Geobiology

Geobiology

Editor:

Quan Cui, Ph.D.

ArclerPress

www.arclerpress.com

Geobiology

Editor: Quan Cui, Ph.D.

708 3rd Avenue, 6th Floor
New York
NY 10017
United States of America
www.arclerpress.com

ISBN: 978-1-68094-480-8

Library of Congress Control Number: 2016951169

About the Editor

Quan Cui, Ph.D.

Quan Cui obtained her PhD from Beijing Normal University in 2015. She worked as environmental impact assessment engineer for years. Aiming at linking science to policy, her study interest spans from ecosystem modelling to water resources management. She published peer-reviewed articles about using machine learning models in inflow forecasting, CO_2 management and watershed ecosystem health assessment.

Contents

List of Contributors

Ceiridwen J. Edwards
Smurfit Institute of Genetics, Trinity College, Dublin 2, Ireland
Research Laboratory for Archaeology, University of Oxford, Dyson Perrins Building, South Parks Road, Oxford OX1 3QY, UK

Carl D. Soulsbury
School of Biological Sciences, University of Bristol, Woodland Road, Bristol BS8 1UG, UK
School of Life Sciences, University of Lincoln, Brayford Pool, Lincoln LN6 7TS, UK

Mark J. Statham
Canid Diversity and Conservation Laboratory, Center for Veterinary Genetics, University of California, One Shields Avenue/Old Davis Road, Davis, CA 95616-8744, USA

Simon Y.W. Ho
School of Biological Sciences, University of Sydney, NSW 2006, Sydney, Australia

Dave Wall
UCD School of Biology and Environmental Science, University College Dublin, Ireland

Gaudenz Dolf
Institute of Genetics, Vetsuisse Faculty, University of Berne, 3001 Berne, Switzerland

Graziella Iossa
School of Biological Sciences, University of Bristol, Woodland Road, Bristol BS8 1UG, UK

Phillip J. Baker
School of Biological Sciences, University of Bristol, Woodland Road, Bristol BS8 1UG, UK

Stephen Harris
School of Biological Sciences, University of Bristol, Woodland Road, Bristol BS8 1UG, UK

Benjamin N. Sacks
Canid Diversity and Conservation Laboratory, Center for Veterinary Genetics, University of California, One Shields Avenue/Old Davis Road, Davis, CA 95616-8744, USA
Department of Population Health and Reproduction, School of Veterinary Medicine, University of California, One Shields Avenue/Old Davis Road, Davis, CA 95616-8744, USA

Daniel G. Bradley
Smurfit Institute of Genetics, Trinity College, Dublin 2, Ireland

Osamu Maeda
Institute for Comparative Research in Human and Social Sciences, University of Tsukuba, Tsukuba 305-8571, Japan

Leilani Lucas
Institute of Archaeology, University College London, London WC1H 0PY, United Kingdom

Fabio Silva
Institute of Archaeology, University College London, London WC1H 0PY, United Kingdom

Ken-Ichi Tanno
Faculty of Agriculture, Yamaguchi University, Yamaguchi 753-8515, Japan

Dorian Q. Fuller
Institute of Archaeology, University College London, London WC1H 0PY, United Kingdom

Helen M. Roe
School of Geography, Archaeology and Palaeoecology, Queen's University of Belfast, Belfast BT7 1NN, UK

G. Russell Coope
Centre for Quaternary Research, Department of Geography, Royal Holloway, University of London, Egham, Surrey TW20 OEX, UK

Robert J.N. Devoy
The Coastal Resources Centre, Department of Geography, University College Cork, Cork, Ireland

Colin J.O. Harrison
Department of Zoology, Natural History Museum at Tring, Akeman Street, Tring, Hertfordshire HP23 6AP, UK

Kirsty E.H. Penkman
BioArCh, Department of Chemistry, University of York, York YO10 5DD, UK

Richard C. Preece
Department of Zoology, University of Cambridge, Downing Street, Cambridge CB2 3EJ, UK

Danielle C. Schreve
Centre for Quaternary Research, Department of Geography, Royal Holloway, University of London, Egham, Surrey TW20 OEX, UK

Stephen J. Brooks
Department of Entomology, Natural History Museum, Cromwell Road, London SW7 5BD, UK

Ian P. Matthews
Centre for Quaternary Research, Department of Geography, Royal Holloway, University of London, Egham, Surrey TW20 0EX, UK

Hilary H. Birks
Department of Biology, University of Bergen, N-5020 Bergen, Norway
Bjerknes Centre for Climate Research, Allégaten 55, N-5007 Bergen, Norway

H.John B. Birks
Department of Biology, University of Bergen, N-5020 Bergen, Norway
Bjerknes Centre for Climate Research, Allégaten 55, N-5007 Bergen, Norway
School of Geography and the Environment, University of Oxford, Oxford OX1 3QY, UK
Environmental Change Research Centre, University College London, London WC1E 6BT, UK

Jörg Schnecker
University of Vienna, Department of Microbiology and Ecosystem Science, Division of Terrestrial Ecosystem Research, Vienna, Austria
Austrian Polar Research Institute, Vienna, Austria

Birgit Wild
University of Vienna, Department of Microbiology and Ecosystem Science, Division of Terrestrial Ecosystem Research, Vienna, Austria
Austrian Polar Research Institute, Vienna, Austria

Mounir Takriti
University of Vienna, Department of Microbiology and Ecosystem Science, Division of Terrestrial Ecosystem Research, Vienna, Austria
Austrian Polar Research Institute, Vienna, Austria

Ricardo J. Eloy Alves
Austrian Polar Research Institute, Vienna, Austria
University of Vienna, Department of Ecogenomics and Systems Biology, Division of Archaea Biology and Ecogenomics, Vienna, Austria

Norman Gentsch
Leibniz Universität Hannover, Institut für Bodenkunde, Hannover, Germany

Antje Gittel
Aarhus University, Center for Geomicrobiology, Department of Bioscience, Aarhus, Denmark
University of Bergen, Centre for Geobiology, Bergen, Norway

Angelika Hofer
University of Vienna, Department of Microbiology and Ecosystem Science, Division of Terrestrial Ecosystem Research, Vienna, Austria

Karoline Klaus
University of Vienna, Department of Microbiology and Ecosystem Science, Division of Terrestrial Ecosystem Research, Vienna, Austria

Anna Knoltsch
University of Vienna, Department of Microbiology and Ecosystem Science, Division of Terrestrial Ecosystem Research, Vienna, Austria
Austrian Polar Research Institute, Vienna, Austria

Nikolay Lashchinskiy
Central Siberian Botanical Garden, Siberian Branch of Russian Academy of Sciences, Novosibirsk, Russia

Robert Mikutta
Leibniz Universität Hannover, Institut für Bodenkunde, Hannover, Germany

Andreas Richter
University of Vienna, Department of Microbiology and Ecosystem Science, Division of Terrestrial Ecosystem Research, Vienna, Austria
Austrian Polar Research Institute, Vienna, Austria

Birgit Wild
University of Vienna, Department of Microbiology and Ecosystem Science, Division of Terrestrial Ecosystem Research, Vienna, Austria
Austrian Polar Research Institute, Vienna, Austria

Jörg Schnecker
University of Vienna, Department of Microbiology and Ecosystem Science, Division of Terrestrial Ecosystem Research, Vienna, Austria
Austrian Polar Research Institute, Vienna, Austria

Ricardo J. Eloy Alves
Austrian Polar Research Institute, Vienna, Austria
University of Vienna, Department of Ecogenomics and Systems Biology, Division of Archaea Biology and Ecogenomics, Vienna, Austria

Pavel Barsukov
Siberian Branch of the Russian Academy of Sciences, Institute of Soil Science and Agrochemistry, Novosibirsk, Russia

Jiří Bárta
University of South Bohemia, Department of Ecosystems Biology, České Budějovice, Czech Republic

Petr Čapek
University of South Bohemia, Department of Ecosystems Biology, České Budějovice, Czech Republic

Norman Gentsch
Leibniz University Hannover, Institute of Soil Science, Hannover, Germany

Antje Gittel
University of Bergen, Centre for Geobiology, Department of Biology, Bergen, Norway
Austrian Polar Research Institute, Vienna, Austria

Georg Guggenberger
Leibniz University Hannover, Institute of Soil Science, Hannover, Germany

Nikolay Lashchinskiy
Siberian Branch of Russian Academy of Sciences, Central Siberian Botanical Garden, Novosibirsk, Russia

Robert Mikutta
Leibniz University Hannover, Institute of Soil Science, Hannover, Germany

Olga Rusalimova
Siberian Branch of the Russian Academy of Sciences, Institute of Soil Science and Agrochemistry, Novosibirsk, Russia

Hana Šantrůčková
University of South Bohemia, Department of Ecosystems Biology, České Budějovice, Czech Republic

Olga Shibistova
Leibniz University Hannover, Institute of Soil Science, Hannover, Germany
Siberian Branch of Russian Academy of Sciences, VN Sukachev Institute of Forest, Krasnoyarsk, Russia

Tim Urich
Austrian Polar Research Institute, Vienna, Austria
University of Vienna, Department of Ecogenomics and Systems Biology, Division of Archaea Biology and Ecogenomics, Vienna, Austria

Margarete Watzka
University of Vienna, Department of Microbiology and Ecosystem Science, Division of Terrestrial Ecosystem

Galina Zrazhevskaya
Siberian Branch of Russian Academy of Sciences, VN Sukachev Institute of Forest, Krasnoyarsk, Russia

Andreas Richter
University of Vienna, Department of Microbiology and Ecosystem Science, Division of Terrestrial Ecosystem Research, Vienna, Austria
Austrian Polar Research Institute, Vienna, Austria

Peter H. Verburg
Environmental Geography Group, VU University Amsterdam, de Boelelaan 1087, 1081 HV Amsterdam, The Netherlands

Neville Crossman
CSIRO Land and Water, Private Bag 2, Glen Osmond, SA 5064, Australia

Erle C. Ellis
Department of Geography and Environmental Systems, University of Maryland, Baltimore County, Baltimore, MD 21250, USA

Andreas Heinimann
Centre for Development and Environment (CDE) and Institute of Geography, University of Bern, Hallerstrasse 10, 3012 Bern, Switzerland

Patrick Hostert
Geography Department and Integrative Research Institute on Transformations of Human-Environment Systems (IRI THESys), Humboldt-Universität zu Berlin, Berlin, Germany

Ole Mertz
Department of Geosciences and Natural Resource Management, University of Copenhagen, Copenhagen, Denmark

Harini Nagendra
School of Development, Azim Premji University, Bangalore, India

Thomas Sikor
School of International Development, University of East Anglia, Norwich, UK

Karl-Heinz Erb
Institute of Social Ecology Vienna (SEC), Alpen Adria University Klagenfurt-Vienna-Graz, Schottenfeldgasse 29, A-1070 Vienna, Austria

Nancy Golubiewski
Ministry for the Environment—Manatu Mo Te Taiao, 23 Kate Sheppard Place, Wellington 6143, New Zealand

Ricardo Grau
Instituto de Ecología Regional, Universidad Nacional de Tucumán-CONICET, CC 34 (4107), Yerba Buena, Tucumán, Argentina

Morgan Grove
Northern Research Station, USDA Forest Service, 5200 Westland Blvd. TRC 171, Baltimore, MD 21227, USA

Souleymane Konaté
IUCN—Central and West Africa, University of Abobo-Adjame, UFR-SN/CRE, 02 BP 801, Abidjan 02, Cote d'Ivoire

Patrick Meyfroidt
F.R.S—FNRS and Université catholique de Louvain, Earth and Life Institute, Georges Lemaître Centre for Earth and Climate Research (TECLIM), Louvain-la-Neuve, Belgium

Dawn C. Parker
School of Planning, University of Waterloo, 200 University Avenue West, Waterloo, Canada

Rinku Roy Chowdhury
Graduate School of Geography, Clark University, 950 Main Street, Worcester, MA 01610, United States

Hideaki Shibata
Field Science Center for Northern Biosphere, Hokkaido University, Kita-9, Nishi-9, Kita-ku, Sapporo 060-0809, Japan

Allison Thomson
Field to Market, The Alliance for Sustainable Agriculture, 777 N Capitol St. NE, Suite 803, Washington, D.C. 20002, United States

Lin Zhen
Institute of Geographic Science and Natural Resources Research, Chinese Academy of Sciences, 11A Datun Road, Chaoyang District, Beijing 100101, China

Goetz M. Richter
Sustainable Soils & Grassland Systems, Rothamsted Research, Harpenden, AL5 2JQ, UK

Francesco Agostini
Sustainable Soils & Grassland Systems, Rothamsted Research, Harpenden, AL5 2JQ, UK

Alexandra Barker
Sustainable Soils & Grassland Systems, Rothamsted Research, Harpenden, AL5 2JQ, UK
National Physical Laboratory, Hampton Road, Teddington, TW11 0LW, UK

Delphine Costomiris
Sustainable Soils & Grassland Systems, Rothamsted Research, Harpenden, AL5 2JQ, UK
SAPEC AGRO en France, 2-12 Chemin des Femmes - Immeuble Odyssee Bat. A, 91300 MASSY, France

Aiming Qi
Sustainable Soils & Grassland Systems, Rothamsted Research, Harpenden, AL5 2JQ, UK

Joris Peters
Department of Veterinary Sciences, Institute of Palaeoanatomy, Domestication Research and the History of Veterinary Medicine, LMU Munich, Kaulbachstr. 37, D-80539 Munich, Germany
SNSB, Bavarian State Collection of Anthropology and Palaeoanatomy, Karolinenplatz 2a, D-80333 Munich, Germany

Ophélie Lebrasseur
Palaeogenomics and Bio-Archaeology Research Network, Research Laboratory for Archaeology, Dyson Perrins Building, South Parks Road, Oxford OX1 3QY, UK

Hui Deng
Department of History and Culture, Shanxi University, No. 92 Wucheng Road, Taiyuan City, Shanxi Province, PR China

Greger Larson
Palaeogenomics and Bio-Archaeology Research Network, Research Laboratory for Archaeology, Dyson Perrins Building, South Parks Road, Oxford OX1 3QY, UK

Lourdes López-Merino
Institute of Environment, Health and Societies, Brunel University London, UB8 3PH Uxbridge, UK

Suzanne A.G. Leroy
Institute of Environment, Health and Societies, Brunel University London, UB8 3PH Uxbridge, UK

Amram Eshel
The Ted Arisson Airborne Allergens Monitoring Laboratory, Tel Aviv University, Tel Aviv 69978, Israel

Valentina Epshtein
The Ted Arisson Airborne Allergens Monitoring Laboratory, Tel Aviv University, Tel Aviv 69978, Israel

Reuven Belmaker
Dr Strauss Department of Marine Geosciences, Charney School of Marine Sciences, University of Haifa, Mount Carmel, Haifa 31905, Israel

Revital Bookman
Dr Strauss Department of Marine Geosciences, Charney School of Marine Sciences, University of Haifa, Mount Carmel, Haifa 31905, Israel

Jonathan Saha
University of Leeds, School of History, Leeds LS2 9JT, UK

K.E.H. Penkman
BioArch, Departments of Biology, Archaeology and Chemistry, Biology S Block, University of York, York YO10 5YW, UK

R.C. Preece
Department of Zoology, University of Cambridge, Downing Street, Cambridge CB2 3EJ, UK

D.H. Keen
Institute of Archaeology and Antiquity, University of Birmingham, Birmingham B15 2TT, UK

D. Maddy
School of Geography, Politics and Sociology, Daysh Building, Newcastle University, Newcastle-upon-Tyne NE1 7RU, UK

D.C. Schreve
Department of Geography, Royal Holloway, University of London, Egham, Surrey TW20 0EX, UK

M.J. Collins
BioArch, Departments of Biology, Archaeology and Chemistry, Biology S Block, University of York, York YO10 5YW, UK

David Beresford-Jones
University of Cambridge, McDonald Institute for Archaeological Research, Downing St., Cambridge CB2 3ER, UK

Alexander G. Pullen
University of Cambridge, McDonald Institute for Archaeological Research, Downing St., Cambridge CB2 3ER, UK

Oliver Q. Whaley
Royal Botanical Gardens Kew, Richmond, Surrey TW9 3AE, UK

Justin Moat
Royal Botanical Gardens Kew, Richmond, Surrey TW9 3AE, UK

George Chauca
Universidad Nacional Mayor de San Marcos, Escuela Profesional de Arqueología, Av. Universitaria s/n., Lima, Peru

Lauren Cadwallader
University of Cambridge, McDonald Institute for Archaeological Research, Downing St., Cambridge CB2 3ER, UK

Susana Arce
Museo Regional de Ica, Ministerio de Cultura, Av. Ayabaca s.n.o, cuadra 8, urb. San Isidro, Ica, Peru

Alfonso Orellana
Proyecto Kew Perú, Conservación, Restauración de Hábitats y Medios de Vida Útiles, Urb. San Antonio B-20, Ica, Peru

Carmela Alarcón
Instituto de Investigaciones Andinas Punku, Calle Los Molles 131, urb. Alto de Los Ficus, Lima 43, Peru

Manuel Gorriti
Proyecto Especial Arqueológico Caral-Supe, Las Lomas de la Molina 327, urb. Las Lomas de la Molina Vieja, Lima 12, Peru

Patricia K. Maita
Colección de Antropología Física, Museo Nacional de Arqueología, Antropología e Historia del Perú, Plaza Bolivar s/n, Pueblo Libre, Lima, Peru

Fraser Sturt
University of Southampton, Department of Archaeology, Avenue Campus, Highfield, Southampton, SO17 1BF, UK

Agathe Dupeyron
University of Cambridge, McDonald Institute for Archaeological Research, Downing St., Cambridge CB2 3ER, UK

Oliver Huaman
Proyecto de Investigación Tambo Colorado, IFEA, Av. Arequipa 4500, Miraflores, Lima, Peru

Kevin J. Lane
University of Cambridge, McDonald Institute for Archaeological Research, Downing St., Cambridge CB2 3ER, UK

Charles French
University of Cambridge, McDonald Institute for Archaeological Research, Downing St., Cambridge CB2 3ER, UK

Birgit Wild
University of Vienna, Department of Microbiology and Ecosystem Science, Division of Terrestrial Ecosystem Research, Althanstrasse 14, 1090 Vienna, Austria
Austrian Polar Research Institute, 1090 Vienna, Austria

Jörg Schnecker
University of Vienna, Department of Microbiology and Ecosystem Science, Division of Terrestrial Ecosystem Research, Althanstrasse 14, 1090 Vienna, Austria
Austrian Polar Research Institute, 1090 Vienna, Austria

Jiří Bárta
University of South Bohemia, Department of Ecosystems Biology, Branišovská 31, 37005 České Budějovice, Czech Republic

Petr Čapek
University of South Bohemia, Department of Ecosystems Biology, Branišovská 31, 37005 České Budějovice, Czech Republic

Georg Guggenberger
Leibniz Universität Hannover, Institut für Bodenkunde, Herrenhäuser Strasse 2, 30419 Hannover, Germany

Florian Hofhansl
University of Vienna, Department of Microbiology and Ecosystem Science, Division of Terrestrial Ecosystem Research, Althanstrasse 14, 1090 Vienna, Austria

Christina Kaiser
University of Vienna, Department of Microbiology and Ecosystem Science, Division of Terrestrial Ecosystem Research, Althanstrasse 14, 1090 Vienna, Austria
International Institute for Applied Systems Analysis (IIASA), Schlossplatz 1, 2361 Laxenburg, Austria

Nikolaj Lashchinsky
Central Siberian Botanical Garden, Siberian Branch of Russian Academy of Sciences, St. Zolotodolinskaya 101, 630090 Novosibirsk, Russia

Robert Mikutta
Leibniz Universität Hannover, Institut für Bodenkunde, Herrenhäuser Strasse 2, 30419 Hannover, Germany

Maria Mooshammer
University of Vienna, Department of Microbiology and Ecosystem Science, Division of Terrestrial Ecosystem Research, Althanstrasse 14, 1090 Vienna, Austria

Hana Šantrůčková
University of South Bohemia, Department of Ecosystems Biology, Branišovská 31, 37005 České Budějovice, Czech Republic

Olga Shibistova
VN Sukachev Institute of Forest, Siberian Branch of Russian Academy of Sciences, Akademgorodok, 660036 Krasnoyarsk, Russia
Leibniz Universität Hannover, Institut für Bodenkunde, Herrenhäuser Strasse 2, 30419 Hannover, Germany

Tim Urich
Austrian Polar Research Institute, 1090 Vienna, Austria
University of Vienna, Department of Ecogenomics and Systems Biology, Althanstrasse 14, 1090 Vienna, Austria
University of Bergen, Department of Biology/Centre for Geobiology, Allégaten 41, 5007 Bergen, Norway

Sergey A. Zimov
Northeast Scientific Station, Pacific Institute for Geography, Far-East Branch of Russian Academy of Sciences, 678830 Chersky, Republic of Sakha, Russia

Andreas Richter
University of Vienna, Department of Microbiology and Ecosystem Science, Division of Terrestrial Ecosystem Research, Althanstrasse 14, 1090 Vienna, Austria
Austrian Polar Research Institute, 1090 Vienna, Austria

R.A. Housley
Department of Geography, Royal Holloway, University of London, Egham Hill, Egham TW20 0EX, UK

C.S. Lane
Research Laboratory for Archaeology and History of Art, University of Oxford, Dyson Perrins Building, South Parks Road, Oxford OX1 3QY, UK

V.L. Cullen
Research Laboratory for Archaeology and History of Art, University of Oxford, Dyson Perrins Building, South Parks Road, Oxford OX1 3QY, UK

M.-J. Weber
Zentrum für Baltische und Skandinavische Archäologie, Stiftung Schleswig-Holsteinische Landesmuseen, Schloss Gottorf, D-24837 Schleswig, Germany

F. Riede
Afdeling for Forhistorisk Arkæologi, Institut for Antropologi, Arkæologi og Lingvistik, University of Aarhus, Moesgård Allé 20, DK-8270 Højbjerg, Denmark

C.S. Gamble
Faculty of Humanities (Archaeology), Building 65A, Avenue Campus, University of Southampton, Southampton SO17 1BF, UK

F. Brock
Research Laboratory for Archaeology and History of Art, University of Oxford, Dyson Perrins Building, South Parks Road, Oxford OX1 3QY, UK

Xiaosen Zhang
Department of Geography, Environment and Earth Sciences, University of Hull, Cottingham Road, Hull HU6 7RX, UK

Jane Reed
Department of Geography, Environment and Earth Sciences, University of Hull, Cottingham Road, Hull HU6 7RX, UK

Bernd Wagner
Institute of Geology and Mineralogy, University of Cologne, Zülpicher Str. 49a, 50674 Cologne, Germany

Alexander Francke
Institute of Geology and Mineralogy, University of Cologne, Zülpicher Str. 49a, 50674 Cologne, Germany

Zlatko Levkov
Institute of Biology, Faculty of Natural Sciences, Gazi Baba bb, 1000 Skopje, Republic of Macedonia

Johann Hofherr
European Commission, Joint Research Centre (JRC), Institute for the Protection and Security of the Citizen (IPSC), Maritime Affairs Unit, E. Fermi 2749, 21027 Ispra (VA), Italy

Fabrizio Natale
European Commission, Joint Research Centre (JRC), Institute for the Protection and Security of the Citizen (IPSC), Maritime Affairs Unit, E. Fermi 2749, 21027 Ispra (VA), Italy

Pablo Trujillo
European Commission, Joint Research Centre (JRC), Institute for the Protection and Security of the Citizen (IPSC), Maritime Affairs Unit, E. Fermi 2749, 21027 Ispra (VA), Italy

Angus Davison
School of Life Sciences, University of Nottingham, Nottingham NG7 2RD, UK

Gary S. McDowell
Center for Regenerative and Developmental Biology, and Department of Biology, Tufts University, Medford, MA 02155, USA

Jennifer M. Holden
School of Life Sciences, University of Nottingham, Nottingham NG7 2RD, UK
[9]Present address: Warwick Medical School, The University of Warwick, Coventry CV4 7AL, UK

Harriet F. Johnson
School of Life Sciences, University of Nottingham, Nottingham NG7 2RD, UK

Georgios D. Koutsovoulos
Institute of Evolutionary Biology, University of Edinburgh, Edinburgh EH9 3JT, UK

M. Maureen Liu
School of Life Sciences, University of Nottingham, Nottingham NG7 2RD, UK

Paco Hulpiau
Department for Biomedical Molecular Biology, Ghent University, and Inflammation Research Center (IRC), VIB, 9052 Ghent, Belgium

Frans Van Roy
Department for Biomedical Molecular Biology, Ghent University, and Inflammation Research Center (IRC), VIB, 9052 Ghent, Belgium

Christopher M. Wade
School of Life Sciences, University of Nottingham, Nottingham NG7 2RD, UK

Ruby Banerjee
Wellcome Trust Sanger Institute, Wellcome Trust Genome Campus, Hinxton, Cambridge CB10 1SA, UK

Fengtang Yang
Wellcome Trust Sanger Institute, Wellcome Trust Genome Campus, Hinxton, Cambridge CB10 1SA, UK

Satoshi Chiba
Community and Ecosystem Ecology, Division of Ecology and Evolutionary Biology, Graduate School of Life Sciences, Tohoku University, Aobayama, Sendai 980-8578, Japan

John W. Davey
Department for Biomedical Molecular Biology, Ghent University, and Inflammation Research Center (IRC), VIB, 9052 Ghent, Belgium
Present address: Department of Zoology, University of Cambridge, Downing Street, Cambridge CB2 3EJ, UK

Daniel J. Jackson
Department of Geobiology, University of Göttingen, Göttingen 37077, Germany

Michael Levin
Center for Regenerative and Developmental Biology, and Department of Biology, Tufts University, Medford, MA 02155, USA

Mark L. Blaxter
Institute of Evolutionary Biology, University of Edinburgh, Edinburgh EH9 3JT, UK
Edinburgh Genomics, School of Biological Sciences, University of Edinburgh, Edinburgh EH9 3JT, UK

Marcia Gibson
MRC Social and Public Health Sciences Unit, Glasgow, UK

Mark Petticrew
Public and Environmental Health Research Unit, Department of Public Health and Policy, London School of Hygiene and Tropical Medicine, 1 Keppel St., London WC1E 7HT, UK

Clare Bambra
Department of Geography, Durham University, Wolfson Research Institute, Durham University Queen's Campus, Stockton on Tees TS17 6BH, UK

Amanda J. Sowden
Centre for Reviews and Dissemination, University of York, York YO10 5DD, UK

Kath E. Wright
Centre for Reviews and Dissemination, University of York, York YO10 5DD, UK

Margaret Whitehead
Division of Public Health, University of Liverpool, Whelan Building, Quadrangle, Brownlow Hill, Liverpool L69 3GB, UK

Preface

Geobiology integrates many disciplines, from genomics to paleoenvironmental modeling to paleontology and geochemical analysis. They can be fruitful individually but, when combined, result in profound insights about life—past, present, future and elsewhere. Acknowledging that earth and life are interacting during the whole process and the earth today are a result of long time of molds to life, geobiology science investigates macro- and micro-scale changes in the present as well as in the fossil record. Methodological approaches used in geobiology varies from lab experiments to field investigations. Geobiology is a new scientific disciplinary, which has just been defined in the last decade. Geobiology is the study of the relations between the Earth and life. It is an area which allows the study of the effects of the cosmic and earthly rays, of the architectural forms, of materials, of colors, of technological products, etc..., on living beings. We all have seen that certain locations spread a feeling of wellbeing, whereas other places let us feel uneasy. For certain families or businesses, only bad luck is to be found; for others, success. Everything around us is vibrations. The energy forces of our environment can be harmonious or in disharmony. Our living space, our place of work and our recreation areas are under the influence of these forces, acting unsuspectingly on our organism. These waves, present in the entire universe, may cause varying pathogenic effects and affect our physical and mental well-being, as well as our behavior. After a long-term exposure at harmful locations, troubles may appear, ranging from simple disagreements to more serious pathologies: loss of vitality, insomnia, cancer, suicide, dementia... according to the sensibility of the individual Given the fact that geobiology can be embodied in any geography or biology related research fields, this book gives a holistic perspective of this science. Subjects in this book include geobiology approaches in environmental problems, natural resource, archeology, archaeology, social science research, and biotechnology. Geobiology is an exciting and rapidly developing research discipline that opens new perspectives in understanding Earth as a system. To determine and to exploit its possibilities, this promising scientific field will benefit from a discussion of its definition as a research discipline, its objectives, and its methodological approaches. Such a spirited discussion is the objective of the book. Geobiology touches various subdisciplines of geology and biology in many ways. The book will serve biogeochemists, paleontologists, biomineralogists, microbiologists and many others as a forum to determine future directions of geobiological research.

Editor
Quan Cui, Ph.D.

INTRODUCTION

Geobiology is the study of the interactions that occur between the biosphere (living organisms and their products) and the geosphere. Therefore, it must include elements of the atmosphere, the hydrosphere (marine and freshwater), and the lithosphere. Geobiology applies the principles and methods of biology and geology to the study of the ancient history of the co-evolution of life and Earth as well as the role of life in the modern world. Geobiologic studies tend to be focused on microorganisms, and on the role that life plays in altering the chemical and physical environment of the lithosphere, atmosphere, hydrosphere and/or cryosphere. It differs from biogeochemistry in that the focus is on processes and organisms over space and time rather than on global chemical cycles.

Geobiological research synthesizes the geologic record with modern biologic studies. It deals with process - how organisms affect the Earth and vice versa - as well as history - how the Earth and life have changed together. Much research is grounded in the search for fundamental understanding, but geobiology can also be applied, as in the case of microbes that clean up oil spills. Geobiology employs molecular biology, environmental microbiology, chemical analyses, and the geologic record to investigate the evolutionary interconnectedness of life and Earth. It attempts to understand how the Earth has changed since the origin of life and what it might have been like along the way. Some definitions of geobiology even push the boundaries of this time frame - to understanding the origin of life and to the role that man has played and will continue to play in shaping the Earth in the Anthropogenic.

Major Geobiological events

Perhaps the most profound geobiological event is the introduction of oxygen into the atmosphere by photosynthetic bacteria. This oxygenation of Earth's primordial atmosphere (the so-called oxygen catastrophe or Great Oxygenation Event) and the oxygenation of the oceans altered surface biogeochemical cycles and the types of organisms that have been evolutionarily selected for. A subsequent major change was the advent of multicellularity. The presence of oxygen allowed eukaryotes and, later, multicellular life to evolve.

More anthropocentric geobiologic events include the origin of animals and the establishment of terrestrial plant life, which affected continental erosion and nutrient cycling, and likely changed the types of rivers observed, allowing channelization of what were previously predominantly braided rivers. More subtle geobiological events include the role of termites in overturning sediments, coral reefs in depositing calcium carbonate and breaking waves, sponges in absorbing dissolved marine silica, the role of dinosaurs in breaching river levees and promoting flooding, and the role of large mammal dung in distributing nutrients.

Life interacting with the Earth

Geobiology is predicated on the observation that biological processes interact with physical processes at and near the Earth's surface. Take, for example, carbon, the defining element of life. Within the biosphere – the sum of all environments that support life on Earth – carbon exists in a number of forms and in several key reservoirs. It is present as CO_2 in the atmosphere; as CO_2, HCO_3^- and CO_3^{2-} dissolved in fresh and marine waters; as carbonate minerals in soils, sediments and rocks; and as a huge variety of organic molecules in organisms, in sediments and soils, and dissolved in lakes and oceans. Physical processes move carbon from one reservoir to another; for example, volcanoes add CO_2 to the atmosphere and chemical weathering removes it. Biological processes do as well. In two notable examples, photosynthesis reduces CO_2 to sugar, and respiration oxidizes organic molecules to CO_2. Since the industrial revolution, humans have oxidized sedimentary organic matter (by burning fossil fuels) at rates much higher than those characteristic of earlier epochs, making us important participants in the Earth's carbon cycle.

Process in geobiology

Geobiologists, then, study how organisms influence the physical Earth and vice versa, and how biological and physical processes have interacted through our planet's long history. Elucidating the changing role of life through Earth history, sometimes called historical geobiology, begins with a basic understanding of geobiological processes, but from there takes on a distinctly geological slant. We would like to interpret the geologic record in terms of active processes and chemical states, but rocks preserve only pattern. Thus, the geobiological interpretation of ancient sedimentary rocks requires that we understand how biological processes and aspects of the ambient environmental state are reflected in the geologically preferable patterns they create. For example, we can use the sulfur isotopic composition of minerals in billion-year-old shale's to constrain the biological workings of the ancient sulfur cycle and sulfate abundance in ancient seawater, but can do so only in light of present day observations and experiments that show how biological and physical processes result in particular isotopic patterns.

Of course, there are at least two features that complicate this linkage of geobiological process to geologic pattern. For one, populations evolve, so biological processes observable today may not been active during the deposition of ancient sedimentary rocks. For this reason, historical geobiology has among its goals the establishment of evolutionary pattern in Earth history.

Most obviously, modern surface environments are permeated with oxygen in ways unlikely to have existed during the first two billion years of our planet's development. Other differences exist, as well. Therefore, the present-day Earth system is far removed from the earliest systems where life evolved and then spread out across the planet; it represents a long accumulation of biological, physical and chemical changes through Earth history.

New horizons in geobiology

It is difficult, if not impossible, to predict the future, and while it would be fun to attempt a forecast of the status of geobiology in say 20 years, and we will avoid this. Rather, we highlight that under all circumstances, geobiology will increasingly look to the heavens. Astrobiology can be thought of as the application of geobiological principles to the study of planets and moons beyond the Earth.

Phylogeny predicts evolutionary history

A phylogenetic tree of living things, based on rRNA data and proposed by Carl Woese, showing the separation of bacteria, archaea, and eukaryotes and linking the three branches of living organisms to the LUCA (the black trunk at the bottom of the tree).

Phylogeny takes genetic sequences from living organisms and compares them to each other to reveal evolutionary relationships, much like a family tree reveals how individuals are connected to their distant cousins. It allows us to decipher modern relationships and infer how evolution happened in the past.

Phylogeny can give some sense of history when combined with a little bit more information. Each difference in the DNA indicates divergence between one species and another. This divergence, whether via drift or natural selection, is representative of some lapse of time. Comparing DNA sequences alone gives a record of the history of evolution with an arbitrary measure of phylogenetic distance "dating" that last common ancestor. However, if information about the rate of genetic mutation is available or geologic markers are present to calibrate evolutionary divergence (i.e. fossils), we have a timeline of evolution. From there, with an idea about other contemporaneous changes in life and environment, we can begin to speculate why certain evolutionary paths might have been selected for.

Caltech Geobiology

Geobiologists study the role of life in shaping the chemical and physical characteristics of the Earth and, conversely, the environmental forces that have shaped the evolution and function of life. These questions span the range of geologic time, from the emergence and evolution of life on the earliest Earth to its role in shaping modern biogeochemical cycles. The tools employed by geobiologists at Caltech include sedimentology and stratigraphy, paleontology, paleomagnetism, organic and trace-element geochemistry, isotope geochemistry, molecular biology, and genetics.

Geobiology/Geopathic Stress and its effect on Human Health

Geobiology is a study of the influence of "earth energies" on all forms of life. The earth is threaded with energy lines or meridians of electromagnetic energy which have a powerful influence on living things. Scientists have proven this invisible energy passes through all matter, affecting every living system on the planet, in either a negative or positive way. Man is an electromagnetic creature with every cell in the living system an electrical battery. The electrical system is reflected even more subtly as the human aura which is a photographable and measurable electrical field. Recognizing these phenomena, geobiologists work to ensure that the area where people live in is clear of "geopathic stress" or zones of energy which conflict with, or cause harm to living organisms. People correctly assume that this work is similar to the oriental practice of "feng shui". The main difference is that the geobiologist works with dowseable negative energy lines and fields which he or she then interrupts or neutralizes to clean up the environment. This may be termed energy balancing and it address only those negative aspects which harmfully interact with the electro- magnetic world around our biomagnetic bodies. So, bodily stress is often caused by a geobiological stimulus; thus the term geopathic stress.

A partial list of Geopathic Stress would include sleep disorders, tingling, numbness and or pain in arms and legs, chronic fatigue, aching and swelling, dizziness, frequent headaches, vision problems, disorientation, memory loss, instability, stress, nausea, inability to heal, allergies, problematic menstrual cycles, anxiety, listlessness, resistance to medical treatment, inability to absorb vitamins and minerals, ADD, hyperactivity, depression and a general feeling of malaise. In order to understand this, simply visualize the "negative to human" energy zones, as areas which upset the electromagnetic and auric field of the body and thus compromise the human mental and or immune systems.

Co-evolution of life and Earth

A core concept in geobiology is that life changes over time through evolution. The theory of evolution postulates that unique populations of organisms or species arose from genetic modifications in the ancestral population which were passed down by drift and natural selection.

Along with standard biological evolution, life and planet co-evolve. Since the best adaptations are those that suit the ecological niche that the organism lives in, the physical and chemical characteristics of the environment drive the evolution of life by natural selection, but the opposite can also be true: with every advent of evolution, the environment changes.

A classic example of co-evolution is the evolution of oxygen-producing photosynthetic cyanobacteria which oxygenated Earth's Archean atmosphere. The ancestors of cyanobacteria began using water as an electron source to harness the energy of the sun and expelling oxygen before or during the early Paleoproterozoic. During this time, around 2.4 to 2.1 billion years ago, geologic data suggests that atmospheric oxygen began to rise in what is termed the Great Oxygenation Event (GOE). It is unclear for how long cyanobacteria had been doing oxygenic photosynthesis before the GOE. Some evidence suggests there were geochemical "buffers" or sinks suppressing the rise of oxygen such as volcanism though cyanobacteria may have been around producing it before the GOE. Other evidence indicates that the rise of oxygenic photosynthesis was coincident with the GOE.

The presence of oxygen on Earth from its first production by cyanobacteria to the GOE and through today has drastically impacted the course of evolution of life and planet. It may have triggered the formation of oxidized minerals and the disappearance of oxidizable minerals like pyrite from ancient stream beds. The presence of banded-iron formations (BIFs) have been interpreted as a clue for the rise of oxygen since small amounts of oxygen could have reacted with reduced ferrous iron (Fe(II)) in the oceans, resulting in the deposition of sediments containing Fe(III) oxide in places like Western Australia. However, any oxidizing environment, including that provided by microbes such as the iron-oxidizing photoautotroph Rhodopseudomonas palustris, can trigger iron oxide formation and thus BIF deposition. Other mechanisms include oxidation by UV light. Indeed, BIFs occur across large swaths of Earth's history and may not correlate with only one event.

Other changes correlated with the rise of oxygen include the appearance of rust-red ancient paleosols, different isotope fractionation of elements such as sulfur, and global glaciations and Snowball Earth events, perhaps caused by the oxidation of methane by oxygen, not to mention an overhaul of the types of organisms and metabolisms on Earth. Whereas organisms prior to the rise of oxygen were likely poisoned by oxygen gas as many anaerobes are today, those that evolved ways to harness the electron-accepting and energy-giving power of oxygen were poised to thrive and colonize the aerobic environment.

The Earth has changed

Earth has not remained the same since its planetary formation 4.5 billion years ago. Continents have formed, broken up, and collided, offering new opportunities for and barriers to the dispersal of life. The redox state of the atmosphere and the oceans has changed, as indicated by isotope data. Fluctuating quantities of inorganic compounds such as carbon dioxide, nitrogen, methane, and oxygen have been driven by life evolving new biological metabolisms to make these chemicals and have driven the evolution of new metabolisms to use those chemicals. Earth acquired a magnetic field about 3.4 Ga that has undergone a series of geomagnetic reversals on the order of millions of years. The surface temperature is in constant fluctuation, falling in glaciations and Snowball Earth events due to ice-albedo feedback, rising and melting due to volcanic outgassing, and stabilizing due to silicate weathering feedback.

And the Earth is not the only one that changed - the luminosity of the sun has increased over time. Because rocks record a history of relatively constant temperatures since Earth's beginnings, there must have been more greenhouse gasses to keep the temperatures up in the Archean when the sun was younger and fainter. All these major differences in the environment of the Earth placed very different constraints on the evolution of life throughout our planet's history. Moreover, more subtle changes in the habitat of life are always occurring, shaping the organisms and traces that we observe today and in the rock record.

Genes encode metabolism

Molecular biology allows scientists to understand a gene's function using microbial culturing and mutagenesis. Searching for similar genes in other organisms and in metagenomic and metatranscriptomic data allows us to understand what processes could be relevant and important in a given ecosystem, providing insight into the biogeochemical cycles in that environment.

For example, an intriguing problem in geobiology is the role of organisms in the global cycling of methane. Genetics has revealed that the methane monooxygenase gene (pmo) is used for oxidizing methane and is present in all aerobic methane-oxidizers, or methanotrophs. The presence of DNA sequences of the pmo gene in the environment can be used as a proxy for methanotrophy. A more generalizable tool is the 16S ribosomal RNA gene, which is found in bacteria and archaea. This gene evolves very slowly over time and is not usually horizontally transferred, and so it is often used to distinguish different taxonomic units of organisms in the environment. In this way, genes are clues to organismal metabolism and identity. Genetics enables us to ask 'who is there?' and 'what are they doing?' This approach is called metagenomics

Metabolic diversity influences the environment

Life harnesses chemical reactions to generate energy, perform biosynthesis, and eliminate waste. Different organisms use very different metabolic approaches to meet these basic needs. While animals such as ourselves are limited to aerobic respiration, other organisms can "breathe" sulfate (SO_4^{2-}), nitrate (NO^{3-}), ferric iron (Fe(III)), and uranium (U(VI)), or live off energy from fermentation. Some organisms, like plants, are autotrophs, meaning that they can fix carbon dioxide for biosynthesis. Plants are photoautotrophs, in that they use the energy of light to fix carbon. Microorganisms employ oxygenic and anoxygenic photoautotrophy, as well as chemoautotrophy. Microbial communities can coordinate in syntrophic metabolisms to shift reaction kinetics in their favor. Many organisms can perform multiple metabolisms to achieve the same end goal; these are called mixotrophs.

Biotic metabolism is directly tied to the global cycling of elements and compounds on Earth. The geochemical environment fuels life, which then produces different molecules that go into the external environment. (This is directly relevant to biogeochemistry.) In addition, biochemical reactions are catalyzed by enzymes which sometimes prefer one isotope over others. For example, oxygenic photosynthesis is catalyzed by RuBisCO, which prefers carbon^{-12} over carbon^{-13}, resulting in carbon isotope fractionation in the rock record.

Life is fundamentally chemistry

The first life arose from abiotic chemical reactions. When this happened, how it happened, and even what planet it happened on are uncertain. However, life follows the rules of and arose from lifeless chemistry and physics. It is constrained by principles such as thermodynamics. This is an important concept in the field because it is represents the epitome of the interconnectedness, if not sameness, of life and Earth.

While often delegated to the field of astrobiology, attempts to understand how and when life arose are relevant to geobiology as well. The first major strides towards understanding the "how" came with the Miller-Urey experiment, when amino acids formed out of a simulated "primordial soup". Another theory is that life originated in a system much like the hydrothermal vents at mid-oceanic spreading centers. In the Fischer-Tropsch synthesis, a variety of hydrocarbons form under vent-like conditions. Other ideas include the "RNA World" hypothesis, which postulates that the first biologic molecule was RNA and the idea that life originated elsewhere in the solar system and was brought to Earth, perhaps via a meteorite.

Methodology

While geobiology is a diverse and varied field, encompassing ideas and techniques from a wide range of disciplines, there are a number of important methods that are key to the study of the interaction of life and Earth that are highlighted here.

1. Laboratory culturing of microbes is used to characterize the metabolism and lifestyle of organisms of interest.
2. Gene sequencing allows scientists to study the relationships between extant organisms using phylogenetics.
3. Experimental genetic manipulation or mutagenesis is used to determine the function of genes in living organisms.
4. Microscopy is used to visualize the microbial world. Microscope work ranges from environmental observation to quantitative studies with DNA probes to high-definition visualization of the microbe-mineral interface by electron microscope (EM).
5. Isotope tracers can be used to track biochemical reactions to understand microbial metabolism.
6. Isotope natural abundance in rocks can be measured to look for isotopic fractionation that is consistent with biologic origin.

7. Detailed environmental characterization is important to understanding what about a habitat might be driving life's evolution and, in turn, how life might be changing that niche. It includes and is not limited to, temperature, light, pH, salinity, concentration of specific molecules like oxygen, and the biologic community.
8. Sedimentology and stratigraphy are used to read the rocks. The rock record stores a history of geobiologic processes in sediments which can be unearthed through an understanding of deposition, sedimentation, compaction, diagenesis, and deformation.
9. The search for and study of fossils, while often delegated to the separate field of paleontology, is important in geobiology, though the scale of fossils is typically smaller (micropaleontology).
10. The biochemical analysis of biomarkers, which are fossilized or modern molecules that are indicative of the presence of a certain group of organisms or metabolism, is used to answer the evidence for life and metabolic diversity questions.
11. Paleomagnetics is the study of the planet's ancient magnetic field. It is significant to understanding magneto fossils, bio mineralization, and global ecosystem changes.

CHAPTER 1 TEMPORAL GENETIC VARIATION OF THE RED FOX

Ceiridwen J. Edwards, Carl D. Soulsbury, Mark J. Statham, Simon Y.W. Ho, Dave Wall, Gaudenz Dolf, Graziella Iossa, Phillip J. Baker, Stephen Harris, Benjamin N. Sacks, Daniel G. Bradley

Keywords: Red fox, Mitochondrial DNA, Cytochrome b gene, Control region, Phylogeography, Europe, British Isles

INTRODUCTION

Studies of ancient DNA have suggested that there was relatively little phylogeographical structuring of species in Europe prior to the last Quaternary glaciation (Hofreiter et al., 2004). Therefore, it appears that cycles of retreat into refugial areas during glacial periods, followed by incomplete dispersal from one refugium into another during interglacial periods, are responsible for the deep genetic divergences now seen between peninsular regions in southern Europe. The Balkans, Iberia and Italy have been identified as refugia where flora and fauna survived during glacial phases (Taberlet et al., 1998; Hewitt, 2000, 2004; Randi, 2007). However, several recent studies have indicated that this 'southern refugial model' (Bhagwat and Willis, 2008) is probably too simplistic and not satisfactory for all taxa. Alternative concepts have been developed involving the existence of northern refugia (Stewart and Lister, 2001; Kotlik et al., 2006; Bhagwat and Willis, 2008; Tougard et al., 2008), microrefugia (small favourable areas outside the main refugium; Rull, 2009), and nunatak versus lowland refugia (specific to mountain species; Holderegger and Thiel-Egenter, 2009). As habitat modelling suggests that boreal and tundra habitat extended across central Europe as far north as England (Hewitt, 2000; Fløjgaard et al., 2009), it is also possible that some ecologically adaptable species (those species that are capable of living in various habitats) were not restricted to refugial areas at all during the Last Glacial Maximum (LGM).

Evidence in a range of species seems to support this absence of refugial populations (see Bhagwat and Willis, 2008). One example is the brown bear (Ursus arctos), which was typically thought to be restricted to peninsular refugia (Taberlet et al., 1998). However, Valdiosera et al. (2007) found a complex phylogeographical history that supported continuous gene flow across southern Europe. Previously published carnivore studies, including those of brown bears (Valdiosera et al., 2007) and grey wolves (Canis lupus; Pilot et al., 2010), have not included material from the British Isles, and so it is still unclear whether some ecologically adaptable species could have survived as far north during the glacial cycles.

The red fox (Vulpes vulpes L.) is the most widely distributed terrestrial mammal, with a natural range extending across the entire Holarctic and an introduced range including Australia (Schipper et al., 2008). Red foxes are one of the most evolutionarily plastic species, being found in habitats ranging from tundra to deserts, although the breadth of the red fox niche could reflect differential local adaptation rather than phenotypic plasticity within populations (Sacks et al., 2010). Despite clear fossil evidence of red fox in continental Europe and England from MIS4 onwards (c. 71,000 years ago) to modern times (Sommer and Benecke, 2005a; Sommer and Nadachowski, 2006), only recently has fox phylogeography been examined. To date, studies in Japan (Inoue et al., 2007) and North America (Aubry et al., 2009) have demonstrated the importance of glacial cycles on the phylogeographic structure of the red fox. In Europe, studies have been restricted to the Mediterranean basin (Frati et al., 1998), or based on small sample sizes from archaeological remains found in northern and central Europe (Teacher et al., 2011). Although it seems clear from these studies that European red foxes appear to comprise a large population that has been relatively stable over the past 40,000 years (Teacher et al., 2011), considerably larger sample sizes are needed to characterise post-Pleistocene phylogenetic patterns. In this study, our principal aim was to examine whether the red foxes in the British Isles were derived

Citation: Ceiridwen J. Edwards, Carl D. Soulsbury, Mark J. Statham, Simon Y.W. Ho, Dave Wall, Gaudenz Dolf, Graziella Iossa, Phillip J. Baker, Stephen Harris, Benjamin N. Sacks, Daniel G. Bradley, Temporal genetic variation of the red fox, Quaternary Science Reviews, Volume 57, 4 December 2012, Pages 95-104, ISSN 0277-3791, http://dx.doi.org/10.1016/j.quascirev.2012.10.010.

from a late Pleistocene colonisation from mainland Europe or whether they might reflect an ancient lineage that persisted in the region throughout the later ice ages. We also examined the phylogeographic and demographic patterns in the western Palaearctic region as a whole, to characterise more fully the effects of Quaternary climatic changes on extant red fox populations throughout Europe.

MATERIALS AND METHODS

DNA Extraction and Amplification

Modern Samples

We collected 331 tissue samples from the following regions across Europe: Denmark, England, Estonia, France, Germany, Holland, Ireland, Italy, Norway, Poland, Serbia, Scotland, Sweden, Switzerland and Wales (Fig. 1). Total genomic DNA was isolated using the DNeasy Blood and Tissue Kit (QIAGEN Ltd.), following the manufacturer's instructions. DNA from 242 modern samples was extracted, amplified and sequenced at Trinity College Dublin (TCD). Two regions of the mitochondrial genome were studied: a 464 bp fragment, including 422 bp of the 5′ end of the cytochrome b (cytb) gene, was amplified using the primers from Frati et al. (1998), and a 402 bp fragment, including 325 bp of the 5′ end of the control region (CR), was amplified using the primer pair JP-VVF (5′-CTC CCT AAG ACT CAA GGA AG-3′) and JP-VVR (5′-CCT GAG GTA AGA ACC AGA TGC-3′), designed as part of this study. The 50 µl reaction mixture contained 5 µl of template DNA, 2.5 Units Platinum®Taq DNA Polymerase (Invitrogen), 2.5 mM $MgCl_2$, 50 mM KCl, 20 mM Tris–HCl (pH 8.4), plus dNTPs at 0.2 mM, and 0.4 mM of each primer. The PCR program consisted of an initial denaturation step of 4 min at 94 °C, followed by 35 cycles of 94 °C for 30 s, 55 °C for 30 s, 72 °C for 30 s, and a final elongation step of 4 min at 72 °C. PCR amplicons were purified using the QIAquick PCR purification kit (QIAGEN Ltd.) according to the manufacturer's instructions, and sequencing was performed by Macrogen Inc. (www.macrogen.com).

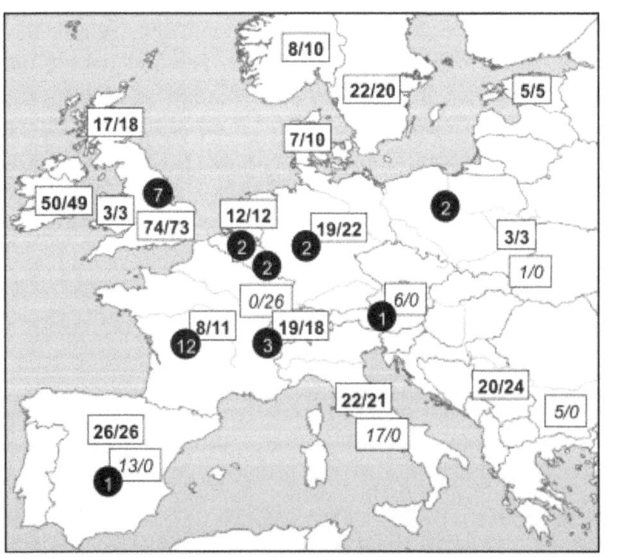

Figure 1: Map of sample locations. Numbers indicate modern cytochrome b data/modern control region data, with numbers in bold denoting sequence data generated as part of this study and numbers in italic showing published data used for analysis. Black circles indicate ancient sequences, with sample sizes.

A subset of 89 samples was analysed at the University of California, Davis. The 5′ portion of cytb was amplified using the primers RF14724 and RF15149 and conditions published by Perrine et al. (2007), and a 409 bp fragment of the CR using primers VVDL1 and VVDL6 and conditions from Aubry et al. (2009). PCR products were purified using ExoSAP-IT (USB/Affymetrix Inc.) and sequenced in both forward and reverse directions on an ABI 3730 capillary sequencer (Applied Biosystems). Sequencher version 4.8 (Gene Codes, Inc.) was used to visualise chromatograms and to edit and align sequences.

Archaeological Samples

Remains from two ancient red foxes were obtained from Carsington Pasture Cave in Derbyshire, England (a left ulna: CPC-98-2251, and a right radius: CPC-98-2252). DNA extractions from ancient material were performed as described in Edwards et al. (2010)and amplification of mitochondrial DNA was performed as described above for the TCD analysis. Appropriate measures were followed to control contamination (Cooper and Poinar, 2000) and multiple controls were used (Spencer and Howe, 2004). No contaminant bands were observed during any part of the analyses and, to identify possible post-mortem damage in the ancient DNA, sequencing was undertaken in both forward and reverse directions from several PCR products derived from two independent extractions. In addition, ancient DNA data from Austria, Belgium, England, France, Germany, Luxembourg, Poland, Spain and Switzerland were included from the study by Teacher et al. (2011): 30 cytb and 27 CR sequences (Fig. 1) were used in the network analysis, while 20 stratigraphically dated samples were included in the Bayesian analysis.

Phylogenetic Analysis

Statistical Analysis

Thirty-six mitochondrial cytb sequences from Frati et al. (1998) were included (eight populations from Austria, Bulgaria, Italy, Sardinia and Spain), as well as four Galician (GenBank accession numbers AY586403–AY586406), and one Ukrainian (Fernandes et al., 2008) sequence. For the analyses of the mitochondrial CR, 26 sequences from France/Switzerland (Valiere et al., 2003) were included. Except for the ancient DNA samples from Teacher et al. (2011), which were shorter fragments, this gave a total long-fragment dataset of 356 cytb sequences and 351 CR sequences. All mitochondrial DNA sequences were aligned by ClustalX (version 1.8; Thompson et al., 1997) and were truncated in order to include the maximum number of published and new sequences, leaving 321 bp of the 5′ end of cytb and 251 bp of the 5′ end of the CR. Median networks were constructed using the median-joining algorithm of Bandelt et al. (1999) in Network (version 4.6; www.fluxus-engineering.com).

Phylogenetic analyses of cytb and CR sequences were conducted separately. The ARLEQUIN software (version 3.5; Excoffier et al., 2005) was used to estimate haplotype and nucleotide diversity, and to generate a matrix of pairwise F_{ST} based on haplotype frequencies (Weir and Cockerham, 1984) and a matrix of pairwise Φ_{ST} to incorporate pairwise differences between haplotypes (Nei and Li, 1979). Pairwise F_{ST} values can be used as an estimate of genetic distances between populations over shallow time depths (Slatkin, 1995). Multidimensional scaling (MDS; Kruskal, 1964) was performed based on the pairwise Φ_{ST} matrices obtained using the SPSS© program (Statistics, version 19). Statistical significance (α = 0.05) of Φ_{ST} was based on 110 permutations, and then corrected for multiple tests using the sequential Bonferroni method (Rice, 1989). The program SAMOVA (version 1.0; Dupanloup et al., 2002) was used to assess population structure in the two datasets. The analysis was run using 100 simulated annealing processes for K values from 2 to 14.

Temporal Network Analysis

To allow comparison of the ancient DNA data taken from Teacher et al. (2011) to our modern results, the cytb sequence was truncated to 201 bp and the CR was truncated to 193 bp. These alignment lengths differ from the Network analysis above as the latter did not include any ancient data. Two alignments were constructed: 386 individuals for cytb, comprising 355 modern and 31 ancient (30 from Teacher et al., 2011; plus our CPC individual); and 378 individuals for the CR, comprising 351 modern and 27 ancient (all from Teacher et al., 2011). For each of the two alignments, sequences were divided into three time categories: modern, Holocene (1500–12,000 years ago), and Pleistocene (>12,000 years ago). To assess the continuity of haplotypes across time, a temporal network was constructed from each sequence alignment using TempNet (Prost and Anderson, 2011).

Bayesian Analysis using Modern and Ancient Data

To investigate phylogeographic patterns and to estimate the timescale of events, Bayesian analysis was performed on a concatenated alignment with BEAST (version 1.6.1; Drummond and Rambaut, 2007). This combined cytb/CR dataset contained 394 bp in total (201 bp of cytb and 193 bp of the CR), with 301 modern and 20 ancient red foxes. First, selection of an appropriate substitution model and partitioning scheme was performed using the Bayesian information criterion (BIC) in PartitionFinder (Lanfear et al., 2012). The best scheme was one in which separate substitution models were assigned to cytb (K80 + G) and to the CR (TIMef + G). This scheme had a better BIC score than schemes in which the two regions were grouped together, or in which the different codon positions in cytb were treated separately. Rate heterogeneity among sites was modelled using a discrete gamma distribution with six rate categories (Yang, 1994). The analysis was repeated without a model of rate variation among sites.

The data were analysed using a discrete phylogeographic model, allowing rates of pairwise spatial diffusion to be estimated (Lemey et al., 2009). Sequences were grouped into seven discrete geographic categories: Ireland, Britain, Iberia, Central Europe, Scandinavia, Italy, and the Balkans. Forward and backward diffusion rates between each pair of locations were allowed to differ, while non-zero diffusion rates were identified using Bayes factors (with a cut-off value of 10.0). All diffusion rates were assumed to be equally likely, such that geographic distance did not inform their prior probabilities.

To place a timescale on the estimates of population history, phylogeny, and phylogeographic history, the molecular clock was calibrated using the stratigraphic dates from Teacher et al. (2011) and the radiocarbon date from the Carsington Pasture Cave specimen. To test whether the structure and spread of these sampling times were sufficient for calibration, five replicates were analysed in which the dates were randomised among the sequences using SiteSampler (Ho and Lanfear, 2010). The 95% credibility intervals of the estimated rates did not include the mean estimate from the original, non-randomised dataset, indicating that the sampling times of the sequences were sufficient for calibration (Firth et al., 2010; Ho et al., 2011). Posterior distributions of parameters were estimated using MCMC sampling. Samples were drawn every 10,000 MCMC steps over a total of 100,000,000 steps, with the first 10% of samples discarded as burn-in. The results of two independent analyses were compared to check for convergence to the stationary distribution, and samples from the posterior were combined. Sufficient sampling was checked using Tracer (version 1.5; http://tree.bio.ed.ac.uk/software/tracer).

Large differences in geographic representation between ancient and modern datasets can lead to biased estimates of substitution rates (Navascués and Emerson, 2009). To investigate this potential problem in our data on our rate estimates, we analysed the geographic subset for which we had sufficient numbers of both ancient and modern samples, specifically Central Europe. This dataset included samples from Austria (1 ancient), Belgium (1 ancient), France (8 modern and 7 ancient), Germany (17 modern and 1 ancient), Holland (12 modern), Poland (3 modern and 2 ancient) and Switzerland (18 modern and 3 ancient).

Modern Demographic Analyses

To test demographic hypotheses and quantify isolation between mainland Europe and Britain and Ireland, Markov chain Monte Carlo (MCMC) simulations were conducted using IMa (Hey and Nielsen, 2007). Both 'isolation with migration' and 'isolation with no migration' models were tested and maximum-likelihood estimates produced, with 90% highest posterior density intervals for splitting times, effective population sizes and, depending on the model, gene flow in each direction. To minimise substructure within the two populations, the mainland European dataset was restricted to sequences from Central Europe, resulting in a comparison between 139 British and Irish sequences and 55 Central Europe sequences. The HKY substitution model was used. These analyses involved mutation-scaled parameters and were, therefore, independent of the mitochondrial mutation rate, which was unknown. However, it was necessary to assume particular mutation rates to translate these scaled estimates into biologically meaningful quantities. This was done using both evolutionary rates consistent with those inferred for, and supported independently in, multiple canid species (for example, Vilà et al., 1997; Wayne et al., 1997; Savolainen et al., 2002; Aubry et al., 2009), and an internal-node based estimate in the present study derived from the Bayesian analysis with ancient and modern samples described above (see also Teacher et al., 2011). In particular, the evolutionary rate for the combined cytb and CR fragment used in this study was assumed to be 9.36% per million years (Myr), which was the weighted average of mutation rates for the cytb (2.8%; 321 bp) and CR (17.75%; 251 bp) portions (Vilà et al., 1997; Wayne et al., 1997; Savolainen et al., 2002; Aubry et al., 2009).

After several initial runs with default parameters, 60 geometrically heated chains (Hey and Nielsen, 2007) were used with parameter '-g1' set to 0.9. Multiple runs of 1000 steps (following 104 or 105 iterations as burn-in as necessary) were conducted to assess mixing and to narrow the range of parameter space. A second independent (different random seed) run of 106 steps in MCMC mode was then conducted. Consistent marginal peak locations (parameter estimates) with unimodal likelihood curves approaching zero on both ends, effective sample sizes of >50, and trend plots free of systematic change indicated good mixing. To produce final parameter estimates based on joint distributions, the 'LoadTree' mode in IMa was employed using trees produced in the longer MCMC run (Hey and Nielsen, 2007).

Inferences regarding the occurrence of past events of population expansion or decline were based on neutrality test estimates, as calculated via Tajima›s D (Tajima, 1996) and Fu›s Fs (Fu, 1997) in ARLEQUIN (version 3.5; Excoffier et al., 2005). Values were calculated for all modern cytb and CR samples, and for each region separately.

RESULTS

We generated 314 modern cytb sequences and 325 CR sequences for this study (Fig. 1). For the cytb analysis, a further 42 sequences from European foxes were taken from published (Frati et al., 1998) and archived (GenBank accession numbers AY586403–AY586406) sources. Of the two archaeological fox samples from Carsington Pasture Cave, CPC-98-2251 yielded a reproducible reliable sequence using the cytb primer set of Frati et al. (1998), but no result from the CR primer set. This sample was radiocarbon dated to 4712 ± 22 uncalibrated radiocarbon years (UBA-14767). This date equates to 5325–5578 years calibrated BP, so this English fox dates to the Neolithic, which arrived in Britain about 4000 cal. years BC (Brown, 2007). No DNA was amplifiable from CPC-98-2252. Twenty-six CR sequences from the region of France/Switzerland were also incorporated into the analysis (Valiere et al., 2003). Thus the long-fragment dataset included 356 individuals with cytb sequences, 351 individuals with CR sequences and 301 individuals with both fragments. The short-fragment dataset included the same concatenated sequences (trimmed) along with 30 additional published ancient sequences (Teacher et al., 2011).

Modern Haplotypes across Europe

Based on the long-fragment dataset, 44 cytb and 69 CR haplotypes were found in Europe, of which only seven and 15, respectively, were found in multiple locations (Fig. 1, Tables 1 and 2). Of the four most common cytb haplotypes, only haplotype A1 showed a wide geographic spread, being found in all regions (Fig. 2a). The remaining haplotypes were generally common in one or two regions: haplotypes B and D (Fig. 2a) were mainly found in the British Isles (Ireland and Britain, at 88.8% and 71.4% respectively), whereas haplotype A2 was predominantly found (76.7%) in Scandinavia and central Europe. In contrast to cytb, the CR haplotypes were more differentiated and no single haplotype was found in all regions. Only four were found in more than two regions, and most were restricted to one or two populations (Fig. 2b).

Table 1: Sample size, number of mtDNA cytb haplotypes including singleton haplotypes, and gene diversity for each population. Tajima's D and Fu's Fs tests of selective neutrality are shown for each region

Population	n	Haplotypes	Unique haplotypes	Singleton haplotypes	Haplotype diversity	Nucleotide diversity	Tajima's D (Pvalue)	Fu's F(Pvalue)
Ireland	50	5	2	2	0.26 ± 0.08	0.0028 ± 0.002	−1.36 (0.07)	−0.42 (0.43)
Britain							−0.47 (0.37)	−0.04 (0.53)
Scotland	17	2	0	0	0.12 ± 0.10	0.0011 ± 0.001		
England	74	7	3	2	0.51 ± 0.06	0.0051 ± 0.003		
Wales	3	2	0	0	0.67 ± 0.31	0.0062 ± 0.006		
Iberia							−1.45 (0.06)	−4.31 (0.01)
Spain	39	9	7	4	0.74 ± 0.05	0.0034 ± 0.003		
Central Europe							−1.17 (0.11)	−5.91 (0.02)
Holland	12	2	0	0	0.41 ± 0.13	0.0038 ± 0.003		
France	8	4	0	0	0.75 ± 0.14	0.0080 ± 0.005		
Switzerland	19	8	5	5	0.82 ± 0.06	0.0088 ± 0.005		
Germany	19	5	3	2	0.71 ± 0.08	0.0030 ± 0.002		
Austria	6	5	3	2	0.93 ± 0.12	0.0166 ± 0.011		
Denmark	7	2	0	0	0.29 ± 0.20	0.0009 ± 0.001		
Scandinavia							−0.67 (0.30)	−1.76 (0.11)
Sweden	21	5	3	2	0.69 ± 0.07	0.0030 ± 0.002		
Norway	8	1	0	0	0.00 ± 0.00	0.0000 ± 0.000		

Estonia	5	2	0	0	0.60 ± 0.18	0.0037 ± 0.003		
Italy	39	8	5	3	0.58 ± 0.09	0.0028 ± 0.002	−1.88 (0.02)	−3.98 (0.00)
Balkans							−0.56 (0.35)	−1.19 (0.27)
Bulgaria	5	4	3	2	0.90 ± 0.16	0.0087 ± 0.006		
Serbia	20	5	3	0	0.80 ± 0.04	0.0062 ± 0.004		
East Europe	4	1	0	0	0.00 ± 0.00	0.0000 ± 0.000		
All samples	**356**	**44**	**37**	**24**	**0.80 ± 0.02**	**0.0072 ± 0.004**	**−1.08 (0.18)**	**−2.52 (0.20)**

Table 2: Sample size, number of mtDNA CR haplotypes including singleton haplotypes, and gene diversity for each population. Tajima's D and Fu's Fs tests of selective neutrality are shown for each region

Population	n	Haplotypes	Unique haplotypes	Singleton haplotypes	Haplotype diversity	Nucleotide diversity	Tajima's D (Pvalue)	Fu's F(Pvalue)
Ireland	49	12	7	5	0.71 ± 0.06	0.0110 ± 0.007	−1.11 (0.14)	−2.08 (0.21)
Britain							−0.47 (0.48)	−1.22 (0.36)
Scotland	18	4	1	1	0.31 ± 0.14	0.0060 ± 0.004		
England	73	11	5	5	0.68 ± 0.05	0.0126 ± 0.007		
Wales	3	2	1	1	0.67 ± 0.31	0.0106 ± 0.010		
Iberia							2.15 (0.98)	1.60 (0.78)
Spain	26	6	5	0	0.83 ± 0.03	0.0124 ± 0.007		
Central Europe							−0.31 (0.45)	−14.23 (0.00)
Holland	12	6	4	2	0.85 ± 0.07	0.0130 ± 0.008		
France	11	6	1	0	0.85 ± 0.09	0.0181 ± 0.011		
Switzerland	18	8	2	2	0.82 ± 0.06	0.0125 ± 0.008		
France/ Switzerland	26	16	10	6	0.95 ± 0.02	0.0153 ± 0.009		
Germany	22	10	4	3	0.92 ± 0.03	0.0130 ± 0.008		
Denmark	10	3	0	0	0.38 ± 0.18	0.0060 ± 0.004		
Scandinavia							−0.63 (0.30)	−2.69 (0.12)
Sweden	20	10	7	3	0.91 ± 0.04	0.0140 ± 0.008		
Norway	10	2	1	1	0.20 ± 0.15	0.0040 ± 0.003		
Estonia	5	2	1	0	0.60 ± 0.18	0.0120 ± 0.009		
Italy	21	5	2	0	0.62 ± 0.09	0.0045 ± 0.003	0.94 (0.83)	0.34 (0.60)
Balkans								
Serbia	24	5	3	2	0.70 ± 0.06	0.0149 ± 0.009	1.23 (0.91)	3.34 (0.94)
East Europe	3	1	0	0	0.00 ± 0.00	0.0000 ± 0.000		
All samples	**351**	**69**	54	31	**0.93 ± 0.01**	**0.0186 ± 0.010**	**0.29 (0.58)**	**−2.14 (0.43)**

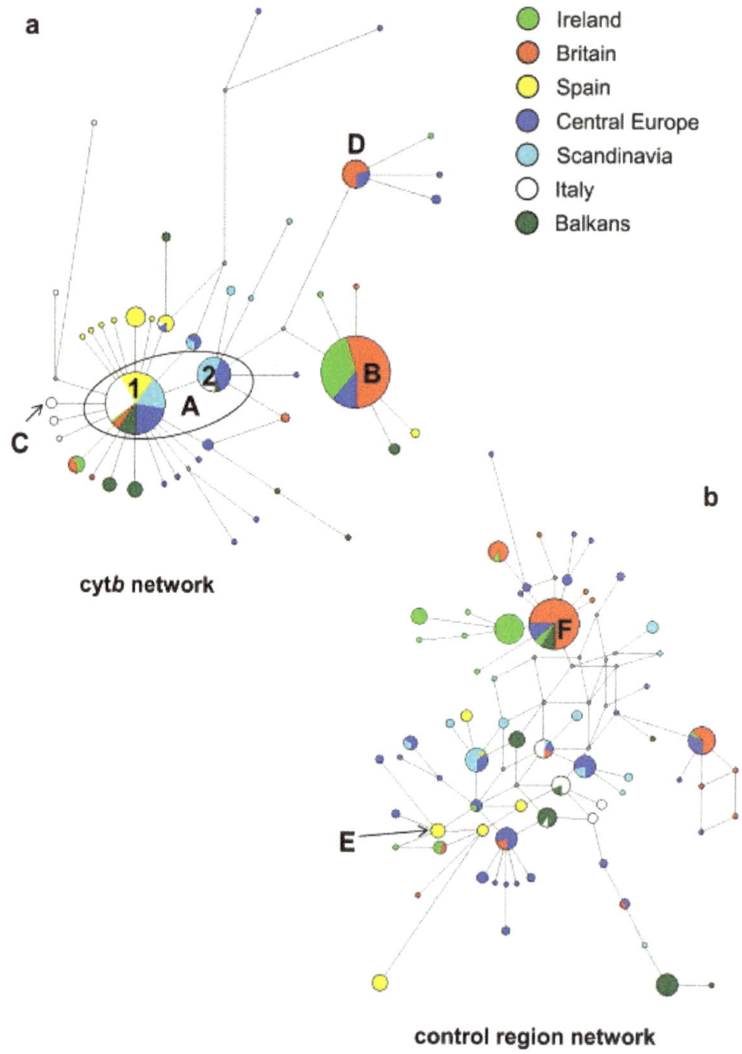

Figure 2: Median networks constructed from (a) 321 bp of cytb from 356 modern red fox sequences, and (b) 251 bp of CR from 351 modern red fox sequences, across Europe. Circles represent sequence haplotypes, the area being proportional to the frequency of the haplotypes. Points are theoretical intermediate nodes introduced by the median-joining algorithm, and branches between haplotypes represent single nucleotide mutations. Haplotypes labelled with a letter are discussed in the main text and equal those shown in Fig. 5and Table 3.

Population Structuring and Glacial Refugia

All Φ_{CT} values produced by SAMOVA were statistically significant (data not shown). The highest Φ_{CT} value for the cytb dataset corresponded to K = 4, but two of these groups contained a single population, that of Austria and Bulgaria, both taken from the study by Frati et al. (1998). The highest configuration with no single population groupings was K = 2. The geographical organisation of these groups showed clustering of the British Isles (Britain and Ireland) with Holland, and all other populations forming the other group. The grouping of Holland and British Isles was consistent across all values of K (Fig. 3a). At K = 7, samples from Estonia, Denmark and Sweden clustered separately from mainland Europe and Norway; this clustering pattern was also supported by pairwise F_{ST} values.

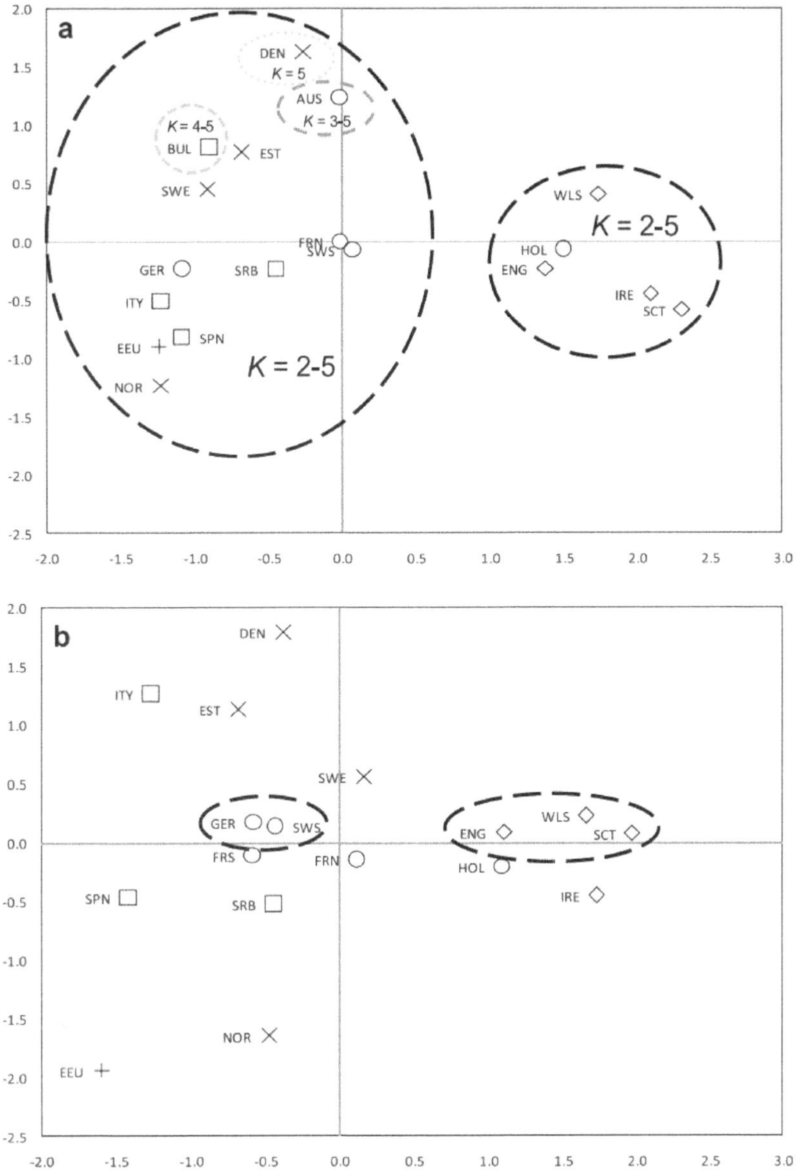

Figure 3: Two-dimensional MDS plots, drawn using data from: (a) 321 bp of cytb, and (b) 251 bp of CR, summarising genetic distances among the modern red fox populations. (a) The proportion of the data explaining the first two principal coordinates, the R^2 value, is 87.0%. Hierarchical groupings as calculated by SAMOVA (Kvalues 2–5) are shown, highlighting the independence of the Holland and British Isles cluster to increasing values of K. (b) The proportion of the data explaining the first two principal coordinates, the R^2 value, is 82.6%. For this rapidly mutating marker, the best hierarchical grouping improves as K is increased, until K = 14, where only neighbouring Germany and Switzerland, and the British Isles cluster together (as shown).

Figure 4: Map describing the seven locations (Balkans, Britain, Central Europe, Iberia, Ireland, Italy and Scandinavia) to which each of the 301

modern and 20 ancient red fox were assigned in the combined cytb/CR phylogeographic analysis. Non-reversible diffusion rates were estimated across the entire distribution of posterior trees and, therefore, reflect average rates of diffusion over time. Pairwise diffusions with Bayes factor >10 are shown; these are considered to be different from zero.

The highest Φ_{CT} value produced by SAMOVA for the CR dataset was K = 14, suggesting a high degree of differentiation based on this more rapidly mutating marker. The highest configuration with no single population groupings was, again, K = 2. However, the best hierarchical grouping improved as K increased, until K = 14, where only neighbouring Germany and Switzerland, and the British Isles (Britain and Ireland) clustered together (Fig. 3b). Pairwise F_{ST} values showed a similar pattern.

Distribution of Haplotypes through Time

Three cytb haplotypes were found in the Pleistocene, Holocene and modern-day eras (Fig. 5a). These were split broadly across samples from Britain and mainland Europe (Table 3). Two of these haplotypes were also the most common ones found in modern samples (Fig. 2a); the geographic distribution of these Pleistocene and Holocene haplotypes is very similar to the modern day haplotype distribution. In particular, haplotype B is found most frequently in the British Isles, Holland and France, and is also found in these regions during the Pleistocene and Holocene, suggesting constant occupation during the LGM. In contrast, no CR haplotypes were found across all eras, although a small number were found in two different eras (Fig. 5b). This may partly reflect high genetic differentiation among locations, but also likely reflects sampling error given the high diversity of this marker in all locations.

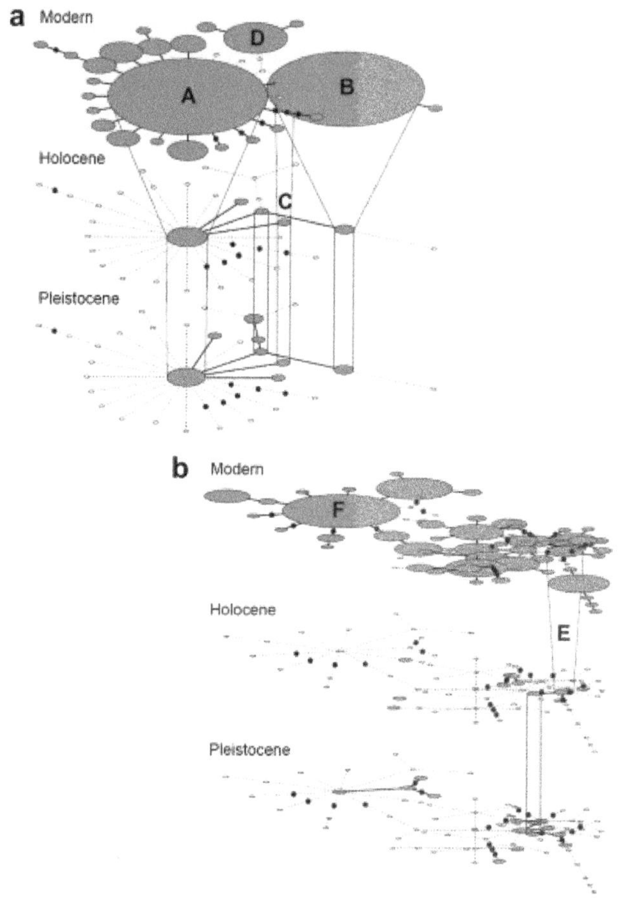

Figure 5: Temporal networks constructed from: (a) 201 bp of cytb from 355 modern and 31 ancient red foxes, and (b) 193 bp of CR from 351 modern and 27 ancient red foxes. Haplotypes labelled with a letter are discussed in the main text and equal those shown in Fig. 2 and Table 3.

Table 3: Table detailing haplotype-sharing across modern, Holocene and Pleistocene time-slices. Haplotype labels correspond to those labelled in Fig. 2

Haplotype label	Modern	Holocene	Pleistocene
A	Italy (31), Spain (21), Sweden (21), Germany (15), Serbia (11), Norway (8), Denmark (7), Switzerland (7), England (4), Estonia (3), Holland (3), Poland (3), Bulgaria (1), France (1), Ireland (1), Scotland (1)	France (7), England (1), Spain (1)	England (2), France (2), Austria (1), Belgium (1), Germany (1), Poland (1)
B	England (50), Ireland (44), Scotland (16), Holland (9), Serbia (3), Switzerland (3), France (2), Spain (2), Wales (2)	England (1), Switzerland (1)	England (1), France (1)
C	Italy (2)	Switzerland (1)	France (1)
D	England (14), Switzerland (4), Austria (1), France (1), Ireland (1), Wales (1)		Luxembourg (2)
E	France (2), Spain (6)	France (2)	
F	England (37), Scotland (15), Holland (8), Ireland (6), France (4), Serbia (4), Wales (2), Switzerland (1)		France (1)

Bayesian Analysis of Modern and Ancient Data

Bayesian phylogenetic analysis using a discrete geospatial model showed that the most highly supported pairwise diffusions were from Britain to Ireland, Central Europe to Britain, and Scandinavia to Central Europe (Fig. 4). The posterior median coalescence time for all lineages was 50,000 years, with a 95% credibility interval of 32,700–77,800 years. The posterior median mutation rates estimated for our 394 bp cytb/CR fragment based on the Central European ancient and modern samples were 33.81% per Myr (95% CI: 13.75–59.38% per Myr) or 26.29% Myr (95% CI: 16.12–41.54% per Myr) depending, respectively, on whether the model included gamma-distributed rates among sites or assumed a constant rate among sites.

Demography

Tests of neutrality based on cytb revealed statistically significant evidence of population expansion only in Iberia and Italy, although both Fu's F and Tajima's D statistics were consistently negative, suggestive of a weak signature of expansion throughout Europe (Table 1). Tests of neutrality based on the more rapidly mutating CR revealed statistically significant evidence of population expansion only in Central Europe, according to a strongly negative Fu›s F value. Otherwise, values varied from weakly and non-significantly negative to positive across regions, revealing no clear trend (Table 2).

Hypotheses of demographic isolation between mainland Europe and the British Isles were tested using IMa (Hey and Nielsen, 2007). Assuming a nucleotide mutation rate of 9.36% per Myr, both the 'isolation with migration' and 'isolation only' models indicated that the red fox populations in the British Isles became isolated from those of mainland Europe around the end of the last ice age (Table 4). Although the model incorporating migration suggested a somewhat earlier split than the model assuming no migration, both models have a lower 90% HPD limit around 6000 years. The isolation with migration model indicated some degree of gene flow since colonisation from mainland Europe to Britain and Ireland but not in the reverse direction, where the posterior density function displayed a mode approaching zero. The higher mutation rates estimated from the Bayesian analyses imply a much more recent splitting time, more in line with a mid-Holocene divergence (Table 4).

Table 4: IMa joint estimates of demographic parameters: N_e (effective population size = $\theta/4\mu$; where θ is genetic diversity and μ is mutation rate), $N_e m$ (effective migration rate = $\theta m/4$; where m is the migration rate per generation), and population splitting times (t/μ) under two models based on concatenated cytb and CR sequences. The 90% highest posterior density (HPD) intervals corresponding to parameter estimates are indicated in parentheses. Unless otherwise indicated, estimates are based on the assumed site-specific mutation rate, 9.36% per Myr for the combined cytb/CR haplotype

Demographic model:	Isolation with migration (90% HPD)	Isolation only (90% HPD)
N_e Britain/Ireland	150,000 (62,000–233,000)	215,000 (139,000–325,000)
N_e Central Europe	481,000 (243,000–1,035,000)	416,000 (225,000–844,000)
N_e Ancestral	112,000 (40,000–374,000)	262,000 (157,000–422,000)
$N_e m$ into Britain/Ireland	3270 (110–20,140)	–
$N_e m$ into Central Europe	–[a]	–
Splitting time	19,000 years (5900–51,500)	10,000 years (5700–14,500)
Alternative splitting time 1[b]	6765 years (2100–18,335)	3560 years (2030–5160)
Alternative splitting time 2[c]	5260 years (1630–14,260)	2770 years (1580–4015)

[a]The posterior density function smoothed peak approached zero for gene flow into Central Europe.

[b]These estimates assume a mutation rate of 26.29% per Myr. This value is based on the Bayesian median posterior estimate assuming homogeneous rates among sites, and does not take into account the 95% CI of the estimate.

[c]These estimates assume a mutation rate of 33.81% per Myr. This value is based on the Bayesian median posterior estimate using a gamma-distributed rate heterogeneity across sites, and does not take into account the 95% CI of the estimate.

DISCUSSION

British Isles Fox Populations

Most British mammals show genetic affinity to the Eurasian mainland populations; for example, badgers (Meles meles; Marmi et al., 2006), red deer (Cervus elaphus; Sommer et al., 2008), and least weasel (Mustela nivalis; Lebarbenchon et al., 2010), but British red foxes are clearly diverged from mainland populations. Fossil evidence suggests that foxes were present in Britain around 28,000 to 35,000 years ago (Sommer and Nadachowski, 2006), and ancient fox samples (Teacher et al., 2011) support the presence of the predominant British cytb and CR haplotypes in England and Central Europe (France, Switzerland and Luxembourg) during the Pleistocene, through into the Holocene. The strongly supported Bayesian diffusion rates suggest that British foxes originated from these Central European populations and, prior to the formation of the English Channel, were part of a contiguous population. Gene flow would have been restricted as the channel was forming but, as evidenced by the SAMOVA grouping between Britain and Holland in modern samples, probably continued between these areas across Doggerland prior to full submersion of this landbridge around 8,200BP. Using the more traditional assumptions about mtDNA mutation rates (that is, based on model-adjusted divergence in closely-related species calibrated to the fossil record), the IMa analysis indicates that British fox populations underwent isolation and subsequent divergence of haplotypes with the closure of the landbridge. This evidence lends support to a founder event and subsequent independent evolution in isolation in Britain.

There is contradictory evidence with respect to the relative origins of Irish and British foxes. As can be seen from the high number of cytb haplotypes shared between the two countries, these populations would seem to share a common origin, but less clear is which population was established first or from the other. On the one hand, the Bayesian geospatial analysis provides support for an origin of Irish foxes from British ones. On the other hand, the CR network indicates that Irish foxes have a high frequency (the majority) of unique haplotypes, most of which form a unique clade with haplotypes up to three mutations removed from the basal haplotype shared with Britain. British foxes have an extremely low frequency of unique haplotypes, comprising three singletons, each one mutation away from a haplotype shared with Ireland and mainland Europe. Independent evidence is also inconclusive. Red foxes are assumed to be native to Ireland, but direct Pleistocene fossil evidence for their presence is lacking (Yalden, 1999). Instead, fossil records indicate that red foxes were first recorded between 5000 and 3000 BP (Sommer and Benecke, 2005a), after the closure of a putative landbridge to Ireland around 9500 years ago (Wingfield, 1995). However, overland colonisation of Ireland did occur in some species, including the brown bear (Sommer and Benecke, 2005b) and stoat (Mustela erminea; Martínková et al., 2007). Thus, despite the lack of fossil evidence for their ancient presence in Ireland, it is possible that foxes colonised Ireland independently during the Pleistocene, and then colonised Britain from there. If, instead, red foxes only arrived in Ireland (from Britain or elsewhere) in the mid-Holocene, the CR data paint a misleading picture. Clearly this question would benefit from the addition of other genetic markers because the mitochondrial markers used in the present study provide only a single genealogical reflection of the demographic history of these populations.

Phylogeographic Structuring of Red Foxes

Our findings were consistent with those of previous studies (Frati et al., 1998; Teacher et al., 2011) in revealing a lack of deep phylogenetic structure, such as on the scale of that seen in North American red foxes, where major lineages diverged on the order of 400,000 years ago (Aubry et al., 2009). Nevertheless, our use of a relatively large dataset with a broad geographic coverage of Europe enabled us to elucidate population structure and historical demography in this region. Both cytb and CR data provided evidence of structuring of populations across Europe. Most notable was the division between most of Europe and Britain, Ireland and Holland, for example as evidenced in the MDS plots. Also evident from Φ_{ST}, MDS, SAMOVA, and Bayesian diffusion analyses was a relatively high degree of isolation of Iberian foxes from those of other European regions and, more generally, less connectivity in the southern than northern regions. This finding was consistent with those of Frati et al. (1998), who found relatively high differentiation across Mediterranean populations. Our findings of population structure, along with palaeontological evidence (Sommer and Benecke, 2005a; Sommer and Nadachowski, 2006) and ancient DNA analyses showing isolation by distance (Teacher et al., 2011), indicate that foxes were present across Europe throughout the last glaciation. A corollary of this inference is that red foxes were not restricted to peninsular refugia during the last ice age. Our results are in concordance with other studies, such as for brown bears (Valdiosera et al., 2007), a species that at one time was (incorrectly) believed to show structuring into peninsular refugia (Taberlet et al., 1998). Further evidence can be taken from the spatio-temporal stability of haplotypes. For example, cytb haplotypes found in areas such as France and England during the Pleistocene were found there subsequently in the Holocene, as well as in modern samples (see also Teacher et al., 2011).

Mitochondrial DNA variation of the red fox in Europe seems to indicate an expansion in population size. The occurrence of a historic bottleneck and subsequent expansion is further supported by the low nucleotide diversity in the cytb region (0.007), and CR (0.019), values similar to those seen in North American red foxes (0.009 and 0.023 respectively; Aubry et al., 2009). Tests of neutrality showed evidence of population expansion in the majority of regions; our data are drawn from a wider geographic area and have a larger sample size than previous studies, providing greater power to detect population change.

In areas covered by glaciers, we would expect to see patterns of colonisation. In Scandinavia, there appears to have been bi-directional recolonisation. Scandinavian populations showed two distinct structures; Denmark and Sweden appeared to cluster together with Estonia (mainly haplotype A2), whilst Norway apparently clustered with mainland European populations (all haplotype A1). However, a larger sample size is needed, especially as Norwegian foxes are from one peninsula in Finnmark, whilst Swedish and Danish foxes were more widely sampled. In addition, significant Bayesian diffusion rates suggest movement from Scandinavia to Central Europe, but not vice versa. Although more genetic evidence is needed from haplotypes in Russia and more easterly populations, this does suggest that foxes colonised Scandinavia from two different directions. This easterly and westerly colonisation into Scandinavia is observed in a range of species, including brown bears (Taberlet et al., 1998; Korsten et al., 2009), field voles (Microtus agrestis; Jaarola et al., 1999) and common shrews (Sorex araneus;Lundqvist et al., 2011), leading to a distinctive suture zone of haplotypes, as would seem evident in our cytb data. CR data are less clear as it shows much greater population differentiation. However, pairwise F_{ST} values also exhibit some evidence of clustering in these regions.

CONCLUSIONS

Our research relates to two broader themes. Firstly, we challenge the standard hypotheses of glacial refugia where mammals were restricted to refugial areas in the ice age. Species that were ecologically adaptable, such as the red fox, could persist outside these areas and now exhibit high diversity and population structuring consistent with a constant occupation of Europe during the LGM and Quaternary glacial cycles. Secondly, our results show the importance of a wide sampling strategy, and highlight the need for inclusion of multiple data sources, including fossil evidence, to strengthen conclusions drawn from modern datasets.

ACKNOWLEDGEMENTS

We thank Andrew Chamberlain for provision of the two ancient fox specimens from Carsington Pasture Cave; Oliver Berry, Janosch Arnold and Christian Gortázar Schmidt for comments on an early draft; and Frank Zachos for both access to samples and constructive criticism of this manuscript. We thank Kate Lynch, Cian Murphy, Gianpiero Cavalleri and Vilena Nazmieva (Trinity College Dublin) for help with laboratory work and the following for access to modern fox tissue samples: Miguel Galiana and Juan Carranza (Spain); Dusko Cirovic (Serbia); Sandro Lovari, Luciano Palazzi, Lucia Burrini, Evidio Bartolini and Giorgia Romeo (Italy); Olavi Grönwall, Sabine Sten and Peter Mortensen (Sweden); Peter Akokan and Ciarán Butler (Ireland); Bruno Keller, Uwe Schaarschmidt, Adrian Vos and Mathias Büttner (Germany); Jaap Mulder (Holland); Marie-Lazarine Poulle and Sandrine Ruette (France); Siw Killengreen (Norway); Trine-Lee Wincentz Jensen (Denmark); Urmas Saarma (Estonia) and Tomasz Pietrzak (Poland). CJE was supported by the Irish Research Council for Science Engineering and Technology Basic Research Grant Scheme (project numbers SC/1999/409 and SC/2002/510), the John Fell Oxford University Press (OUP) Research Fund (2009–10), and a Wellcome Trust VIP Award (2010–11).

REFERENCES

1. Aubry, K.B., Statham, M.J., Sacks, B.N., Perrine, J.D., Wisely, S.M., 2009. Phylogeography of the North American red fox: vicariance in Pleistocene forest refugia. Mol. Ecol. 18, 2668e2686.
2. Bandelt, H.-J., Forster, P., Röhl, A., 1999. Median-joining networks for inferring intraspecific phylogenies. Mol. Biol. Evol. 16, 37e48.
3. Bhagwat, S.A., Willis, K.J., 2008. Species persistence in northerly glacial refugia of Europe: a matter of chance or biogeographical traits? J. Biogeogr. 35, 464e482.
4. Brown, A., 2007. Dating the onset of cereal cultivation in Britain and Ireland: the evidence from charred cereal grains. Antiquity 81, 1042e1052.
5. Cooper, A., Poinar, H., 2000. Ancient DNA: do it right or not at all. Science 289, 1139. Drummond, A.J., Rambaut, A., 2007. BEAST: Bayesian evolutionary analysis by sampling trees. BMC Evol. Biol. 7, 214.
6. Dupanloup, I., Schneider, S., Excoffier, L., 2002. A simulated annealing approach to define the genetic structure of populations. Mol. Ecol. 11, 2571e2581.
7. Edwards, C.J., Magee, D.A., Park, S.D., McGettigan, P.A., Lohan, A.J., Murphy, A., Finlay, E.K., Shapiro, B., Chamberlain, A.T., Richards, M.B., Bradley, D.G., Loftus, B.J., MacHugh, D.E., 2010. A complete mitochondrial genome sequence from a Mesolithic wild aurochs (Bos primigenius). PLoS One 5, e9255.
8. Excoffier, L., Laval, G., Schneider, S., 2005. Arlequin ver. 3.0: an integrated software package for population genetics data analysis. Evol. Bioinform. Online 1, 47e50.
9. Fernandes, C.A., Ginja, C., Pereira, I., Tenreiro, R., Bruford, M.W., Santosreis, M., 2008. Species-specific mitochondrial DNA markers for identification of non-invasive samples from sympatric carnivores in the Iberian peninsula. Conserv. Genet. 9, 681e690.
10. Firth, C., Kitchen, A., Shapiro, B., Suchard, M.A., Holmes, E.C., Rambaut, A., 2010. Using time-structured data to estimate evolutionary rates of double-stranded DNA viruses. Mol. Biol. Evol. 27, 2038e2051.
11. Fløjgaard, C., Normand, S., Skov, F., Svenning, J.C., 2009. Ice age distributions of European small mammals: insights from species distribution modelling. J. Biogeogr. 36, 1152e1163.
12. Frati, F., Hartl, G.B., Lovari, S., Delibes, M., Markov, G., 1998. Quaternary radiation and genetic structure of the red fox Vulpes vulpes in the Mediterranean Basin, as revealed by allozymes and mitochondrial DNA. J. Zool. 245, 43e51.
13. Fu, Y.X., 1997. Statistical tests of neutrality of mutations against population growth, hitchhiking and background selection. Genetics 147, 915e925.
14. Hewitt, G.M., 2000. The genetic legacy of the Quaternary ice ages. Nature 405, 907e 913.
15. Hewitt, G.M., 2004. Genetic consequences of climatic oscillations in the Quaternary. Philos. Trans. R. Soc. Lond. B Biol. Sci. 359, 183e195.
16. Hey, J., Nielsen, R., 2007. Integration within the Felsenstein equation for improved Markov chain Monte Carlo methods in population genetics. Proc. Natl. Acad. Sci. U. S. A. 8, 2785e2790.
17. Ho, S.Y.W., Lanfear, R., 2010. Improved characterisation of among-lineage rate variation in cetacean mitogenomes using codon-partitioned relaxed clocks. Mitochondrial DNA 21, 138e146.
18. Ho, S.Y.W., Lanfear, R., Phillips, M.J., Barnes, I., Thomas, J.A., Kolokotronis, S.-O., Shapiro, B., 2011. Bayesian estimation of substitution rates from ancient DNA sequences with low information content. Syst. Biol. 60, 366e375.
19. Hofreiter, M., Serre, D., Rohland, N., Rabeder, G., Nagel, D., Conard, N., Münzel, S., Pääbo, S., 2004. Lack of phylogeography in European mammals before the last glaciation. Proc. Natl. Acad. Sci. U. S. A. 101, 12963e12968.
20. Holderegger, R., Thiel-Egenter, C., 2009. A discussion of different types of glacial refugia used in mountain biogeography and phylogeography. J. Biogeogr. 36, 476e480.
21. Inoue, T., Nonaka, N., Mizuno, A., Morishima, Y., Sato, H., Katakura, K., Oku, Y., 2007. Mitochondrial DNA phylogeography of the red fox (Vulpes vulpes) in northern Japan. Zool. Sci. 24, 1178e1186.
22. Jaarola, M., Tegelström, H., Fredga, K., 1999. Colonization history in Fennoscandian rodents. Biol. J. Linn. Soc. Lond. 68, 113e127.
23. Korsten, M., Ho, S.Y.W., Davison, J., Pähn, B., Vulla, E., Roht, M., Tumanov, I.L., Kojola, I., Andersone-Lilley, Z., Ozolins, J., Pilot, M., Mertzanis, Y., Giannakopoulos, A., Vorobiev, A.A., Markov, N.I., Saveljev, A.P., Lyapunova, E.A., Abramov, A.V., Männil, P., Valdmann, H., Pazetnov, S.V., Pazetnov, V.S., Rõkov, A.M., Saarma, U., 2009. Sudden expansion of a single brown bear maternal lineage across northern continental Eurasia after the last ice age: a general demographic model for mammals? Mol. Ecol. 18, 1963e1979.
24. Kotlik, P., Deffontaine, V., Mascheretti, S., Zima, J., Michaux, J.R., Searle, J.B., 2006. A northern glacial refugium for bank voles (Clethrionomys glareolus). Proc. Natl. Acad. Sci. U. S. A. 103, 14860e14864.
25. Kruskal, J.B., 1964. Multidimensional scaling by optimising goodness to fit to a nonmetric hypothesis. Pyschometrika 29, 1e27.
26. Lanfear, R., Calcott, B., Ho, S.Y.W., Guindon, S., 2012. PartitionFinder: combined selection of partitioning schemes and substitution models for phylogenetic analyses. Mol. Biol. Evol.. http://dx.doi.org/10.1093/molbev/mss020. First published online: January 20, 2012.
27. Lebarbenchon, C., Poitevin, F., Arnal, V., Montgelard, C., 2010. Phylogeography of the weasel (Mustela nivalis) in the western-Palaearctic region: combined effects of glacial events and human movements. Heredity 105,

449e462.

28. Lemey, P., Rambaut, A., Drummond, A.J., Suchard, M.A., 2009. Bayesian phylogeography finds its roots. PLoS Comput. Biol. 5, e1000520.

29. Lundqvist, A.-C., Alström-Rapaport, C., Tegelström, H., 2011. Fennoscandian phylogeography of the common shrew Sorex araneus. Postglacial recolonisationdcombining information from chromosomal variation with mitochondrial DNA data. Acta Theriol. 56, 103e116.

30. Marmi, J., López-Giráldez, F., Macdonald, D.W., Calafell, F., Zholnerovskaya, E., Domingo-Roura, X., 2006. Mitochondrial DNA reveals a strong phylogeographic structure in the badger across Eurasia. Mol. Ecol. 15, 1007e1020.

31. Martínková, N., McDonald, R.A., Searle, J.B., 2007. Stoats (Mustela erminea) provide evidence of natural overland colonization of Ireland. Proc. Biol. Sci. 274, 1387e 1393.

32. Navascués, M., Emerson, B.C., 2009. Elevated substitution rate estimates from ancient DNA: model violation and bias of Bayesian methods. Mol. Ecol. 18, 4390e4397.

33. Nei, M., Li, W.H., 1979. Mathematical model for studying genetic variation in terms of restriction endonucleases. Proc. Natl. Acad. Sci. U. S. A. 76, 5269e5273.

34. Perrine, J.D., Pollinger, J.P., Sacks, B.N., Barrett, R.H., Wayne, R.K., 2007. Genetic evidence for the persistence of the critically endangered Sierra Nevada red fox in California. Conserv. Genet. 8, 1083e1095.

35. Pilot, M.M., Branicki, W., Je˛drzejewski, W., Goszczynski, J., Je ˛drzejewska, B., Dykyy, I., Shkvyrya, M., Tsingarska, E., 2010. Phylogeographic history of grey wolves in Europe. BMC Evol. Biol. 10, 104.

36. Prost, S., Anderson, C.N.K., 2011. TempNet: a method to display statistical parsimony networks for heterochronous DNA sequence data. Methods Ecol. Evol. 2, 663e667.

37. Randi, E., 2007. Phylogeography of south European mammals. In: Weiss, S., Ferrand, N. (Eds.), Phylogeography of Southern European Refugia. Springer, The Netherlands, pp. 101e126.

38. Rice, W.R., 1989. Analyzing tables of statistical tests. Evolution 43, 223e225.

39. Rull, V., 2009. Microrefugia. J. Biogeogr. 36, 481e484.

40. Sacks, B.N., Statham, M.J., Perrine, J.D., Wisely, S.M., Aubry, K.B., 2010. North American montane red foxes: expansion, fragmentation, and the origin of the Sacramento Valley red fox. Conserv. Genet. 11, 1523e1539.

41. Savolainen, P., Zhang, Y.P., Luo, J., Lundeberg, J., Leitner, T., 2002. Genetic evidence for an East Asian origin of domestic dogs. Science 298, 1610e1613.

42. Schipper, J., Chanson, J.S., Chiozza, F., Cox, N.A., Hoffmann, M., Katariya, V., Lamoreux, J., Rodrigues, A.S., Stuart, S.N., Temple, H.J., Baillie, J., Boitani, L., Lacher Jr., T.E., Mittermeier, R.A., Smith, A.T., Absolon, D., Aguiar, J.M., Amori, G., Bakkour, N., Baldi, R., Berridge, R.J., Bielby, J., Black, P.A., Blanc, J.J., Brooks, T.M., Burton, J.A., Butynski, T.M., Catullo, G., Chapman, R., Cokeliss, Z., Collen, B., Conroy, J., Cooke, J.G., da Fonseca, G.A., Derocher, A.E., Dublin, H.T., Duckworth, J.W., Emmons, L., Emslie, R.H., Festa-Bianchet, M., Foster, M., Foster, S., Garshelis, D.L., Gates, C., Gimenez-Dixon, M., Gonzalez, S., GonzalezMaya, J.F., Good, T.C., Hammerson, G., Hammond, P.S., Happold, D., Happold, M., Hare, J., Harris, R.B., Hawkins, C.E., Haywood, M., Heaney, L.R., Hedges, S., Helgen, K.M., Hilton-Taylor, C., Hussain, S.A., Ishii, N., Jefferson, T.A., Jenkins, R.K., Johnston, C.H., Keith, M., Kingdon, J., Knox, D.H., Kovacs, K.M., Langhammer, P., Leus, K., Lewison, R., Lichtenstein, G., Lowry, L.F., Macavoy, Z., Mace, G.M., Mallon, D.P., Masi, M., McKnight, M.W., Medellín, R.A., Medici, P., Mills, G., Moehlman, P.D., Molur, S., Mora, A., Nowell, K., Oates, J.F., Olech, W., Oliver, W.R., Oprea, M., Patterson, B.D., Perrin, W.F., Polidoro, B.A., Pollock, C., Powel, A., Protas, Y., Racey, P., Ragle, J., Ramani, P., Rathbun, G., Reeves, R.R., Reilly, S.B., Reynolds 3rd, J.E., Rondinini, C., Rosell-Ambal, R.G., Rulli, M., Rylands, A.B., Savini, S., Schank, C.J., Sechrest, W., Self-Sullivan, C., Shoemaker, A., Sillero-Zubiri, C., De Silva, N., Smith, D.E., Srinivasulu, C., Stephenson, P.J., van Strien, N., Talukdar, B.K., Taylor, B.L., Timmins, R., Tirira, D.G., Tognelli, M.F., Tsytsulina, K., Veiga, L.M., Vié, J.C., Williamson, E.A., Wyatt, S.A., Xie, Y., Young, B.E., 2008. The status of the worlds land and marine mammals: diversity, threat, and knowledge. Science 322, 225e230. Slatkin, M., 1995. A measure of population subdivision based on microsatellite allele frequencies. Genetics 139, 457e462.

43. Sommer, R., Benecke, N., 2005a. Late-Pleistocene and early Holocene history of the canid fauna of Europe (Canidae). Mamm. Biol. 35, 227e241.

44. Sommer, R., Benecke, N., 2005b. The recolonization of Europe by brown bears Ursus arctos Linnaeus, 1758 after the Last Glacial Maximum. Mamm. Biol. 35, 156e164.

45. Sommer, R.S., Nadachowski, A., 2006. Glacial refugia of mammals in Europe: evidence from fossil records. Mamm. Rev. 36, 251e265.

46. Sommer, R.S., Zachos, F.E., Street, M., Jöris, O., Skog, A., Benecke, N., 2008. Late Quaternary distribution dynamics and phylogeography of the red deer (Cervus elaphus) in Europe. Quat. Sci. Rev. 27, 714e733.

47. Spencer, M., Howe, C., 2004. Authenticity of ancient DNA results: a statistical approach. Am. J. Hum. Genet. 75, 240e250.

48. Stewart, J.R., Lister, A.M., 2001. Cryptic northern refugia and the origins of the modern biota. Trends Ecol. Evol. 16, 608e613.

49. Taberlet, P., Fumagalli, L., Wust-Saucy, A.G., Cosson, J.F., 1998. Comparative phylogeography and postglacial colonization routes in Europe. Mol. Ecol. 7, 453e464.

50. Tajima, F., 1996. The amount of DNA polymorphism maintained in a finite population when the neutral mutation rate varies among sites. Genetics 143, 1457e 1465.

51. Thompson, J.D., Gibson, T.J., Plewniak, F., Jeanmougin, F., Higgins, D.G., 1997. The ClustalX windows interface: flexible strategies for multiple sequence alignment aided by quality analysis tools. Nucleic Acids Res. 24, 4876e4882.

52. Teacher, A.G.F., Thomas, J.A., Barnes, I., 2011. Modern and ancient red fox (Vulpes vulpes) in Europe show an unusual lack of geographical and temporal structuring, and differing responses within the carnivores to historical climatic change. BMC Evol. Biol. 11, 214.

53. Tougard, C., Renvoisé, E., Petitjean, A., Quéré, J.P., 2008. New insight into the colonization processes of common voles: inferences from molecular and fossil evidence. PLoS One 3, e3532.

54. Valdiosera, C.E., García, N., Anderung, C., Dalén, L., Crégut-Bonnoure, E., Kahlke, R.D., Stiller, M., Brandström, M., Thomas, M.G., Arsuaga, J.L., Götherström, A., Barnes, I., 2007. Staying out in the cold: glacial refugia and mitochondrial DNA phylogeography in ancient European brown bears. Mol. Ecol. 16, 5140e5148.

55. Valiere, N., Fumagalli, L., Gielly, L., Miquel, C., Lequette, B., Poulle, M.-L., Weber, J.-M., Arlettaz, R., Taberlet, P., 2003. Long-distance wolf recolonization of France and Switzerland inferred from non-invasive genetic sampling over a period of 10 years. Anim. Conserv. 6, 83e92.

56. Vilà, C., Savolainen, P., Maldonado, J.E., Amorim, I.R., Rice, J.E., Honeycutt, R.L., Crandall, K.A., Lundeberg, J., Wayne, R.K., 1997. Multiple and ancient origins of the domestic dog. Science 276, 1687e1689.

57. Wayne, R.K., Geffen, E., Girman, D.J., Koepfli, K.P., Lau, L.M., Marshall, C.R., 1997. Molecular systematics of the Canidae. Syst. Biol. 46, 622e653.

58. Weir, B.S., Cockerham, C.C., 1984. Estimating F-statistics for the analysis of population structure. Evolution 38, 1358e1370.

59. Wingfield, R.T.R., 1995. A model of sea-levels in the Irish and Celtic seas during the end-Pleistocene to Holocene transition. In: Preece, R.C. (Ed.), Island Britain: a Quaternary Perspective. Geological Society Special Publication No. 96, pp. 209e242. London.

60. Yalden, D.W., 1999. The History of British Mammals. Poyser, London.

61. Yang, Z., 1994. Maximum likelihood phylogenetic estimation from DNA sequences with variable rates over sites: approximate methods. J. Mol. Evol. 39, 306e314.

CHAPTER 2

NARROWING THE HARVEST: INCREASING SICKLE INVESTMENT AND THE RISE OF DOMESTICATED CEREAL AGRICULTURE IN THE FERTILE CRESCENT

Osamu Maeda, Leilani Lucas, Fabio Silva, Ken-Ichi Tanno, Dorian Q. Fuller

Keywords: Neolithic, Southwest Asia, Archaeobotany, Paleoethnobotany, Early Holocene, Sickle blade, Conformist bias

INTRODUCTION

The origins of agriculture transformed the social and natural worlds. Key plant species were transformed into domesticated crops, their micro-habitats became new ecosystems for weeds and pests (Willcox, 2012), and their annual seed produce was co-opted into cycles of harvest, storage and planting by people (Fuller, 2007 and Harlan et al., 1973). During the same process human societies, economies and technologies changed, as sedentism increased (Belfer-Cohen and Goring-Morris, 2011), population densities increased (Kuijt, 2000) and subsistence shifted increasingly towards more intensive management of cultivated environments and land-ownership/private property (Bowles and Choi, 2013), as opposed to extensive hunting and gathering in communal territories. The development of agriculture was an entangled process where human society and the natural environment interplayed. Much of this process, however, remains poorly understood or actively debated. For example, there has been active debate between a hypothesis of rapid domestication driven by conscious human selection (Abbo et al., 2010a and Honne and Heun, 2009), and a protracted process in which selection pressures were lower and domestication traits were the unintended outcomes of human behavioural strategies of environmental management and species propagation (Allaby et al., 2010, Asouti and Kabukcu, 2014 and Fuller et al., 2014). While the rapid expansion of archaeobotanical data over the past decade has provided more robust and quantitative models of crop evolution, which strongly favours slower domestication processes and weaker coefficients of selection of crops (Fuller et al., 2014, Purugganan and Fuller, 2011, Riehl et al., 2013 and Tanno and Willcox, 2012), the integration of these data with evidence for human technologies, settlement structure and activity patterns has been more limited (Asouti and Fuller, 2013 and Watkins, 2008). In the current contribution we present a comprehensive and quantitative reanalysis of non-obsidian stone harvesting tools, i.e. sickle blades, from across the Fertile Crescent zone of Southwest Asia from ca. 12,000 to 5000 cal. BCE in order to trace the development of early agricultural practices (Fig. 1). For the first time large-scale, regional quantitative data on sickle use is directly compared to quantitative archaeobotanical evidence for the transformation of cereal crops towards morphologically domesticated forms, and the transformation in plant economies towards increasing reliance on crops over gathered wild foods. Our analysis, using quantile regression (detailed below) statistically examines a correlation between increasing use of, and investment in, sickle blades and increasing reliance on cereal crops, while at the same time questions a connection between sickle harvesting and evolution of non-shattering cereals (contrary to what has often been proposed, e.g., Wilke et al., 1972 and Hillman and Davies, 1990). The paper below elucidates these patterns through a detailed consideration of regional lithic and archaeobotanical variation anchored to a uniformly applied chronometric timescale.

Citation: Osamu Maeda, Leilani Lucas, Fabio Silva, Ken-Ichi Tanno, Dorian Q. Fuller, Narrowing the harvest: Increasing sickle investment and the rise of domesticated cereal agriculture in the Fertile Crescent, Quaternary Science Reviews, Volume 145, 1 August 2016, Pages 226-237, ISSN 0277-3791, http://dx.doi.org/10.1016/j.quascirev.2016.05.032.

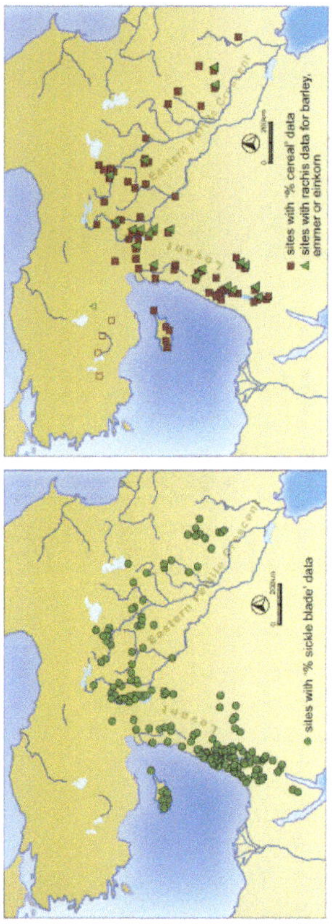

Figure 1: Map of the Fertile Crescent showing the sites with the data for sickle blades and cereals.

HARVESTING AND PROTRACTED DOMESTICATION AS ENTANGLEMENT OF TECHNOLOGY AND ECOLOGY

The transformations of domestication can best be appreciated as a co-evolutionary entanglement between plant adaptations, human socioeconomic systems and technology. Many of the key traits that define the domestication syndrome in cereal crops can be tied directly to selection pressures caused by human actions, which modified the seed dispersal and germination parts of the plants' life cycle (Fuller and Allaby, 2009 and Harlan et al., 1973). Thus, for example, harvesting methods that retain the entire cereal ear, such as uprooting or cutting by sickle, will favour non-shattering morphotypes, which have lost the wild type adaptation of spikelet dehiscence (Fuller et al., 2014, Harlan et al., 1973 and Hillman and Davies, 1990). Larger seeds that lack germination inhibition will better compete at becoming established parts of crops planted in freshly tilled fields in contrast to smaller-seeded and dormant forms (Cunniff et al., 2014, Fuller, 2007, Harlan et al., 1973 and Kluyver et al., 2013). There are, however, trade-offs from a human point of view. Non-shattering ears may be easier to harvest if one first invests in making tools, such as sickles for this purpose. However, once harvested, spikelets must be separated by threshing, which is not necessary for gathered mature spikelets; dehusking is required of both wild and domesticated cereals. Therefore, there are some additional labour investments, in sickle production and threshing, which become incumbent on the cultivators of domesticated cereals (Fuller et al., 2010). In addition, cultivated fields likely require some nutritional supplementation if they are to remain in use and support larger-seeded, more erect and tightly spaced domesticated plants. Much recent discussion has cast agricultural origins as a particular form of ecological niche construction (Smith, 2011 and Zeder, 2015). In this sense human activities of soil preparation, planting and harvesting created a new environment for crop life cycles, especially with regards to seed dispersal and germination. At the same time this economic environment, that included cultivation, formed a new niche to which humans had to adapt through labour investment, technology and cultivation practices. While for the plant, evolution can be seen as adaptation through natural selection for genetic changes, the evolution for the humans was phenotypic in the sense that it was cultural. As argued by Larson et al. (2013) such evolution fits the definition of exaptation (Gould and Vrba, 1982), in that existing techniques and technologies were co-opted into a new use and then gradually improved and modified for that purpose. One likely example of this is sickles (Fuller, 2007). In the case of Chinese rice domestication in the Yangtze, clear sickle tools only appear a millennium after morphological domestication (Fuller et al., 2007 and Fuller et al., 2009), and millennia after the start of cultivation. By contrast, in Southwest Asia sickles were in use for millennia before accepted evidence for pre-domestication cultivation (Willcox et al., 2008), suggesting that they were used for other tasks or a wider range of plant cutting needs. This contradicts the hypothesis that sickles were developed as part and parcel of domesticating cereals (Hillman and Davies, 1990 and Wilke et al., 1972). Instead, there is a more complex relationship between the adoption of sickles for cereal harvesting with the evolution of domesticated cereals and the increasing importance of cereals economically.

THE USE OF SICKLES AND SICKLE BLADES IN THE FERTILE CRESCENT

The use of sickles for harvesting cereals can be attested in the archaeological record by the presence of sickle blades, a common stone tool used in prehistory. Sickle blades are usually made of flint or chert and used by being inserted in a wooden or bone handle, as evidenced in rare examples of complete sickles which have survived to date (Borrell and Molist, 2007 and Edwards, 2007) or inferred from the trace of fixative remaining on their surface (Copeland and Verhoeven, 1996 and Milevski et al., 2013). They are distinguished from other stone tools by having characteristic gloss, i.e. sickle gloss or sickle sheen, visible along the edges of blades. The gloss indicates their function as cereal harvesting tools since it is formed when used for cutting silica-rich plants, such as sedges or grasses, including wheat and barley (Fullagar, 1991).

The use of sickle blades in association with the development of agriculture has been investigated by various researchers particularly through typological studies (Cauvin, 1983, Gopher, 1989 and Gopher et al., 1996) and microscopic use-wear analysis (Anderson, 1999, Clemente and Gibaja, 1998, Goodale et al., 2010, Unger-Hamilton, 1985 and Vardi et al., 2010). It has been assumed that increasing use of sickle blades and their typological change throughout the Neolithic reflect an important role of sickle blades which contributed to increasing productivity of cereal crops (e.g. Borrell, 2015 and Vardi and Gilead, 2013). However, such an assumption has never been tested against archaeobotanical data and often ignores regional diversity in agricultural development. The full-scale quantitative analysis of sickle blades, alongside archaeobotanical evidence for their harvested plant products, has not previously been undertaken. Here we do so on a chronological and geographical scale comparable to macro-regional processes of crop domestication and the transition to agriculture, facilitating comparison with regional archaeobotanical data.

ARCHAEOBOTANICAL TRENDS IN CEREAL DOMESTICATION AND CONSUMPTION

Rather than seeing domestication and the origins of agriculture as a single "revolution", recent research has indicated a protracted episode of evolution, in which both morphological domestication evolved slowly and the reduction in use of diverse wild foods declined gradually (Asouti and Fuller, 2012, Asouti and Fuller, 2013 and Fuller et al., 2014). Non-shattering rachises is usually seen as a key trait making grain crops dependent on humans for reproduction by removal of the natural seed-dispersal mechanisms (Hillman and Davies, 1990 and Zohary, 1969). In archaeological cereal remains this difference is documented by the preserved abscission scar on the base of the spikelets of the rachis segments (Fuller, 2007, Tanno and Willcox, 2012, Tanno and Willcox, 2006 and Weide et al., 2015). Rather than the rapid replacement of shattering with non-shattering predicted under a strong selection model that assumed application of sickle harvesting from the beginnings of cultivation (Hillman and Davies, 1990), archaeobotanical evidence that has become available mostly in the past decade, for two wheats (*Triticum monococcum* and *Triticum dicoccum*) and barley (*Hordeum vulgare*) indicates a much more gradual transition to dominance in non-shattering rachises, taking 2000–2500 years (Fuller, 2007, Fuller et al., 2014 and Tanno and Willcox, 2012).

MATERIALS AND METHODS

Chronology

The chronology of the lithic and botanical assemblages comes from available published radiocarbon dates, and for a number of the sites, calibration information can be found in the supplementary material of Asouti and Fuller (Asouti and Fuller, 2013 and Fuller et al., 2012). The dates were converted to a point estimate of age based on the median of summed calibrated radiocarbon dates, as described in Fuller et al. (Fuller et al., 2012 and Fuller et al., 2014). By this method all available dates for a site, or phase, were recalibrated and summed with OxCal 3.10 (Bronk Ramsey, 1995), using the IntCal13 calibration curve (Reimer et al., 2013). The median of the 1-sigma probability distribution was used.

This has the advantage of readily identifying and downplaying outlier dates in sites that have multiple dates, as the summed probability emphasizes the overlaps in like chronological range. The resulting median generally represents an estimate of the calibrated range that is most likely to fall during the site occupation and towards the middle of the timespan over which our studied assemblages formed. The summed probability approach has been used to combine multiple dates from single sites in regional studies of summed radiocarbon probability that use dates as data to infer demographic trends (e.g. Shennan and Edinborough, 2007, Shennan et al., 2013, Borrell et al., 2015 and Flohr et al., 2015), and has been found to be a robust means of representing individual sites. While there has been debate over the use of summed probability plots at the regional level as proxies of past population (e.g. Williams, 2012), the approach has been robustly statistically tested (Timpson et al., 2014). In any case, our purpose is not demographic trends but merely to arrive at the best possible point estimate for placing assemblages in time, and therefore to understand general chronological trends in the use of cereals and sickle blades rather than to establish a very precise chronology for the start date, end date or occupation duration of any particular site. For this the use of summed probability is convenient and effective and does not require the time-consuming assessment and argumentation about quality of various ^{14}C-dates that is necessary to address some other questions (e.g. Flohr et al., 2015). As a result, a total of 1556 radiocarbon dates were used for the calibration of 181 archaeological sites/phases. For the other 119 sites/phases, for which radiocarbon dates were not available, the median age was either interpolated between later and earlier dates in the same sequence, or the median of the estimated date range provided in the publications, based on cultural materials, was used.

Sickle Blades

Sickle blades recovered from 277 archaeological contexts across the Fertile Crescent (Fig. 1), covering about seven millennia, have been studied typologically and quantitatively. First, typology of sickle blades has been examined since it reflects the degree of investment on their production not necessarily in terms of labour and time consumption but in an attempt towards specialized production of these tools. Secondly, the frequency of sickle blades at each site/phase has been quantitatively evaluated to understand increasing importance of sickle use over time.

The use patterns of sickle blades, based on their typology and quantitative trends, can be classified into 12 groups (Fig. 2). Typological classification is based on their shape, type of blank and manner of retouch, referring to previous studies (e.g. Cauvin, 1983 and Gopher, 1989). It has been suggested that the shape of sickle blades reflect the shape of sickle handles and the way in which they were hafted (Bar-Yosef, 1987 and Cauvin, 1983). However, recent discoveries of complete sickles demonstrate that the way sickle blades were hafted was not so easily predictable (e.g. Borrell and Molist, 2007). Therefore, types of handles and hafting are not considered here in detail. The definition of 12 groups is detailed in supplementary information and summarized.

Quantitative data of sickle blades used at each site were calculated as the percentage of sickle blades to all retouched flint/chert tools by count and plotted in the chronological order provided by (Fig. 3a, b). The figures were calculated from published data with the exception of new data from Hasankeyf Höyük. Any glossed pieces (including those made on flakes) or those described as "with sheen" in published lithic data were counted as sickle blades while retouched tools involve any pieces which have edge modification, including retouched and used blades and flakes. It is known that gloss can be produced on the edge of lithic tools by activities such as reed-cutting (Winter 1994) or use of threshing sledges (Anderson,

1999 and Gurova, 2013) so glossed blades do not exclusively indicate cereal harvesting. Nevertheless, we assume that the scale of these other activities during the Neolithic would be outweighed by cereal cutting when cereal cultivation is a significant part of the economy. Therefore, in this study all the glossed tools reported from Epi-Palaeolithic and Neolithic sites are regarded as plausible cereal harvesting tools for the purpose of exploring regional and chronological quantitative patterns and how these correlate with data on cereal use and domestication. Experimental studies have demonstrated that only a few hours work of cereal cutting develops gloss on the edge (Quintero et al., 1997) and thus the misidentification of sickle blades due to the lack of gloss should be minimal.

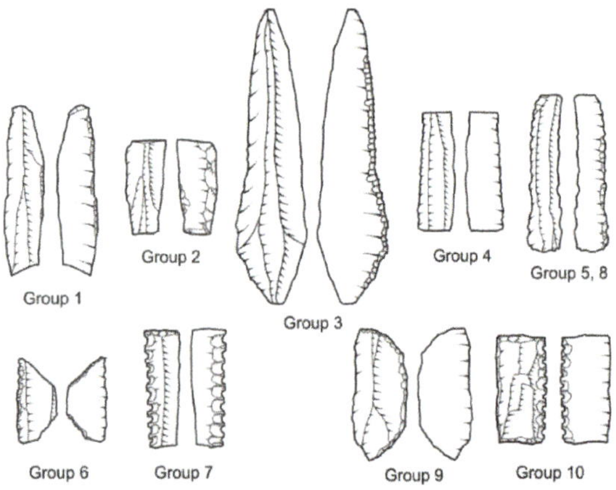

Figure 2: Schematic drawing of typological groups of sickle blades. Group 11 is defined for the Cyprus area where the typological features of sickle blades vary between different sites and periods and cannot be represented by a single type. Group 12 is defined for the Levantine arid area where the use of sickle blades is very infrequent and their typology is not standardised.

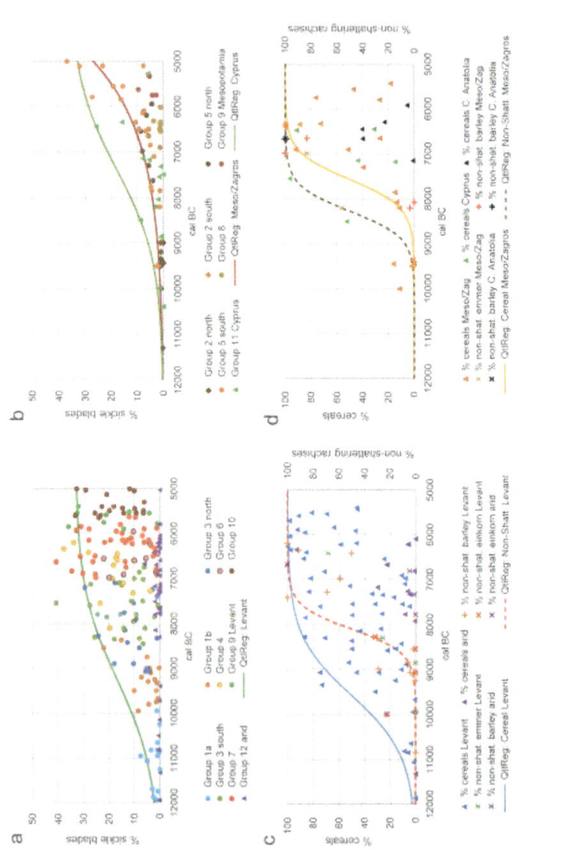

Figure 3: Changes in sickle blade frequency and cereal frequency over time, with trends indicated by 90% quantile regression. a. Sickle blade frequencies in the Levant. b. Sickle blade frequencies in Mesopotamia, the Zagros and Cyprus. c. Percentages of cereals and percentages of non-shattering cereal rachises at sites in the Levant. d. Percentages of cereals and percentages of non-shattering cereal rachises at sites in Mesopotamia, the Zagros and Cyprus.

When the amount of sickle blades has been reported as "rare" or "virtually absent", a numerical value of 1% was chosen for the purposes of quantitative analysis. For some sites the percentage of sickle blades are presented as a range value (e.g. 15–20%). This is for sites where the quantity of sickle blades are only reported as a range of percentage and their numerical counts are not available, or in instances where there are two or more datasets obtained from different excavations at one site but correlations between them are difficult. In this case the median values were used for the quantitative analysis in Fig. 3.

The percentage of sickle blades can be biased by various factors, such as a number of blades inserted to one sickle and longevity of each blade but it is unlikely to affect the general pattern in our overall dataset. Another factor which may bias quantification of sickle blades is the use of obsidian blades for cereal harvesting. Since distinctive sickle gloss is not formed on obsidian (Fullagar, 1991), it is not possible to identify obsidian sickle blades with the naked eye. This could misrepresent the percentage of sickle blades used at sites where obsidian was a major raw material. Although recent progress in use wear analysis on a microscopic level allows us to identify the use of obsidian for harvesting purposes, those methods have not been widely enough applied to be useful for the quantitative analysis in the present paper. For this reason, sites in Central Anatolia, where obsidian accounts for the majority of lithic assemblages, are not included in this study. Elsewhere, obsidian is usually a minority, ca. 0–30% of all lithics (Ibáñez et al., 2015), and therefore has a minor effect on the estimated percentage of sickle blades based on flint/chert examples.

Plant Data: Non-Shattering Rachises and Percentage of Cereals

In order to address the relationship between sickles and economic change, we have taken two straightforward indices from the archaeobotanical record, the proportion of non-shattering (domesticated morphotype) cereals, from 9148 rachises, and the overall proportion of cereals in archaeobotanical assemblages, totaling 742,860 charred specimens from all assemblages across 113 sites. While these data are not always available from all sites, nor from the same sites, they can nevertheless be plotted chronologically (by median age) and grouped by region as defined by the sickle blade groups presented in Section 5.2 (Fig. 3c, d).

The percentage of non-shattering spikelet bases in cereals recovered from archaeological contexts has been used previously (Fuller et al., 2014) to document the evolution of the domestication syndrome. By calculating the percentage of non-shattering in an archaeological assemblage and plotting against an estimate of time, it is possible to document the slow progression over time to the dominance of non-shattering rachises. The evidence for non-shattering rachises in barley (*Hordeum vulgare*) comes from previously published datasets presented earlier by Fuller et al. (2014) . The limited rachis datasets for emmer (*Triticum dicoccum*) and einkorn (*T. monococcum*) has been expanded from that presented earlier (Fuller et al., 2014) with additional emmer rachis data from Chogha Golan (Weide et al., 2015) included. Differences between data presented earlier by Fuller et al. (2014) and the present dataset are rachises that were not determined to be either shattering or non-shattering (i.e. 'indeterminates') were not taken into consideration in this study, although presented in the tables, and the estimated standard deviations are not presented.

While the broader trends in the evolution of non-shattering are similar across the Fertile Crescent the importance of cereals in the diet varied greatly across sites and regions. To see the growth in cereal production over time, we have taken a percentage of the number of cereal specimens (including grains and chaff of barley, wheats and rye, whether wild or domestic) out of all counted charred plant remains. In other words this is the total number of identifiable specimens of all cereals combined, rather than attempting to estimate the number of specimens by combining chaff and grain from the same species. Although chaff to grain ratios differ across cereal crops, and the proportion of chaff to grain differ between different products and by-products of crop-processing stages (Hillman, 1984 and Stevens, 2003), it is unrealistic to attempt to account for this across all assemblages. For one thing, numbers and ratios are subject to variations in sampling, sorting and reporting strategies of different archaeobotanists. In addition, different patterns in which crop-processing stages were exposed to fire and preserved may structure the proportions of chaff and grains, and this may differ across cereal types (e.g. between glume wheats and barley), systematic variation in this regard is as likely to differ across contexts within an individual site as it is between sites. It is the case that sites include a range of contexts and context types, therefore providing some balance between intra-site variations. Nevertheless, biases will tend to balance out when all cereal types are considered. Thus, for example, a greater input from dehusking by-product will be expected to increase glume wheat representation, while less dehusking will favour more barley or rye and fewer glume wheats. It has also been experimentally demonstrated that chaff rich assemblages before charring will be made to look grain-rich through the charring process (Boardman and Jones, 1990). Taking into account all of these concerns we took a simple summed frequency approach as this requires the least assumptions, except that archaeobotanical evidence reflects some recurrent aspects of plant use on a site, and it is straightforward in terms of using existing archaeobotanical evidence, i.e. closer to the primary data as reported.

Quantile Regression and the Analysis of Covariance

In order to objectively identify trends, we used quantile regression of the 90th percentile. Unlike an ordinary least squares regression, which approximates to the conditional mean, a quantile regression can approximate to any quantile of the dataset being analysed (Koenker, 2005). This is quite useful when trying to understand, as in our case, the evolution not of the mean but of the top-end of the dataset, that is driving the process of adoption (for more on this see Section 6.2). The shape of the regression curve chosen was logistic as this best mimics the data. It is also the case that logistic curves are the predicted form on many evolutionary processes, including selection for domestication traits and the transition to agricultural economies over foraging (Rindos, 1984 and Hillman and Davies, 1990). In addition, a logistic curve replicates on theoretical grounds the process of development/adoption of an innovation (Fitzhugh, 2001). The regressions were done using the quantreg package (Koenker, 2013) for R (R Core Team, 2014). It is worth noting that quantile regressions focusing on such trends are robust to minor variation to both horizontal and vertical axis values, which means minor differences in the observational error between sites should not affect the overall trends. For example, even if median dates are misestimated somewhat, by a couple of centuries in either direction, for example, the logistic quantile regression is unaffected.

To then quantitatively compare the different identified logistic curves, particularly in what relates to their covariance with time, an analysis of covariance (ANCOVA) was applied (e.g. Garcia-Berthou, 2001). This type of analysis evaluates whether a series of given trends are equal across different levels of a categorical independent variable, in our case, between cereals and sickle blades across the same, or different regions. Because ANCOVA is a general linear model that combines ANOVA (analysis of variance) with regression, it requires that the relationship between the dependent and independent variables be linear. As mentioned above, a logistic curve is the best representation of our data, and this required the data to be logit-transformed so as to recover a linear relationship with time. The ANCOVA test that uses quantile regression, available from the quantreg package mentioned above, was then applied to compare the different trends. A significant effect of the categorical variable, given by a low p-value, is then interpreted as a lack of covariance – in other words, the two trends being compared are significantly different from each other.

RESULTS AND DISCUSSION

Diverse Investment: Sickle Blade Types and Sickle Blade Frequency in the Neolithic Tool-Kits

Our study demonstrates the increasing investment in production and the use of sickle blades as well as their regional diversity. While Fig. 3a and b shows a general quantitative trend of sickle blade use, the chronological and geographical distributions of different types of sickle blades defined in section 5.2 (Fig. 2) can be represented in four time slices (Fig. 4). Cyprus is dealt with separately because sickle blade typology in Cyprus does not exactly match the situation in the mainland, although a close relationship between Cyprus and Levantine coastal regions is inferred from lithic industry as a whole (McCartney and Todd, 2005). The following time slices show the chronological change of sickle blade typology, which reflect the change in degree of their specialized manufacture.

(1)During the last phase of the Epi-Palaeolithic to the PPNA (ca. 12,000 to 8700 cal. BCE), the use of sickle blades is limited across all the area. Sickle blades used in the northern and southern Levant (Group 1) are of a simple rectangular type made of less-standardised blades from which other retouched tools were also manufactured. The percentage of sickle blades are mostly less than 5% (Fig. 3a). In the Eastern Fertile Crescent

(Mesopotamia and Zagros) sickle blades are extremely rare (Group 2). When present they are made of simple blades or rather irregular flakes.

(2)From the last phase of the PPNA to the PPNB (ca. 9100 to 6700 cal. BCE), sickle blades begin to increase at some sites in the Levant and regional difference in their typology and quantity becomes clearer between the Levant and Eastern Fertile Crescent (Fig. 3a and b). In the northern and southern Levant, sickle blades made of bi-directional blades produced from opposed platform cores become common (Group 3) and their percentage increases through time up to about 30% at some sites but remains low at others (Fig. 3a). The same type of sickle blades is distributed in the arid zone but in a very small quantity. Although the production of bi-directional blades are highly specialized (Abbès, 2003), these blades were used for various types of tools and it indicates the production of sickle blades was not specialized with respect to other lithic types. On the other hand, specialized production of sickle blades is witnessed in the north-western Levant in the Late PPNB (Group 4), where sickle blades were made of standardised uni-directional blades produced from single-platform prismatic cores (Arimura, 2011 and Borrell, 2015). It is assumed that this type of blade was specifically produced to be used for sickle blades while other tools were usually made of bi-directional blades through a separate production sequence. In the Eastern Fertile Crescent the occurrence of sickle blades increase from the previous period up to 10% (Fig. 3b) and becomes more standardised (Group 5). They are made of uni-directional blades produced from bullet-shaped cores using pressure detachment technique (Inizan and Lechevallier, 1994), as are other tools. Except for the presence of gloss these sickle blades are basically identical to ordinary retouched blades used for various purposes. Thus, although the pressure blade detachment requires higher skill, the production of sickle blades was not particularly specialized.

(3)In the later phase of Late PPNB and the Pottery Neolithic (PN) (ca. 7000 to 5700 cal. BCE), Group 4 continued in the north-western Levant, and sickle blade types changed in other areas of the Levant. The maximum percentages of sickle blades in an assemblage levels off (Fig. 3a). In the north Levant crescent-shaped sickle blades became popular (Group 6). Although their production from irregular blades/flakes is not highly skilled, their distinctive shape and the orientation of gloss oblique to the blade edge indicates that they were apparently different from ordinary retouched blades and produced to be inserted into sickles in a specific manner (Cauvin, 1973 and Nishiaki, 1997). Another distinctive type is seen in the southern Levant (Group 7), where sickle blades made of uni-directional blade/flake segments with coarsely denticulated retouch become common. The higher degree of retouch indicates higher investment in their production (Gopher et al., 1996). In the Eastern Fertile Crescent, Group 5 continues in the Zagros foothills and the percentage of sickle blades remains more or less the same. On the other hand, sites in northern Mesopotamia experienced a drop in the use of sickle blades at this time (Group 8).

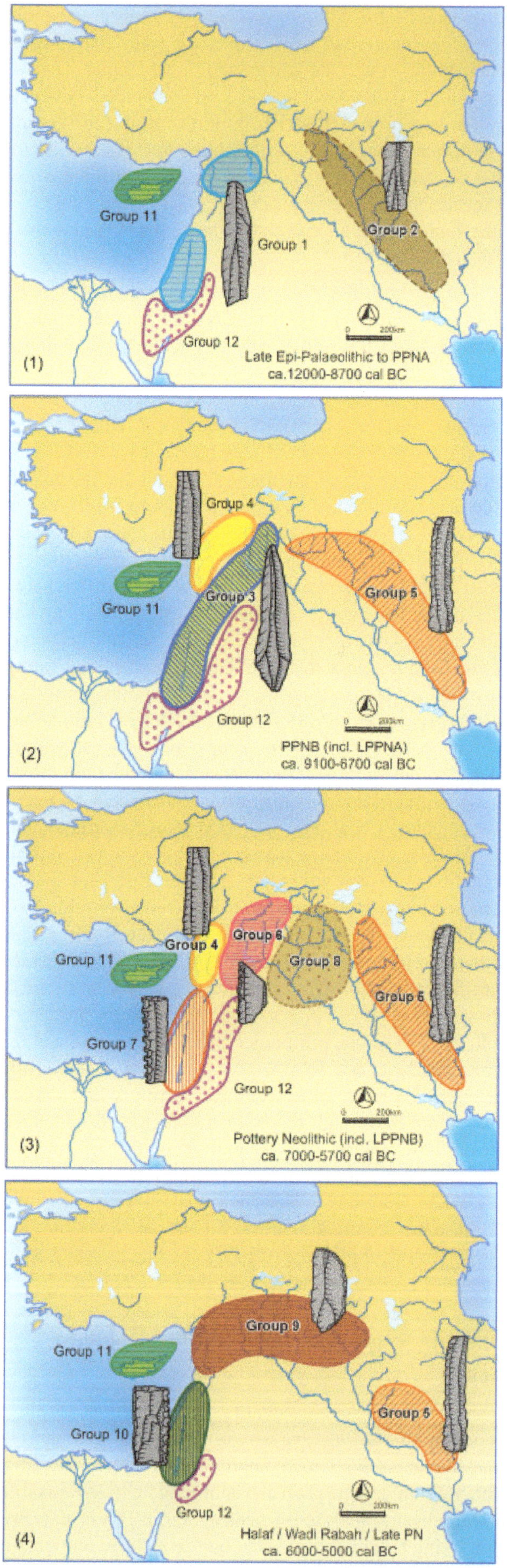

Figure 4: Map of the Fertile Crescent showing the regional typological groups of sickle blades in four time slices.

(4)In the Late PN to the beginning of the Chalcolithic (ca. 6000 to 5000 cal. BCE), the Halaf culture expands across the northern half of both the Eastern Fertile Crescent and the Levant. In this area crescent sickle blades were used alongside rectangular ones (Group 9) (Arimura, 1999). Despite the typological commonality from east to west, the quantitative differences of sickle blades remained with a low frequency in the Eastern Fertile Crescent (Fig. 3a, b). In the southern half of Eastern Fertile Crescent where Group 5 continues, however, the percentage of sickle blades begins to increase and catches up with Levantine assemblages by the end of this period. Sickle blades used in the southern Levant were carefully made into standardised rectangular segments (Kadowaki, 2005 and Vardi and Gilead, 2013) and those with backed retouch and truncated ends prevail (Group 10).

We have then classed the assemblages into four grades that reflect increasing levels of effort in production, and plotted their occurrence over time (Fig. 5). The production of Group 1 sickle blades were not separated from that of ordinary retouched blades, which are morphologically identical to sickle blades. It is thus likely that sickle blades were simply selected from ordinary retouched blades and inserted into handles to be used for harvesting. The same is true for Group 3 sickle blades. The appearance of the Group 3 sickle blades made on fine bi-directional blades at the end of the PPNA should be regarded as an outcome of the change in overall blade production rather than specialized production of sickle blades. The specific production of sickle blades targeted to be only used as sickle blades comes to the fore in the Pottery Neolithic in the Levant. Sickle blades of Groups 4, 6, 7, 9 and 10 are those produced for the specific use as harvesting tools. This suggests a trend towards increasing interest in sickle production through time (Fig. 5). Taken together with the trend of increased numbers of sickles, this typological change highlights the increasing importance of this tool category over the course of the Neolithic.

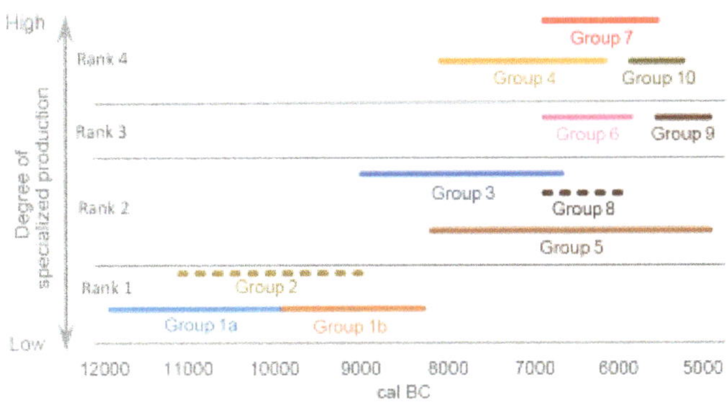

Figure 5: The occurrence of sickle typological groups over time, classed into 4 ranks of increasingly specialized production and investment.

Comparing Trends of Sickle Blades and Cereals

Fig. 3c and d shows the chronological trends in the proportion of non-shattering cereals (barley, emmer and einkorn) and the percentage of cereals (barley, *Triticum* spp. and rye) out of all charred remains. While a few sites in the Levant have large minorities of non-shattering einkorn or emmer early in the Neolithic (10,000-9000 cal. BCE), sites with a majority of domesticated rachis remains or approaching 100% all date to the later PPNB (after 7500 cal. BCE) (Fig. 3c). There is less data from the Eastern Fertile Crescent, which indicates both a late appearance of non-shattering barley (after 8000 cal. BCE), one early appearance of a high proportion (large minority) of non-shattering emmer at Chogha Golan (Riehl et al., 2013 and Weide et al., 2015), and otherwise the fixation of non-shattering in populations after 7000 cal. BCE (Fig. 3d).

What is evident is that cereals only start to occur as a majority of the archaeobotanical assemblages in a few sites in the Levant just before 9000 cal. BCE, but lower proportions are more common until after 7000 cal. BCE when assemblages low in cereals disappear, except for in the arid area. Data are more limited in the eastern Fertile Crescent and Cyprus but the trend towards cereal consumption appears later than in the Levant (Fig. 3d). In the arid zone of the southern Levant cereals remain rare throughout the period considered and at most of these sites wild cereals persist, raising the possibility that either cereal collecting persisted or that selection for non-shattering did not occur in this region.

We tested whether on individual sites we could see a strong correlation between cereal statistics and sickle frequency, but there is no clear regression by which proportion of sickle blades predicts the amount of cereal use or domesticated cereals (Fig. 6). Nevertheless, there are some tendencies for sites with more specialized sickle types to have more cereals and sites with few cereals to have fewer sickles.

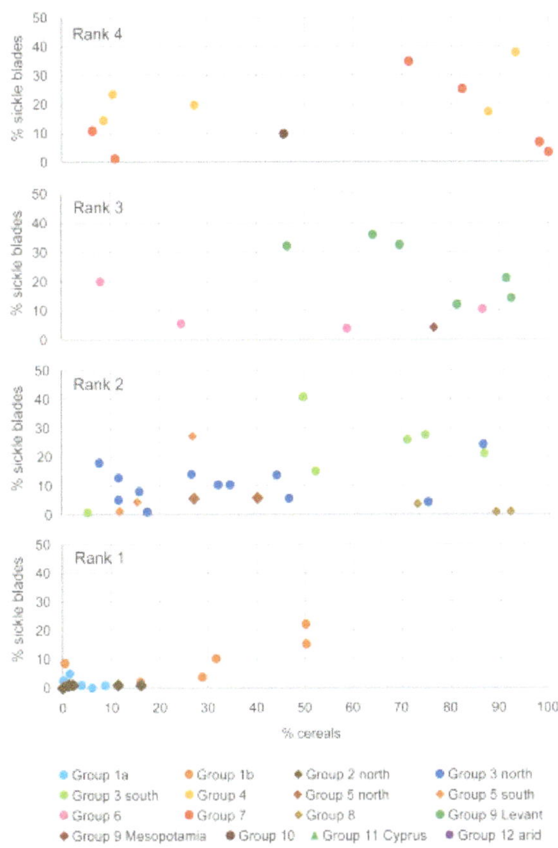

Figure 6: Sickle blade frequency and cereal frequency at individual sites. Each dot represents an individual site for which data for % sickle blade and % cereal are both available.

Thus patterns emerge only at the regional level, but even at this level there are sites that are further in the innovation adoption process than others, as Fig. 3 makes clear. As we are more interested in these "innovators" we cannot simply look at how the conditional mean evolves through time. Instead, one should look at the top quantiles of the conditional distribution or, in other words, focus our analysis on what is driving the process regionally. We have, as noted in Section 5.4, conducted quantile regression to the top 90th percentile in order to recover the top-end of the range, adopting a logistic regression curve (curved lines in Fig. 3), and subsequently did ANCOVA analyses on the identified trends.

The resulting curves in Fig. 3 allow for a direct comparison of these maximum trends between the variables and regions analysed, which can be seen in Fig. 7, and is supported by the ANCOVA tests of covariance whose results are shown. We can observe that, in the Levant, the logistic growth curves for % sickle blades and % cereal seem to co-vary with time (Fig. 7a), however they are significantly different (p-value of 0.03393). Cereals reach the turning point (50%) around 9600 cal. BCE, whereas sickle blades do so three centuries later. Non-shattering cereals, on the other hand, only reach the turning point fifteen hundred years later, around 8100 cal. BCE. The trend of the curves for % sickle blades and % cereal is very different in Mesopotamia and Zagros (Fig. 7b), and this is again supported by the ANCOVA tests (highly significant p-value of 0.0002187).

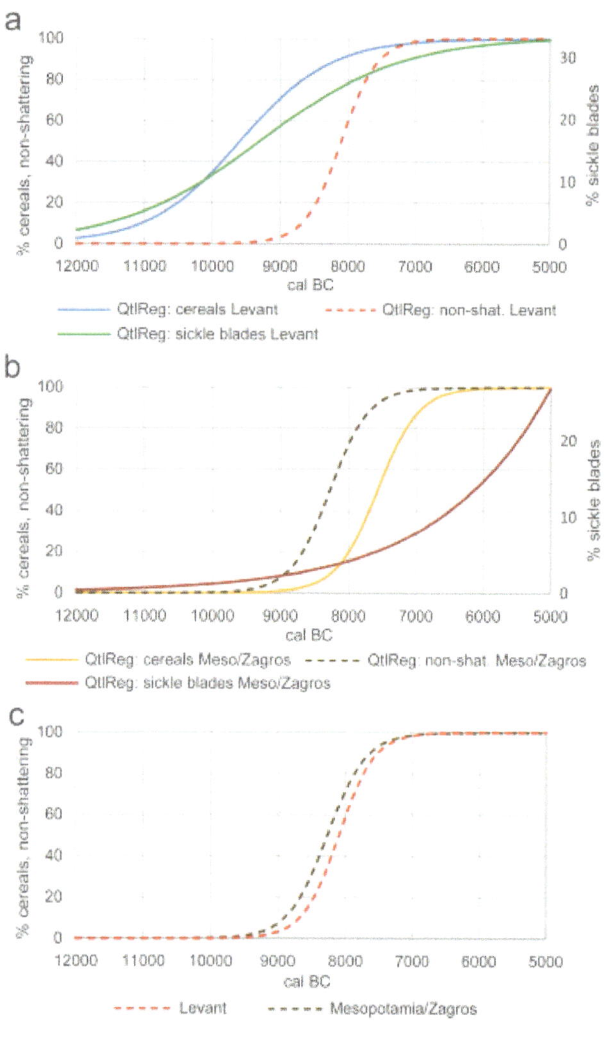

Figure 7: A comparison of the quantile regression trends in sickle blade frequency, cereal frequency and domesticated cereal frequency. a. For the Levant, b. For Mesopotamia and the Zagros, c. A comparison of the cereal domestication trends (non-shattering rachides) for the Levant and Mesopotamia plus the Zagros.

Another interesting observation is the similarity between the % non-shattering regression curves for both the Levant and the Mesopotamia/Zagros regions with a point difference sum of merely 171.14 (Fig. 7c). All three parameters for these curves are very similar, their turning points differ by a mere 170 years. The ANCOVA test returned a non-significant p-value of 0.25753, meaning that the two trends are not significantly different. However, the curve for the Mesopotamia/Central Zagros regions is based on only eight data points and the error margins for this curve are too elevated. Their similarity should therefore be taken as merely suggestive.

Nevertheless, these data may suggest cereals across the whole region shared in the same general trend, as assumed in recent studies of domestication rate (Fuller et al., 2014 and Purugganan and Fuller, 2011), and can be regarded as a single meta-population. This means that although early cereal cultivation patches were geographically dispersed and separated, between them there was enough gene flow that new alleles could spread between cultivated patches through the bridge of intervening wild populations as well as through human exchanges of seed (Allaby, 2010). Such a pattern is consistent with the view that cereal domestication evolved as a geographical mosaic drawing on several local populations in parallel (Allaby, 2015). Such a pattern is strongly indicated in recent genomic analyses on domesticated and wild barley (Poets et al., 2015), and consistent with available analyses of wheat genetics (Kilian et al., 2007 and Civáň et al., 2013).

The close correspondence between the timing of the increases in tough rachis cereals in both the Levant and the Mesopotamia/Zagros regions raises the possibility that tough rachis cereals were introduced into the Mesopotamia/Zagros region from the west, and this could have taken place in several waves. At Chogha Golan, for example, after a shift in the plant assemblage that suggests some subsistence reorganization at ca. 8200 BCE, emmer with a high proportion of non-shattering rachises (~56%) appears, and this could indicate introduction of cultivars further along the evolutionary path to non-shattering, rather than in situ selection for this trait. Further archaeobotanical evidence is needed to test whether the hypothesis of independent cereal domestication in the eastern Fertile Crescent (e.g. Riehl et al., 2013 and Riehl et al., 2015) can be sustained.

Alternative Regional Investments in Sickles and Cereals

The trends in quantile regression show the correlation between sickle use and cereal use. In both the northern and southern Levant a clear trend towards increasing sickle production, compounded by the appearance of highly specialized sickles, correlates with a rise in consumption of cereals, with the trend in sickles lagging slightly behind that for cereal consumption. This suggests that the increasing investment on sickle blades was driven by the growth of cereal production. However, at the same time, the use of sickle blades clearly demonstrates regional diversity in these trends. In the Eastern Fertile Crescent and Cyprus both trends begin later. While sickle blades increase after 9000 cal. BCE and reach the peak around 7000 cal. BCE at many sites in the Levant, those in the Eastern Fertile Crescent begin to increase only after 6500 cal. BCE. In Cyprus, although the early introduction of cereals may be associated with elevated rates of becoming domesticated (Lucas et al., 2012), the late adoption of sickles appears unconnected to this process. These regional differences demonstrate that different technological trajectories occurred in the development of agriculture across Fertile Crescent regions. This is also supported by the regional differences of sickle blade typology as discussed in section 6.1. Therefore, even if cereal domestication can be seen as a region-wide process affecting cereal meta-populations, the accompanying cultural evolution was regionally varied with a mosaic of cultural pathways. Indeed a number of authors have argued for a Neolithic cultural mosaic, despite geographical proximity and long-distance trade (e.g. Asouti and Fuller, 2013, Belfer-Cohen and Goring-Morris, 2011 and Watkins, 2008). It is definitely worth exploring what causes these regional differences in agricultural trajectories. Multiple factors can be considered, such as, resource availability, including the degrees to which legumes or animal resources contributed to protein consumption, climate differences, population density, settlement pattern, and social tradition relating to land tenure or ownership. But enquiry into these potential causes requires further substantial research beyond the scope of this paper.

The lag between the rise of sickles and the rise towards fixation in non-shattering cereal ears contradicts a long-standing hypothesis of sickle pressure, i.e. that use of sickles drove the evolution of non-shattering rachises (Harlan, 1992, Hillman and Davies, 1990 and Wilke et al., 1972). This hypothesis presupposes strong selection of domestic-type cereals during sickle reaping because shattering (wild-type) spikelets more likely drop when the ears are shaken by contact with sickles (or people holding sickles), whereas non-shattering forms have a high likelihood of being harvested via sickle and sown the next year. Proponents of the hypothesis of domestication via sickle pressure have argued that sickle reaping prompted rapid evolution of non-shattering rachises, taking cultivated cereal populations from morphologically wild to genotypically fixed for non-shattering morphotypes in less than 200 years and perhaps as fast as a few decades (Hillman and Davies, 1990). Our study, on the contrary, supports the contention that sickles were not necessary for non-shattering cereals to evolve, particularly when their evolution was a slow, prolonged process (Fuller, 2007, Fuller et al., 2009 and Sauer, 1958). It is assumed that the use of sickle did not prevent the evolution of non-shattering rachises but it did not prompt it either.

It is more likely that the evolution of non-shattering types was caused by other sources of weak selection (Fuller, 2007 and Fuller et al., 2010), through annual harvesting and sowing cycles. First, simple by virtue of being cultivated and re-sown by people, the wild state in which natural selection favours shattering is overturned thus allowing for tough rachis mutants to accumulate in cultivated populations. The timing of harvesting, i.e. late harvests, may then select for non-shattering. As wild cereals are non-synchronous in their ripening, their exploitation, including under early cultivation, would have necessitated multiple harvests, for example weekly over a period of 4–6 weeks of cereal ripening. The latest of such harvests would have included larger proportions of non-shattering mutants. Therefore if this last harvest was saved for sowing, even by rare households, something like 1 household in 20, there would be a basis of selection for non-shattering to increase overall in regional populations. By similar reasoning it is also plausible that occasional households experimented with sickle harvesting of cereals, long before this became a standard practice, and this would have contributed to selection for non-shattering. This rare experimentation with sickles for cereal harvests in due course became a fad and a more widely accepted standard practice.

We should regard sickles as having been developed as a cutting tool for raw materials such as reeds and sedges for basketry, matting or thatch. The extensive use of grass culms, reeds and sedges for such purposes, including for lining burials, has been inferred from phytolith analyses at some Natufian sites (e.g. Portillo et al., 2010 and Power et al., 2014). We suggest that sickles were later, over the course of the PPNB, transferred to agricultural harvesting, in cultural evolutionary terms an exaptation (Fuller, 2007 and Larson et al., 2013). Attributing a role to sickles in morphological evolution of cereal domestication appears to be an anachronism, and the imposition of modern "agronomic" thinking on early forager-cultivator societies. This conclusion also makes sense in comparative global terms, as archaeological evidence argues against a role for sickles or other harvesting knives in the domestication of rice, where sickles were instead adopted in the later Fourth millennium cal. BCE, after *Oryza sativa* was already domesticated and non-shattering (Fuller et al., 2007 and Fuller et al., 2009).

This disconnection between sickles and cereal domestication in evolutionary terms, then, raises the question as to why sickles were adopted into agriculture. While this can be considered to increase efficiency, especially if measured in terms of harvest yield per unit of area (Bar-Yosef, 1998 and Fuller, 2007), the regional differences indicate that there were also cultural factors. For one thing many sites have low sickle percentages throughout the period of study even if quantile regression shows a steady trend of an increase in the upper limits of frequency. On a regional level, the use of sickle blades remains humble in northern Mesopotamia during the Pottery Neolithic (Group 8 in Fig. 4) while the use of cereals had already begun to rise at this time (Fig. 3b,d). This means that sickles were not always requisite for cereal harvesting. In fact, sickle harvesting is not necessarily more efficient than other methods (e.g. Hillman and Davies, 1990). It has also been reported that hand plucking of wheat and barley by Bedul Bedouin is as efficient as harvesting with lithic and metal sickles (Simms and Russell, 1997). Even if sickle harvesting has a slight advantage in harvest yield per unit, the difference in time investment to harvest a certain area must be only few hours when compared with other methods. Sickles also require investment in manufacture, and as the data indicate this investment in each sickle also tended to increase over time (Fig. 5). It is indeed not our intention to exclude an economic factor promoting sickle use in cereal harvesting. However, we believe that the economic advantage of sickles is subtle and was not a sufficient motivation for their adoption into agriculture. Our study clearly shows that the growth in cereal consumption and sickle use are correlated but it also shows that it was a process which cannot be simply explained by cause-and-effect relations in economic terms. In this sense those regions like the Levant, where sickles became increasingly prominent, followed historically contingent trajectories in which sickle use made sense within the cultural and ecological logic of that cultural tradition. We would suggest that as a focus of labour investment sickles were adopted as a character of the cultural package of farming. Cultural transmission processes such as conformist bias (Acerbi and Bentley, 2014), could contribute to explaining this pattern in the Levant, beyond mere content biases such as efficiency. Conformist bias means that people mostly copy the behaviours that they see as more common or frequent in the society around them, on the assumption that what is common is "right" or "better". In a sense then the adoption of sickles was as much stylistic, conforming to styling oneself to be a "farmer", as it was about efficiency, and that cultural evolution was not a straight-forward adaptive process, but developed by cultural exaptation, a solution that was efficient enough and developed from retooling an existing technology (blades) for a new purpose (cereal harvesting), i.e. exaptation. The stylistic aspects of sickles as signalling the cultural identity of a farmer may provide part of the explanation for the curious later development of fired clay sickles in the Ubaid period of Mesopotamia (Benco, 1992). While this has sometimes been attributed to the lack of lithic raw materials in the floodplains of Mesopotamia, such an explanation fails to explain why such tools have been found in the northern Mesopotamia, e.g. Tepe Gawra (Oates, 2010), where lithic materials for more efficient stone-blade sickles would have been readily available. Therefore, we infer that the protracted transition to agriculture in the Near East was in part a matter of developing the cultural logic of being a farmer, which in the Levant included the use of sickles.

The long-term co-evolutionary trajectory that we have highlighted between cereal cultivation and sickle use has a number of implications for re-framing explanations for the origins of agriculture. Clearly there is no single event, or step change, apparent as we might expect in relation to hypotheses of simple climatic triggers for the origins of agriculture. There is certainly no indication that the Younger Dryas or its termination was a dividing line between foraging and cultivation that precipitated rapid change (e.g. Bar-Yosef, 1998; Hillman et al., 2001; Blockley and Pinhasi, 2011). Nevertheless, the onset of the Holocene and the PPNA, after ca. 10,000 cal. BCE (as per Blockley and Pinhasi, 2011) does correlate with start of the upward trend in cereal use and sickle blades, and the first reports of a few non-shattering wheat or barley on very few sites, but this is only the beginning of a 2000 year trend that levels off after 8000 cal. BCE. Borrell et al. (2015)argue that around 8000 cal. BCE there was a climate driven break in population in the northern Levant and this transition led to the establishment of agriculture and new blade production patterns after this dry event. While the abandonment of sites in this region at this period is compelling (Borrell et al., 2015), the trend illustrated by our study from archaeobotanical and lithic data across the wider region fit with what still appears a gradual and macro-regional pattern of change. Elsewhere, such as in western Iran this same dry event (ca. 8200-8000 cal. BCE) correlated with other subsistence shifts, such as the brief appearance and then abandonment of non-shattering barley at Chogha Golan, followed by its replacement by semi-domesticated emmer wheat (Riehl et al., 2015). As Riehl et al. (2015) argue, the protracted process of domestication would have included a period of stasis, or even reversal, at least on a local level, facilitated by the resilience that was inherent in mixed strategies of foraging and landscape management. A further implication of the reviewed evidence is that we need to broaden the study of early cereal domestication from the traditional focus on non-shattering rachises. There is a long tradition of regarding non-shattering ears as the single most important trait differentiating domesticated crops from their wild ancestors. This has been called "the main diagnostic character that serves for distinction of wild cereals from their cultivated counterparts (Zohary, 1969, p. 157–158), or "the most conspicuous… crucial in maintaining the disruptive selection that effectively maintains separation of the two kinds of populations" (Harlan, 1992, p. 118). Non-shattering rachises remain important for at least two reasons: first, they are an archaeobotanically recoverable trait that is less ambiguous than most in studying cereal evolution under cultivation. Second, non-shattering helps to maintain a dependence of domesticated species on humans for sowing, and subsequent selection and maintenance of varietal differences in cereals is facilitated by co-dependence of humans and cereals that do not readily reseed themselves. Nevertheless, and contrary to Harlan's (1992) inference, this trait may not have been so crucial during domestication. Instead, traits like even germination and ripening, reliability of yield (e.g. Abbo et al., 2010b), and increased food value brought about by better grain filling and larger grains (Fuller et al., 2010) may have been at least as important, alongside the storability of cereals and how they fit with traditions of food preparation and consumption, such as the flour-based foods that were characteristic of the Near East (e.g. Haaland, 2007 and Fuller and Rowlands, 2011).

CONCLUSIONS

It is suggested that to the extent that change was directional— on a regional level but not in every locality— it was directed by an internal logic of cultural evolutionary processes, including conformist bias and exaptation of pre-existing technologies to new uses, that in some cases may have produced marginal efficiency gains or, in others, non-significant efficiency losses. This would then predict considerable scope of variation between regions as cultural traditions drifted in alternative directions. Variations in traditions of harvesting, such a tool choice, harvesting height, when and how many times during ripening, and which temporal portion of the harvested cereals were stored for re-sowing would have all contributed to creating varied levels of selection for domestication traits (as per Fuller, 2007 and Fuller et al., 2010). Nevertheless the slow accumulation of domestication traits amongst cereals across a macro-regional meta-population could have promoted but not required becoming a more committed sickle-harvester of cereals. Trends towards human population increase and sedentism would have similarly provided more opportunities to commit to this trend towards a cereal producer identity, even if local collapses and climate shock may have sometimes gone the other way. What is nevertheless clear is that the net regional trend was to establish a high commitment to cereals, sickles and fixation on non-shattering domesticated forms throughout the Fertile Crescent by the sixth millennium cal. BCE.

ACKNOWLEDGMENTS

This work was supported by JSPS KAKENHI Grant Number 25370899 to OM and carried out while he was a visiting scholar at UCL. Research by LL and DF was supported by a European Research Council grant (#323842) on "Comparative Pathways to Agriculture" to DF. FS was supported by a grant from UK NERC (NE/K003402/1) to DF. We thank peer reviewers for their useful suggestions.

REFERENCES

1. Abbes, F., 2003. Les outillages n eolithiques en Syrie du Nord. Archaeopress, Oxford .
2. Abbo, S., Lev-Yadun, S., Gopher, A., 2010a. Agricultural origins: Centers and noncenters; a near eastern reappraisal. Crit. Rev. Plant Sci. 29 (5), 317e328.
3. Abbo, S., Lev-Yadun, S., Gopher, A., 2010b. Yield stability: an agronomic perspective on the origin of near Eastern agriculture. Veg. Hist. Archaeobot 19 (2), 143e150.
4. Acerbi, A., Bentley, R.A., 2014. Biases in cultural transmission shape the turnover of popular traits. Evol. Hum. Behav. 35 (3), 228e236.
5. Allaby, R.G., 2010. Integrating the processes in the evolutionary system of domestication. J. Exp. Bot. 61 (4), 935e944.
6. Allaby, R.G., 2015. Barley domestication: the end of a central dogma? Genome Biol. 16 (176). http://dx.doi.org/10.1186/s13059-015-0743-9.
7. Allaby, R.G., Brown, T.A., Fuller, D.Q., 2010. A simulation of the effect of inbreeding on crop domestication genetics with comments on the integration of archaeobotany and genetics: a reply to Honne and Heun. Veg. Hist. Archaeobot 19 (2), 151e158.
8. Anderson, P.C., 1999. Experimental cultivation, harvest and threshing of wild cereals: the relevance for interpreting the use of Epipaleolithic and Neolithic Artifacts. In: Anderson, P.C. (Ed.), Prehistory of Agriculture: New Experimental and Ethnographic Approaches. Institute of Archaeology, University of California, Los Angeles, pp. 118e144.
9. Arimura, M., 1999. The lithic industry of the final Pottery Neolithic: New evidence from tell Ain el-Kerkh, northwest Syria. Neo-Lithics 3 (99), 7e10.
10. Arimura, M., 2011. LPPNB blade caches at Tell Ain el-Kerkh, north-west Syria. In: Healey, E., Campbell, S., Maeda, O. (Eds.), The State of the Stone: Terminologies, Continuities and Contexts in Near Eastern Lithics. Ex oriente, Berlin, pp. 373e383.
11. Asouti, E., Fuller, D.Q., 2012. From foraging to farming in the southern Levant: the development of Epipalaeolithic and Pre-Pottery Neolithic plant management strategies. Veg. Hist. Archaeobot 21 (2), 149e162.
12. Asouti, E., Fuller, D.Q., 2013. A contextual approach to the emergence of agriculture in Southwest Asia. Curr. Anthropol. 54 (3), 299e345.
13. Asouti, E., Kabukcu, C., 2014. Holocene semi-arid oak woodlands in the IranoAnatolian region of Southwest Asia: Natural or anthropogenic? Quat. Sci. Rev. 90, 158e182.
14. Bar-Yosef, O., 1987. Direct and indirect evidence for hafting in the Epi-Palaeolithic and Neolithic of the southern Levant. In: Stordeur, D. (Ed.), La Main et l'Outil. Manches et emmanchements prehistoriques. Travaux de la Maison de l 'Orient, Lyon, pp. 155e164.
15. Bar-Yosef, O., 1998. The Natufian culture in the levant, threshold to the origins of agriculture. Evol. Anthropol. 6 (5), 159e177.
16. Belfer-Cohen, A., Goring-Morris, A.N., 2011. Becoming farmers. Curr. Anthropol. 52 (S4), S209eS220.
17. Benco, N.L., 1992. Manufacture and use of clay sickles from the Uruk mound, Abu Salabikh, Iraq. Paleorient 18 (1), 119 e134.
18. Blockley, S.P.E., Pinhasi, R., 2011. A revised chronology for the adoption of agriculture in the southern Levant and the role of Lateglacial climatic change. Quat. Sci. Rev. 30, 98e108.
19. Borrell, F., 2015. Further remarks about lithic production at Akarçay tepe (middle Euphrates valley) during the late PPNB. In: Affanni, G., et al. (Eds.), Broadening Horizons 4: A Conference of Young Researchers Working in the Ancient Near East, Egypt and Central Asia. Archaeopress, Oxford, pp. 265e277.
20. Borrell, F., Molist, M., 2007. Projectile points, sickle blades and glossed points tools and hafting systems at Tell Halula (Syria) during the 8th millennium cal. BCE. Paleorient 33 (2), 59 e77.
21. Borrell, F., Junno, A., Barcelo, J.A., 2015. Synchronous environmental and cultural change in the emergence of agricultural economies 10,000 years ago in the Levant. PLoS One 10 (8), e0134810.
22. Bowles, S., Choi, J.K., 2013. Coevolution of farming and private property during the early Holocene. Proc. Natl. Acad. Sci. U. S. A. 110 (22), 8830e8835.
23. Bronk Ramsey, C., 1995. Radiocarbon calibration and analysis of stratigraphy: the OxCal program. Radiocarbon 37 (2), 425e430.
24. Cauvin, M.-C., 1973. Problemes d 'emmanchement des faucilles du Proche-Orient les documents de Tell Assouad (Djezireh, Syrie). Paleorient 1 (1), 101 e106.
25. Cauvin, M.-C., 1983. Les faucille prehistoriques du Proche-Orient donn ees mor- phologiques et fonctionelles. Paleorient 9, 63 e79.
26. Civan, P., Ivani cova, Z., Brown, T.A., 2013. Reticulated origin of domesticated emmer wheat supports a dynamic model for the emergence of agriculture in the Fertile Crescent. PLoS One 8 (11), e81955.
27. Clemente, I., Gibaja, J.F., 1998. Working processes on cereals: an approach through microwear analysis. J. Archaeol. Sci. 25 (5), 457e464.
28. Copeland, L., Verhoeven, M., 1996. Bitumen-coated sickle-blade elements at Tell Sabi Abyad II, Northern Syria. In: Caneva, I., et al. (Eds.), Beyond Tools: Redefining the PPN Lithic Assemblages of the Levant. Ex oriente, Berlin, pp. 327e330.
29. Cunniff, J., Wilkinson, S., Charles, M., Jones, G., Rees, M., Osborne, C.P., 2014. Functional traits differ between cereal crop progenitors and other wild grasses gathered in the Neolithic Fertile Crescent. PLoS One 9 (1), e87586.
30. Edwards, P.C., 2007. A 14 000 year-old hunter-gatherer's toolkit. Antiquity 81, 865e876.

31. Fitzhugh, B., 2001. Risk and invention in human technological evolution. J. Anthropol. Archaeol. 20 (2), 125e167.

32. Flohr, P., Fleitmann, D., Matthews, R., Matthews, W., Blacka, S., 2015. Evidence of resilience to past climate change in Southwest Asia: early farming communities as the 9.2 and 8.2 ka events. Quat. Sci. Rev. in press http://dx.doi.org/10.1016/j. quascirev.2015.06.022.

33. Fullagar, R.L.K., 1991. The role of silica in polish formation. J. Archaeol. Sci. 18, 1e24.

34. Fuller, D.Q., 2007. Contrasting patterns in crop domestication and domestication rates recent archaeobotanical insights from the old world. Ann. Bot. Lond. 100 (5), 903e924.

35. Fuller, D.Q., Allaby, R., 2009. Seed dispersal and crop domestication: shattering, germination and seasonality in evolution under cultivation. In: Østergaard, L. (Ed.), Annual Plant Reviews: Fruit Development and Seed Dispersal, 38. WileyBlackwell, Oxford, pp. 238e295.

36. Fuller, D.Q., Rowlands, M., 2011. Ingestion and Food Technologies: Maintaining Differences over the Long-term in West, South and East Asia. In: Bennet, J., Sherratt, S., Wilkinson, T.C. (Eds.), Interweaving Worlds - Systematic Interactions in Eurasia, 7th to 1st Millennia BC. Essays from a Conference in Memory of Professor Andrew Sherratt. Oxbow Books Ltd, Oxford, pp. 37e60.

37. Fuller, D.Q., Allaby, R.G., Stevens, C., 2010. Domestication as innovation: the entanglement of techniques, technology and chance in the domestication of cereal crops. World Archaeol. 42 (1), 13e28.

38. Fuller, D.Q., Asouti, E., Purugganan, M.D., 2012. Cultivation as slow evolutionary entanglement: comparative data on rate and sequence of domestication. Veg. Hist. Archaeobot 21 (2), 131e145.

39. Fuller, D.Q., Qin, L., Zheng, Y., Zhao, Z., Chen, X., Hosoya, L.A., Sun, G.-P., 2009. The domestication process and domestication rate in rice: spikelet bases from the Lower Yangtze. Science 323 (5921), 1607e1610.

40. Fuller, D.Q., Denham, T., Arroyo-Kalin, M., Lucas, L., Stevens, C.J., Qin, L., Allaby, L.G., Purugganan, M.D., 2014. Convergent evolution and parallelism in plant domestication revealed by an expanding archaeological record. Proc. Natl. Acad. Sci. U. S. A. 111 (17), 6147e6152.

41. Fuller, D.Q., Harvey, E., Qin, L., 2007. Presumed domestication? Evidence for wild rice cultivation and domestication in the fifth millennium BC of the Lower Yangtze region. Antiquity 81, 316e331.

42. Garcia-Berthou, E., 2001. On the misuse of residuals in ecology: testing regression residuals vs. the analysis of covariance. J. Anim. Ecol. 70 (4), 708e711.

43. Goodale, N., Otis, H., Andrefsky Jr., W., Kuijt, I., Finlayson, B., Bart, K., 2010. Sickle blade life-history and the transition to agriculture: an early Neolithic case study from Southwest Asia. J. Archaeol. Sci. 37, 1192e1201.

44. Gopher, A., 1989. The Flint Assemblages of Munhata: Final Report. Association

45. Paleorient, Paris.

46. Gopher, A., Barkai, R., Asaf, A., 1996. Trends in Sickle Blades Production in the Neolithic of the Hula Valley, Israel. In: Kozlowski, S.K., Gebel, H.G.K. (Eds.), Neolithic Chipped Stone Industries of the Fertile Crescent, and Their Contemporaries in Adjacent Regions. Ex oriente, Berlin, pp. 411e425.

47. Gould, S.J., Vrba, E.S., 1982. Exaptation: a missing term in the science of form. Paleobiology 8 (1), 4e15.

48. Gurova, M., 2013. Tribulum inserts in ethnographic and archaeological perspective: Case studies from Bulgaria and Israel. Lithic Technol. 38 (3), 179e201.

49. Haaland, R., 2007. Porridge and pot, bread and oven: food ways and symbolism in African and the Near East from the Neolithic to the present. Cambridge Archaeol. J. 17, 165e182.

50. Harlan, J.R., 1992. Crops and Man, second ed. American Society of Agronomy, 236 O. Mae Madison, WI.

51. Harlan, J.R., De Wet, J.M.J., Price, E.G., 1973. Comparative evolution of cereals. Evolution 27 (2), 311e325.

52. Hillman, G., 1984. Interpretation of archaeological plant remains: the applications of ethnographic models from Turkey. In: Van Zeist, W., Casperie, W.A. (Eds.), Plants and Ancient Man. Balkema, Rotterdam, pp. 1e41.

53. Hillman, G., Davies, M.S., 1990. Domestication rates in wild wheats and barley under primitive cultivation. Biol. J. Linn. Soc. Lond 39 (1), 39e78.

54. Honne, B.I., Heun, M., 2009. On the domestication genetics of self-fertilizing plants. Veg. Hist. Archaeobot 18 (3), 269e272.

55. Ibanez, J.J., Ortega, D., Campos, D., Khalidi, L., M ~ endez, V., 2015. Testing complex networks of interaction at the onset of the Near Eastern Neolithic using modelling of obsidian exchange. J. R. Soc. Interface 12 (107), 1e11.

56. Inizan, M., Lechevallier, M., 1994. L'adoption du debitage laminaire par pression au Proche-Orient. In: Gebel, H.G., Kozlowski, S.K. (Eds.), Neolithic Chipped Stone Industries of the Fertile Crescent. Ex oriente, Berlin, pp. 23e32.

57. Kadowaki, S., 2005. Designs and production technology of sickle elements in Late Neolithic Wadi Ziqlab, northern Jordan. Paleorient 31 (2), 69 e85.

58. Kilian, B., Ozkan, H., Walther, A., Kohl, J., Dagan, T., Salamini, F., Martin, W., 2007. € Molecular diversity at 18 loci in 321 wild and 92 domesticate lines reveal no reduction of nucleotide diversity during Triticum monococcum (einkorn) domestication: implications for the origin of agriculture. Mol. Biol. Evol. 24, 2657e2668.

59. Kluyver, T.A., Charles, M., Jones, G., Rees, M., Osborne, C.P., 2013. Did greater burial depth increase the seed size of domesticated legumes? J. Exp. Bot. 64 (13), 4101e4108.

60. Koenker, R., 2005. Quantile Regression. Cambridge University Press, Cambridge.

61. Koenker, R., 2013. Quantreg: Quantile Regression. R Package Version 5.05. Available. http://CRAN.R-project.org/package¼quantreg.

62. Kuijt, I., 2000. People and space in early agricultural villages: exploring daily lives, community size, and architecture in the Late Pre-Pottery Neolithic. J. Anthropol. Archaeol. 19 (1), 75e102.

63. Larson, G., Stephens, P.A., Tehrani, J.J., Layton, R.H., 2013. Exapting exaptation. Trends Ecol. Evol. 28 (9), 497e498.

64. Lucas, L., Colledge, S., Simmons, A., Fuller, D.Q., 2012. Crop introduction and accelerated island evolution: archaeobotanical evidence from 'Ais Yiorkis and pre-Pottery Neolithic Cyprus. Veg. Hist. Archaeobot 21 (2), 117e129.

65. McCartney, C., Todd, I.A., 2005. Chipped stone. In: Todd, I.A. (Ed.), Vasilikos Valley Project 7: Excavations at Kalavasos-tenta. Volume II. Studies in Mediterranean Archaeology, vol. 71(7). Paul Åstroms F € orlag, S € avedalen, pp. 177 € e264.

66. Milevski, I., Vardi, J., Gilead, I., Eirikh-Rose, A., Birkenfeld, M., Mienis, H.K., Horwitz, L.K., 2013. Excavations at Horbat 'Illit B: a Chalcolithic (Ghassulian) site in the Haelah valley. J. Isr. Prehist. Soc. 43, 73e147.

67. Nishiaki, Y., 1997. Neolithic sickle elements from the Balikh Valley, north Syria. AlRafidan 18, 59e67.

68. Oates, J., 2010. More thoughts on the Ubaid. In: Carter, R.A., Graham, P. (Eds.), Beyond the Ubaid: Transformation and Integration in the Late Prehistoric Societies of the Middle East. Ancient Oriental Civilization 63. The Oriental Institute of the University of Chicago, Chicago, pp. 45e49.

69. Poets, A.M., Fang, Z., Clegg, M.T., Morell, P.L., 2015. Barley landraces are characterized by geographically heterogeneous genomic origins. Genome Biol. 16, 173.

70. Portillo, M., Rosen, A.M., Weinstein-Evron, M., 2010. Natufian plant uses at el-Wad terrace (Mount Carmel, Israel): the phytolith evidence. Eurasian Prehist. 7 (1), 99e112.

71. Power, R.C., Rosen, A.M., Nadel, D., 2014. The economic and ritual utilization of plants at the Raqefet Cave Natufian site: the evidence from phytoliths. J. Anthropol. Archaeol. 33 (1), 49e65.

72. Purugganan, M.D., Fuller, D.Q., 2011. Archaeological data reveal slow rates of evolution during plant domestication. Evolution 65 (1), 171e183.

73. Quintero, L.A., Wilke, P.J., Waines, J.G., 1997. Pragmatic studies of Near eastern Neolithic sickle blades. In: Gebel, H.G.K., Kafafi, Z., Rollefson, G.O. (Eds.), The Prehistory of Jordan, II. Perspectives from 1997. Ex oriente, Berlin, pp. 263e286. R Core Team, 2014. R: a Language and Environment for Statistical Computing. R Foundation for Statistical Computing, Vienna. Available. http://www.R-project. org/.

74. Reimer, P.J., et al., 2013. IntCal13 and Marine13 radiocarbon age calibration curves 0e50,000 years cal BP. Radiocarbon 55 (4), 1869e1887.

75. Riehl, S., Zeidi, M., Conard, N.J., 2013. Emergence of agriculture in the foothills of the Zagros mountains of Iran. Science 341 (6141), 65e67.

76. Riehl, S., Asouti, E., Karakaya, D., Starkovich, B.M., Zeidi, M., Conard, N.J., 2015. Resilience at the transition to agriculture: the long-term landscape and resource development at the Aceramic Neolithic tell site of Chogha Golan (Iran). Biomed. Res. Int. 2015 (532481). http://dx.doi.org/10.1155/2015/532481.

77. Rindos, D., 1984. The Origins of Agriculture: an Evolutionary Perspective. Academic Press, New York.

78. Sauer, C.O., 1958. Jericho and composite sickles. Antiquity 32, 187e189.

79. Shennan, S., Edinborough, K., 2007. Prehistoric population history: from the late glacial to the late Neolithic in central and northern Europe. J. Archaeol. Sci. 34, 1339e1345.

80. Shennan, S., Downey, S.S., Timpson, A., Edinborough, K., Colledge, S., Kerig, T., Manning, K., Thomas, M.G., 2013. Regional population collapse followed initial agriculture booms in mid-Holocene Europe. Nat. Commun. 4 (2486). http://dx. doi.org/10.1038/ncomms3486.

81. Simms, S.R., Russell, K.W., 1997. Bedouin hand harvesting of wheat and barley: implications for early cultivation in Southwestern Asia. Curr. Anthropol. 38 (4), 696e702.

82. Smith, B.D., 2011. General patterns of niche construction and the management of 'wild' plant and animal resources by small-scale pre-industrial societies. Philos. Trans. R. Soc. Lond. B Biol. Sci. 366 (1566), 836e848.

83. Stevens, C.J., 2003. An investigation of agricultural consumption and production models for prehistoric and Roman Britain. Environ. Archaeol. 8 (1), 61e76.

84. Tanno, K., Willcox, G., 2006. How fast was wild wheat domesticated? Science 311 (5769), 1886e1886.

85. Tanno, K., Willcox, G., 2012. Distinguishing wild and domestic wheat and barley spikelets from early Holocene sites in the Near East. Veg. Hist. Archaeobot 21 (2), 107e115.

86. Timpson, A., Colledge, S., Crema, E., Edinborough, K., Kerig, T., Manning, K., Thomas, M.G., Shennan, S., 2014. Reconstructing regional population fluctuations in the European Neolithic using radiocarbon dates: a new case-study using an improved method. J. Archaeol. Sci. 52, 549e557.

87. Unger-Hamilton, R., 1985. Microscopic striations on flint sickle-blades as an indication of plant cultivation: preliminary results. World Archaeol. 17 (1), 121e126.

88. Vardi, J., Gilead, I., 2013. Keeping the razor sharp: hafting and maintenance of sickles in the southern Levant during the 6th and 5th millennia BC. In: Borrell, F., Ib anez, J.J., Molist, M. (Eds.), Stone Tools in Transition: from Hunter- ~ Gatherers to Farming Societies in the Near East. Universidad Autonoma de Barcelona, Barcelona, pp. 377e394.

89. Vardi, J., Golan, A., Levy, D., Gilead, I., 2010. Tracing sickle blade levels of wear and discard patterns: a new sickle gloss quantification method. J. Archaeol. Sci. 37, 1716e1724.

90. Watkins, T., 2008. Supra-regional networks in the

Neolithic of Southwest Asia. J. World Prehist. 21 (2), 139e171.

91. Weide, A., Riehl, S., Zeidi, M., Conard, N.J., 2015. Using new morphological criteria to identify domesticated emmer wheat at the aceramic Neolithic site of Chogha Golan (Iran). J. Archaeol. Sci. 57, 109e118.

92. Wilke, P.J., Bettinger, R., King, T.F., O'Connell, J.F., 1972. Harvest selection and domestication in seed plants. Antiquity 46, 203e209.

93. Willcox, G., 2012. Searching for the origins of arable weeds in the Near East. Veg. Hist. Archaeobot 21, 163e167.

94. Willcox, G., Fornite, S., Herveux, L., 2008. Early Holocene cultivation before domestication in northern Syria. Veg. Hist. Archaeobot 17 (3), 313e325.

95. Williams, A.N., 2012. The use of summed radiocarbon probability distributions in archaeology: a review of methods. J. Archaeol. Sci. 39, 578e589.

96. Winter, H., 1994. Reed harvesting experiment. In: Lechevallier, M., Ronen, A. (Eds.), Le gisement de Hatoula en Judee occidentale, Isra el: Rapport des fouilles, 1980- € 1988. Editions Faton, Quetigny Cedex, France, pp. 271e275.

97. Zeder, M.A., 2015. Core questions in domestication research. Proc. Natl. Acad. Sci. U. S. A. 112 (11), 3191e3198.

98. Zohary, D., 1969. The Progenitors of Wheat and Barley in Relation to Domestication and Agriculture Dispersal in the Old World. In: Ucko, P.J., Dimbleby, G.W. (Eds.), The Domestication and Exploitation of Plants and Animals. Duckworth, London, pp. 47e66. O. Mae

CHAPTER 3

DIFFERENTIATION OF MIS 9 AND MIS 11 IN THE CONTINENTAL RECORD: VEGETATIONAL, FAUNAL, AMINOSTRATIGRAPHIC AND SEA-LEVEL EVIDENCE FROM COASTAL SITES IN ESSEX, UK

Helen M. Roe, G. Russell Coope, Robert J.N. Devoy, Colin J.O. Harrison, Kirsty E.H. Penkman, Richard C. Preece, Danielle C. Schreve

INTRODUCTION

The differentiation of MIS 9 from MIS 11 in the continental record is a topic of ongoing debate. In Britain there is a consensus in favour of correlation of the Anglian Stage with MIS 12 and of the ensuing Hoxnian Stage with MIS 11 (e.g. Bridgland, 1994, Bridgland et al., 1999, Rowe et al., 1999, Grün and Schwarcz, 2000, Schreve, 2001 and Preece et al., 2007). However, in continental Europe, much uncertainty surrounds both the correlation of the Elsterian and the Holsteinian with other continental or deep-sea records. Like the consensus in Britain, many European workers favour correlation of the Holsteinian with at least part of MIS 11 (Sarntheim et al., 1986, Sibrava, 1986, Reille et al., 2000, deBeaulieu et al., 2001, Rousseau, 2003, Antoine and Limondin-Lozouet, 2004, Nitychhoruk et al., 2005 and Nitychhoruk et al., 2006). However, a recent study by Geyh and Müller (2005) has promoted correlation with MIS 9 based on some new TIMS ^{230}Th/U dates from two peat layers from the Holsteinian reference site at Bosel in northern Germany and reappraisal of uranium series dates from two Hoxnian sites in England. They also reconsider the correlation of certain critical pollen sequences from maar lakes in France to support their case and they consequently dismiss the potential conflation of two "Holstein-like Interglacials" as "a problem confined to the Quaternary of the British Isles" (Geyh and Müller, 2006: p. 3072). Such divergent views hamper assessment of events within the late Middle Pleistocene.

The Thames Valley in southern England is one region where a range of deposits of late Middle Pleistocene age have been distinguished, largely because the critical interglacial sediments occur at different elevations in the terrace staircase (e.g. Bridgland, 1994). Pollen analysis, the technique most frequently used for biostratigraphy, often fails to discriminate between late Middle Pleistocene interglacials, especially in fluvial contexts, where the records are notoriously fragmentary and frequently undiagnostic (cf. Thomas, 2001). However, different stages can be distinguished, often fairly easily, using molluscan (Preece, 1995 and Preece, 1999) and especially mammalian biostratigraphy (Schreve, 2001). By integrating terrace and faunal biostratigraphical data with aminostratigraphy, Bridgland (1994) identified four post-Anglian temperate stages in the Thames terrace record in the Lower Thames region downstream of central London, which he correlated with MIS 11, 9, 7 and 5e respectively. Moreover, new aminostratigraphic techniques, involving the analysis of multiple amino acids contained within the crystalline lattice of fossil shells (especially the calcitic opercula of the freshwater gastropod *Bithynia*), have provided additional independent support for such differentiation (Penkman, 2005 and Penkman et al., 2007a).

Citation: Helen M. Roe, G. Russell Coope, Robert J.N. Devoy, Colin J.O. Harrison, Kirsty E.H. Penkman, Richard C. Preece, Danielle C. Schreve, Differentiation of MIS 9 and MIS 11 in the continental record: vegetational, faunal, aminostratigraphic and sea-level evidence from coastal sites in Essex, UK, Quaternary Science Reviews, Volume 28, Issues 23–24, November 2009, Pages 2342-2373, ISSN 0277-3791, http://dx.doi.org/10.1016/j.quascirev.2009.04.017.

This paper reviews evidence from eastern Essex, on the western margin of the North Sea basin, that has a direct bearing on these issues. New data are presented from Cudmore Grove, a coastal site on Mersea Island, where a deeply-incised channel cut into London Clay (Eocene) has been infilled by interglacial estuarine silts and clays and now lies buried beneath fluvial gravels. The sediments, which cover a substantial part of the interglacial, can be linked into the terrace sequence of the Thames. Cudmore Grove is located just 10 km to the west of Clacton-on-Sea, the type site of the Lower Palaeolithic Clactonian industry (Oakley and Leakey, 1937), which has been correlated with the Hoxnian interglacial (Fig. 1; Pike and Godwin, 1953, Warren, 1955, Turner and Kerney, 1971, Miller et al., 1979, Bowen et al., 1989 and Bridgland et al., 1999). The pollen records from these two sites bear a number of similarities and they were initially attributed to the same interglacial, namely the Hoxnian (cf. Bridgland, 1988, Bridgland, 1994 and Roe, 1994). However, several lines of evidence suggest that this correlation is not secure.

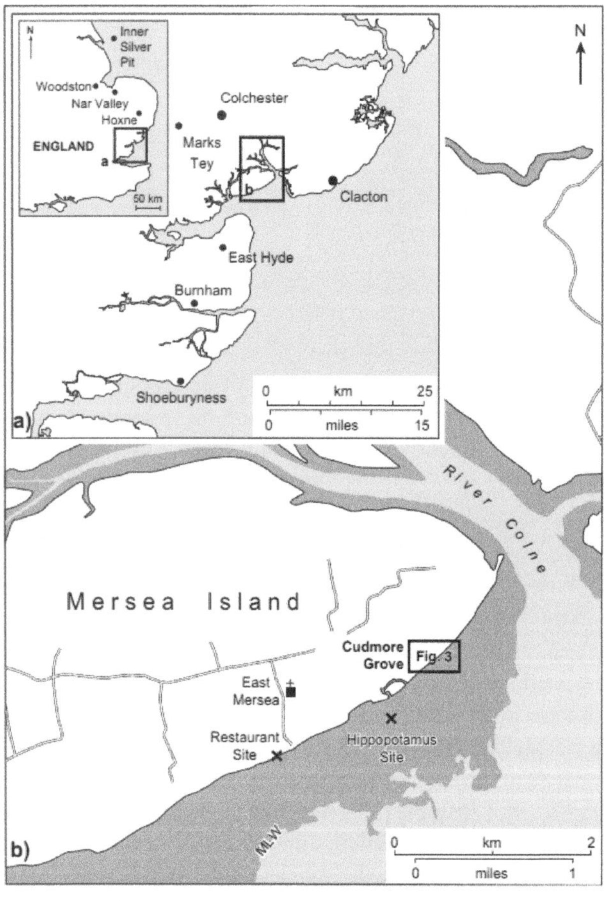

Figure 1: Location map of Cudmore Grove and the East Mersea Restaurant Site. The location of key interglacial sites in East Anglia referred to in the text is shown.

Despite their proximity, we show that the faunal records from the two sites differ significantly, as does their relative sea-level history. These differences indicate that the sites are not of the same age, a conclusion supported by the new aminostratigraphic data presented here that indicate that the sediments at Clacton formed during MIS 11 and those at Cudmore Grove during MIS 9. On the same foreshore, 2 km from the site at Cudmore Grove, another interglacial channel occurs now buried by modern beach shingle. This is the 'East Mersea Restaurant Site', which has yielded a limited fauna of non-marine molluscs and vertebrates, including *Hippopotamus* indicative of an Ipswichian age (correlated with MIS 5e) (Fig. 1). Thus channel sediments belonging to four different interglacials (including the Holocene), occur along this limited stretch of coastline. This provides an opportunity to examine faunal, sea level and other differences between the stages within a small geographical region and in particular to define the characteristics of MIS 9 in greater detail.

REGIONAL SETTING

The coastal fringe of eastern Essex is dominated by a series of low-lying (<ca 30 m O.D.) Pleistocene terrace gravels that extend in a north-easterly direction between Southend and Clacton (Fig. 2). These deposits are largely of Thames origin and post-date the Anglian glaciation, when the Thames was diverted by ice from an earlier, more northerly course across central and northern Essex into the valley of an existing tributary, the Essex Medway (cf. Gruhn et al., 1974, Bridgland, 1983 and Bridgland, 1994, Fig. 5.5; Gibbard, 1985 and Gibbard, 1994). Older, pre-Anglian diversion Thames deposits, assigned to the 'Kesgrave Member' of the Lower Thames Formation (Gibbard, 1999) occur to the north of Mersea Island in the Tendring Plateau area, whilst pre-diversion gravels of the Essex Medway, the 'High-level East Essex Gravels' (Bridgland, 1994), survive as isolated remnants in the west of the coastal region (Fig. 2). The study area lies beyond the Anglian ice limits; the nearest glacial deposits occur a few kilometres to the west of Colchester (Fig. 2).

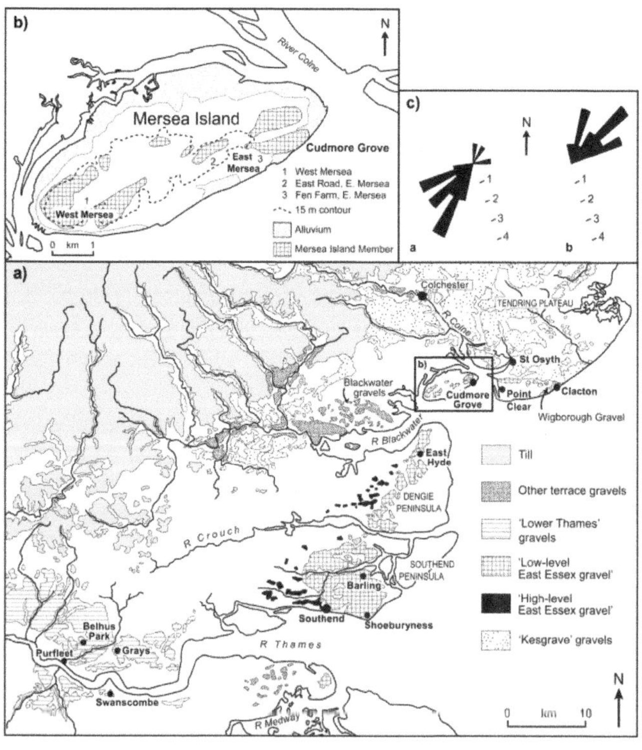

Figure 2: Distribution of sand and gravels associated with the Thames and its tributaries in eastern Essex and the eastern area of the Lower Thames region (after Bridgland, 1988 and Bridgland, 1994). The distribution of Anglian till and the pre-Anglian gravels of the Thames ('Kesgrave' gravels) are indicated. For ease of interpretation, and to enable the various deposits to be related to the diversion of the Thames into the Southend and Dengie Peninsulas during the Anglian Stage, the terminology applied to the gravels follows Bridgland (1994). The

pre-diversion deposits of the Essex Medway ('High-level East Essex gravels'), the post-diversion Thames–Medway deposits ('Low-level East Essex gravels') and the Thames gravels of the Lower Thames area are all identified separately (cf. Bridgland, 1988 and Bridgland, 1994). A more complete overview of the formal nomenclature applied to the gravels is given in the text and in Gibbard (1999). Inset (b) shows the distribution of the Mersea Island Member outcrops on Mersea Island (after Bridgland, 1983); inset (c) shows the palaeocurrent measurements from Units 6a (a) and 6b (b). The location of key sites mentioned in the text is indicated.

The terrace gravels of south-eastern Essex are locally underlain by remnants of a series of palaeo-channels cut into the London Clay (Lake et al., 1977, Bridgland, 1983, Bridgland, 1988, Bridgland, 1994, Roe, 1994 and Roe, 1999). Of these, the channel-fill sequence at Cudmore Grove is the best preserved and most accessible. These features represent incised river channels of the diverted Essex Thames (or 'Thames–Medway') and its associated tributaries as the river migrated southwards across the region during successive warm stages of the late Middle Pleistocene (Bridgland et al., 1988, Bridgland et al., 1999, Bridgland, 1994, Roe, 1994 and Roe, 2001). Palaeontological analyses have confirmed that the southeast Essex channel deposits accumulated under fully temperate conditions at times of high relative sea level (Warren, 1955, Bridgland, 1994, Roe, 1994, Roe, 1999, Roe, 2001 and Bridgland et al., 1999). The terrace gravels and the channel-fill sediments are truncated locally by Holocene estuarine sediment, particularly around the east-trending estuaries of the Crouch, Blackwater and Colne (Fig. 2).

FIELD INVESTIGATIONS

The sedimentary sequence at Cudmore Grove was examined using a combination of surface mapping, section logging and coring techniques. The exposed parts of the sequence, which include sand and gravel cliffs and foreshore exposures, were mapped along a 0.3 km section extending between Cudmore Grove wood [TM 065 144] and the sea wall [TM 069 147] (Fig. 3, Fig. 4 and Fig. 5). The subsurface stratigraphy was investigated using a series of shallow (ca 1–9 m) borings taken with a hand-auger and a 'Minute Man' portable drilling apparatus. A deep (22 m) borehole (CG1) was also excavated with a 10 cm diameter percussion corer to sample the complete Pleistocene sequence (Fig. 3 and Fig. 4). This was positioned on the cliff top at a point where the channel-fill sequence was estimated to be at maximum thickness. Bulk samples were obtained from a trench dug into the foreshore with a mechanical excavator (Fig. 3; see also Bridgland, 1994, Fig. 5.25).

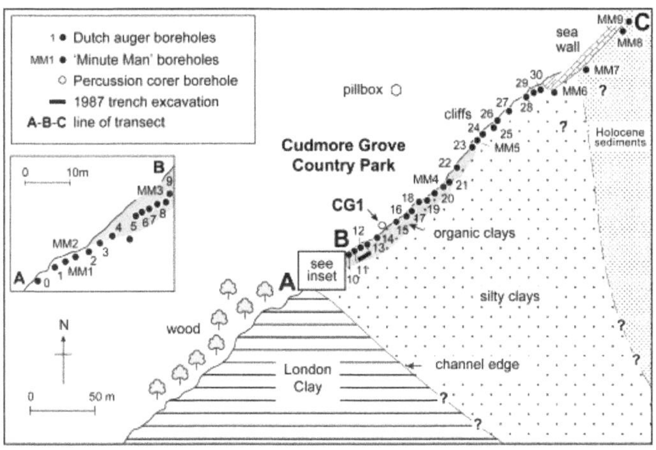

Figure 3: Location of boreholes at Cudmore Grove. The channel edge is indicated. The cross sections A–C and A–B refer to Fig. 4 and Fig. 5 respectively.

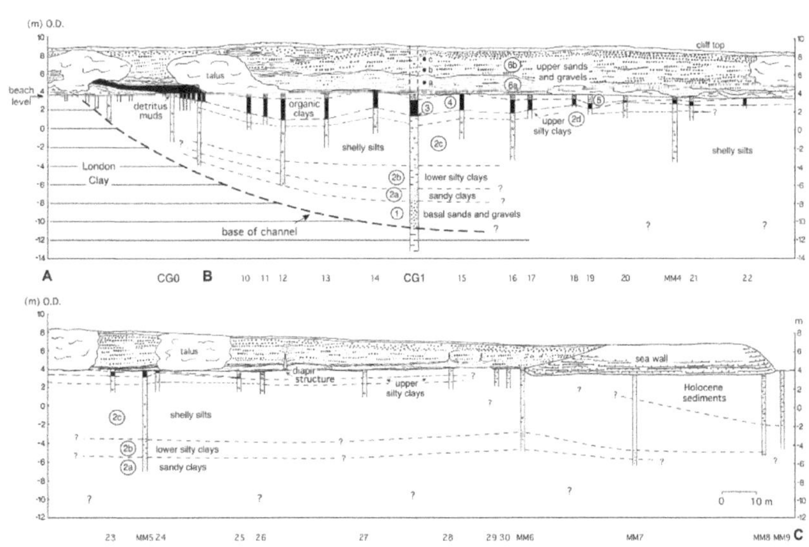

Figure 4: Section A–C showing the lithostratigraphy of the Cudmore Grove channel-fill sequence and overlying sands and grav-

els (Unit 6). The location of the boreholes is shown in Fig. 3. The position of Unit 3, only a few centimetres thick, at the base of the organic sequence (depicted in black) is indicated. The stratigraphical position of this unit is shown more clearly in Fig. 5. The lateral continuity of sub-units 2c and 2d near boreholes MM6–MM7 is unclear; here the Pleistocene beds are replaced by Holocene estuarine deposits (Fig. 3) which are compositionally similar to the upper part of the Pleistocene channel-fill sequence and locally indistinguishable. Samples for clast lithological analysis (Table 1) were collected from the cliff sections near borehole CG1 (sampling points a, b, c).

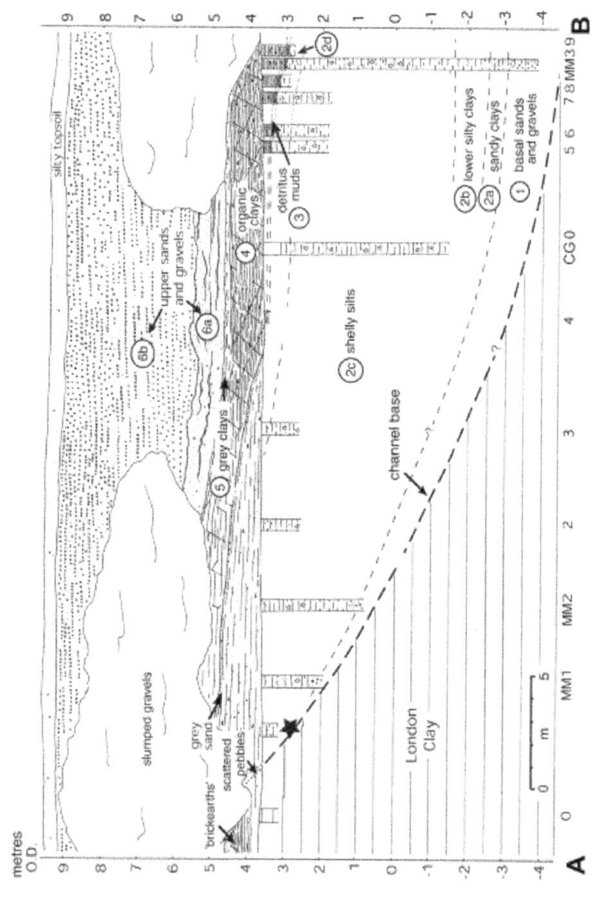

Figure 5: Channel margin, section A–B. The star symbol marks the position of the single struck flake found *in situ* in the Cudmore Grove Channel gravels (Unit 1).

LITHOSTRATIGRAPHY

With the exception of the sands and gravels exposed in the cliffs, which extend for several kilometres along the coast of Mersea Island (Fig. 3), the entire Pleistocene sequence at Cudmore Grove fills a steep-walled, channel-like depression cut into the London Clay, the 'Cudmore Grove Channel' (Roe, 1994) (Fig. 3, Fig. 4 and Fig. 5). This feature is at least 0.28 km wide and 0.25 km long and is aligned in a northwest–southeast direction. The southern edge of the channel is well exposed at the base of the cliffs near Cudmore Grove wood and on the foreshore (Fig. 3). Its maximum observed depth was recorded in borehole CG1 at −10.40 m O.D. (Fig. 4). The depth further to the east is unknown, where the Pleistocene beds are truncated by Holocene sediment.

The Pleistocene sequence at Cudmore Grove can be divided into six units:

Unit 1: Basal sands and gravels (Cudmore Grove Channel Gravel)

The floor of the channel is lined with sands and gravels that form the basal member of the Pleistocene sequence and directly overlie London Clay. These reach a maximum recorded thickness (3.10 m) in borehole CG1 (Fig. 4), then thin progressively south-westwards towards the channel margin, where they taper into a 1–2 cm pebble lag (Fig. 4 and Fig. 5), the 'Cudmore Grove Channel lag' (Bridgland et al., 1988). The persistence of Unit 1 to the northeast of borehole CG1 is unknown.

Unit 2: Silty clays

The basal gravels grade upwards into a thick sequence of silts and clays, which occupy most of the channel and are exposed over wide areas of the foreshore. This unit can be divided into four sub-units:

a. Sandy clays

Massive, dark grey sandy clays occupy the basal 0.5–1 m. These reach maximum thickness (1.2 m) in borehole CG1, and thin towards the channel margin (Fig. 4). They can also be traced at the eastern end of the site below −5 m O.D. The continuity of the sub-unit beyond this point is unknown.

b. Lower silty clays

The sandy clays pass gradually upwards into brown-grey silty clays (Munsell: 10YR 4/1), commonly interbedded with thin, horizontal and gently inclined sand bands. Shell fragments occur occasionally. The lower silty clays reach maximum thickness (ca 3 m) between boreholes CG1 and MM5 and can be traced as far north as the sea wall (Fig. 4).

c. Shelly silts

The lower silty clays grade upwards into a thick sequence of massive, dark grey silts with scattered shell debris. These vary in thickness, from ca 5 m near borehole MM5 to <1 m in the southwest. The shelly silts can also be traced to the end of the sea wall in the northeast, where they extend from −3 m to 2 m O.D. (Fig. 4). The sub-unit is truncated further to the east by Holocene estuarine sediment.

d. Upper silty clays

Green-grey silty clays, (ca 0.5 m) with occasional horizontal sand beds form the uppermost member of this unit, which contains occasional scattered wood fragments and shell debris.

Unit 3: Detritus muds

The silty clays are truncated at the southwestern part of the site by a thin bed of shelly detritus muds, composed of abundant shell debris in a matrix of dark brown (7.5YR 2/0) silty clay. This deposit, which when sampled in the early 1990s was only a few centimetres thick, is richly fossiliferous and has yielded a wealth of faunal remains (Holman et al., 1990; see below). The detritus muds were formerly more extensive and up to 30 cm thick (Bridgland et al., 1988) but have been eroded significantly by ongoing cliff retreat. The contact between the detritus muds and the silty clays of Unit 2 is sharp and is marked by a thin bed of coarse sand. The maximum thickness of the detritus muds (15 cm) was recorded near borehole MM3 (Fig. 5). From here the unit thins rapidly westwards, grading into 2–3 cm layer of grey sand and eventually disappearing near borehole MM5.

Unit 4: Organic clays

Organic clays with abundant wood fragments overlie the detritus muds. The contact between the two units is abrupt and is marked by the sudden disappearance of shell debris. Like the detritus muds, the organic clays achieve maximum thickness (2.5 m) in the southwestern part of the site, where two facies can be recognised, a lower ca 20–50 cm bed of highly compressed wood interbedded with organic clay, and an upper 1–2 m bed of massive, organic clay with occasional wood fragments. The lower deposit is the more heterogeneous and is partly laminated. Three tree trunks have been exhumed from this horizon, one with its roots still intact (Bridgland, 1994, Fig. 5.26). Both beds show signs of compaction and local deformation. The organic clays thin steadily south-westwards, where they finally meet the rising surface of the London Clay. They also thin north-eastwards and eventually become truncated by the Upper sands and gravels (Unit 6) (Figs. 5 and 6A).

Figure 6: A) Photograph of an outcrop of the organic clays (Unit 4) at the base of the Cudmore Grove cliffs (facing north west), taken near borehole 14 (Fig. 4). The overlying sands and gravels (Unit 6) are sandier at the base (sub-unit 6a) and become clast-dominated in the upper part of the cliff exposure (sub-unit 6b). B) Photograph of the organic clays (Unit 4) and overlying grey clays (Unit 5) taken in the cliff sections near borehole 4 (Fig. 5).

Unit 5: Grey clays

Massive, grey clays, 5–30 cm thick, form the uppermost member of the channel-fill sequence (Fig. 6B). The contact between the two deposits is variable, sometimes it is marked by a transitional zone of brown-grey clay, but elsewhere it is abrupt. The unit extends laterally towards the eastern part of the site where it rests directly above Unit 2. It finally disappears near borehole 25 (Fig. 4 and Fig. 5).

Units 1–5 together complete the infill sequence of the Cudmore Grove Channel, which we formally define here as the 'Cudmore Grove Member' of the Lower Thames Formation (cf. Gibbard, 1999). Borehole CG1, which includes the thickest representation of most of the sedimentary units, is proposed as the stratotype [TM 0681 1456].

Unit 6: Upper sands and gravels (Mersea Island Member, Lower Thames Formation)

The fine-grained beds are capped by 4–5 m of sands and gravels, which are exposed in the Cudmore Grove cliffs (Figs. 4, 5 and 6A). These were originally assigned to the 'Mersea Island Gravel' of the Low-level East Essex Gravel Formation (Bridgland, 1983, Bridgland, 1988 and Bridgland, 1994), which comprised all the diverted Thames gravel aggradations in eastern Essex. This has since been re-termed the 'Mersea Island Member' of the Lower Thames Formation (Gibbard, 1999) with Cudmore Grove the designated type locality. Accelerated cliff erosion in recent years has provided excellent exposures of the sands and gravels that comprise two main facies: a lower 2–3 m suite of cross-bedded sands (Unit 6a); and an upper 3 m of mainly clast-supported gravels (Unit 6b) (Fig. 4 and Fig. 5). The lower unit, which forms a sharp contact with the underlying grey clays and organic clays, is best developed in the western part of the cliff exposure, where it comprises large-scale, gently dipping foresets of sand and subsidiary gravel. Individual foresets are 15–50 cm thick and dip at angles of 5–20°. Palaeocurrent measurements show that the prevailing flow direction was towards the north-north-east (Fig. 2c). In the lower part of this unit the sand-rich facies are locally interbedded with a series of thin (1–5 cm) grey clay horizons (Fig. 6B). The sandy beds are truncated by the clast-rich facies of Unit 6b, which thicken progressively towards the sea wall. These too are cross-stratified, particularly in the lower 2 m, although the foresets are smaller than in Unit 6a. In contrast to the lower facies, palaeocurrent measurements show that flow was towards the southeast (Fig. 2c).

Unit 6b shows several signs of periglacial deformation, including a number of ice-wedge casts. At the southwestern end of the exposure the gravelly facies are also interbedded with large lobes of massive, heavily oxidized silty clay ('brickearth') (Bridgland, 1983 and Bridgland et al., 1988; Fig. 5), thought to comprise soliflucted London Clay (Bridgland, 1994).

CLAST LITHOLOGICAL ANALYSIS

Clast lithological analysis has been an important tool in this region for distinguishing deposits of the Thames, which became confluent with the Essex Medway after its diversion into the region during the Anglian (forming the 'Thames–Medway'), from older, pre-diversion Medway aggradations (Bridgland, 1983 and Bridgland, 1988). The former contain clasts sourced from the wider catchment of the Thames, including the Weald, whereas the latter comprise lithologies originating from the London Basin (Bridgland, 1988). To ensure compatibility with previous work, samples were analysed using the technique described by Bridgland (1983: p. 25). Two samples were examined from the Cudmore Grove Channel Gravel (Unit 1, samples 1a and 1b) and three from Unit 6, the Upper sands and gravels (samples 6a–6c) (Fig. 4; Table 1). At least 400 stones were counted per sample. During counting, the pebbles were divided into three main categories: i) the 'local' group, which consists of rocks outcropping in the London Basin; ii) the 'southern' group comprising rocks occurring in the area to the south of the North Downs (including Lower Greensand chert); and iii) the 'exotic' group, comprising rocks from any other area and including exotic types reworked from within the London Basin (cf. Bridgland, 1983 and Bridgland, 1988).

Table 1: Clast lithological composition of gravels from Mersea Island. Samples 1a and 1a are from Unit 1 (basal sands and gravels) sampled in borehole CG1 at Cudmore Grove; samples 6a–6c are from Unit 6 (upper sands and gravels) (Fig. 4). Other sample numbers follow Bridgland, 1983 and Bridgland, 1994

				Local			Southern		Exotics						
Sub-division	Location	Sample no.	Height (m O.D.)	Non-Tertiary flint (%)	Tertiary flint (%)	Total (%)	Total greensand chert (%)	Total (%)	Quartz & quartzite (%)	Total (%)	Ratio Tertiary: non-Tertiary	Ratio Southern:local	Ratio Southern:exotics	Total count	Grid reference
Mersea Island Member	Cliff section Cudmore Grove	6a	5.1	54.8	33.3	88.1	7.3	7.3	4.6	4.6	1.65	12.07	0.63	562	TM 0683 1457
	Cliff section Cudmore Grove	6b	6.3	57.8	28.3	86.1	10.3	10.3	3.6	3.6	2.04	8.35	0.35	417	TM 0683 1457
	Cliff section Cudmore Grove	6c	7.8	57.3	31.7	89.0	8.5	8.5	2.2	2.2	1.80	10.39	0.28	409	TM 0683 1457
	West Mersea[a]	1	19.3	43.8	38.6	82.4	14.2	14.2	2.8	3.5	1.13	5.80	0.24	578	TM 0134 1361
	West Mersea[a]	2	19.4	42.9	44.8	87.7	10.0	10.0	1.4	2.3	0.95	8.79	0.23	431	TM 0144 1373
	Fen Farm[a] East Mersea	1	10.6	39.6	47.6	87.2	10.7	10.7	1.8	2.2	0.83	8.17	0.20	553	TM 0590 1444
	Fen Farm[a] East Mersea	2	11.8	37.7	52.3	90.0	7.6	7.6	1.8	2.3	0.72	11.82	0.31	512	TM 0583 1437
Cudmore Grove Member	Borehole CG1 Cudmore Grove Channel Gravel	1a	−8.30 to −9.40	70.7	13.2	83.9	2.9	2.9	12.0	13.2	5.36	29.00	4.57	483	TM 0681 1456
	Borehole CG1 Cudmore Grove Channel Gravel	1b	−8.30 to −9.40	73.5	11.8	85.3	5.0	5.0	9.3	9.7	6.22	5.86	1.94	525	TM 0681 1456
	Cudmore Grove Channel lag gravel[b]	1	ca 1–2 m	40.1	44.9	85.0	14.0	14.0	1.0	1.0	1.12	6.07	0.07	301	TM 0682 1451

[a]Counts reported by Bridgland (1983).

[b]Counts reported by Bridgland (1994).

The basal sands and gravels (Unit 1) recovered from borehole CG1 were found to comprise the same general lithologies (southern, exotic and local) as typical Thames–Medway gravels (Bridgland, 1988; Table 1). However, proportions of the three types differed significantly. First, the samples included much higher levels of angular (non-Tertiary flint), which gave Tertiary:non-Tertiary flint ratios of 5.36 and 6.22 (ratios from Thames–Medway gravels rarely exceed 1). These ratios are comparable to those recorded by Bridgland (1983) in the 'Upper Holland Gravel' near Clacton, a deposit that is thought to be enriched with distal outwash material (Bridgland, 1994: p. 287). The Unit 1 samples also yielded much higher frequencies of exotics (9.7% and 13.2% of the total) than local Thames–Medway gravels, which typically contain 0.5–2.5% (Bridgland, 1983). In this case, the most probable source is the exotic-rich Kesgrave Member deposits that outcrop widely to the north and west of Mersea Island (Fig. 2).

Overall, the range of lithologies in the Unit 1 gravels indicates a composition intermediate between Thames–Medway and outwash, with an additional element of Kesgrave material. The gravels may, therefore, have been aggraded by the Thames–Medway itself, downstream of a confluence with a river re-working fluvio-glacial sediments. Alternatively, they could have been deposited by a tributary that was fed by streams re-working older Thames–Medway, fluvio-glacial and Kesgrave deposits.

A single analysis from the Cudmore Grove Channel lag differed from the counts described above in showing a typical Thames–Medway composition (cf. Bridgland et al., 1988; Table 1). Unfortunately, no counts were obtained from this marginal deposit in the present study because pebble numbers were too low to obtain a valid sample size. Nevertheless, general observations confirmed that the deposit was indeed much richer in Tertiary flint than the gravels in borehole CG1. It therefore appears that the basal gravel is a composite deposit, the thinner, marginal part of which probably represents a lag derived from older Thames–Medway sediments.

Unit 6

The upper sands and gravels (Unit 6) show some compositional similarities to the Thames–Medway gravels of the Southend and Dengie Peninsulas (cf. Bridgland, 1983 and Gibbard et al., 1996). They differ, however, in their marginally higher proportions of exotics, which reach 2.4–4.6% (for comparison, frequencies in the Asheldham Member on the Dengie Peninsula rarely exceed 2.0%). The assemblages nevertheless compare closely with those of gravels found elsewhere on the island, which have also been assigned to the Mersea Island Member (Bridgland, 1983; Table 1). Bridgland (1983) attributed their slight enrichment of exotics to the input of additional quartzose material from a tributary, possibly the proto-Blackwater, which joined the Thames–Medway between the Dengie Peninsula and Mersea Island.

Whatever the origin of the exotics, the close similarities shown between the counts from Units 6a and 6b suggest that the same materials were being sourced throughout the aggradation, in spite of the inferred changes in flow direction (Table 1; Fig. 2c). This is somewhat surprising, since the upper gravel-rich facies appear to have been deposited by a river flowing from the northeast, a direction that is inconsistent with a Thames–Medway origin. It is therefore possible that these upper gravels were reworked by a smaller river (perhaps a proto-River Colne) from older Thames–Medway material to the north.

REGIONAL DISTRIBUTION OF THE SEDIMENT SEQUENCE

The thickness of the sand and gravel overburden in the Cudmore Grove cliffs, (Fig. 4) renders it impossible to trace the Cudmore Grove Channel or its associated infill inland or immediately beyond the foreshore area. However, sand and gravel outcrops occur elsewhere on the island, which have been correlated with the gravel in the Cudmore Grove cliffs (Bridgland, 1983 and Bridgland, 1988). Compositionally these gravels are similar to those at Cudmore Grove, although they differ considerably in height. The outcrops at West Mersea, for example, have basal elevations of ca 10–13 m O.D. (Bridgland, 1983), some 10 m higher than the gravels at Cudmore Grove (Fig. 2 and Fig. 4). These outcrops rest directly on London Clay. The outcrops near East Mersea are less well documented, although Bridgland (1983) established that the small outcrop to the west of the village (Fig. 2b) has a base of ca 10 m O.D. Sand and gravel outcrops at Point Clear, 4 km to the east of Cudmore Grove on the Tendring Plateau (Fig. 2), have also been assigned to the Mersea Island Member (cf. Bridgland, 1994).

Bridgland (1994) showed that the various outcrops of the Mersea Island Member correspond well, both in thickness and height, with the sands and gravels of the Asheldham Member (cf. Gibbard, 1999) on the Dengie Peninsula, which he considered its upstream equivalent. The Wigborough Gravel, ca 8 km to the east in the Clacton area (Fig. 2), was thought to represent the downstream equivalent on the Tendring Plateau (see Bridgland, 1994: p. 294 for a full account of the correlations). Like the gravels in the Cudmore Grove cliffs these also include slightly enriched levels of exotics (Bridgland, 1994: p. 322).

In spite of these long profile projections, it remains difficult to establish whether all the outcrops of the Mersea Island Member represent part of a single aggradation or whether they are composite in age. It is possible, for example, that some of the more elevated outcrops are older than those in the Cudmore Grove cliffs and may even pre-date the Cudmore Grove Channel and its infill. Relative age determinations based on long profile projections are also hampered, in this instance, by the fact that at least part of the Cudmore Grove cliff gravels sequence appears to have been deposited by a tributary river or rivers (the proto-Blackwater and/or proto-Colne) and not by the Thames–Medway (Roe, 1994). The stratigraphical and palaeogeographical implications of these findings are considered further below.

POLLEN AND PLANT MACROFOSSILS

Sampling and Preparation

Pollen analyses were undertaken on the sediments of borehole CG1 and were prepared using the method of Berglund and Ralska-Jasiewiczowa (1986). Pollen was recovered from the sediments between 6.27 and 14.80 m in this borehole, spanning Units 2–4 (Fig. 7). The sandy facies at the base of Unit 2 (14.80–16 m) yielded only degraded *Pinus*pollen at sparse concentrations, precluding detailed analyses. Plant macrofossil remains retained at the sieving stage of pollen preparation are given in Table 2.

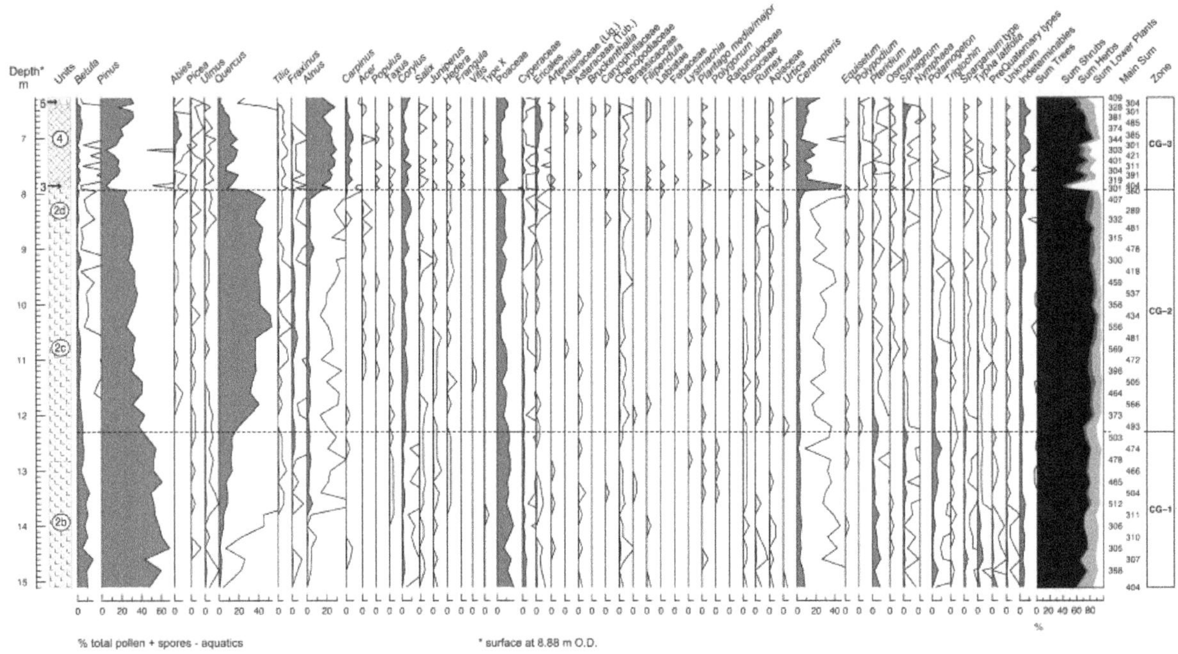

Figure 7: Pollen percentage diagram from core CG1. 10× exaggeration lines are indicated.

Table 2: Plant macrofossils preserved after the sieving stage of pollen analysis in samples from borehole CG1. None of the other pollen samples analysed from this core yielded identifiable macrofossils

Pollen zone	Depth (m)	Unit	Terrestrial	Waterside & damp ground			Aquatic					Unclassified	
			*Urtica*sp(p). seeds	*Eupatorium cannabinum*L. achenes	*Typha*sp(p). fruit	*Alisma*sp. fruit	*Elatine hydropiper*L. fruit	*Ranunculus*Batrachian sp. achenes	cf. *Chara*sp(p). oospores	*Azolla filiculoides*Lam. megaspores	*Najas minor*All. fruit	Bryophyte fragments	*Sphagnum*leaves
CG-3	6.62	4	1										
CG-3	6.82	4		1									
CG-3	6.94	4			2				2			1	1
CG-3	7.02	4		1						3	1		
CG-3	7.12	4	1	1	1				1				
CG-3	7.21	4								3	1		
CG-3	7.31	4							1	5			1
CG-3	7.40	4								1			
CG-3	7.50	4			1	1			1				
CG-3	7.58	4				1							
CG-3	7.66	4			1								
CG-3	7.75	4	1										
CG-3	7.85	3			1							1	
CG-2	9.80	2c										1	
CG-2	12.20	2c			1								
CG-1	12.80	2b			1								
CG-1	13.00	2b			2								
CG-1	13.40	2b										1	
No pollen	15.10	2b			2								
No pollen	15.30	2a	1										1
No pollen	15.87	2a					1	1				1	
No pollen	16.00	2a					1						

Pollen Assemblage Biozones

The pollen percentage diagram (Fig. 7) is divided into three pollen assemblage biozones, labelled with the prefix ‘CG’.

CG-1 (14.80–12.30 m)

This basal zone occurs in the sediments of sub-units 2b–2c. The spectra are dominated *Pinus* pollen, which rises to 69% at 14.40 m. *Betula* pollen also peaks in the lower section, then falls gradually near the upper zone boundary. *Quercus* pollen increases steadily through the zone. Of the other tree taxa, *Alnus* pollen frequencies are most significant, rising in the middle of the zone. Shrub taxa are poorly represented, with *Corylus* the most persistent. Poaceae and Cyperaceae pollen dominate the herbaceous taxa. Lower plant and aquatic pollen types are well represented, attaining peak frequencies of ca 15%.

With the exception of several *Typha* fruits, plant macrofossils were rare. However, several macrofossils were recovered from the underlying sandy sediments, which contained little or no pollen. These included two fruits of *Elatine hydropiper* (at 15.87 and 16.00 m) and an achene of *Ranunculus*, subgenus Batrachian (at 15.87 m) (Table 2).

CG-2 (12.30–7.93 m)

Sub-units 2c and 2d continue to the top of this zone. The base of the zone is marked by a sudden rise in the frequency of *Quercus* and a small increase in *Corylus* pollen. These trends continue to the middle of the zone, where *Quercus* achieves peak frequencies (53%). *Pinus* and *Betula* pollen frequencies decline simultaneously. The frequencies of most taxa stabilise in the upper half of the zone, with *Quercus* and *Pinus* dominant. *Alnus*and *Corylus* pollen undergo slight frequency increases. A single *Typha* fruit (at 12.20 m) and a bryophyte fragment (at −10.0 m) were the only macrofossils recorded.

CG-3 (7.93–6.27 m)

The base of this zone coincides with the appearance of the organic clays (Unit 4). The basal few centimetres contain sand and shell debris, representing the thin remnant in this core of Unit 3 (Fig. 7). The pollen spectra undergo many changes, which begin with a sharp peak in *Ceratopteris* spores and a steep rise in *Alnus* pollen. These changes coincide with a sharp fall in *Pinus*, *Quercus*, *Corylus* and Poaceae pollen. *Alnus*frequencies continue to rise in the central part of the zone, whereas *Ceratopteris* spores decline. *Pinus*, *Quercus* and *Corylus* pollen frequencies recover slightly, together with Poaceae and *Betula*. *Carpinus*, *Abies* and Ericales pollen frequencies also increase in the middle part of the zone. With the exception of *Pinus* and *Alnus*, frequencies of all the major arboreal taxa decrease at the top of the zone to <3%. These changes coincide with another sedimentary change, as the organic clays are replaced at 6.4 m by the grey clays (Unit 5). *Ceratopteris* spore frequencies show another sharp increase at this point. Only two samples from this unit yielded pollen.

The samples from this zone yielded abundant macrofossils, including several megaspores of *Azolla filiculoides* and three achenes of *Eupatorium cannabinum* (Table 2). A tree trunk (over 2 m long and probably *in situ*) and several other large pieces of *Alnus* wood were recovered from the basal part of the organic sequence elsewhere at the site.

VEGETATION AND LOCAL ENVIRONMENT

The absence of pollen in the basal, sand-rich sediments of sub-unit 2a suggests that conditions were initially unfavourable for pollen sedimentation or preservation. However, the fact that these deposits yielded waterside herb macrofossils shows that marshland communities were already established. The occurrence of fruits of *E. hydropiper* is noteworthy in implying that interglacial conditions had already become established when Unit 2a accumulated. This species flourishes today in eutrophic, sandy-bottomed, shallow-water habitats with a thin cover of silt and organic detritus (Brinkkemper et al., 2008).

CG-1

The sediments and other microfossils (see below) associated with this zone suggest that a fairly substantial, tidally-influenced river channel now existed. The pollen source area is likely to have been large, including inputs from fluvial and limited marine sources. Boreal-type woodland prevailed in the regional catchment at the start of this phase; *Betula* and *Pinus* were both widespread, although the latter is likely to be overrepresented due to the buoyancy properties and high pollen productivity of *Pinus* grains (cf. Fægri and Iversen, 1989 and Roe and van de Plassche, 2005). Oak, elm, ash and hazel persisted on a limited scale.

The composition of the forest changed subsequently as oak, and then ash, began to expand, out-competing birch in some areas. With this expansion came the colonisation of *Hedera* into the forest understorey. The unidentified palynomorph Type X first appears in this zone. This taxon has been reported from several Hoxnian sites (Thomas, 2001) and is believed to have been produced by a forest shrub, possibly a member of the Oleaceae (Turner, 1970). *Alnus* too became established, probably near the river margins.

Herb communities were fairly diverse throughout this period and point to the existence of lightly shaded forest-floor habitats or more open areas of grassland. *Pteridium* and *Sphagnum* confirm the development of damp, slightly acidic soils in parts of the region. Waterside and marsh plants also formed an important part of the flora during this time, probably flourishing near the river. These included sedges, ferns and several marshland herbs (*Typha latifolia*, *Filipendula*, *Sparganium* and *Lysimachia*). The *Potamogeton*pollen record shows that relatively calm, freshwater habitats prevailed locally.

CG-2

The sediments and other microfossil datasets suggest that marine influence became more pronounced as saltwater inflow gradually increased (see Sections 8, 9 and 10below). An enlargement of the pollen catchment can thus be inferred, as pollen was now carried in from estuarine and marine sources. The sudden increase in *Quercus* pollen at the start of the zone may partly reflect this taphonomic change, as *Quercus* pollen was introduced from coastal waters. In spite of this overrepresentation, oak was clearly becoming well established in the region, initiating a decline in both birch and pine. Pine probably persisted in some areas but on a reduced scale. *Fraxinus* and *Tilia* also expanded in the middle of the zone, reflecting the development of well-drained, base-rich soils within the catchment. This was accompanied by the gradual expansion of *Corylus*and the first arrival of *Vitis*, a shrub of damp, woodland habitats (Turner, 1968 and Turner, 1970).

Some open areas persisted during this zone, supporting such genera as *Polygonum* and Caryophyllaceae. Some of the Chenopodiaceae and *Plantago* pollen may have been carried in from local saltmarshes. Waterside and aquatic plants declined, probably reflecting the continued salinity rise.

CG-3

The onset of zone CG-3 coincided with a marked change in depositional environment, as the estuarine channel regime of the previous zone was replaced by a much quieter phase of organic clay deposition, associated with a lagoonal backwater. This was associated with the deposition of Unit 4. A thin bed of sand (Unit 3, the detritus muds) accumulated in the intervening period, probably as a lag deposit following a short phase of denudation. This sequence of events appears to have initiated a dramatic change in pollen source area, as the input of regionally derived estuarine pollen (mostly *Quercus* and *Pinus*) ceased and was replaced by a strong local pollen signal. The influx of *Ceratopteris*spores at the start of zone shows that fern communities colonised the site, probably spreading over the damp, exposed surface of the infilled channel. Some of these resistant spores may also have been reworked.

Alder now dominated the local woodland and even became rooted at the site, showing that local soils were fairly wet. Drier areas, probably fringing the alder carr, still supported mixed oak woodland during the early part of the zone, with the same suite of taxa that flourished previously. This soon changed, however, as firstly *Carpinus* and then boreal trees (*Abies*, *Pinus* and *Betula*) began to expand. *Picea* was the only conifer not to invade, possibly because of the high clay content of the local soils (cf. Turner, 1970). The ground vegetation also changed, as Ericaceous plants became more widespread, probably in response to soil acidification.

Wetter areas continued to support *Alnus* in the latter half of the zone, although there was no longer any input of *Alnus* wood to the sediment, possibly because the water was now too deep for the tree to grow. Conditions were now more favourable to waterside, aquatic and semi-aquatic herb communities, including *E. cannabinum*, *A. filiculoides*, *Chara* and *Alisma*. All these plants prefer fairly quiet, eutrophic, freshwater conditions, although with the exception of *Azolla*, all could tolerate low levels of salt. *Sphagnum* too became more widespread, reflecting increasing dampness.

Conditions changed again near the end of the zone as the input of organic material ceased and inorganic clays and sands (Unit 5) began to accumulate. *Pinus*, *Alnus* and *Ceratopteris* were still present during this phase, although *Abies* appears to have declined in the immediate region. The thermophilous trees also diminished, possibly in response to a progressive shift towards boreal woodland. Because only the more resistant pollen grains are represented in the two uppermost samples it is possible that some of the pollen has been reworked, probably from the underlying organic beds.

Pollen Biostratigraphy

The pollen sequence from Cudmore Grove shows two clear phases of vegetational development. The first, in biozones CG-1 and CG-2, records the regional establishment and expansion of mixed oak woodland, during the mesocratic phase of an interglacial, when closed forests developed under temperate conditions and soils became more fertile (cf. Birks and Birks, 2004). Biozone CG-3 in contrast, records the local development of alder carr woodland with a peripheral expansion of *Carpinus* and *Abies*. In northwest Europe the expansion of these two trees occurred in the oligocratic phase, when soils deteriorated and forest cover declined. In Britain, this broad pattern of vegetational succession has commonly been divided into four zones: the 'pre-temperate', 'early-temperate', 'late-temperate' and 'post-temperate sub-stages' (Turner and West, 1968). More complete palynological records from central and southern Europe (e.g. Roucoux et al., 2008) indicate that some interglacials have more complex histories of forest change, but we retain these terms here in order to facilitate comparisons of the pollen records with those of other Middle Pleistocene sites and as a framework for the discussion of the relative sea-level history.

The representation of pollen spectra of both early and late-temperate sub-stage character at Cudmore Grove is highly unusual within the context of the Thames region, where interglacial pollen records are typically fragmentary. A hiatus probably occurs between biozones CG-2 and CG-3, but there are strong grounds for believing that both of these phases date from the same interglacial. There is, for example, no evidence for a period of weathering or a major unconformity between the two biozones, or any evidence for an intermediate deterioration of climate. Alder also appeared to have been expanding locally at the end of zone CG-2 and continued to expand in CG-3, suggesting that the two biozones date from the same temperate episode. The profile from Cudmore Grove thus represents one of the most complete, non-lacustrine interglacial pollen records from Britain and is arguably the most important from the Thames system. However, given the changes in pollen source area and the estuarine/lagoonal context of the sediments, correlations with other sites, particularly records from lakes, must be made with caution.

Despite this difficulty, a number of palaeobotanical elements strongly support correlation with a 'Hoxnian-type' interglacial (cf. Scourse et al., 1999 and Thomas, 2001). This evidence is strongest in the upper part of the sequence (biozone CG-3), where the presence of Type X and *Abies* pollen and megaspores of *A. filiculoides* together preclude correlation with any other interglacial sequences of the British Late-, Early and Middle Pleistocene. All three floristic elements were absent, for example, in the Ipswichian (=Eemian/MIS 5e) (cf. West, 1980). These elements are also absent at sites that have been correlated on the basis of their terrace stratigraphy, faunal assemblages or aminostratigraphy with MIS 7, for example, Aveley and Ilford in the Lower Thames region (Bridgland, 1994), although the pollen records from these sites are far less complete (West et al., 1964 and West, 1969). Sparse *Abies* pollen (<0.5–1% of the total land pollen) has been reported from fluvial deposits which have been correlated with MIS 7 at Stoke Goldington (Green et al., 1996) and Marsworth (Murton et al., 2001) in Buckinghamshire, although these can be attributed to long-distance inputs (cf. Green et al., 1996). The spectra from both of these sites differ from those of Cudmore Grove in several other respects, particularly in their absence of Type X and their inclusion of greater frequencies of Poaceae pollen, which point to an incomplete regional tree cover in southern Britain at least during part of this interglacial.

More detailed correlations between the Cudmore Grove pollen assemblages and those from other 'Hoxnian-type' sites in Britain and continental Europe are hampered by several problems, not least the taphonomic issues outlined above, and the fact that more than one 'Hoxnian-type' temperate episode may have occurred with similar floristic characteristics (cf. Scourse et al., 1999, Thomas, 2001 and Scourse, 2006). It is nevertheless appropriate to make brief comparisons with the records from three key sites in the immediate vicinity (<20 km) that have also yielded pollen assemblages of 'Hoxnian' character: i) the Hoxnian parastratotype succession at Marks Tey, Essex, a lacustrine sequence infilling a depression cut in Anglian till (Turner, 1970); and ii) the pollen spectra from the neighbouring channel-fill sequences at Clacton (Pike and Godwin, 1953 and Bridgland et al., 1999) and Tillingham (East Hyde) (Roe, 2001) (Fig. 1). The last two channel-fills were deposited by the Thames in the interglacial that immediately followed the diversion of the river into eastern Essex during the late Anglian, i.e. the Hoxnian (cf. Bridgland, 1994, Bridgland et al., 1999 and Roe, 2001). The sediments at all three sites have been correlated with MIS 11 (Bridgland et al., 1999 and Rowe et al., 1999).

The Cudmore Grove pollen spectra share several attributes with those from Marks Tey, only 16 km to the northwest (Turner, 1970: p. 396; Fig. 7). Biozones CG-1 and CG-2, for example, compare closely with the spectra assigned to Ho IIa at this site, which are characterized by a sharp rise in *Quercus* pollen, a gradual expansion of *Alnus* and *Corylus* and persistent presence of *Fraxinus* pollen (Turner, 1970), although this sub-stage is not especially diagnostic (Thomas, 2001). The Cudmore Grove pollen profile differs from that of Marks Tey in the high frequencies of *Pinus*, although this might be explained by taphonomic differences between the sites.

Biozone CG-3 at Cudmore Grove is also broadly similar to the late-temperate sequence (Ho III) at Marks Tey, particularly in its record of *Abies*, *Carpinus* and Type X pollen. However, all three taxa achieve much greater abundances at Marks Tey than at Cudmore Grove, with *Abies* in particular peaking at frequencies of 45%, as opposed to ca 6% at Cudmore Grove (Fig. 7). This attribute alone does not provide sufficient basis for assuming that the sites are of different ages; the comparatively modest *Abies* profile at Cudmore Grove could be an artifact of the local dominance of *Alnus*, which may have obscured some of the regional vegetational changes upon which zonation of this sub-stage has been based (cf. West, 1956 and Turner, 1970). *Abies* prefers drier soils and is thus unlikely to have flourished around Cudmore Grove at any time during the high *Alnus*phase. Moreover, even if *Abies* was widespread in the hinterland, it is doubtful whether much of its pollen would have reached the site, given its low rates of pollen productivity and restricted patterns of dispersal (Reille, 1990). *Taxus* pollen is also far more strongly represented at Marks Tey than at Cudmore Grove, achieving frequencies of ca 10% (Turner, 1970), although varying frequencies of yew pollen could reflect local differences in soil conditions (cf. Kelly, 1964). Unlike Marks Tey, there is also no record of *Pterocarya*in the late-temperate sub-stage at Cudmore Grove (cf. Turner, 1970).

The pollen profiles from Clacton and East Hyde (Tillingham) also include far greater abundances of *Abies* and *Carpinus* pollen than Cudmore Grove, although in other respects the assemblages are broadly similar, and all three sites have yielded Type X pollen and remains of *Azolla*. At Clacton, *Abies* peaks at ca 65% (Pike and Godwin, 1953), whilst at East Hyde a peak of 40% is recorded (Roe, 2001). This difference might again reflect taphonomic variations between the sites; during the late-temperate sub-stage of the interglacial represented at Cudmore Grove a sheltered lagoon appears to have existed, whereas the channels at East Hyde and Clacton were more open to estuarine influence during the corresponding interval (Pike and Godwin, 1953, Bridgland et al., 1999 and Roe, 2001). This might have provided suitable conditions for the selective accumulation of bisaccate grains.

The palaeobotanical data from Cudmore Grove thus display broad similarities with the neighbouring records from Marks Tey, East Hyde and Clacton, but there are a number of subtle differences, particularly the differing proportions of *Abies* recorded in the late-temperate sub-stage. These may either reflect taphonomic differences between the sites, or may imply that the Cudmore Grove profile belongs to a younger interglacial also characterized by 'Hoxnian-type' vegetation, and possibly co-eval with MIS 9 (cf. Scourse et al., 1999 and Thomas, 2001). Wider comparisons between the Cudmore Grove pollen record and sites in southern Britain and elsewhere are considered further below in the light of the other bio- and aminostratigraphic data.

DIATOMS

Sampling and Preparation

Samples for diatom analysis were taken from boreholes CG0 and CG1. Borehole CG0 (Fig. 5), collected in 1986, recovered ca 5.5 m of sediment spanning sub-units 2b–2d, below 0.5 m of modern beach sand. Samples from the eroded Units 3–4 at CG0 were taken from a separate borehole ca 8 m away, near the position of borehole 9 (Fig. 5). Borehole CG1, ca 50 m east of CG0 in the deeper part of the channel, which was collected in 1990, provided a thicker and more complete sedimentary sequence (Units 2–4) through the interglacial (Fig. 4). Diatom samples from this core came from 6.7 m to below 16 m, providing overlap and continuation of the record from CG0. The diatom assemblages from the cores have been integrated. Preparation of the sediments for diatoms was based upon standard techniques (Battarbee, 1986). Identification of diatoms, together with interpretations of their taphonomy and palaeoecology were based on established floras and other texts (e.g., van der Werff and Huls, 1957–1974 and Admiraal, 1984). Results are presented as a percentage frequency diagram showing key species within salinity groupings (Fig. 8). The diatoms were generally well preserved and included >220 taxa.

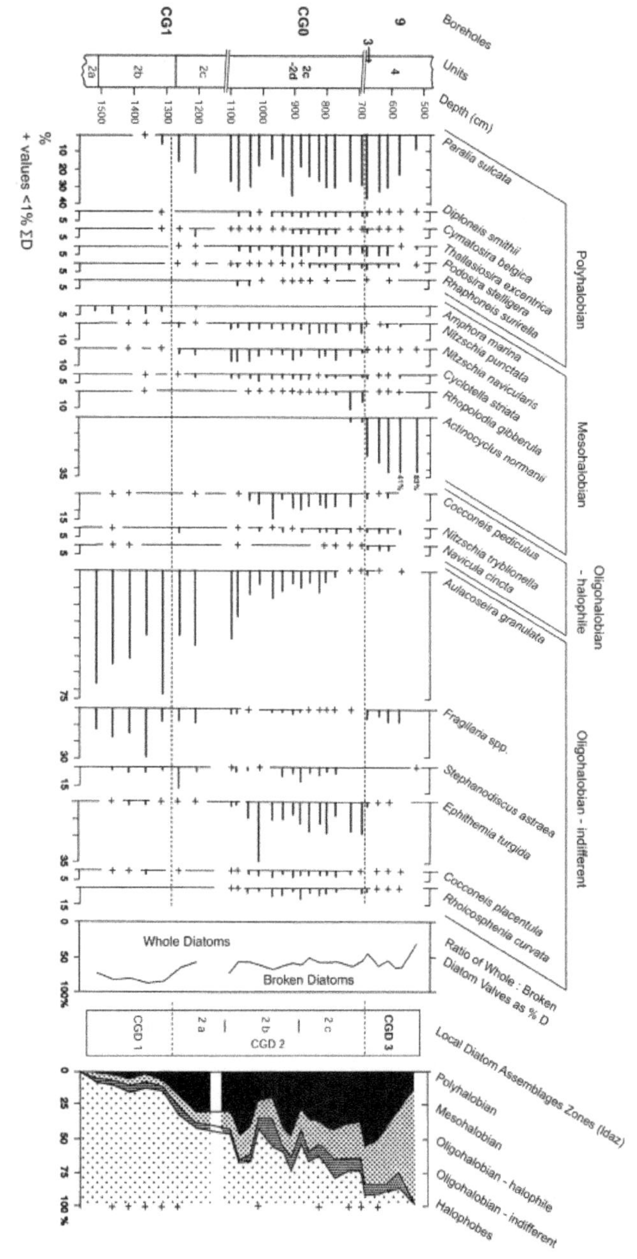

Figure 8: Composite diatom percentage diagram (selected taxa) based on samples from cores CG0, CG1 and a core collected from the organic clays and detritus muds (Units 3 and 4) near the point of borehole 9 (Fig. 5). The break recorded between 1100 cm and 1200 cm is an artifact of the change in cores at this point. Salinity groupings follow Hustedt (1957).

Diatom Assemblages

Three main local diatom assemblage zones (ldaz) are recognised (CGD1–CGD3; Fig. 8). In ldaz CGD2 the assemblage is complex and the zone has been divided into three subzones ldaz CGD2a–2c.

ldaz CGD1 (15.10–12.70 m)

This zone, which spans the lower silty clays (sub-unit 2b), is characterized by a fining upward sedimentary sequence. Diatoms are absent below ca 15 m. Freshwater (oligohalobous–indifferent) taxa, dominated by *Aulacoseira granulata* (Ehr.) Simonsen reach >50% of the total assemblage, with *Fragilaria pinnata* Ehr., *Fragilaria brevistriata*Grun. and *Fragilaria lapponica* Grun. together at >15%. Brackish-water (mesohalobous and oligohalobous–halophilous) species are present throughout the zone at low values (<10%). Marine diatoms (polyhalobous) occur erratically, with the littoral species *Paralia sulcata* (Ehr.) Cleve making an early appearance.

ldaz CGD2a (12.70–11.00 m)

This subzone begins close to the contact of sub-units 2b and 2c in borehole CG1 and marks the major expansion of polyhalobous taxa to >28%, with *P. sulcata* rising to >20%. Other marine taxa present at lower frequencies include *Diploneis smithii* (Bréb.) Cleve, *Cymatosira belgica* Grun. and *Thalassiosira excentrica* (Ehr.) Cleve. The expansion of marine diatoms is accompanied by a increase in brackish species, dominated by *Nitzschia punctata* (W. Smith) Grun., *Nitzschia navicularis* (Bréb.) Grun. and *Cyclotella striata* (Kütz.) Grun. Freshwater taxa remain dominant in ldaz CGD2a at >50%.

ldaz CGD2b (11.00–8.60 m)

Significant fluctuations occur between freshwater and marine diatoms in this subzone. Marine and brackish-water taxa continue to increase in abundance and species diversity, together reaching >60%. Freshwater, halophilous species also increase in importance, especially *Cocconeis pediculus* Ehr. and *Nitzschia tryblionella* Hantzsch. Freshwater taxa (oligohalobous–indifferent) generally represent <35% of the total assemblage. *Fragilaria* spp. fall through ldaz CGD2a to low values in ldaz CGD2b (<3%). *A. granulata*declines to 10–15% and is joined at similar values by *Epithemia turgida* (Ehr.) Kütz, accompanied by lower frequencies of *Stephanodiscus astraea* (Ehr.) Grun.

ldaz CGD2c (8.60–6.90 m)

The 'peak and trough' behaviour of the major marine and freshwater groups ceases. Marine and brackish-water taxa expand to >70%, with the mesohalobous and halophilous groups together reaching ca 35%. Freshwater diatoms continue to decline resulting in a community now characterized by *E. turgida*, and lower values of *Cocconeis placentula* Ehr. and *Rhoicosphenia curvata* (Kütz) Grun.

ldaz CGD3 (6.90–5.30 m)

The base of this zone marks the boundary between the silty clays (Unit 2d) and the organic beds (Units 3 and 4). Mesohalobous and halophilous diatoms expand in frequency to >75% of the total assemblage, dominated by the rapid rise of *Actinocyclus normanii* (Greg.) Hust., which reaches >80%. At the base of the zone, marine taxa, dominated by *P. sulcata* and *Th. excentrica*, rise to maximum values (ca 55%), before falling to ca 12% at the top of the zone. Freshwater diatoms show low, erratic frequencies and are dominated by the return of *Fragilaria* spp. to values of ca 8%.

Local Environment Inferred from the Diatoms

Changes in the diatom assemblages show the probable long-term landward progression of the tidal head and accompanying salt wedge of a palaeo-estuary, with the development of full brackish-water conditions by ldaz CGD2c. Shorter term changes in the diatoms may represent i) major fluctuations in freshwater discharge to the estuary; ii) rises of relative sea level; and iii) sedimentary and/or estuarine–coastal geometry changes. Together these eventually led to the replacement of an estuarine tidal channel and mudflat environment by more isolated conditions, possibly an estuarine embayment or lagoon. Detailed diatom records of this kind are rare from British interglacial sites, although Mitlehner (1992) provides a skeletal record from the Nar Valley Clay in Norfolk (see Section 16.3 below).

During the early phase of channel infilling, associated with ldaz CGD1, the dominance of *Aulacoseira* spp. and *Fragilaria* spp. and the high values of whole diatoms (ca 80%), suggest that eutrophic, freshwater conditions prevailed locally, indicative of a riverine environment with adjacent alkaline freshwater areas. The presence of brackish and marine taxa reflects the penetration of saltwater into the river. *F. pinnata* and *F. lapponica*(each rising to >12%), have a northerly distribution (de Wolf and Cleveringa, 1994) and the diatom community is consistent with stressed conditions during the early part of an interglacial. *Fragilaria* spp. are pioneer diatoms that thrive under conditions of environmental stress, particularly where increasing salinity is involved (Stabell, 1985 and Denys, 1990), as is *A. granulata* (Stachura and Witkowski, 1997). The assemblage as a whole is representative of the upper reaches of a river estuary close to the tidal limit, with salinities of ca 0.5–5‰.

The rapid rise in marine species in ldaz CGD2a, dominated by *P. sulcata* and other littoral and lower estuary taxa, shows the influence of tidal action. The rise in broken diatoms over the ldaz CGD1/2a boundary coupled with an increase in species diversity, together suggest the transport and inwashing of diatoms from a range of salinity regimes. The fining in sediment within Unit 2 may be associated with increasing salinity, resulting in flocculation. The continuing dominance of freshwater diatoms indicates that the site remained close to the tidal head in the river.

Above ca 11 m (CGD2b) the diatom flora reflects the effects of the developing estuary, shown by the expansion of the number and frequencies of brackish-water taxa. In the freshwater diatom community, the dominance of *A. granulata* and other planktonic taxa is replaced by benthic, epipelic taxa, especially, *E. turgida*. This indicates an increase in the extent of shallower, estuarine margin habitats. The fluctuations in marine to freshwater diatoms in ldaz CGD2b also indicate a period of variability in the environmental controls operating in the estuary. The occurrence of such fluctuating freshwater: marine diatom ratios in some estuarine records have been attributed to the effects of storms, river floods and marine surges, combined with the impact of longer term relative sea-level rises (Devoy et al., 1994). The record here shows that the ‘cycles’ in the diatom ratios each represent ca 0.5 m of sediment accumulation with no apparent hiatuses. This sediment thickness, together with its fine texture, suggest that these ‘cycles’ are unlikely to be the result of storms, which might be expected to have produced coarser layers, although it is possible that the channel margin location (for CG0) may have influenced the diatom pattern concentrating diatoms from many habitats and sources. Longer term controls (10^1–10^2 years) are more likely to have caused the observed changes. These include climate variability, leading to changes in freshwater river discharge, and alterations in the tidal prism as the estuary shape changed with rising relative sea level and sediment deposition.

The disappearance of these ‘cycles’ from the diatom record above ca 850 cm coincides with the steady expansion in the importance of brackish-water and marine diatoms. The maintenance of *N. punctata*, *N. navicularis*, *C. pediculus* and other epipelic – epiphytic mesohalobous and halophilous diatoms, indicates the occurrence of mudflat and intertidal conditions. The rising values of *C. belgica*, accompanied by many other marine planktonic and tychoplanktonic diatoms at the base of ldaz CGD2c, point to wave and tidal scouring (Vos and de Wolf, 1993) and the proximity of open coastal sandy areas. The site at this time was probably in a lower estuarine setting. The observed changes probably reflect the landward shift in estuary zones with long-term sea-level rise, causing increased exposure to marine influence.

The base of ldaz CDG3 is marked by an increase in broken diatoms and a peak in marine diatoms. The sediments also become more sandy, suggesting an increase in available energy. These samples came from Unit 3, which has yielded rich mammal and molluscan assemblages (see below). Given that these sediments appear to represent a lag or a high-energy flood deposit, erosion and/or re-working of diatoms from proximal marine sources seems likely.

The subsequent collapse of both the marine and freshwater diatom frequencies through ldaz CGD3, suggests a distinct change in conditions, as supported by the development of the organic clays (Unit 4). The rapid rise in *A. normanii* coupled with *Fragilaria* spp., indicates increased water eutrophication and environmental stress (cf. Stabell, 1985 and Denys, 1990). The diatom flora represents the maintenance of fresh-brackish to brackish-water, indicative of isolated or semi-enclosed areas of standing water. This interpretation is consistent with the isolation of the site from the open estuary. The sediment infilling and shallowing of the estuary through the interglacial may now have led to the development of channel cut-offs, sediment barriers and estuarine lagoons.

OSTRACODA

Sampling and Preparation

Ostracods were recovered from the silty clays (Unit 2) and detritus muds (Unit 3) penetrated in borehole CG1. Depths given below refer to this borehole. The organic clays (Unit 4) and the basal sands and gravels (Unit 1) were barren. After wet-sieving, all identifiable fragments retained on 125 μm sieves were counted. Two local assemblage biozones, 1 and 2 were identified in the ostracod diagram (Fig. 9).

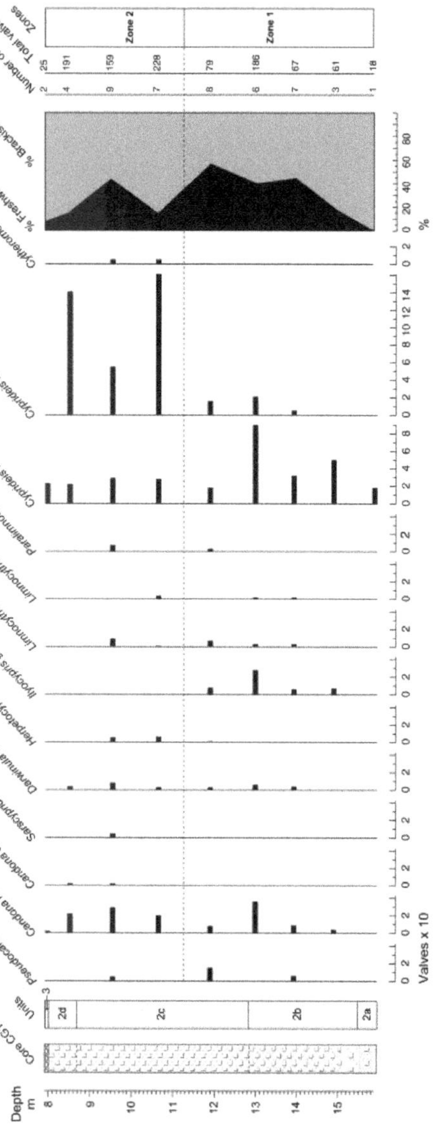

Figure 9: Ostracod diagram for core CG1. Nomenclature follows Meisch (2000).

Assemblages

Biozone 1: 15.87–11.25 m

This lowermost zone extends from the sandy clays to the shelly silts (sub-units 2a–2c). The sandy clays below 14.00 m yielded a sparse ostracod fauna dominated by juveniles of the euryhaline species *Cyprideis torosa* (Jones). Many of the specimens of *C. torosa*showed signs of nodal development, a characteristic that has been attributed to salinities (<6‰) at the lowest limit of the species' range (cf. Meisch, 2000). Freshwater species increase in frequency and become more diverse in the clay-rich sediments at 13.00 m. These include the bottom-burrowing *Candona neglecta* Sars and *Pseudocandona marchica* Hartwig and the plant-crawling *Ilyocypris gibba* (Ramdohr) and *Limnocythere inopinata* (Baird). Two of these, *C. neglecta* and *I. gibba*, show particularly large increases in abundance at 13.00 m (Fig. 9). Noded valves of *C. torosa* also increase at this level.

Biozone 2: 11.25–7.90 m

This upper zone extends from the middle of sub-unit 2c to Unit 3 and is characterized by fluctuations in the frequencies of both brackish and freshwater taxa. At 10.63 m, for example, numbers of smooth-valved *C. torosa* increase dramatically, whilst the diversity and frequency of freshwater species decline. Numbers of *C. torosa* fall, however, at 9.50 m, with a corresponding rise in the frequencies of freshwater taxa. *C. neglecta*dominates the latter, with well-balanced numbers of adults and juveniles and males and females (Roe, 1994). The free-swimming *Sarscypridopsis aculeata* (Lilljeborg) also occurs. *C. torosa* valve frequencies rise again near the upper part of the unit. The thin detritus muds overlying the silty clays (Unit 3) yielded only a few noded *C. torosa* and the occasional broken valve of *C. neglecta*.

Local Environment Inferred from the Ostracoda

The mixed ostracod faunas are typical of a highly transitional inner estuarine environment in which the euryhaline *C. torosa* was able to flourish locally alongside fluctuating numbers of freshwater species. The early phase of silty clay deposition (sub-unit 2a) was characterized by conditions poorly suited to ostracod life, probably resulting from the sandy nature of the substrate or current activity. The presence of large numbers of noded *Cyprideis* suggests a salinity at ca 5‰ or less, if they are *in situ*.

In contrast, conditions during the accumulation of the overlying clay and silt dominated beds (sub-units 2b–2d) allowed *C. torosa* to thrive. Salinities were also low enough locally to enable *C. neglecta* and *I. gibba* to survive at all levels of development (Roe, 1994), the latter suggesting that aquatic or bankside vegetation existed within or in proximity to the channel (Meisch, 2000).

Salinities must have increased subsequently, as reflected by an expansion in smooth-valved *C. torosa* and a fall in local freshwater populations. Some of the sparse freshwater species were probably transported from sites further upstream. The well-balanced freshwater populations from the overlying sediments at 9.50 m appear to mark a brief decline in saltwater influence. Salinity levels increased again when the uppermost part of the silty clays accumulated.

Overall, the fauna of the silty clays typifies an estuarine section of river, which gradually became more open to marine influence through time. The brief lowering of salinities that took place in the latter part of this episode (at 9.50 m) may reflect a short-lived increase in river discharge, a change in tidal or depositional regime or possibly a small fall in relative sea level. This parallels the changes noted in the diatom floras at the same level.

The limited fauna of the detritus muds (Unit 3), points to a marked change in conditions, which were generally unfavourable for ostracods. The currents which carried sand and shell debris into the site at this time may have restricted their survival. Re-working from the lower silty clays also cannot be excluded.

MOLLUSCA

Sampling and Preparation

Small samples for molluscan analysis were taken at regular intervals throughout borehole CG1. Units 1, 2a, 4, 5 and 6 were barren but molluscan assemblages were recovered from sub-units 2c and 2d and Unit 3, whereas a few comminuted fragments were recovered from sub-unit 2b. The detritus muds (Unit 3) were easily the most shelly horizon. Consequently, the residues of large bulk samples from Unit 3 previously analysed for vertebrate remains by John Clayden (see Holman et al., 1990), were sorted and the frequencies of molluscan taxa estimated (Table 3).

Table 3: Estimated abundance of molluscan taxa in Units 3 and sub-units 2c and 2d. Abundant = >200, very common = 76–200, common = 21–75, occasional = 6–20, rare = 1–5

	Sub-units 2c–2d	Unit 3
Freshwater taxa		
Valvata piscinalis (Müller)	Occasional	Common
Borysthenia naticina (Menke)	–	Occasional
Bithynia tentaculata (Linnaeus)	–	Very common
Radix balthica (Linnaeus) (=*Lymnaea peregra* (Müller))	–	Rare
Ancylus fluviatilis Müller	–	Common
Anodonta sp.	–	Rare
Unionidae	–	Occasional fragments
Corbicula fluminalis (Müller)	–	Abundant
Pisidium amnicum (Müller)	–	Common
Pisidium clessini Neumayr	–	Rare
Pisidium casertanum f. *ponderosa* Stelfox	–	Common
Pisidium henslowanum (Sheppard)/*P. supinum* Schmidt	–	Very common
Pisidium subtruncatum Malm	–	Rare
Brackish-water taxa		
Hydrobia acuta (Draparnaud) (=*Heleobia neglecta*(Muus))	Abundant	Abundant
Peringia ulvae (Pennant, 1777)	Rare	–
Cerastoderma glaucum (Poiret)	Common	Abundant
Cerastoderma sp. (probably *glaucum*)	Common fragments	Common

Assemblages and Environmental Interpretations

The assemblages from the silty clays (sub-units 2c–2d) are dominated by brackish species, including *Hydrobia acuta* (=*Heleobia* cf. *neglecta*), a hydrobiid previously listed as '*Paladilhia radigueli*', and the cockle *Cerastoderma glaucum*. Attribution of these hydrobiids to *H. acuta* is tentative but they possess tall slender shells with deep sutures that match reference material. *Peringia ulvae*, another hydrobiid tolerant of higher salinities, also occurs occasionally. This fauna is consistent with a quiet estuarine environment, which was sheltered from the influence of strong tidal activity. A brackish fluviatile setting appears more likely than a lagoonal or saltmarsh habitat, an interpretation that is further supported by the occasional presence of the freshwater gastropod *Valvata piscinalis*.

The faunas of the detritus muds (Unit 3) are more difficult to interpret because of the highly broken and probably allochthonous nature of the shells. Fourteen taxa have been recovered (Table 3). The preservation is generally poor and many of the shells are corroded. *H. acuta* and *C. glaucum* dominate the assemblage but freshwater species are also common, especially *Bithynia tentaculata*, *V. piscinalis*, *Ancylus fluviatilis*, *Corbicula fluminalis*, *Pisidium amnicum*, *Pisidium casertanum* f. *ponderosa* and *Pisidium henslowanum*/*supinum*. Other species include *Borysthenia naticina*, *Radix balthica*(=*Lymnaea peregra*) and *Pisidium clessini*. Fragments of freshwater mussel (*Anodonta*sp.) have also been recorded (Bridgland et al., 1988). No land snails were recovered. This fauna suggests an episode of inwashing from a nearby freshwater source, an event characterized by a marked change in depositional energies.

COLEOPTERA

Introduction

Samples taken throughout the organic clays (Unit 4) mostly yielded sparse, fragmentary and indeterminable insect remains. However, bulk samples (~5 kg wet weight) from the detritus muds (Unit 3) exposed in a foreshore trench (Fig. 3) produced fossil Coleoptera that were well enough preserved to enable some taxa to be identified. The insect remains were recovered from the sediment using the method outlined by Coope (1986). Altogether 65 coleopteran taxa were recognised of which 47 could be determined to species level (Table 4). Seven taxa no longer live in the British Isles.

Table 4: Coleoptera recorded in a bulk sample of Unit 3 (detritus muds) collected from the foreshore trench section (Fig. 3). Nomenclature follows Lucht (1987).

Carabidae		**Helodidae**	
Elaphrus cupreus Duft[b]	1	Gen. et sp. indet	1
Loricera pilicornis (F.)[b]	1		
Bembidion (*Peryphus*) sp.	1	**Dryopidae**	
Bembidion unicolor Chaud.[b]	1	*Drypos* sp.	1
Patrobus atrorufus (Ström)[b]	1	*Oulimnius tuberculatus* (Müll.)	1
Pterostichus strenuus (Panz.)[b]	1	*Macronychus quadrituberculatus* Müll.	2
Pterostichus nigrita (Payk.)[b]	2		
Pterostichus gracilis (Dej.)[b]	1	**Phalacridae**	
Platynus ruficornis (Goeze)	1	*Phalacrus caricis* Sturm	1
Oodes gracilis Villa a [and] b	2		
Odacantha melanura (L.)[b]	1	**Anobiidae**	
		Gastrallus immarginatus (Müll.)	1
Dytiscidae		*Anobium punctatum* (Geer)	2
Agabus bipustulatus (L.)[b]	1		
Ilybius sp.	1	**Scarabaeidae**	
Acilius sp.	1	*Onthophagus* spp.	3
Dytiscus sp.	1	*Aphodius* sp.	1
		Valgus hemipterus (L.) a [and] b	1
Rhysodidae		*Triodonta* sp.[a]	1
Rhysodes sulcatus (F.)[a]	1	*Osmoderma eremita* (Scop.)[a]	1
Hydraenidae		**Lucanidae**	
Hydraena sp.	2	*Dorcus parallelopipedus* (L.)	1
Ochthebius sp.	1		
		Cerambididae	
Hydrophilidae		*Prionus coriarius* (L.)	2
Coelostoma orbiculare (F.)[b]	1		
Cercyon convexiusculus Steph.	2	**Chrysomelidae**	
Hydrobius fuscipes (L.)[b]	1	*Macroplea appendiculata* (Panz.)	2
Chaetarthria seminulum (Hbst.)[b]	1	*Donacia semicuprea* Panz.	1
		Donacia cinerea Hbst.	1
Histeridae		*Plateumaris sericea* (L.)	1
Hister (*sensu lato*) sp.	1	*Plateumaris braccata* (Scop.)	3
		Agelastica alni (L.)	1
Silphidae			
Silpha tristis Ill.	1	**Scolytidae**	
Phosphuga atrata (L.)	1	*Scolytus scolytus* (F.)	1
		Hylesinus crenatus (F.)	1
Staphylinidae			
Aploderus caesus (Er.)[a]	1	**Curculionidae**	
Trogophloeus sp.	1	*Apion* sp.	1
Stenus sp.	1	*Rhyncolus elongatus* (Gyll.)[a]	1

Philonthus sp.	1	*Stenoscelis (Brachytemnus) submuricatus* (Schoen.)[a]	4
		Tanysphyrus lemnae (Payk.)	1
Pselaphidae		*Notaris scirpi* (F.)	1
Pselaphus heisei Hbst.	1	*Notaris acridulus* (L.)	1
		Anthonomus sp.	1
Elateridae		*Hylobius abietis* (L.)	1
Adelocera murina (L.)	2	*Limnobaris pilistriata* (Steph.)	1
		Phytobius sp.	1
Dascillidae		*Rhynchaenus testaceus* (Müll.)	2
Dascillus cervinus (L.)	1		

[a]Species no longer living in Britain.

[b]Species present in the MCR database.

Environmental Implications of the Coleopteran Assemblage

The large numbers of species represented by only one individual indicates that the assemblage is only a small subsample of the contemporary beetle fauna. The ecologically heterogeneous assemblage suggests that it was probably brought together in flood debris, some of it swept off the adjacent land surface, and some originating from a slowly moving river. This faunal variety means that the various ecological requirements of the species can be built up into a mosaic picture of the local environment.

Aquatic Habitats

Oulimnius tuberculatus, represented by a single specimen, is the only beetle indicative of running water, where it lives amongst stones and moss in shallow well-aerated riffles (Holland, 1972). However, the caddisfly *Hydropsyche contubernalis* McL. is usually found in large slow-flowing rivers (Edington and Hildrew, 1995). All the rest of the aquatic beetles are typical of standing or very slowly moving. The large carnivorous dytiscid taxa *Agabus bipustulatus*, *Ilybius* sp., *Acilius* sp. and *Dytiscus* sp. live in well-vegetated ponds. The hydrophilids *Coelostoma orbiculare*, *Cercyon convexiusculus* and *Hydrobius fuscipes* occur in similar habitats where they feed on decomposing plant material. *Chaetarthria seminulum* lives in wet mud, usually beside eutrophic pools or on the muddy banks of slow-flowing streams (Hansen, 1987). *Macroplea appendiculata* lives permanently underwater feeding mostly on *Potamogeton* but also on *Myriophyllum spicatum* and other pond-weeds. *Macronychus quadrituberculatus* lives in the fissures of submerged logs or rocks in fairly large, slow-moving rivers, where it probably feeds on algae. It is an extremely rare species in Britain today probably due to the reduction of its habitat by riverine management activities. *Tanysphyrus lemnae* is a minute weevil that feeds on the duckweed *Lemna* on the surface of stagnant water.

Marginal Damp Habitats

In this category are included marsh habitats of varying degrees of humidity. By far the largest number of beetle species in this fauna fall into this group. The Carabidae are ground beetles that are either predators or general scavengers. All the recovered species live close to still or slowly flowing eutrophic water where the soil is rich in humus and where the vegetation is rather lush and composed of *Carex*, *Scirpus* or *Phragmites*. *Bembidion unicolor* occurs in deciduous forests and is especially characteristic of the drier parts of *Alnus* swamps. Some species, such as *Loricera pilicornis* and *Platynus ruficornis*, require bare patches of soil. Of particular interest are the two carabids *Oodes gracilis* and *Odacantha melanura*, which inhabit accumulations of dead reeds, such as *Phragmites*, on the swampy borders of eutrophic lakes (Lindroth, 1992).

Plateumaris braccata, monophagous on *Phragmites communis* (Koch, 1992), was represented by numerous fragments. *Donacia cinerea* feeds on reeds and sedges, such as *Typha*, *Phragmites* and *Carex*. *Donacia semicuprea* and *Notaris acridulus* feed principally on the aquatic grass *Glyceria*. The weevil *Limnobaris pilistriata* chiefly feeds on Cyperaceae and similar plants; the larvae usually eating the roots. *Phalacrus caricis*feeds on the smutted inflorescences of various species of *Carex* (Thompson, 1958). *Dascillus cervinus* lives in damp meadows where the larvae feed on the roots of grasses.

Meadow-like Habitats

Hardly any xerophilous species were recovered except for the 'click beetle' *Adelocera murina*, which is common today in dry grassland where its larvae feed on the roots of various plants.

Woodland Habitats

The Coleoptera provide abundant evidence for the local presence of trees both living and dead. The Elm bark beetle *Scolytus scolytus*, the vector for Dutch Elm disease, is not restricted to *Ulmus* but also attacks *Fraxinus*, *Carpinus* and other deciduous trees. *Hylesinus crenatus* chiefly attacks *Fraxinus* and only exceptionally other broad-leaved trees. Both the adults and larvae of *Agelastica alni* (alder leaf beetle) eat the leaves of *Alnus*. The larvae of the weevil *Rhynchaenus testaceus* mine its leaves. *Stenoscelis submuricatus*, the commonest species in this assemblage, bores into the wood of diseased broad-leaved trees. *Rhyncolus elongatus* feeds in the trunks and stumps of partly rotting conifers, both *Pinus* and *Abies*, whereas *Hylobius abietis* (large pine weevil) attacks both pines and spruces.

Evidence of dead or dying trees is provided by a number of

species. *Gastrallus immarginatus* burrows under dead bark of old oak trees but only rarely other deciduous trees. *Rhysodes sulcatus* lives in decaying wood, both deciduous and conifer, mainly in fallen trunks in moist places (Lindroth, 1986). It is extremely rare in Europe having a relict distribution from the times when the old natural forests were more widespread. *Valgus hemipterus*, *Osmoderma eremita* (hermit beetle) and *Dorcus parallelopipedus* (lesser stag beetle) develop in decaying wood of deciduous trees. *Prionus coriarius* is a large beetle (up to 45 mm) that lives in rotten trunks and major branches of various broad-leaved trees, less often on conifers; its larvae mostly feed on tree roots. *Anobium punctatum* (woodworm beetle) attacks dead wood that is moderately dry but which has often already been attacked by decomposing fungae (Hickin, 1963).

Specialist Habitats

A number of species occur with other specialised habitats. Thus *Onthophagus* and *Aphodius* are dung feeders. Species of *Hister* are most often found in carcasses where the larvae live on maggots. *Silpha tristis* is another carrion beetle that can feed on other dead insects (Koch, 1989). Their presence indicates that both dung and carcases must have existed on dry ground rather than in the marsh itself. *Phosphuga atrata* is a specialist predator on snails and slugs.

Summary

This beetle assemblage indicates a sluggish river surrounded by a marsh dominated by *Phragmites*. Alder trees grew alongside the river. Nearby the mature forest consisted mainly of deciduous trees, including oak and ash, as well as coniferous trees such as pine. There is much evidence of dead and rotting timber. Some of the trunks were submerged in water. The presence of beetles that feed on dung and carrion indicates that the area was visited by large herbivorous mammals, perhaps attracted by the presence of *Glyceria* (sweet grass). There is no evidence of saline conditions, as would be indicated by obligate halobiont species.

Climatic Inferences of the Coleoptera

This coleopteran assemblage is entirely composed of temperate species some of which have present-day geographical ranges that do not extend as far north as Britain. Of particular importance is the presence of *O. gracilis*, which has a strong preference for habitats that heat up rapidly in the summer (Lindroth, 1943). Several of the wood-dependent species (e.g. *R. sulcatus*, *V. hemipterus*, *R. elongatus* and *S. submuricatus*) are also predominantly 'southern' but these inhabitants of old mature forests may now be restricted by anthropogenic reduction of their habitats rather than by climate. However, there is no doubt that this fauna indicates a climate rather warmer than that in southern England today.

Using the Mutual Climatic Range (MCR) method (Atkinson et al., 1987) it is possible to give quantitative estimates of the thermal climate based on the coleopteran assemblage. Fourteen species in the Cudmore Grove fauna were also present on the MCR database (see Table 4). This method gave the following estimates, where T_{max} is the mean temperature of the warmest month (July) and T_{min} is the mean temperature of the coldest month(s), namely January and February:

T_{max} lay somewhere between 16 °C and 22 °C

T_{min} lay somewhere between −7 °C and 4 °C

Within this range of temperatures it is possible to give a best estimate (Coope et al., 1998). This gives the following figures for the most likely values:

$$T_{max} = 19\,°C, \quad T_{min} = 1\,°C$$

Thus the mean July temperatures were about 2 °C warmer than those of the present day in southeast England and mean January temperatures may have been very slightly colder.

VERTEBRATES

Introduction

Fossiliferous deposits were first reported at East Mersea in the early 1900s by Dalton (1908), although no specific mention was made of vertebrate remains. Later accounts by Warren, 1917 and Warren, 1933 and Cornwall (1958) mention bones of *Hippopotamus*and therefore probably refer to the deposits of Ipswichian age that are known from the nearby 'Hippopotamus Site' (Bridgland and Sutcliffe, 1995) and the 'East Mersea Restaurant Site' (Bridgland et al., 1995) a short distance away on the foreshore (Fig. 1). The detritus muds (Unit 3) at Cudmore Grove have yielded an extensive vertebrate assemblage including mammals, birds and herpetiles (Bridgland et al., 1988 and Holman et al., 1990) and fish. Unit 1 has yielded a fragmentary molar of an indeterminate elephant (G.R. Ward, personal communication) and a single bone of *Bison*or *Bos* was recovered from Unit 2 (Roe, 1994). The specimens from Unit 3 are extremely well preserved, showing no significant signs of rolling or abrasion. The presence of fragile herpetile vertebrae and small mammal mandibles with the dentition still *in situ*suggests that there has been little post-depositional disturbance.

At least 18 mammalian taxa have been recognized amongst the 1508 bones and teeth recovered from the detritus muds (Unit 3) (Table 5). In addition, 11 fish taxa, 14 herpetofaunal taxa (364 specimens) and 11 avian taxa (41 specimens) have been recorded.

Table 5: Vertebrates recorded in a bulk sample of Unit 3 (detritus muds). (NIS = number of identified specimens, MNI = minimum number of individuals). Percentage values represent the abundance of each taxonomic group (e.g. Aves and Mammalia). Amphibia and Reptilia percentage totals are grouped and the identifications follow Holman et al. (1990)

Species	Present	Common	Abundant
PISCES			
Acipenseridae			
Acipenser sp., sturgeon	+		
Salmonidae			
Salmo sp., salmon or trout	+		
Esocidae			

Esox lucius L., pike			+++
Cyprinidae			
Tinca tinca L., tench	+		
Abramis brama L., common bream	+		
Leuciscus leuciscus L., dace	+		
Rutilus rutilus L., roach		++	
Scardinius erythropthalmus L., rudd	+		
Anguillidae			
Anguilla anguilla L., eel		++	
Gasterosteidae			
Gasterosteus aculeatus L., three-spined stickleback	+		
Percidae			
Perca fluviatilis L., perch	+		

Species	N.I.S.	% of Total within each group	M.N.I.
AMPHIBIA			
Salamandridae			
Triturus cristatus (Laurenti), warty newt	145	39.83	3
Triturus vulgaris L., smooth newt	2	0.54	1
Triturus sp., indeterminate newt	5	1.37	1
Bufonidae			
Bufo bufo L., common toad	12	3.29	6
Hylidae			
Hyla sp., tree frog	1	0.27	1
Ranidae			
Rana arvalis Nilsson, moor frog	5	1.37	3
Rana ridibunda Pallas, or *R. esculenta* (L.), marsh or edible frog	20	5.49	12
Rana sp., indeterminate frog	45	12.36	13
Pelophylax lessonae Camerano, pool frog	10	2.74	5
REPTILIA			
Emydidae			
Emys orbicularis (L.), European pond terrapin	4	1.09	1
Lacertidae			
Lacerta vivipara Jacquin, viviparous lizard	2	0.54	1
Anguidae			
Anguis fragilis L., slow worm	9	2.47	1
Colubridae			
Zamenis longissimus (Laurenti), Aesculapian snake	3	0.82	1
Natrix natrix (L.), grass snake	84	23.07	1
Natrix tessellata (Laurenti) or *N. maura* (L.), dice or viperine snake	7	1.92	1
Viperidae			

Vipera berus (L.), adder	10	2.74	1
AVES			
Anatidae			
Anas platyrhynchos L., mallard	13	30.95	3
Anas querquedula L., garganey duck	1	2.38	1
Anas strepera L., gadwall duck	3	7.14	1
Cygnus cygnus (L.), whooper swan	2	4.76	1
Mergus albellus L., smew	2	4.76	1
Laridae			
Sterna sandvicensis (Latham), Sandwich tern	1	2.38	1
Rallidae			
Porzana parva (Scopoli), little crake	2	4.76	1
Turdidae			
Turdus philomelos Brehm, song thrush	6	14.28	1
Turdus pilaris L., fieldfare	6	14.28	1
Paridae			
Parus major L., 1758, great tit	2	4.76	1
Parus caeruleus L. or *P. ater* L., blue or coal tit	2	4.76	1
Sylviidae			
Phylloscopus trochilus (L.) or *P. collybita* (Viellot), willow warbler or chiffchaff	2	4.76	1
MAMMALIA			
Insectivora			
Sorex cf. *araneus* L., common shrew	9	0.59	4
Sorex cf. *minutus* L., pygmy shrew	4	0.26	2
Neomys cf. *browni* Hinton, water shrew	38	2.51	18
Crocidura cf. *leucodon* Hermann, bicoloured white-toothed shrew	1	0.06	1
Chiroptera			
Eptesicus serotinus (Schreber), serotine bat	1	0.06	1
Primates			
Macaca sylvanus (L.), Barbary macaque	1	0.06	1
Rodentia			
Sciurus vulgaris L., red squirrel	2	0.13	1
Castor fiber L., European beaver	19	1.25	5
Clethrionomys glareolus (Schreber), bank vole	191	12.66	46
Arvicola terrestris cantiana (Hinton), water vole	563	37.33	60
Microtus agrestis (L.), field vole	31	2.05	19
Microtus agrestis (L.) or *M. arvalis* (Pallas), field vole or common vole	82	5.43	47
Microtus sp., indeterminate vole	503	33.35	106
Apodemus cf. *sylvaticus* (L.), wood mouse	42	2.78	9
Carnivora			
Canis lupus L., wolf	1	0.06	1
Ursus arctos L., brown bear	4	0.26	1
Meles meles (L.), badger	1	0.06	1
Mustela cf. *putorius* L., polecat	3	0.19	2

Perissodactyla			
Equus ferus Boddaert, horse	1	0.06	1
Artiodactyla			
Capreolus capreolus (L.), roe deer	11	0.72	2

Climatic and Environmental Interpretation of the Vertebrate Remains

Fish

The 11 fish taxa recovered from Unit 3 belong to seven different families (Table 5). All are typical of freshwater rivers and streams, although a few can tolerate more saline habitats. The fish remains would therefore seem to have been transported from purely freshwater habitats upstream of tidal influence. The most interesting species is sturgeon (*Acipenser*sp.) represented by distinctive fragments of scute, the presence of which indicates that Unit 3 includes material transported by a sizeable stream or river. Interestingly, sturgeon (*Acipenser sturio*) was recovered from the interglacial beds at Purfleet in the Lower Thames region (Schreve et al., 2002; Fig. 2), a site which is believed to occupy a similar position in the Thames terrace sequence to Cudmore Grove (Bridgland, 1994 and Schreve et al., 2002).

Herpetofauna

The herpetofaunal assemblage, one of the richest reported from Britain (Holman et al., 1990), provides valuable palaeoclimatic information, since the geographical distributions of these animals are closely linked to temperature. All herpetiles are directly reliant upon external heat for their survival, and in the case of oviparous reptiles, require certain minimum temperatures for the hatching of their eggs. Six of the 14 herpetile taxa reported from Cudmore Grove by Holman et al. (1990) are not native in Britain: water frog or edible frog (*Rana ridibunda* or *Rana esculenta*), moor frog (*Rana arvalis*) tree frog (*Hyla* sp.), viperine or dice snake (*Natrix tessellata* or *Natrix maura*), aesculapian snake (*Zamenis longissimus*) and European pond terrapin (*Emys orbicularis*). There is some debate over the degree of oceanicity or continentality indicated by these exotic taxa. Based upon their present-day ranges, Holman et al. (1990) suggested that the palaeoclimate at Cudmore Grove was slightly warmer than at present in Britain, with mean July temperatures above 17–18 °C and with mild winters. Comparable modern analogues have been suggested for central or southern France (Holman et al., 1990) or northern Italy (Holman, 1993) but Gleed-Owen (1999) proposed more continental conditions. *Z. longissimus* and *N. tessellata* are distributed throughout most of eastern Europe, where they experience cold continental winters (Gasc et al., 1997). Taken in association with the occurrence of *R. arvalis* at Cudmore Grove, Gleed-Owen (1999) suggested that a more realistic modern analogue would be around the southeast German border with north-east Austria and the western Czech Republic.

The Cudmore Grove herpetofaunal assemblage is indicative of still or slowly flowing water with abundant aquatic vegetation, adjacent damp grassland and nearby well-vegetated, damp habitats with plenty of ground cover. Some drier, open shrub or woodland habitats are also indicated by *Bufo bufo* and *Z. longissimus*, whereas *Vipera berus* and *Lacerta vivipara* occur in a wide range of open terrestrial habitats, including heaths, dunes, woodland, marshes and meadowland (Arnold and Burton, 1980).

Birds

Avian remains, which are usually under-represented in open sites, are represented by 42 identifiable bones or bone fragments. The presence of garganey (*Anas querquedula*), gadwall (*Anas strepera*) and little crake (*Porzana parva*) suggests summers as warm as that of central Europe at the present-day, in which case the northern boreal breeding whooper swan (*Cygnus cygnus*) and smew (*Mergellus albellus*) would probably have been present as winter visitors. These species suggest the presence of an open body of freshwater, with a well-vegetated margin and shallow-water plantlife. A range of passerines, including song thrush (*Turdus philomelos*), fieldfare (*Turdus pilaris*), great tit (*Parus major*), blue or coal tit (*Parus caeruleus* or *Parus ater*) and willow warbler or chiffchaff (*Phylloscopus trochilus* or *Phylloscopus collybita*) indicate the proximity of woodland. The Sandwich tern (*Sterna sandvicensis*) suggests the proximity of the coast.

Mammals

The mammalian remains suggest a similar range of environments, with woodland and aquatic species particularly well represented. The presence of deciduous or mixed woodland with thick ground cover is suggested by the abundance of bank vole (*Clethrionomys glareolus*) and wood mouse (*Apodemus sylvaticus*), with smaller numbers of pygmy shrew (*Sorex minutus*) and serotine bat (*Eptesicus serotinus*), for which there are few British Pleistocene records. This is supported by larger mammal taxa such as red squirrel (*Sciurus vulgaris*), macaque (*Macaca sylvanus*), beaver (*Castor fiber*), brown bear (*Ursus arctos*), badger (*Meles meles*) and roe deer (*Capreolus capreolus*). Locally open vegetation is indicated by field vole (*Microtus agrestis*) and horse (*Equus ferus*), whereas aquatic habitats are indicated by the abundance of water vole (*Arvicola terrestris cantiana*), which represents 37% of the mammalian assemblage (MNI = 60), and the high frequency of water shrew (*Neomys* cf. *browni*), which frequent the well-vegetated banks of rivers and lakes with still or slow-flowing water. The presence of bicoloured white-toothed shrew (*Crocidura* cf. *leucodon*), which today has a predominantly southern European distribution, supports the inference that the climate may have been slightly warmer than at present.

FAUNAL BIOSTRATIGRAPHY

The faunal datasets provide a wealth of information about the local environment and climate associated with the accumulation of the Cudmore Grove interglacial beds, but few groups are biostratigraphically significant. Neither the

Coleoptera nor the Ostracoda provide definitive stratigraphical information. Amongst the Mollusca, however, the occurrence of *B. naticina* and *C. fluminalis* is significant, since neither species has been reported in the Lower Gravel or Lower Loam at Swanscombe in Kent (Fig. 1) or in the Freshwater Beds at Clacton. These horizons are thought to relate to the early-temperate sub-stage of the Hoxnian (Kerney, 1971 and Turner and Kerney, 1971). During this interglacial in Britain *Corbicula* is known only from the late-temperate sub-stage, where it occurs in the Middle Gravel at Swanscombe, the Estuarine Beds at Clacton and in tidally-influenced sediments at East Hyde near Tillingham (Roe and Preece, 1995, Meijer and Preece, 2000 and Roe, 2001). *B. naticina* is similarly known only from the upper levels (Middle Gravel and equivalents) at Swanscombe (Kerney, 1971). The occurrence of these species in brackish-water sediments belonging to a late-temperate sub-stage at Cudmore Grove would therefore be consistent with deposition during the Hoxnian. Other evidence presented in this paper suggests, however, that the organic sediments at Cudmore Grove belong to a later temperate stage, probably equivalent to MIS 9 (see below). Sites that have been correlated with this stage, such as Barling in Essex and Hackney Downs in central London, have yielded *Corbicula* during the early-temperate sub-stage (Bridgland et al., 2001 and Green et al., 2006). It would appear that *Corbicula*occurred throughout much of MIS 9, in contrast to its more restricted occurrence during MIS 11. In Britain, *B. naticina* has not previously been recorded after the Hoxnian, and thus its occurrence at Cudmore Grove represents its first record from the late-temperate sub-stage of the ensuing interglacial.

The best faunal evidence for the age of the interglacial beds comes from the mammal remains. The assemblage lacks any of the 'indicator species' that have been found to characterise the Hoxnian (Schreve, 1997, Schreve, 2001 and Preece et al., 2007) at sites such as Hoxne (Stuart et al., 1993), Barnham (Parfitt, 1998), Beeches Pit, West Stow (Preece et al., 2007), all in Suffolk, Swanscombe (Sutcliffe, 1964 and Schreve, 1996) and Clacton in Essex (Singer et al., 1973). Thus, in the case of the small mammals, the extinct small mole (*Talpa minor*), rabbit (*Oryctolagus cuniculus*), extinct giant beaver (*Trogontherium cuvieri*) and European pine vole (*Microtus* (*Terricola*) *subterraneus*) are all absent at Cudmore Grove, despite extensive sieving that led to the recovery of over 1500 mammalian remains. In Britain this assemblage of species only co-existed during the Hoxnian so their absence at Cudmore Grove strongly suggests that this site dates from a younger temperate episode. This is supported in the large mammal assemblage by the occurrence at Cudmore Grove of brown bear (*U. arctos*), as opposed to cave bear (*Ursus spelaeus*) known from the Hoxnian.

The distinction first made by Currant (1989) between the vertebrate assemblages from Hoxnian sites and those from Grays and Cudmore Grove was initially dismissed by Bridgland (1994), who suggested that variations in sample size might explain the apparent differences. However, Schreve (1997) counted 1575 specimens from 27 taxa at Grays, 1622 specimens from 33 taxa at Swanscombe and 952 specimens from 19 taxa at Clacton, compared with 1508 specimens from 18 taxa at Cudmore Grove. These counts demonstrate that all four sites are comparably rich, indicating that the absence of the Hoxnian 'indicator species' at Cudmore Grove is biostratigraphically significant, although it is true that the number of large mammals recovered from the other sites is far larger than the comparable total from Cudmore Grove.

On the basis of mammalian biostratigraphical evidence, in particular the presence of white-toothed shrews (*Crocidura* spp.), the Cudmore Grove deposits were originally grouped not only with those of Grays Thurrock (Essex) but also with the lower part of the sequence at Aveley (Essex) and Itteringham (Norfolk) (Bridgland et al., 1988 and Currant, 1989). However, *Crocidura* has now been recovered from deposits of Hoxnian age at Barnham (Parfitt, 1998) and of Last (Ipswichian) Interglacial age at Tornewton Cave, Devon (A.P. Currant, personal communication) so, pending further refinement to species level, the genus can no longer be used as an indicator of age.

The Cudmore Grove mammalian assemblage compares most closely with those faunas known from Purfleet and Grays Thurrock in the Lower Thames region (Fig. 2). These localities have previously been attributed to a wide range of ages but have most recently been assigned to MIS 9 on the basis of terrace stratigraphy (Bridgland, 1994) and mammalian biostratigraphy (Schreve, 1997, Schreve, 2001 and Schreve et al., 2002). These assemblages are unified not only by the species present but also characterized by some that are absent, where such absence is not related to collection failure. Moreover, all three sites contain remains of water shrew, attributed to *N.* cf. *browni* on the basis of size. *N. browni* was first described by Hinton (1911) from Grays Thurrock and was considered by him to represent a transitional form between the small early Middle Pleistocene water shrew, *Neomys newtoni* and the larger, modern species *Neomys fodiens*. Whether this represents a valid species is open to question but relative size may nevertheless serve as a useful biostratigraphic marker (Schreve, 2001).

The remains of water vole (*A. terrestris cantiana*) from Cudmore Grove are also of biostratigraphical significance, since they show differences from those recovered from Hoxnian sites. *A. terrestris cantiana* is characterized by relatively small and permanently growing rootless molars, which display a '*Mimomys*' enamel differentiation (i.e. the enamel of the trailing (convex) edges is markedly thicker than that of the leading (concave) edges). The modern NW European *Arvicola terrestris terrestris*, on the other hand, has relatively larger molars that display a '*Microtus*' enamel differentiation, with thicker enamel on the leading (concave) edges of the salient angles. Throughout the Middle and Late Pleistocene, an evolutionary trend is apparent in the water vole lineage as the size of the molars increases and the thickened enamel on the convex sides of the molars reduces (Sutcliffe and Kowalski, 1976, Heinrich, 1982, Heinrich, 1987, Stuart, 1982, van Kolfschoten, 1990 and von Koenigswald and van Kolfschoten, 1996). A separation of the Cudmore Grove sample from that of Hoxnian sites has been demonstrated on the basis of this enamel differentiation, using a quotient (SDQ) to express the relative thickness of the enamel on the trailing and leading edges of the molars. The SDQ values for Hoxnian *Arvicola* samples of first lower molars fluctuate around 140 (Parfitt, 1998, Schreve, 2001 and Preece et al., 2007, Fig. 27), whereas those from Cudmore Grove have a mean value of 133.36 ($n = 48$, range = 105–147), implying a younger age. The SDQ values for Cudmore Grove accord well with those from other sites attributed to MIS 9, such as Grays

Thurrock and Purfleet (Schreve, 2001 and Preece et al., 2007).

A further trend is also visible in *Arvicola*, namely the progressive loss of the '*Mimomys*fold' in the first lower molar. The earliest forms of *Arvicola* from the end of the 'Cromerian Complex' retain the '*Mimomys* fold' and this trait consistently occurs in up to one third of individuals from Hoxnian deposits (Parfitt, 1998 and Preece et al., 2007, Fig. 27). In later interglacials, that number has dwindled or even disappeared. The evidence from Cudmore Grove is consistent with a post-Hoxnian age, since of the 112 first lower molars recovered, the '*Mimomys* fold' is present in only 5 specimens (4.46% of the sample).

An upper limit for the age of the Cudmore Grove interglacial is suggested by the presence of *M. sylvanus*, which is unknown in Britain from deposits attributed to MIS 7 and later. The length of the Cudmore Grove water vole first lower molars may also be significant, since they are smaller (mean = 3.49 mm, n = 33) than samples from localities attributed to MIS 7, such as Pontnewydd Cave (Green, 1984) (3.69 mm, n = 3, Schreve, 1997) and to the Ipswichian (see Preece et al., 2007, Fig. 27 for details). The distinctive mammalian assemblage from Cudmore Grove therefore differs from those described from several Hoxnian sites that have produced reasonable samples of mammalian remains. The absence of critical Hoxnian 'indicator species', the presence of both brown bear and the occurrence of *Arvicola* molars with relatively lower SDQ values, all indicate a post-Hoxnian age for the Cudmore Grove interglacial.

ARCHAEOLOGY

At least three unstratified flint flakes have been found on the beach at Cudmore Grove towards the southern channel margin (G.R. Ward and P. Spencer, personal communication). All these finds probably originate from the feather-edge of the Cudmore Grove Channel Gravel (Unit 1), from which a single struck flake was recovered *in situ* (Bridgland, 1994) (Fig. 5). A scraper has also been found on the foreshore to the immediate east of the Cudmore Grove sea wall (TM 06891466; Fig. 3 and Fig. 4). The presence of iron-stained sand within cavities of the scraper suggests it may have been derived from the adjacent Upper sands and gravels (Unit 6). A broken hand-axe has also been recovered from a foreshore exposure of gravel near the 'Hippopotamus Site' (TM 06541424; P. Spencer, personal communication). Its relationship to the infill sequence at this site is uncertain.

AMINO ACID GEOCHRONOLOGY

Introduction

A new technique of amino acid analysis has been developed for geochronological purposes (Penkman, 2005, Penkman et al., 2007a and Penkman et al., 2008a), combining a reverse-phase high-pressure liquid chromatography (RP-HPLC) method of analysis (Kaufman and Manley, 1998) with the isolation of an 'intra-crystalline' fraction of amino acids by bleach treatment (Sykes et al., 1995). This combination of techniques results in the analysis of D/L (dextro- and laevo-rotatory optical isomers) values of multiple amino acids from the chemically protected protein within the biomineral, thereby enabling both decreased sample sizes and increased reliability of the analysis. Amino acid racemization (AAR) analyses were undertaken on shells and opercula from bulk samples from Unit 3: four shells of *B. tentaculata* (NEaar 0859, 0888–0890; published in Penkman et al., 2007a), four shells of *V. piscinalis* (NEaar 0855–0858; published in Penkman et al., 2007a) and four opercula of *B. tentaculata* (NEaar 0891–0894). Seven *B. tentaculata* opercula (NEaar 2407–8, 3728–3730, 3820–2821) have also been analysed from the neighbouring 'East Mersea Restaurant Site' (Sample 3) (Fig. 1); five previously published by Penkman et al. (2007b) (NEaar 3728–3730, 3820–3821) and two new analyses (NEaar 2407–8). All samples were prepared using the procedures of Penkman (2005) and Penkman et al. (2008a) to isolate the intra-crystalline protein by bleaching. Two subsamples were then taken from each shell; one fraction was directly demineralised and the free amino acids analysed (referred to as the 'Free' amino acids, FAA, F), and the second was treated to release the peptide-bound amino acids, thus yielding the 'total' amino acid concentration, referred to as the 'Total Hydrolysable amino acid fraction' (THAA, H*). Samples were analysed in duplicate by RP-HPLC.

The DL ratios of aspartic acid/asparagine, glutamic acid/glutamine, serine, alanine and valine (D/L Asx, Glx, Ser, Ala, Val) as well as the [Ser]/[Ala] value provide an overall estimate of protein decomposition (Fig. 10; Supplementary Online Appendix 1). Serine is one of the most geochemically unstable amino acids, with one of its decomposition products being alanine (Bada et al., 1978). This enables the ratio of the concentration of serine ([Ser]) to the concentration of alanine ([Ala]) to be used as a useful indication of the extent of protein decomposition. The D/L of an amino acid will increase with increasing time, whilst the [Ser]/[Ala] value will decrease. The D/L of Ser is less useful as a geochronological tool for samples of this age, but is presented here as aberrant values are useful indications of contamination.

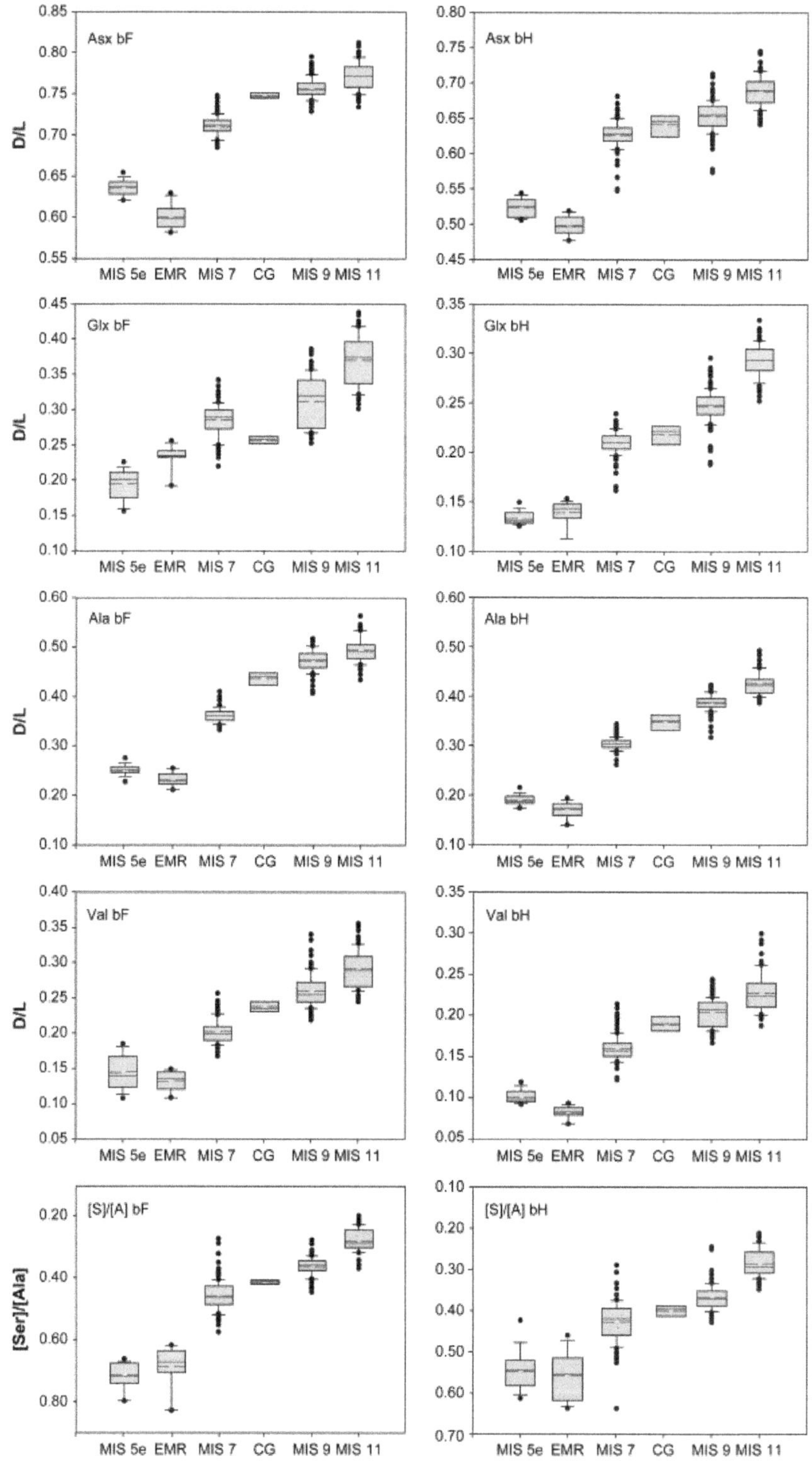

Figure 10: D/L values of Asx, Glx, Ala, Val; and [Ser]/[Ala] for the Free (FAA; F) and Total Hydrolysable amino acid (THAA; H) fractions of bleached *Bithynia tentaculata* opercula from East Mersea Restaurant Site (EMR) and Cudmore Grove (CG), compared with shells from sites correlated with MIS 5e (Bobbitshole, Trafalgar Square), MIS 7 (Aveley, Lion Pit), MIS 9 (Barling, Grays, Hackney, Purfleet) and MIS 11 (Swanscombe, Hoxne, Elveden, Barnham, Clacton and Beeches Pit). For each group, the base of the box indicates the 25th percentile. Within the box, the solid line plots the median and the dashed line shows the mean. The top of the box indicates the 75th percentile. Where more than 9 data points are available, the 10th and 90th percentiles can be calculated (shown by lines below and above the boxes respectively). The results of each duplicate analysis are included in order to provide a statistically significant sample size. The *y*-axes for the [Ser]/[Ala] data are plotted in reverse, so that the direction of increased protein degradation for each of the indicators remains the same. Note different scales on the *y*-axes.

Results

In a closed system, the amino acid ratios of the FAA and the THAA subsamples should be highly correlated, enabling the recognition of compromised samples. The extent of protein decomposition in both the FAA and THAA increases with time, with increased levels of protein breakdown during warm stages and decreased rates of degradation in cold stages. Data from British interglacial deposits using the intra-crystalline fraction of these three biominerals (*V. piscinalis* shell, *B. tentaculata* shell and opercula) tend to fall into clusters of amino acid ratios (Penkman, 2005 and Penkman et al.,

2007a). Given a similar temperature history, this then allows a regional aminostratigraphic framework to be developed, with independent geochronology allowing correlation with marine Oxygen Isotope Stages (Penkman, 2005). The extent of protein degradation in the *V. piscinalis*and *B. tentaculata* shells indicates an age for the Cudmore Grove shells between MIS 7 and MIS 11 (Penkman et al., 2007a).

It has become apparent that the protein fraction within gastropod shells has greater variability in samples of this age, resulting in difficulties in discrimination between different interglacial warm stages (Penkman et al., 2007a). Amino acid data obtained from the intra-crystalline fraction of the calcitic *Bithynia* opercula indicate that this biomineral is a particularly robust repository for the original protein (Penkman et al., 2007a, Penkman et al., 2008a and Penkman et al., in press). The *B. tentaculata* opercula from Cudmore Grove (Fig. 10) have protein that is significantly more degraded than that from Aveley and Lion Pit, which are in the Mucking Formation of the Thames, correlated with MIS 7 (Bridgland, 1994). Aveley also has OSL dates consistent with an age within MIS 7 (E. Rhodes, personal communication). The opercula protein from Cudmore Grove is also less degraded than that from sites correlated with MIS 11, such as Hoxne (Ashton et al., 2008), Swanscombe, Clacton (Bridgland et al., 1999), Barnham (Preece and Penkman, 2005), Elveden (Ashton et al., 2005) and Beeches Pit (Preece et al., 2007). Both Hoxne (Grün and Schwarcz, 2000) and Beeches Pit (Preece et al., 2007) have independent geochronology (U series/ESR and TL) confirming attribution to MIS 11. The opercula show similar levels of protein degradation to sites correlated with MIS 9, including Hackney (Green et al., 2006), Barling (Bridgland et al., 2001) Grays (Bridgland, 1994) and Purfleet (Schreve et al., 2002). Purfleet also has OSL data consistent with an MIS 9 age (E. Rhodes, personal communication). The intra-crystalline protein decomposition is slightly lower than that observed at Purfleet, which is consistent with an age within a later part of MIS 9.

Due to the low levels of degradation with cold stages and because of the natural variability in biological samples, there is a potential difficulty of discriminating between the end of one warm stage and the beginning of the next, particularly with some amino acids. We have therefore used statistical tests on the *B. tentaculata* opercula to establish whether it is possible to use the extent of protein breakdown to discriminate between Cudmore Grove and selected sites that have multiple lines of evidence correlating them with either MIS 11, MIS 9 or MIS 7. Together with the problem of small sample sizes, amino acid data have upper and lower limits (0 and 1) and therefore statistical tests must be applied with caution. However, they can provide useful insights alongside interpretations based on the graphical data alone. For each amino acid within each fraction (FAA and THAA), a 2-tailed *t*-test (assuming normally distributed data, as shown to be usual in Penkman et al., 2007a) was performed on the data. If the result of the *t*-test produces a $p < 0.05$, this enables discrimination between the two groups tested at a 95% confidence level. The 2-tailed *t*-test assumes no prior knowledge of the stratigraphy and therefore provides a useful guide of the resolving power of this technique when applied to sites of unknown age. The results are presented in Supplementary Online Appendix 2 but are also summarised in Table 6.

Table 6: Summary of statistical test results (Minitab v.15), using 2-tailed *t*-tests (assuming normal distribution). The results of each duplicate analysis are included in order to provide a statistically significant sample size. Number in "Yes" column represents the number of amino acid (AA) fractions that enable discrimination between the two groups in question at a 95% confidence level. Number in "No" column represents the number of AA fractions that do not enable discrimination between the two sites at a 95% confidence level. For example, in the 2-tailed *t*-tests, analysis of *Bithynia* opercula results in ten out of ten AA fractions support the hypothesis that Cudmore Grove (CG) is distinguishable and younger than the sites correlated with MIS 11. The sites were selected because they have multiple lines of evidence that enable them to be correlated to a specific MIS. Full details of the statistical tests and further stratigraphical information and references associated with the sites are given in Appendix 2. EMR = East Mersea Restaurant Site

	Sites included in 2-tailed *t*-test	Yes	No
Is CG younger than sites correlated with MIS 11?	Swanscombe, Clacton, Hoxne, Barnham, Elveden, Beeches Pit	10	0
Is CG younger than sites correlated with MIS 9?	Barling, Grays, Hackney, Purfleet	9	1
Is CG older than sites correlated with MIS 7?	Lion Pit, Aveley	8	2
Is EMR younger than sites correlated with MIS 7?	Lion Pit, Aveley	10	0

The amino acid data from *B. tentaculata* and *V. piscinalis* shells and *B. tentaculata*opercula indicate an age for Cudmore Grove lying between MIS 7 and MIS 11 but the enhanced chronological resolution from the *Bithynia* opercula data indicates a probable mid-late MIS 9 age.

The *B. tentaculata* opercula protein analysed from the East Mersea Restaurant Site (Sample 3) was significantly less degraded than those from Cudmore Grove (Fig. 10). Although one sample has slightly lower ratios in Glx and Val (NEaar 2408), the samples have very similar values to opercula from sites correlated with the Ipswichian (MIS 5e), such as Trafalgar Square in central London and Bobbitshole, the Ipswichian type site near Ipswich (Suffolk) (Penkman et al., 2008b). The extent of protein decomposition is significantly less than that observed in opercula from sites correlated with MIS 7 (Appendix 1). The amino acid data therefore support the correlation of the Restaurant Site with the Ipswichian.

SUMMARY AND DISCUSSION

Environment and Climate Associated with the Interglacial Beds: Synthesis

The channel-fill sequence at Cudmore Grove has yielded an impressive suite of thermophilous plant and animal remains clearly indicative of interglacial conditions. Particularly important in this respect are the beetle assemblages, from which a temperature range T_{max} 16–22 °C and T_{min} −7 °C to +4 °C has been estimated, and the herpetofauna that is indicative of a climatic regime equivalent to that found today in parts of central Europe. Moreover, the Cudmore Grove channel sediments span much of the interglacial and have provided a long pollen record extending from the early-temperate stage, when oak and other thermophilous forest elements were expanding, to the late-temperate stage, when late immigrating trees such as *Abies* and *Carpinus* were becoming established (Table 7). The post-temperate sub-stage, during which pollen values of *Quercus*, *Carpinus* and *Abies* decline and those of *Pinus*, Poaceae and Ericales increase, may also be represented in the grey clays (Unit 5) in the upper part of the sequence.

Table 7: Summary of the environmental history of the Pleistocene succession at Cudmore Grove showing details of the fossil groups that have been used in the interpretation of the depositional environments and climate. The deposits with *Hippopotamus* at the 'East Mersea Restaurant Site' and the 'Hippopotamus Site' (both thought to be MIS 5e in age) are not included

Unit	Name	Local pollen assemblage zones	Diatom zones ldaz	Ostracoda and Mollusca	Vertebrates and Coleoptera	Depositional environment	Climate	Palynological sub-stage (after Turner & West, 1968)
6	Upper sands and gravels (Mersea Island Member)	barren	barren	barren	barren	Fluviatile – braided regime. Post-depositional periglacial processes, including ice-wedge formation	?Cold and periglacial	?
5	Grey clays	*Alnus, Pinus* and Ericales. Pollen may be re-worked				Slow-moving water, favouring clay flocculation, possibly marine	Late interglacial?	Post-temperate (IV)?
4	Organic clays	*Alnus-Quercus-Carpinus, Pinus* and *Abies*. Ericales expands **lpaz CG-3**	ldaz CGD3			Lagoonal backwater with episodic saltwater penetration	Interglacial	Late-temperate (III)
3	Detritus muds			Sparse ostracod fauna with low numbers of *C. torosa* Freshwater molluscan fauna with some brackish taxa	Rich mammalian, avian, herpetofaunal and fish assemblage (54 taxa) 65 coleopteran taxa	'Lag' deposit with allochthonous material derived from proximal freshwater and estuarine sources, probably following erosive episode	Full interglacial. Temperatures ~2°C warmer than in southern England today	
2d	Silty clays: *Upper silty clays*	**lpaz CG-2** *Quercus-Pinus*, some *Corylus*	**ldaz CGD2b-c**	**Ostracod Zone 2** Brackish-water Mollusca	barren	Outer estuary	Interglacial	Early-temperate (II)
2c	Silty clays: *Shelly silts*		**ldaz CGD2a**			Significant estuarine channel		
2b	Silty clays: *Lower silty clays*	**lpaz CG-1** *Pinus-Betula-Quercus*	**ldaz CGD1**	**Ostracod Zone 1** Low numbers of *C. torosa*		Tidally-influenced river channel	Early interglacial with water temperatures similar to Scandinavian rivers today	
2a	Silty clays: *Sandy clays*	No pollen. *Elatine* macrofossils reflect temperate conditions	barren		*Bos* or *Bison* (femur)	Low-energy river channel, eutrophic with sandy substrate and marginal saltwater influence	Early interglacial	?
1	Basal sands and gravels	barren		barren	Elephant (molar)	Fluviatile	Early interglacial or cold stage?	barren

Further insights into the depositional context of the interglacial beds have been provided by the diatom, ostracod, molluscan and coleopteran assemblages, which show that the channel-fill sequence accumulated in a dynamic coastal environment that underwent a number of changes through time (Table 7). The diatoms and ostracods together show that the silty clays (Unit 2) were deposited in a tidally-influenced river channel, which became increasingly open to marine influence as the oak-dominated early-temperate sub-stage progressed. Outer estuarine conditions prevailed by the end of this phase. Conditions changed markedly when the detritus muds (Unit 3) accumulated; this richly fossiliferous deposit was the product of high-energy riverine or coastal processes that carried shell debris, insect remains, bones and other detritus into the site from the surrounding hinterland and adjacent estuarine areas. Sheltered lagoonal conditions became established subsequently as the organic muds (Unit 4) accumulated and the site became isolated from the open estuary. The persistent presence of marine diatoms in these sediments, which date from the late-temperate sub-stage, confirms that relative sea levels remained high, with saltwater episodically penetrating the lagoon. The grey clays (Unit 5) capping the channel-fill sequence may even reflect a return to more open, estuarine or fully marine conditions towards the end of the interglacial, although the absence of diatoms or ostracods precludes confirmation.

The Age of the Cudmore Grove Interglacial Beds

The biostratigraphical datasets have provided varying degrees of insight into the age of the interglacial beds. The pollen and plant macrofossil data are important not only for constraining the length of interglacial time recorded, but they additionally include elements that have traditionally been regarded as uniquely 'Hoxnian', including the presence of Type X pollen, modest levels of *Abies* and *Carpinus* pollen in the late-temperate sub-stage and megaspores of *A. filiculoides* (Turner, 1970, Phillips, 1976 and Horton et al., 1992). A last interglacial (MIS 5e) or an MIS 7 age for the Cudmore Grove interglacial beds can therefore be firmly ruled out. However, given the uncertainty over the number and status of 'Hoxnian-type' interglacials represented in the British Pleistocene (Roe, 1994, Scourse et al., 1999 and Thomas, 2001), there are potential flaws in undertaking correlations with spectra from other 'Hoxnian' sites. This problem is further compounded by the fact that most of the sites that have been attributed to the younger of the two 'Hoxnian-type' interglacials, which was possibly co-eval with MIS 9, have only yielded fragmentary pollen records, reflecting their fluvial origin (Thomas, 2001 and Nitychhoruk et al., 2006). Notwithstanding this, the pollen spectra from Cudmore Grove zone CG-1 show some similarities to the short pollen profiles from Hackney (Green et al., 2006) and Barling (Bridgland et al., 2001) in the Thames system, which have both been correlated with MIS 9 on the basis of their faunal biostratigraphy and position in the Thames terrace sequence. These profiles, which span part of the early-temperate sub-stage of an interglacial (cf. Turner and West, 1968), also include a phase of *Quercus–Pinus* dominance with subsidiary levels of mixed oak forest taxa, e.g., *Ulmus* and *Fraxinus*. They have also both yielded macrofossils of *A. filiculoides*. Unfortunately, like many other early-temperate sub-stage spectra (cf. Turner and West, 1968), the observed palynological changes are biostratigraphically uninformative and do not provide independent grounds for correlation. Pollen biostratigraphical correlations with longer regional pollen records from lacustrine sequences are complicated by the fact that the Cudmore Grove pollen assemblages accumulated in a coastal environment where different taphonomic processes operate.

With regard to the molluscs, the occurrence of the invasive bivalve *C. fluminalis* in the early-temperate sub-stage is significant because during the Hoxnian it does not occur until the late-temperate sub-stage, at least in the Thames system (cf. Meijer and Preece, 2000). For example, at both

Swanscombe and Clacton (Fig. 2), it only appears with the so-called 'Rhenish fauna' in the aggradations which are attributed to Ho IIIb (Kerney, 1971). Its early appearance at Cudmore Grove is again more consistent with its distribution at Barling (Bridgland et al., 2001) and Hackney Downs (Green et al., 2006) which both are attributed to MIS 9.

The mammalian faunas and the AAR data provide the strongest insights into the age of the Cudmore Grove interglacial beds and firmly support a correlation with MIS 9. In the case of the mammalian data, this correlation is not only based on the absence of several 'indicator' species of Hoxnian age, but on the presence of *U. arctos* (brown bear), which is unknown from the Hoxnian, and also on the '*Mimomys*' enamel differentiation of the *Arvicola* molars (cf. Preece et al., 2007). The fact that only 4.46% of the *Arvicola*specimens from Cudmore Grove display the '*Mimomys* fold' further points to a post-Hoxnian age (cf. Parfitt, 1998 and Preece et al., 2007). The AAR datasets are derived from the shells of two species of molluscs (*V. piscinalis* and *B. tentaculata*), as well as the opercula of the latter. These data independently support a correlation with MIS 9. Despite their proximity and palynological similarities, the differences in the amino acid ratios between Cudmore Grove and Clacton are particularly striking. Not surprisingly, the Cudmore Grove amino acid ratios also differ from those of the East Mersea Restaurant Site (Bridgland et al., 1988 and Bridgland et al., 1995), which lies only 2 km to the south-west of Cudmore Grove (Fig. 1). This interglacial deposit, which comprises ca 1.2 m of silts and gravel channelled into the London Clay, has yielded a freshwater molluscan fauna and a rich vertebrate assemblage (10 species) that includes the water vole, *A. terrestris terrestris* and *Hippopotamus amphibius*; a combination unknown in Britain before the Late Pleistocene (cf. Sutcliffe and Kowalski, 1976 and Sutcliffe, 1995). The mammal assemblages compare closely with those from several other Ipswichian (MIS 5e) sites, including Joint Mintor Cave, Devon (Sutcliffe, 1960) and Trafalgar Square, London (Bridgland et al., 1995).

Further independent support for a separation of the Cudmore Grove interglacial beds from those at Clacton is provided by the marked differences in the sea-level histories of the two sites. The Clacton Channel deposits do not record the onset of estuarine/marine conditions until relatively late in the interglacial, during the late-temperate sub-stage (Ho IIIb), whereas the earlier part of the channel-fill sequence, representing the early-temperate sub-stage, accumulated in a freshwater environment (Turner and Kerney, 1971 and Bridgland et al., 1999). A similar onset of marine conditions, again during Ho IIIb, is recorded in the channel-fill sequences at East Hyde near Tillingham (Roe, 2001) (Fig. 11) and at Swanscombe (Kerney, 1971) and all three channels are believed to have infilled during MIS 11 (Bridgland et al., 1999 and Roe, 2001; Fig. 12a). At Cudmore Grove, in contrast, estuarine conditions were already established in the early-temperate sub-stage of the interglacial represented, and relative sea levels remained high until at least the late-temperate sub-stage. It is difficult to reconcile the presence of outer estuarine conditions at Cudmore Grove during the early-temperate sub-stage with the freshwater conditions at Clacton; the channels lie at approximately the same height and are only 10 km apart and are thus likely to have been flooded by the sea around the same time had they formed during the same interglacial (Roe, 1994). To accommodate these differences within the same temperate event requires a complex model of coastal evolution, which would have allowed the sea to penetrate the eastern part of Mersea Island, but not neighbouring Clacton (Roe, 1999). Whilst this cannot be ruled out, it is not supported by any evidence or the bio- and aminostratigraphical data.

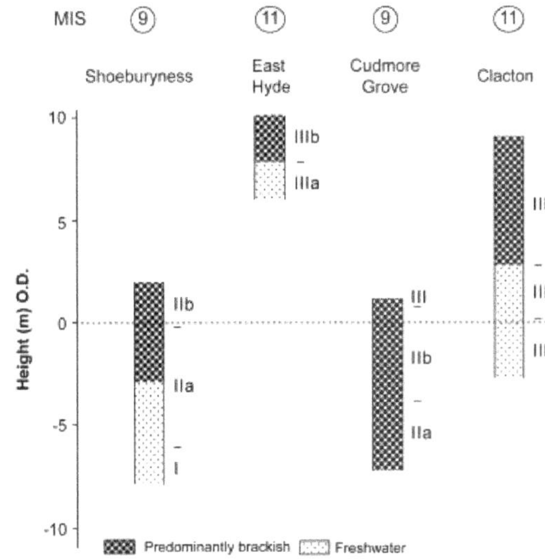

Figure 11: Altitudinal plot of interglacial channel-fill sediments in eastern Essex showing the salinity characteristics inferred from diatoms and ostracods and their associated pollen zones. The assemblage zones follow the terminology of Turner and West (1968). Early-temperate sub-stage = zone II; late-temperate sub-stage = zone III. The record from East Hyde follows Roe (2001); Clacton (Pike and Godwin, 1953 and Turner and Kerney, 1971), Shoeburyness (Roe, 1999). The correlations proposed in the text with MIS 9 and 11 are indicated.

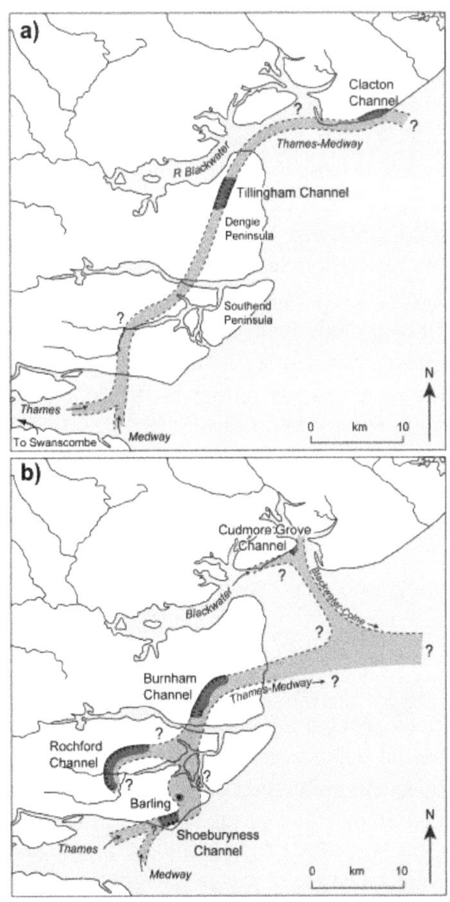

Figure 12: Palaeodrainage maps of eastern Essex during MIS 11 (a)

and MIS 9 (b). Map a) shows the tidally-influenced Thames–Medway flowing between Tillingham (East Hyde) and Clacton during the late-temperate sub-stage of the inferred MIS 11 interglacial. The sequence at Swanscombe is considered to be the upstream equivalent of this channel system (Bridgland, 1994 and Roe and Preece, 1995). Map b) depicts a larger, more complex palaeo-estuary inferred for the end of the early-temperate sub-stage of the younger interglacial, when lower estuarine conditions prevailed at Cudmore Grove (after Roe, 1994 and Schreve et al., 2002). The Cudmore Grove palaeo-channel is placed in the more northerly part of this estuarine system and is assumed to have formed a tributary of the main Thames–Medway estuary. The dark shading shows the distribution of channel-fill sediments of estuarine character which have been examined for pollen and other microfossils (Roe, 1994).

The clast lithological and palaeocurrent data presented here further re-enforce the interpretation that the Cudmore Grove Channel is not an upstream equivalent of the Clacton Channel as was once suggested (e.g. Bridgland, 1988 and Bridgland, 1994). This interpretation was based on the assumption that both interglacial channel-fills were overlain by what were assumed to be Thames–Medway gravels: the Mersea Island Member aggradation above the Cudmore Grove Channel and the Wigborough Gravel at Clacton (Fig. 2). The fine-grained channel-fills were thus thought to represent the interglacial course of the Thames–Medway during the interglacial that preceded the Mersea Island–Wigborough Gravel aggradation: the Hoxnian (Bridgland, 1994, Fig. 5.5). The Hoxnian correlation was based on the biostratigraphy of the Clacton deposits and their position in the Thames terrace sequence (cf. Turner and Kerney, 1971, Bridgland, 1988 and Bridgland, 1994, Fig. 5.5). The clast lithological data presented here cannot confirm that either the basal sands and gravels lining the Cudmore Grove Channel (Unit 1) or the Unit 6 gravels in the Cudmore Grove cliffs were deposited by the Thames–Medway. Indeed, the Unit 1 gravels have a composition that is more consistent with aggradation by a smaller river, possibly the Essex Blackwater, re-working Thames–Medway deposits, local Kesgrave Member gravels and fluvio-glacial gravels from the area to the west of Cudmore Grove. Moreover, the palaeocurrent directions from Unit 6 confirm that the upper part of this aggradation (sub-unit 6b) was deposited by a river flowing from the northeast (possibly a proto-River Colne), which is inconsistent with a Thames–Medway origin. The Unit 6 gravels also display compositional differences from Thames–Medway gravels further to the south, for example, on the Dengie Peninsula (cf. Bridgland, 1994), which suggest that they too may have been aggraded by tributary rivers (the Colne, Blackwater or both) re-working older Thames–Medway deposits. The terrace gravels above the two channels cannot therefore be used to confirm that the underlying interglacial beds are either contemporaneous or part of the same palaeo-channel system.

There is therefore compelling biostratigraphical, aminostratigraphical and palaeo-sea level evidence, together with evidence from the regional terrace stratigraphy to suggest that interglacial deposits from three separate interglacials are represented in this small area of eastern Essex: i) the Clacton and Tillingham Channel deposits, which are of Hoxnian age and formed during part of MIS 11 (Bridgland et al., 1999 and Roe, 2001); ii) the Cudmore Grove Channel deposits, which we suggest formed within MIS 9; and iii) the East Mersea Restaurant Site deposits, which are of last interglacial (Ipswichian) age (MIS 5e). We reject the possibility that the Cudmore Grove sequence dates from a younger temperate sub-stage within MIS 11 (for example, sub-stage 11c) because the sequence records a full interglacial forest succession and temperatures several degrees warmer than present, characteristics that are inconsistent with other records attributed to this warm interval (cf. Reille et al., 2000, Desprat et al., 2005 and Preece et al., 2007). The Cudmore Grove mammal faunas are also markedly different to those recorded in the upper part of the Hoxnian type-sequence at Hoxne (strata C and B), which has recently been re-assigned to a later sub-stage within MIS 11 (Ashton et al., 2008).

The palaeogeographies of eastern Essex were quite different during the two late Middle Pleistocene interglacials (Fig. 12). During MIS 11, the Thames–Medway appears to have flowed north-eastwards between Tillingham and Clacton, maintaining its early post-diversion route, which had been established late in the Anglian (cf. Bridgland, 1994; Fig. 12a). During MIS 9, a larger fluvio-estuarine drainage network appears to have existed in the region, of which the Cudmore Grove Channel sequence was only a small remnant (cf. Bridgland, 1994, Roe, 1994 and Schreve et al., 2002) (Fig. 12b). On the basis of the clast lithological data, we assume that this section of the palaeo-estuary was associated with a southward-draining tributary system (the 'proto-Colne or proto-Blackwater'), which may have become confluent with the Thames–Medway somewhere in the current offshore area, possibly east of the Dengie Peninsula. Interglacial deposits filling buried channels at Shoeburyness, Canewdon (Roe, 1999) and Barling (Bridgland et al., 2001) on the Southend Peninsula, and Burnham (North Wick) on the Dengie Peninsula (Roe, 1999) are all considered to have formed part of this palaeo-estuary (Roe, 1999 and Schreve et al., 2002; Fig. 1 and Fig. 12). Like Cudmore Grove, all these sites have yielded brackish-water assemblages indicative of the early-temperate sub-stage of an interglacial (Roe, 1994 and Roe, 1999). The spectra from Shoeburyness and North Wick show striking similarities to those from zone CG-2 at Cudmore Grove, although the late-temperate sub-stage is missing at these sites (Roe, 1999). The exact geography of this palaeo-estuary is speculative, although bathymetric and seismic data from the offshore area to the east of the Dengie Peninsula confirm the existence of a buried channel system with associated submerged terrace gravels in the postulated vicinity of this inferred MIS 9 estuary (Bridgland et al., 1993, Bridgland and D›Olier, 1995 and Bridgland, 2002). It is worth noting that the occurrence of channel-fills in close proximity that can demonstrably be shown to date from different interglacials is not unique in Britain; a similar series of channel-fills of Middle to Late Pleistocene age occur in close juxtaposition in the eastern Solent on the south coast of England, incised into the soft, Eocene Bracklesham Beds (Preece et al., 1990).

Correlations with other Regional Sites of Inferred MIS 9 Age and Wider Significance

The data presented here form part of a growing body of evidence that supports the existence of a second interglacial in the British late Middle Pleistocene, which included some 'Hoxnian' palaeovegetation elements (cf. Scourse et al., 1999, Thomas, 2001 and Scourse, 2006). This interglacial has been correlated with MIS 9 on the basis of the widely held assumption that the Anglian can be equated with MIS 12 and the Hoxnian with part of MIS 11 (cf. Bridgland, 1994, Rowe et al., 1999, Grün and Schwarcz, 2000, Schreve, 2001 and Preece et al., 2007). In addition to Cudmore Grove and the aforementioned interglacial estuarine deposits in southern Essex (Shoeburyness, Barling, North Wick and Canewdon), the interglacial beds at Purfleet and Grays (Bridgland, 1994 and Schreve et al., 2002; Fig. 2) and Hackney Downs (Green et al., 2006) have all been assigned to this temperate event. These deposits have been separated from the Hoxnian and later temperate stages using aminostratigraphy, mammalian biostratigraphy and by their position in the Thames terrace sequence; in no case has separation been possible on the basis of pollen biostratigraphy alone. Of these sites, the sequence at Cudmore Grove is by far the most complete, and, we argue, the most important: it spans a considerable part of the interglacial, provides detailed insights into the climate and local environment associated with this temperate stage, and evidence of high relative sea levels in the southern North Sea region. Moreover, the record shows that the marine transgression of this interglacial differed in its timing in this region from that of the preceding Hoxnian, taking place during the oak-dominated, early-temperate sub-stage. The maximum height attained by this inferred MIS 9 marine transgression is less easy to ascertain, as the Cudmore Grove interglacial beds are locally compacted and do not necessarily record the maximum flooding surface, which may or not be represented in the grey clays (Unit 5). Evidence from the interglacial channel-fills on the Southend Peninsula suggests that relative sea levels attained a height of at least ca 5–7 m O.D. in this region during the early-temperate sub-stage of this interglacial (Roe, 1999 and Bridgland et al., 2001). This compares with the estimate of ca >20 m O.D. inferred for the Hoxnian RSL highstand during Ho IIIb from Swanscombe (Kerney, 1971 and Bridgland, 1994). However, the possibility of differential post-depositional movements associated with long-term tectonic subsidence in the Thames Estuary (cf. D›Olier, 1972) means that comparisons of the highstand maxima must be made with caution.

Flint artefacts indicate that hominins were present during the infilling of the Cudmore Grove Channel, as they were at contemporary sites such as Grays and Purfleet in the Lower Thames (Schreve et al., 2002). A full archaeological excavation of the margins of the channel may therefore be worthwhile.

Perhaps the most important attribute of the Cudmore Grove record are the insights that it provides into the vegetation history of MIS 9 in southern Britain, particularly its record of the late-temperate sub-stage, which is missing from all the other regional sites presumed to be of this age (cf. Thomas, 2001). Interestingly, one possible exception is the interglacial sequence in the Nar Valley, Norfolk, which has long been correlated with the Hoxnian (e.g. Stevens, 1960, Ventris, 1986, Ventris, 1996 and Gibbard et al., 1992). Here, a complex succession of freshwater and marine deposits, locally over 10 m thick and extending up to 23 m O.D., overlies Anglian glacial sediments (Woodlands Farm Member of the Lowestoft Formation) (Ventris, 1986, Gibbard et al., 1992 and Lewis, 1999). The pollen spectra from the marine beds include well defined *Abies*, *Carpinus* and *Empetrum* peaks, which have been ascribed to the Hoxnian late-temperate sub-stage (Stevens, 1960, Ventris, 1986 and Ventris, 1996), whereas the same unit has also yielded a diatom flora indicative of a nearshore, shelf environment (Mitlehner, 1992). This Hoxnian correlation has recently been challenged by $^{230}Th/^{238}U$ dates from peats within the Nar Valley freshwater beds, which have produced a mean age of 317 ± 14 ka, suggesting a correlation with MIS 9 (Rowe et al., 1997). Conventional isoleucine epimerization determinations on the foraminiferid *Ammonia beccarii* from the Nar Valley marine beds add independent support to this interpretation (Scourse et al., 1999). The mean aIle/Ile ratio obtained (0.135) is much lower than ratios (ca 0.2) based on the same species from the offshore interglacial Sand Hole Formation in the Inner Silver Pit, ca 20 km off the coast of East Yorkshire (Fig. 1). This deposit also yielded high levels of *Abies* and *Picea* and has similarly been ascribed to the Hoxnian late-temperate sub-stage (Scourse et al., 1998). These differences in ratios might reflect contrasting post-depositional thermal histories between the two sites, but it seems more likely that the Nar Valley and the Sand Hole Formation interglacial beds date from two temperate stages (MIS 9 and MIS 11 respectively) characterized by similar palaeovegetation (Scourse et al., 1999). If this interpretation is correct, then the widely held correlation of the Anglian glacial sediments in the Nar Valley with MIS 12 is thrown into question, although Scourse et al. (1999: p. 554) suggest that a substantial hiatus may exist between the Woodlands Farm Member and the Nar Valley freshwater beds. This warrants further investigation.

The stratigraphic separation of the Cudmore Grove interglacial deposits from those at neighbouring Clacton directly thus directly parallels the separation that has been made between the Nar Valley interglacial beds and the Sand Hole Formation on the basis of conventional aminostratigraphy and uranium series dating. Together these findings imply that: i) *Abies* was indeed a constituent of the regional forests in southern Britain during the late-temperate sub-stages of two successive late Middle Pleistocene interglacials, which are suggested to be equivalent to MIS 9 and 11; and ii) these interglacials were both characterized by high relative sea levels in the Southern North Sea region. This postulated distribution of *Abies* in the British late Middle Pleistocene tallies with the inferred distribution of the tree from the maar lake records of the Velay region, central France, where *Abies* appears to have been a significant forest component in both the latter stages of the Praclaux (MIS 11) and Landos (MIS 9) interglacials (Reille et al., 2000 and deBeaulieu et al., 2001), although it was also present on a significantly reduced scale in the Bouchet 2 and 3 interstadials, which are correlated with MIS 7 (Reille et al., 2000). Such long-distance comparisons must be made with caution, however, as the composition of the interglacial forests of central France may have differed to those prevailing in southern Britain. Certainly the collective evidence from Britain implies that the biostratigraphical significance formerly attributed to *Abies* as a Hoxnian 'indicator taxon' in the late Middle Pleistocene should now be abandoned.

On a wider European scale, the evidence for two marine highstands associated with interglacial deposits of 'Hoxnian-type' character in the Southern North Sea region has more far-reaching implications, for it follows that the 'Holsteinian' marine beds of continental Europe may also record more than one interglacial high sea-level event. In short, we refute the contention that the issue of "two Holstein-like Interglacials" is "a problem confined to the Quaternary of the British Isles" as advocated by Geyh and Müller (2006: p. 3072). Whether polleniferous marine sediments of the inferred MIS 9 and MIS 11 highstands are widely present in other areas of the Southern North Sea and western Baltic regions remains to be seen; this paper has certainly highlighted their rather patchy distribution in southern Britain. What has become clear is the need to continue to adopt a multidisciplinary approach, supplementing pollen biostratigraphy with other more rigorous stratigraphical and geochronological techniques, to separate and better constrain these interglacials with 'Hoxnian-like' affinities.

ACKNOWLEDGEMENTS

This work was initiated while HMR was at Cambridge University in receipt of an NERC studentship, which is gratefully acknowledged. She thanks Phil Gibbard for his help during the tenure of this studentship and for many valuable discussions. Steve Boreham also provided field assistance and help with clast lithological analysis. We also thank David Bridgland for discussions on the age and origin of the Mersea Island Gravel and artefacts, Graham Ward and Patrick Spencer for their comments on the archaeology, and Eric Robinson and John Whittaker for advice about ostracods. We are grateful to John Clayden and David Harrison for supplying a substantial part of the vertebrate material and Simon Parfitt for kindly providing some of the *Arvicola* SDQ measurements, data on the fish and for useful discussion. We thank Matthew Collins, Rebecca Briant and Francis Wenban-Smith for support for the amino acid analyses, which were funded by NERC, English Heritage and the Wellcome Trust and Eddie Rhodes for access to unpublished OSL data from Aveley and Purfleet. Jane Bunting commented on an early draft of the manuscript and Maura Pringle provided cartographic support. Borehole CG1 was funded by the Department of Plant Sciences, University of Cambridge and Essex County Council. This paper represents a contribution to the Ancient Human Occupation of Britain (AHOB 2) project.

REFERENCES

1. Admiraal, W., 1984. The ecology of estuarine sediment-inhabiting diatoms. Progress in Phycological Research 3, 269–322.
2. Antoine, P., Limondin-Lozouet, N., 2004. Identification of MIS 11 interglacial tufa deposit in the Somme Valley (France): new results from the Saint-Acheul fluvial sequence. Quaternaire 15, 41–52.
3. Arnold, E.N., Burton, J.A., 1980. A Field Guide to the Reptiles and Amphibians of Britain and Europe. Collins, London.
4. Ashton, N.M., Lewis, S.G., Parfitt, S.A., Candy, I., Keen, D.H., Kemp, R., Penkman, K., Thomas, G., Whittaker, J., White, M.J., 2005. Excavations at the Lower Palaeolithic site at Elveden, Suffolk, UK. Proceedings of the Prehistoric Society, 7, pp. 11–61.
5. Ashton, N., Lewis, S.G., Parfitt, S.A., Penkman, K.E.H., Coope, G.R., 2008. New evidence for complex climate change in MIS 11 from Hoxne, Suffolk, UK. Quaternary Science Reviews 27, 652–668.
6. Atkinson, T.C., Briffa, K.R., Coope, G.R., 1987. Seasonal temperatures in Britain during the past 22,000 years, reconstructed using beetle remains. Nature 325, 587–592.
7. Bada, J.L., Ho, M.Y., Man, E.H., Schroeder, R.A., 1978. Decomposition of hydroxy amino acids in foraminiferal tests: kinetics, mechanism and geochronological implications. Earth Planetary Science Letters 41, 67–76.
8. Battarbee, R.W., 1986. Diatom analysis. In: Berglund, B.E. (Ed.), Handbook of Holocene Palaeoecology and Palaeohydrology. John Wiley & Sons Ltd., Chichester, pp. 527–570.
9. Berglund, B.E., Ralska-Jasiewiczowa, M., 1986. Pollen analysis and pollen diagrams. In: Berglund, B.E. (Ed.), Handbook of Holocene Palaeoecology and Palaeohydrology. Wiley, Chichester, pp. 455–484. deBeaulieu, J.L., de Andrieu-Ponel, V., Reille, M., Gru¨ ger, E., Tzedakis, C., Svobodova, H., 2001. An attempt at correlation between the Velay pollen sequence and the Middle Pleistocene stratigraphy from central Europe. Quaternary Science Reviews 20, 1593–1602.
10. Birks, H.J.B., Birks, H.H., 2004. The rise and fall of forests. Science 305, 484–485.
11. Bowen, D.Q., Hughes, S., Sykes, G.A., Miller, G.H., 1989. Land–sea correlations in the Pleistocene based on isoleucine epimerization in non-marine molluscs. Nature 340, 49–51.
12. Bridgland, D.R., 1983. The Quaternary Fluvial Deposits of North Kent and Eastern Essex. Unpublished Ph.D. thesis, City of London Polytechnic.
13. Bridgland, D.R., 1988. The Pleistocene fluvial stratigraphy and palaeogeography of Essex. Proceedings of the Geologists' Association 99, 291–314.
14. Bridgland, D.R., 1994. Quaternary of the Thames. In: Geological Conservation Review Series 7. Chapman & Hall, London.
15. Bridgland, D.R., 2002. Fluvial deposition on periodically emergent shelves in the Quaternary: examples from the shelf around Britain. Quaternary International 92, 25–34.
16. Bridgland, D.R., D'Olier, B., 1995. The Pleistocene evolution of the Thames and Rhine drainage systems in the southern North Sea basin. In: Preece, R.C. (Ed.), Island Britain: a Quaternary Perspective. Geological Society Special Publication 96, pp. 27–45.
17. Bridgland, D.R., Sutcliffe, A.J., 1995. Cudmore Grove Hippopotamus site (TM 065142). In: Bridgland, D.R., Allen, P., Haggart, B.A. (Eds.), The Quaternary of the Lower Reaches of the Thames. Field Guide. Quaternary Research Association, Durham, pp. 275–276.
18. Bridgland, D.R., Allen, P., Currant, A.P., Gibbard, P.L., Lister, A.M., Preece, R.C., Robinson, J.E., Stuart, A.J., Sutcliffe, A.J., 1988. Report of Geologists' Association field meeting in north-east Essex, May 22nd–24th, 1987. Proceedings of the Geologists' Association 99, 315–333.
19. Bridgland, D.R., D'Olier, B., Gibbard, P.L., Roe, H.M., 1993. Correlation of Thames terrace deposits between the Lower Thames, eastern Essex and the submerged offshore continuation of the Thames–Medway Valley. Proceedings of the Geologists' Association 104, 51–57.

20. Bridgland, D.R., Currant, A.P., Preece, R.C., 1995. East Mersea Restaurant site (TM 053136). In: Bridgland, D.R., Allen, P., Haggart, B.A. (Eds.), The Quaternary of the Lower Reaches of the Thames. Field Guide. Quaternary Research Association, Durham, pp. 271–274.
21. Bridgland, D.R., Field, M.H., Holmes, J.A., McNabb, J., Preece, R.C., Selby, I., Wymer, J.J., Boreham, S., Irving, B.G., Parfitt, S.A., Stuart, A.J., 1999. Middle Pleistocene interglacial Thames–Medway deposits at Clacton-on-Sea, England: reconsideration of the biostratigraphical and environmental context of the type Clactonian Palaeolithic industry. Quaternary Science Reviews 18, 109–146.
22. Bridgland, D.R., Preece, R.C., Roe, H.M., Tipping, R.M., Coope, G.R., Field, M.H., Robinson, J.E., Schreve, D.C., Crowe, K., 2001. Middle Pleistocene interglacial deposits at Barling, Essex, UK: evidence for a longer chronology for the Thames terrace sequence. Journal of Quaternary Science 16, 813–840.
23. Brinkkemper, O., Weeda, E.J., Bohnke, S.J.P., Kuijper, W.P., 2008. The past and present occurrence of Elatine and implications for palaeoenvironmental reconstructions. Vegetation History and Archaeobotany 17, 15–24.
24. Coope, G.R., 1986. Coleoptera analysis. In: Berglund, B.E. (Ed.), Handbook of Holocene Palaeoecology and Palaeohydrology. J. Wiley & Sons, Chichester, pp. 703–713.
25. Coope, G.R., Lemdahl, G., Lowe, J.J., Walkling, A., 1998. Temperature gradients in northern Europe during the last glacial–interglacial transition (14–9 14C kyr BP.) interpreted from coleopteran assemblages. Journal of Quaternary Science 13, 419–433.
26. Cornwall, I.W., 1958. Soils for the Archaeologist. Phoenix House, London, 230 pp.
27. Currant, A.P., 1989. The Quaternary origins of the modern British mammal fauna. Biological Journal of the Linnean Society 38, 22–30.
28. Dalton, W.H., 1908. Post-glacial beds in Mersea, Essex. Essex Naturalist 15, 136–137.
29. Denys, L., 1990. Fragilaria blooms in the Holocene of the western coastal plain of Belgium. In: Simola, H. (Ed.), The Proceedings of the 10th Diatom Symposium 1988. Koeltz Scientific Books, Koenigstein, pp. 397–406.
30. Desprat, S., Sa´nchez Gon˜i, M.F., Turon, J.L., McManus, J.F., Loutre, M.F., Duprat, J., Malaize, B., Peyron, O., Peypouquet, J.P., 2005. Is vegetation responsible for glacial inception during periods of muted insolation changes? Quaternary Science Reviews 24, 1361–1374.
31. Devoy, R.J.N., Dodson, J.R., Thom, B.G., Nichol, S., 1994. Holocene environments in the Hawkesbury Valley, New South Wales: a comparison of terrestrial and marine records. Quaternary Science Reviews 13, 241–256.
32. D'Olier, B., 1972. Subsidence and sea-level rise in the Thames Estuary. Philosophical Transactions of the Royal Society of London A 272, 121–130.
33. Edington, J.M., Hildrew, A.G., 1995. Caseless Caddis Larvae: a Key with Ecological Notes. Freshwater Biological Association Scientific Publications 53. pp. 1–134.
34. Fægri, K., Iversen, J., 1989. Textbook of Pollen Analysis, fourth ed. Wiley, New York, 328 pp.
35. Gasc, J-P., Cabela, A., Crnobrnja-Isailovic, J., Dolmen, D., Grossenbacher, K., Haffner, P., Lescure, J., Martens, H., Martı´nez Rica, J.P., Maurin, H., Oliviera, M.E., Sofianidou, T.S., Veith, M., Zuiderwijk, A. (Eds.), 1997. Atlas of Amphibians and Reptiles in Europe. Societas Europaea Herpetologica and Museum National d'Histoire Naturelle, Paris.
36. Geyh, M.A., Mu¨ller, H., 2005. Numerical 230Th/U dating and a palynological review of the Holsteinian/Hoxnian interglacial. Quaternary Science Reviews 24, 1861–1872.
37. Geyh, M.A., Mu¨ller, H., 2006. Missing evidence for two Holstein-like interglacials. Reply to the comments of J.D. Scourse on Numerical 230Th/U dating and a palynological review of the Holsteinian/Hoxnian interglacial. Quaternary Science Reviews 25, 3072–3073.
38. Gibbard, P.L., 1985. Pleistocene History of the Middle Thames Valley. University Press, Cambridge, 155 pp.
39. Gibbard, P.L., 1994. The Pleistocene History of the Lower Thames Valley. Cambridge University Press, Cambridge, 221 pp.
40. Gibbard, P.L., 1999. The Thames Valley, its tributary, valleys and their former courses (Chapter 4). In: Bowen, D.Q. (Ed.), A Revised Correlation of Quaternary Deposits in the British Isles, pp. 56–58.
41. Gibbard, P.L., West, R.G., Andrew, R., Pettit, M., 1992. The margin of a Middle Pleistocene ice advance at Tottenhill, Norfolk, England. Geological Magazine 129, 59–76.
42. Gibbard, P.L., Boreham, S., Roe, H.M., Burger, A.W., 1996. Middle Pleistocene lacustrine deposits in eastern Essex, England and their palaeogeographical implications. Journal of Quaternary Science 11, 281–298.
43. Gleed-Owen, C., 1999. The palaeoclimatic and biostratigraphic significance of herpetofaunal remains from the British Quaternary. In: Andrews, P., Banham, P. (Eds.), Late Cenozoic Environments and Hominid Evolution: a Tribute to Bill Bishop. Geological Society, Bath, pp. 201–215.
44. Green, C.P., Coope, G.R., Jones, R.L., Keen, D.H., Bowen, D.Q., Currant, A.P., Holyoak, D.T., Ivanovich, M., Robinson, J.E., Rogerson, R.J., Young, R.C., 1996. Pleistocene deposits at Stoke Goldington in the valley of the Great Ouse. Journal of Quaternary Science 11, 59–87.
45. Green, C.P., Branch, N.P., Coope, G.R., Field, M.H., Keen, D.H., Wells, J.M., Schwenninger, J.-L., Preece, R.C., Schreve, D.C., Canti, G., Gleed-Owen, C.P., 2006. Marine Isotope Stage 9 environments of fluvial deposits at Hackney, North London, UK. Quaternary Science Reviews 25, 89–113.
46. Green, H.S., 1984. Pontnewydd Cave. A Lower Palaeolithic Hominid Site in Wales. The first report.

National Museum of Wales Quaternary Studies Monograph 1.

47. Gruhn, R., Bryan, A.L., Moss, A.J., 1974. A contribution to Pleistocene chronology of south-east Essex, England. Quaternary Research 4, 53–71.

48. Gru¨n, R., Schwarcz, H.P., 2000. Revised open system U-series/ESR age calculations for teeth from Stratum C at the Hoxnian interglacial type locality, England. Quaternary Science Reviews 19, 1151–1154.

49. Hansen, M., 1987. The Hydrophiloidea (Coleoptera) of Fennoscandia and Denmark. In: Fauna Entomological Scandinavica 18. E.J. Brill, Leiden, pp. 1–254.

50. Heinrich, W.D., 1982. Zur Evolution und Biostratigraphie von Arvicola (Rodentia, Mammalia) im Pleistoza¨n Europas. Zeitschrift fu¨ r Geologische Wissenschaften 10, 683–735.

51. Heinrich, W.D., 1987. Neue Ergebnisse zur Evolution und Biostratigraphie von Arvicola (Rodentia, Mammalia) im Quarta¨ r Europas. Zeitschrift fu¨ r Geologische Wissenschaften 15, 389–406.

52. Hickin, N.E., 1963. The Insect Factor in Wood Decay. Hutchinson,, London, 336 pp.

53. Hinton, M.A.C., 1911. The British fossil shrews. Geological Magazine 8, 529–539.

54. Holland, D.G., 1972. A Key to the Larvae, Pupae and Adults of the British Species of Elminthidae. In: Freshwater Biological Association Scientific Publication No. 26. The Ferry House, Ambleside, pp. 1–58.

55. Holman, J.A., 1993. British Quaternary herpetofaunas: a history of adaptations to Pleistocene disruptions. British Herpetological Society Bulletin 50, 37–39.

56. Holman, J.A., Stuart, A.J., Clayden, J.D., 1990. A Middle Pleistocene herpetofauna from Cudmore Grove, Essex, England, and its paleogeographic and paleoclimatic implications. Journal of Vertebrate Paleontology 10, 86–94.

57. Horton, A., Keen, D.H., Field, M.H., Robinson, J.E., Coope, G.R., Currant, A.P., Graham, D.K., Green, C.P., Phillips, L.M., 1992. The Hoxnian interglacial deposits at Woodston, Peterborough. Philosophical Transactions of the Royal Society of London B 338, 131–164.

58. Hustedt, F., 1957. Die Diatomeenflora des Fluss-systems der Weser im Gebeit der Hansestadt Bremen. Abhandlungen Naturwissenschaftlichen Verein zu Bremen 34, 181–440.

59. Kaufman, D.S., Manley, W.F., 1998. A new procedure for determining DL amino acid ratios in fossils using reverse phase liquid chromatography. Quaternary Science Reviews 17, 987–1000.

60. Kelly, M.R., 1964. The Middle Pleistocene of North Birmingham. Philosophical Transactions of the Royal Society of London B 247, 533–592.

61. Kerney, M.P., 1971. Interglacial deposits in Barnfield pit, Swanscombe, and their molluscan fauna. Journal of the Geological Society of London 127, 69–93.

62. Koch, K., 1989. Die Ka¨fer Mitteleuropas: O¨ kologie. 1. Goecke & Evers, Krefeld, pp. 1–440.

63. Koch, K., 1992. Die Ka¨fer Mitteleuropas: O¨ kologie. 3. Goecke & Evers, Krefeld, pp. 1–389.

64. von Koenigswald, W., van Kolfschoten, T., 1996. The Mimomys–Arvicola boundary and the enamel thickness quotient (SDQ) of Arvicola as stratigraphic markers in the Middle Pleistocene. In: Turner, C. (Ed.), The Early Middle Pleistocene in Europe. Balkema, Rotterdam, pp. 211–226.

65. van Kolfschoten, T., 1990. The evolution of the mammal fauna in the Netherlands and the middle Rhine area (Western Germany) during the late Middle Pleistocene. Mededelingen Rijks Geologische Dienst 43 (3), 1–69.

66. Lake, R.D., Ellison, R.A., Hollyer, S.E., Simmons, M., 1977. Buried channel deposits in south-east Essex area: their bearing on Pleistocene palaeogeography. Report 77/ 21. Institute of Geological Sciences, 13 pp.

67. Lewis, S.G., 1999. Eastern England (Chapter 2). In: Bowen, D.Q. (Ed.), A Revised Correlation of Quaternary Deposits in the British Isles. Geological Society Special Report No. 23, pp. 10–27.

68. Lindroth, C.H., 1943. Oodes gracilis Villa. Eine thermophile Carabidae Schwedens. Notulae Entomologia 22, Helsingfors.

69. Lindroth, C.H., 1986. The Carabidae (Coleoptera) of Fennoscandia and Denmark (2). In: Fauna Entomologica Scandinavica 15. E.J. Brill, Leiden, pp. 223–497.

70. Lindroth, C.H., 1992. Ground Beetles (Carabidae) of Fennoscandia: a Zoogeographical Study. Part 1. Intercept, Andover, UK, pp. 1–630 (English translation of 1945 German original text).

71. Lucht, W.H., 1987. Die Ka¨fer Mitteleuropas: Katalog. Goecke & Evers, Krefeld, pp. 1–342.

72. Meijer, T., Preece, R.C., 2000. A review of the occurrence of Corbicula in the Pleistocene of North-west Europe. Netherlands Journal of Geosciences 79, 241–255.

73. Meisch, C., 2000. Freshwater Ostracoda of Western and Central Europe. In: Schwoerbel, J., Zwick, P. (Eds.), Süßwasserfauna von Mitteleuropa, Band 8/3. Spektrum Akademischer Verlag, Heidelberg and Berlin, 522 pp.

74. Miller, G.H., Hollin, J.T., Andrews, J.T., 1979. Aminostratigraphy of UK Pleistocene deposits. Nature 281, 539–543.

75. Mitlehner, A.G., 1992. Palaeoenvironments of the Hoxnian Nar Valley Clay, Norfolk, England: evidence from an integrated study of diatoms and ostracods. Journal of Quaternary Science 7, 335–341.

76. Murton, J.B., Baker, A., Bowen, D.Q., Caseldine, C.J., Coope, G.R., Currant, A.P., Evans, J.G., Field, M.H., Green, C.P., Hatton, J., Ito, M., Jones, R.L., Keen, D.H., Kerney, M.P., McEwan, R., McGregor, D.F.M., Parish, D., Robinson, J.E., Schreve, D.C., Smart, P.L., 2001. A late Middle Pleistocene temperate–periglacialtemperate sequence (Oxygen Isotope Stages 7–5e) near Marsworth, Buckinghamshire, UK. Quaternary Science Reviews 20, 1787–1825.

77. Nitychhoruk, J., Bin′ ka, K., Hoefs, J., Ruppert, H., Schneider, J., 2005. Climate reconstruction for the Holsteinian interglacial in eastern Poland and its

comparison with isotopic data from Marine Isotope Stage 11. Quaternary Science Reviews 24, 631–644.

78. Nitychhoruk, J., Bin´ ka, K., Ruppert, H., Schneider, J., 2006. Holsteinian interglacial ¼ Marine Isotope Stage 11? Quaternary Science Reviews 25, 2678–2681.

79. Oakley, K.P., Leakey, M., 1937. Report on excavations at Jaywick Sands, Essex (1934), with some observations on the Clactonian industry and on the fauna and geological significance of the Clacton Channel. Proceedings of the Prehistoric Society 3, 217–260.

80. Parfitt, S., 1998. The interglacial mammalian fauna from Barnham. In: Ashton, N., Lewis, S.G., Parfitt, S. (Eds.), Excavations at the Lower Palaeolithic Site at East Farm, Barnham, Suffolk, 1989–94. Occasional Paper of the British Museum 125, pp. 111–147.

81. Penkman, K.E.H., 2005. Amino Acid Geochronology: a Closed System Approach to Test and Refine the UK Model. Unpublished Ph.D. thesis, University of Newcastle.

82. Penkman, K.E.H., Preece, R.C., Keen, D.H., Maddy, D., Schreve, D.C., Collins, M.J., 2007a. Testing the aminostratigraphy of fluvial archives: the evidence from intra-crystalline proteins within freshwater shells. Quaternary Science Reviews 26, 2958–2969.

83. Penkman, K.E.H., McGrory, S., Collins, M.J., 2007b. Amino acid racemization analysis: Medway project. English Heritage Research Department Report Series 73/2007.

84. Penkman, K.E.H., Kaufman, D.S., Maddy, D., Collins, M.J., 2008a. Closed-system behaviour of the intra-crystalline fraction of amino acids in mollusc shells. Quaternary Geochronology 3, 2–25.

85. Penkman, K., Preece, R.C., Keen, D.H., Collins, M., 2008b. British aggregates: an improved chronology using amino acid racemization and degradation of intracrystalline amino acids (IcPD). English Heritage Research Department Report Series 6/2008.

86. Penkman, K.E.H., Preece, R.C., Keen, D.H., Collins, M.J. Amino acid geochronology of the type Cromerian of West Runton, Norfolk, UK. Quaternary International, in press.

87. Phillips, L., 1976. Pleistocene vegetational history and geology in Norfolk. Philosophical Transactions of the Royal Society of London B 275, 215–286.

88. Pike, K., Godwin, H., 1953. The interglacial at Clacton-on-Sea. Quarterly Journal of the Geological Society of London 108, 11–22.

89. Preece, R.C., 1995. Mollusca from interglacial sediments at three critical sites in the Lower Thames. In: Bridgland, D.R., Allen, P., Haggart, B.A. (Eds.), The Quaternary of the Lower Reaches of the Thames. Field Guide. Quaternary Research Association, pp. 53–60.

90. Preece, R.C., 1999. Mollusca from last interglacial fluvial deposits of the River Thames at Trafalgar Square, London. Journal of Quaternary Science 14, 77–89.

91. Preece, R.C., Penkman, K.E.H., 2005. New faunal analyses and amino acid dating of the Lower Palaeolithic site at East Farm, Barnham, Suffolk. Proceedings of the Geologists' Association 116, 363–377.

92. Preece, R.C., Scourse, J.D., Houghton, S.D., Knudsen, K.L., Penney, D.N., 1990. The Pleistocene sea-level and neotectonic history of the eastern Solent, southern England. Philosophical Transactions of the Royal Society of London B 328, 425–477.

93. Preece, R.C., Parfitt, S.A., Bridgland, D.R., Lewis, S.G., Rowe, P.J., Atkinson, T.C., Candy, I., Debenham, N.C., Penkman, K.E.H., Rhodes, E.J., Schwenninger, J.-L., Griffiths, H.I., Whittaker, J.E., Gleed-Owen, C., 2007. Terrestrial environments during MIS 11: evidence from the Palaeolithic site at West Stow, Suffolk, UK. Quaternary Science Reviews 26, 1236–1300.

94. Reille, M., 1990. Relations pluie Pollinique-Vegetation, leur interet en Analyse Pollinique. L'eçons de Palynologie et d'Analyse Pollinique. E´ditions du Centre National de la Recherche Scientifique, Paris, pp. 45–53 (Chapitre C).

95. Reille, M., de Beaulieu, J.L., Svobodova, H., Andrieu-Ponel, V., Coeury, C., 2000. Pollen analytical biostratigraphy of the last five climatic cycles from a long continental sequence from the Velay region (Massif Central, France). Journal of Quaternary Science 15, 665–685.

96. Roe, H.M., 1994. Pleistocene Buried Channels in Eastern Essex. Unpublished Ph.D. thesis, University of Cambridge.

97. Roe, H.M., 1999. Late Middle Pleistocene sea-level change in the southern North Sea: the record from eastern Essex, UK. Quaternary International 55, 115–128.

98. Roe, H.M., 2001. The late Middle Pleistocene biostratigraphy of the Thames Valley, England: new data from eastern Essex. Quaternary Science Reviews 20,1603–1619.

99. Roe, H.M., Preece, R.C., 1995. A new discovery of the Middle Pleistocene 'Rhenish' fauna in Essex. Journal of Conchology 35, 272–273.

100. Roe, H.M., van de Plassche, O., 2005. Modern pollen distribution in a Connecticut salt marsh: implications for studies of sea-level change. Quaternary Science Reviews 24, 2030–2049.

101. Roucoux, K.H., Tzedakis, P.C., Frogley, M.R., Lawson, I.T., Preece, R.C., 2008. Vegetation history of the Marine Isotope Stage 7 interglacial complex at Ionnina, NW Greece. Quaternary Science Reviews 27, 1378–1395.

102. Rousseau, D.-D., 2003. The continental record of stage 11: a review. Earth's climate and orbital eccentricity: the Marine Isotope Stage 11 question. Geophysical Monograph Series 137, 213–222.

103. Rowe, P.J., Richards, D.A., Atkinson, T.C., Bottrell, S.H., Cliff, R.A., 1997. Geochemistry and radiometric dating of a Middle Pleistocene peat. Geochimica et Cosmochimica Acta 61, 4201–4211.

104. Rowe, P.J., Atkinson, T.C., Turner, C., 1999. U-series dating of Hoxnian interglacial deposits at Marks Tey, Essex, England. Journal of Quaternary Science 14, 693–702.

105. Sarntheim, M., Stremme, H.E., Mangini, A., 1986. The Holstein interglaciation: timestratigraphic position and

correlations to the stable-isotope stratigraphy of deep-sea sediments. Quaternary Research 29, 75–79.

106. Schreve, D.C., 1996. The mammalian fauna from the Waechter excavations, Barnfield Pit, Swanscombe. In: Conway, B., McNabb, J., Ashton, N. (Eds.), Excavations at Barnfield Pit, Swanscombe, 1968–1972. Occasional Paper of the British Museum 94, pp. 149–162.
107. Schreve, D.C., 1997. Mammalian Biostratigraphy of the Later Middle Pleistocene in Britain. Unpublished Ph.D. thesis, University of London.
108. Schreve, D.C., 2001. Differentiation of the British late Middle Pleistocene interglacials: the evidence from mammalian biostratigraphy. Quaternary Science Reviews 20, 1693–1705.
109. Schreve, D.C., Bridgland, D.R., Allen, P., Blackford, J.J., Gleed-Owen, C.P., Griffiths, H.I., Keen, D.H., White, M.J., 2002. Sedimentology, palaeontology and archaeology of late Middle Pleistocene River Thames terrace deposits at Purfleet, Essex, UK. Quaternary Science Reviews 21, 1423–1464.
110. Scourse, J.D., 2006. Comment on: numerical 230Th/U dating and a palynological review of the Holsteinian/ Hoxnian interglacial by Geyh and Mu¨ller. Quaternary Science Reviews 25, 3070–3071.
111. Scourse, J.D., Ansari, M.H., Wingfield, R.T.R., Harland, R., Balson, P.S., 1998. A Middle Pleistocene shallow marine interglacial sequence, Inner Silver Pit, Southern North Sea: pollen and dinoflagellate cyst stratigraphy and sea-level history. Quaternary Science Reviews 17, 871–900.
112. Scourse, J.D., Austin, W.E.N., Sejrup, H.P., Ansari, M.H., 1999. Foraminiferal isoleucine epimerization determinations from the Nar Valley Clay, Norfolk, UK: implications for Quaternary correlations in the Southern North Sea basin. Geological Magazine 136, 543–560.
113. Sibrava, V., 1986. Correlations of European glaciations and their relation to the deep sea record. Quaternary Science Reviews 5, 433–442.
114. Singer, R., Wymer, J.J., Gladfelter, B.G., Wolff, R.G., 1973. Excavation of the Clactonian industry at the golf course, Clacton-on-Sea, Essex. Proceedings of the Prehistoric Society 39, 6–74.
115. Stabell, B., 1985. The development and succession of taxa within the diatom genus Fragilaria Lyngbye as a response to basin isolation from the sea. Boreas 14, 273–286.
116. Stachura, K., Witkowski, A., 1997. Response of the Gulf of Gdansk diatom flora to the sewage run-off from the Vistula River. Fragmenta Floristica et Geobotanica 42, 517–545.
117. Stevens, L.A., 1960. The interglacial of the Nar Valley. Quarterly Journal of the Geological Society of London 115, 291–316.
118. Stuart, A.J., 1982. Pleistocene Vertebrates in the British Isles. Longman, London and New York.
119. Stuart, A.J., Wolff, R.G., Lister, A.M., Singer, R., Egginton, J., 1993. Fossil vertebrates. In: Singer, R., Gladfelter, B.G., Wymer, J.J. (Eds.), The Lower Paleolithic Site at Hoxne, England. University of Chicago Press, pp. 163–206.
120. Sutcliffe, A.J., 1960. Joint Mintor Cave, Buckfastleigh. In: Transactions and Proceedings of the Torquay Natural History Society 13. pp. 1–26.
121. Sutcliffe, A.J., 1964. The mammalian fauna. In: Ovey, C.D. (Ed.), The Swanscombe Skull: a Survey of Research on a Pleistocene Site. Occasional Paper No. 20. Royal Anthropological Institute, pp. 85–111.
122. Sutcliffe, A.J., 1995. Insularity of the British Isles 250,000–30,000 years ago: the mammalian, including human, evidence. In: Preece, R.C. (Ed.), Island Britain: a Quaternary Perspective. Geological Society Special Publication 96, pp. 127–140.
123. Sutcliffe, A.J., Kowalski, K., 1976. Pleistocene rodents of the British Isles. Bulletin of the British Museum (Natural History) Geology 27 (2).
124. Sykes, G.A., Collins, M.J., Walton, D.I., 1995. The significance of a geochemically isolated intracrystalline organic fraction within biominerals. Organic Geochemistry 23, 1039–1065.
125. Thomas, G.N., 2001. Late Middle Pleistocene pollen biostratigraphy in Britain: pitfalls and possibilities in the separation of interglacial sequences. Quaternary Science Reviews 20, 1621–1630.
126. Thompson, R.T., 1958. Coleoptera, Phalacridae. part 5b. In: Handbooks for the Identification of British Insects, vol. 5. Royal Entomological Society of London, pp. 1–16.
127. Turner, C., 1968. A note on the occurrence of Vitis and other new plant records from the Pleistocene deposits at Hoxne, Suffolk. New Phytologist 67, 333–334.
128. Turner, C., 1970. The Middle Pleistocene deposits at Marks Tey, Essex. Philosophical Transactions of the Royal Society of London B 257, 373–437.
129. Turner, C., Kerney, M.P., 1971. A note on the age of the freshwater beds of the Clacton Channel. Journal of the Geological Society of London 127, 87–93.
130. Turner, C., West, R.G., 1968. The subdivision and zonation of interglacial periods. Eiszeitalter und Gegenwart 19, 93–101.
131. Ventris, P.A., 1986. The Nar Valley. In: West, R.G., Whiteman, C.A. (Eds.), The Nar Valley and North Norfolk. Field Guide. Quaternary Research Association, Coventry, pp. 6–55.
132. Ventris, P.A., 1996. Hoxnian interglacial freshwater and marine deposits in northwest Norfolk, England and their implications for sea-level reconstruction. Quaternary Science Reviews 15, 437–450.
133. Vos, P.C., de Wolf, H., 1993. Reconstruction of sedimentary environments in Holocene coastal deposits of the southwest Netherlands; the Poortvliet boring, as a case study of palaeoenvironmental diatom research. Hydrobiologia 269/270, 297–306.
134. Warren, S.H., 1917. The study of pre-history in Essex, as recorded in the publications of the Essex Field Club. Essex Naturalist 18, 145–152.
135. Warren, S.H., 1933. The Palaeolithic industries of the

Clacton and Dovercourt districts. Essex Naturalist 24, 1–29.

136. Warren, S.H., 1955. The Clacton (Essex) Channel deposits. Quarterly Journal of the Geological Society of London 111, 283–307.

137. van der Werff, A., Huls, H., 1957–1974. Reprinted 1976. Diatomee¨nflora van Nederland, vols. 1–10. Otto Koeltz Science Publishers, Koenigstein.

138. West, R.G., 1956. The Quaternary deposits at Hoxne, Suffolk. Philosophical Transactions of the Royal Society of London B 239, 265–356.

139. West, R.G., 1969. Pollen analyses from interglacial deposits at Aveley and Grays, Essex. Proceedings of the Geologists' Association 74, 147–186.

140. West, R.G., 1980. Pleistocene forest history in East Anglia. New Phytologist 85, 571–622.

141. West, R.G., Lambert, C.A., Sparks, B.W., 1964. Interglacial deposits at Ilford, Essex. Philosophical Transactions of the Royal Society of London B 247, 185–212.

142. de Wolf, H., Cleveringa, P., 1994. Eemian Diatom Floras in the Amsterdam Glacial Basin. Report 582. Rijks Geologische Dienst, Haarlem, 19 pp.

CHAPTER 4

HIGH RESOLUTION LATEGLACIAL AND EARLY-HOLOCENE SUMMER AIR TEMPERATURE

Stephen J. Brooks, Ian P. Matthews, Hilary H. Birks, H.John B. Birks

Keywords: Lake sediments, Tephra chronology, Tephra age-depth model, Climate change, Chironomid-inferred temperatures

INTRODUCTION

Climatic changes during the last glacial–interglacial transition (LGIT) between approximately 16–8 ka BP are of considerable interest to climate scientists because they provide insights into atmosphere-ocean-terrestrial climate linkages during a period of rapid, high amplitude climate change. Changes in the ratio of stable oxygen isotopes from Greenland ice-cores (e.g. Johnsen et al., 1992; Rasmussen et al., 2006; Svensson et al., 2008) provide high resolution records of temperature trends but records from elsewhere are required to determine independently the magnitude and timing of climatic trends at other locations, and to provide a framework for understanding the terrestrial environmental responses to these forcing factors. The climate of the British Isles is strongly influenced by the North Atlantic current and so is especially sensitive to changes in ocean circulation. Quantitative climate records for the LGIT from the British Isles have mostly been inferred from beetle sequences (e.g. Coope et al., 1998) but these are not as highly resolved temporally as the ice-core records, because (i) large quantities of sediment, which may have been laid down over many decades, are required to acquire sufficient numbers of beetle remains; (ii) the inferred temperatures are prone to large range-estimates and (iii) beetle records frequently lack precise and accurate independently derived chronologies. For these reasons most are not suitable to confirm whether or not the detailed climatic trends apparent in the ice-core record also occur in Britain. However, low error, quantitative climate records can be inferred at high temporal resolution from chironomid remains, which are preserved in large numbers as larval head capsules in lake sediments. At present, the only published Lateglacial to early-Holocene chironomid-inferred temperature records from Britain and Ireland are from Whitrig Bog, south-east Scotland (Brooks and Birks, 2001), Lough Nadourcan, western Ireland (Watson et al., 2010), Fiddaun in western Ireland (van Asch et al., 2012), and five sites in the English Lake District (Bedford et al., 2004; Lang et al., 2010). These records show general similarities between each other and with the ice-core records but there are differences in detail, which may reflect regional climatic differences or may be related to the low precision of the chronological data available to anchor and correlate each record temporally. In addition, lake size, depth and morphometry may influence the chironomid-based temperature reconstructions (Lang et al., 2010). Thus, more high resolution records with well-founded, independently derived, chronologies are required to quantify the magnitude, timing and rate of climate change in the British Isles during LGIT and establish the extent of regional and local differences in these trends and in other records across Europe.

We use chronology and temperature trends derived from our sites to compare with the Greenland ice-core eventstratigraphy proposed by Lowe et al. (2008). Radiocarbon chronologies frequently have large uncertainties which make correlation difficult with the climatostratigraphic units defined for Scandinavia (Mangerud et al., 1974) and for Greenland (Rasmussen et al., 2006). Here, for convenience, we adopt a simplified LGIT climatostratigraphic terminology of Lateglacial Interstadial-Younger Dryas-Holocene as these are widely used terms in the literature relating to well-defined climatic events. This scheme is only used for the broad

Citation: tephen J. Brooks, Ian P. Matthews, Hilary H. Birks, H.John B. Birks, High resolution Lateglacial and early-Holocene summer air temperature, Quaternary Science Reviews, Volume 41, 18 May 2012, Pages 67-82, ISSN 0277-3791, http://dx.doi.org/10.1016/j.quascirev.2012.03.007.

sub-divisions of stratigraphical units and not to infer any precise chronological associations to each unit. Ultimately, comparisons are made to the Greenland event-stratigraphy outlined in Lowe et al. (2008).

There have been many investigations into vegetational change in response to climate change in the Lateglacial and early-Holocene in Britain, and regional trends are now well-established (Walker et al., 1994). Certain localities having detailed pollen sequences have been nominated as reference sites (Walker, 1993). They include Abernethy Forest near the Cairngorm mountains in north-east Scotland (Birks and Mathewes, 1978), and Loch Ashik on the Isle of Skye in north-west Scotland (Walker and Lowe, 1990, 1991). These records provide evidence of rapid vegetational changes to climate warming at the beginning of the Lateglacial Interstadial (equivalent to the Bølling-Allerød interstadial in mainland Europe and GI-1 in Greenland), to cooling during the Younger Dryas (GS-1), and finally to rapid warming at the start of the Holocene. Less severe cooling events during the Lateglacial Interstadial may also be suggested by vegetation responses. However, these important pollen sequences cannot be placed in an independently derived climatic context until quantitative, high resolution climate reconstructions with well-founded and precise chronologies are available.

This is the focus of the present study. We provide high temporal resolution chironomid-inferred mean July air temperature records for Loch Ashik and Abernethy Forest and compare them against other high resolution temperature records from Britain, mainland north-west Europe and Greenland to determine regional and local trends. We briefly compare the chironomid-inferred estimates with summer temperature estimates derived from beetles and also with changes in pollen assemblages. The addition of these detailed chironomid sequences and their accompanying mean July air temperature estimates to the pollen records from these two Scottish sites considerably enhances their value as reference sites for the LGIT.

STUDY SITES

Loch Ashik (57°14′32.73″N; 05°49′39.89″W) (Fig. 1) is located in the south-east of the Isle of Skye, western Scotland, about 5 km east of Broadford. Mean July air temperature at Kinlochewe (57°36′016.2″N, 5°19′59″W; 145 m asl) is 14.4 °C (Meteorological Office data for 1971–2000 [www.metoffice.gov.uk]). The loch lies on Torridonian sandstone, which is slightly calcareous in parts. The lake is about 150 m long by about 130 m wide and is situated at about 50 m asl. Streams enter at both the western and eastern ends and there is an out-flowing stream at the north end. Our core was taken from below soft peaty ground at the in-filled western end of the loch near the water's edge, close to the site from where Walker and Lowe (1990) described a LGIT pollen sequence and from where the Vedde Ash was described by Davies et al. (2004). Lake sediments corresponding to the LGIT were situated below peat at a core depth of 500–675 cm (Table 1).

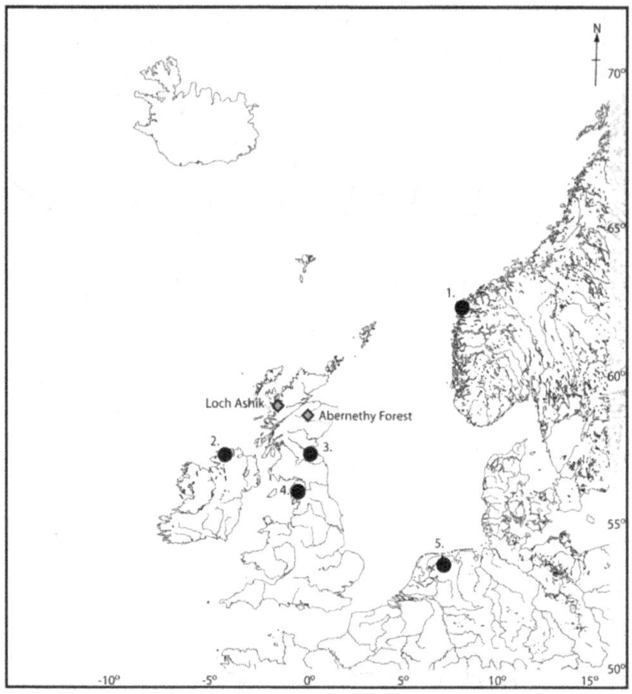

Figure 1: Location of Loch Ashik and Abernethy Forest and other sites referred to in the text (1: Kråkenes; 2: Lough Nadourcan; 3: Whitrig Bog; 4: Hawes Water; 5: Hijkermeer).

Table 1: Lithostratigraphical description of the sediment sequence recovered from Loch Ashik

Depth (cm from mire surface)	Composition
500–569	Dark brown gyttja with coarse detritus at top getting gradually finer
569	Sharp sloping boundary to clay
569–578	Variable pinkish brown or pale grey clay with dark flecks
578.0–578.5	Black Vedde Ash
578.5–587	Variable pinkish brown or pale grey clay with dark flecks
587	Sharp boundary between clay above and gyttja below
587–600	Olive-brown firm fine-detritus gyttja with scattered dark flecks of moss, about 25% silt
600–625	Olive-brown firm fine-detritus gyttja with scattered dark flecks of moss, about 50% silt
625–635	Grey firm silty gyttja with moss inclusions
635–637	Mid-brown silty gyttja, moss absent
637–648.5	Yellow-brown firm silty gyttja, becoming greyer with depth. Moss infrequent
648.5–651.5	Pinkish firm clay-gyttja. Few moss flecks
651.5–672	Finely laminated brownish grey hard clay-silt with pink and black layers. Layers more distinct below pink band at 665 cm.

Abernethy Forest (57°14′13.08″N; 03°42′25.65″W) (Fig. 1) is in the Cairngorms region of north-east Scotland, about 10 km north-west of Aviemore. Mean July air temperature at Braemar (57°00′20.88″N, 3°23′53″W; 340 m asl) is 13.4 °C (Meteorological Office data for 1971–2000 [www.metoffice.gov.uk]). The coring site is in the in-filled south-western end of Loch Garten (about 1100 m long by about 670 m wide, altitude about 230 m asl) and is situated to the south of a channel linking the loch with Loch Mallachie. The base of the Lateglacial lake-sediment was situated at a core depth of about 700 cm. Our coring site is close to the sites used by Birks (1970) and Birks and Mathewes (1978) to describe LGIT pollen and plant macrofossil sequences.

MATERIAL AND METHODS

Coring

Six overlapping core sequences, less than 1 m apart, were taken at Loch Ashik using a 7.5-cm diameter square-rod Livingstone piston corer (Wright, 1967). The early-Holocene and Lateglacial sequence was covered in two drives in each core. The longest sequence (series VI) was 172 cm. The cores were correlated by lithological changes and loss-on-ignition analysis (LOI).

Three overlapping core sequences were taken from Abernethy Forest using the same corer. The LGIT sediments were recovered in three drives in each sequence. The longest sequence (285 cm) was obtained from core series III but core series II was used to provide overlap at the core junction between the second and third drives of series III (at 600 cm). The cores were correlated by sand layers and LOI analysis.

All the cores were sliced at 1 cm intervals. The cores were stored at 4 °C and Abernethy Forest macrofossil samples were stored at −20 °C.

Chronology

Fourteen radiocarbon AMS dates were obtained from Abernethy Forest based on terrestrial plant macrofossils. An age-depth model was constructed using a Bayesian-based *P_ Sequence* within the OxCal v4.1 program in combination with the IntCal09 calibration set (Reimer et al., 2009). The *P_ Sequence* is a Poisson–process depositional model which allows for changing depositional rates within a sediment sequence. This flexibility is achieved through altering the rigidity of the model via parameter '*k*'. High '*k*' values assume close to linear sedimentation while lower values allow greater flexibility. Additionally, these models permit the inclusion of depth and stratigraphical information within the model. The calculated likelihood of the models was checked via the OxCal agreement index using a value of 60% as a lower cut off (for a full explanation of the OxCal terminology and modelling procedures see Bronk Ramsey, 2008, 2009). More details of the final age model, which was used to make best-estimates of the ages of the tephras in the sediments, are presented in Matthews et al. (2011).

Too few plant macrofossils were found in the Loch Ashik sequence for radiocarbon dating. Bulk sediments are not suitable for dating because of the old carbon reservoir effect caused by carbonate in the sediments derived from the bedrock. These problems are common when attempting to derive chronologies for Lateglacial sequences in Britain. Therefore, the age-depth model for Loch Ashik is based solely on tephrochronology using the age-estimates derived from the Abernethy Forest age-depth model for the Borrobol and Penifiler tephras (Matthews et al., 2011).

For the tephrostratigraphical analysis, 0.5 cm^3 volumes from each 1 cm sample were sieved between 80 and 25 µm meshes to remove coarse sand particles (>80 µm) and obscuring silts and clays (<25 µm). The samples were then processed following Blockley et al. (2005). Shards were picked from samples containing shard maxima, mounted on resin stubs and polished for geochemical analysis by WDS–EPMA (Wavelength-Dispersive X-Ray Spectroscopy–Electron Probe Micro-Analyser). Tephra layers were detected in the same stratigraphical positions as those previously reported for Loch Ashik (Pyne-O›Donnell, 2011). Correlations between these layers and previously recognised ashes across Scotland were made by geochemical comparisons to published datasets (Pyne-O›Donnell, 2007). The tephra investigations at Abernethy are reported by Matthews et al. (2011) and the Loch Ashik tephra investigations are reported in Pyne-O›Donnell (2011). Besides providing a chronological framework for the Loch Ashik sequence, we use tephra layers here to provide stratigraphical and chronological links between the Loch Ashik and Abernethy Forest chironomid data and to allow direct comparison between Loch Ashik, Abernethy Forest and other regional climate archives where these tephras have been identified.

Loss-on-Ignition Analysis (LOI)

Following Heiri et al. (2001), LOI analysis was carried out on contiguous samples at 1 cm intervals on cores II (overlap) and III from Abernethy Forest and on core VI from Loch Ashik, which are the same cores used for chironomid analysis. LOI at 550 °C as a percentage of dry weight was used as a proxy for organic content of the sediments.

Chironomid Analysis

Altogether 83 samples were analysed from Loch Ashik, between core depths 500 and 652 cm, and 126 samples from Abernethy Forest, between core depths 410 and 691 cm. At both sites the cores were analysed initially at 4 cm intervals but then resolution was increased to 2 cm or 1 cm intervals at points in the sequence which showed critical changes in the chironomid assemblages or in the inferred temperatures.

Sample preparation for chironomid analysis followed Brooks et al. (2007) using warm 10% KOH and Euparal slide mountant. The head capsules were identified using a compound microscope at 100–400× magnification, with reference to Cranston (1982), Wiederholm (1983), Rieradevall and Brooks (2001), Brooks et al. (2007) and the chironomid collection housed at the Natural History Museum, London.

Data Analysis

The chironomid results are presented stratigraphically using TGVIEW v. 2.0.2 (Grimm, 2004). The stratigraphies were zoned by optimal partitioning using the program ZONE

v. 1.2 (Juggins, 1991) and statistically significant zones were identified by comparison with a broken stick model (Bennett, 1996) using the program BSTICK v. 1 (JM Line and HJB Birks, unpublished). CANOCO v. 4.5 (ter Braak and Šmilauer, 2002) was used for multivariate analyses. In all these analyses percentage abundance data were square-root transformed to stabilise variances and rare taxa were down-weighted. Detrended canonical correspondence analysis (DCCA), with detrending by segments, non-linear rescaling, and constrained by sample order was used to determine the amount of compositional turnover (Birks and Birks, 2008). Use of DCCA is essential as taxon turnover estimates based on detrended correspondence analysis (DCA) are not constrained to the stratigraphical sequence (Birks, 2007). Canonical correspondence analysis (CCA) was used to assess the environmental variables that may have influenced the response of the fossil chironomid assemblages (Birks et al., 1990; Velle et al., 2005). The fossil samples were plotted passively in CCA ordination space of a modern Norwegian calibration dataset (Brooks and Birks, 2000a; 2001, 2004; unpublished) which included the four environmental variables that had a statistically significant influence on the distribution and abundance of the modern chironomid taxa. These are mean July air temperature (which explains 48% of the total variance explained by the subset of four environmental variables), depth (14.0%), conductivity (7.6%) and alkalinity (7.6%). The CCA used Hill›s scaling and inter-sample distances to provide estimates of the turnover distances of the samples (Birks et al., 1990).

Temperature Reconstruction

Mean July air temperature estimates were derived from the square-root transformed fossil chironomid percentage abundance data using a modern Norwegian chironomid-based temperature calibration dataset expanded from the previously published 109-lake calibration set (Brooks and Birks, 2001). It consists of 157 Norwegian lakes spanning mean July air temperatures of 3.5–16 °C, latitudes from 80°N to 58°N, altitudes from 0 to 1600 m, and 142 chironomid taxa. After removing four outlier lakes from the dataset to improve the fit of the temperature model, the minimal adequate model with the best predictive power is a 2-component weighted-averaging partial least squares (WA-PLS) inverse regression model that has a root mean squared error of prediction (RMSEP) of 1.06 °C, a coefficient of determination (r^2_{jack}) of 0.91, and a maximum $bias_{jack}$ of 1.05 °C, all estimated by leave-one-out cross-validation. Sample specific errors of the inferred temperatures were estimated by 1000 bootstrap cycles using WA-PLS (Juggins and ter Braak, 2001). The Norwegian and North British chironomid faunas are similar, so modern Norwegian data can be used to reconstruct temperatures in the British Isles, especially as temperatures were colder in the LGIT than today.

Goodness-of-fit to temperature was evaluated by passively positioning the fossil samples on a CCA of the modern training set constrained solely against July temperature. Any fossil samples that had a squared residual distance value within the most extreme 10% of values in the modern training set were considered to have a poor fit-to-temperature. The modern analogue technique (MAT) was used to detect fossil samples that lacked good analogues in the modern calibration dataset using squared chord distance as a measure of dissimilarity. Samples with a dissimilarity larger than the 95% threshold in the modern data were considered as having no good analogues in the modern calibration dataset (Birks et al., 1990; Birks, 1995, 1998; Velle et al., 2005). The taxon percentage abundance data were untransformed prior to this analysis.

RESULTS

Chronology

Abernethy Forest

The age-depth model was based on 12 radiocarbon dates, after two dates were rejected (Fig. 2a) (Matthews et al., 2011). This model is generally well-constrained through the first part of the Lateglacial Interstadial where the uncertainties are comparable or better than the counting errors produced for NGRIP (Lowe et al., 2008). The uncertainties increase higher in the sequence. During the YD the uncertainties are as high as +/− 500 years for some samples, as a consequence of the scarcity of dateable material coupled with a varying deposition rate.

a

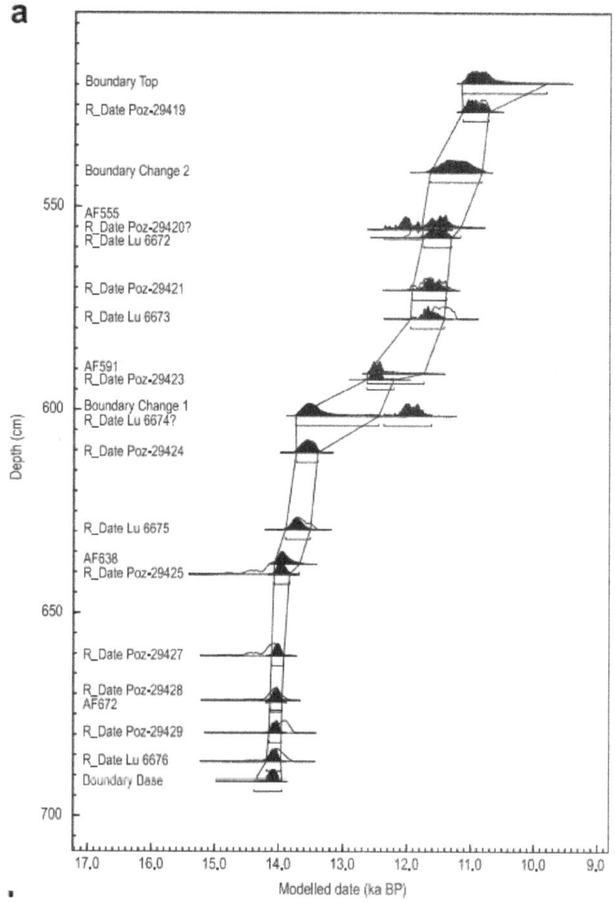

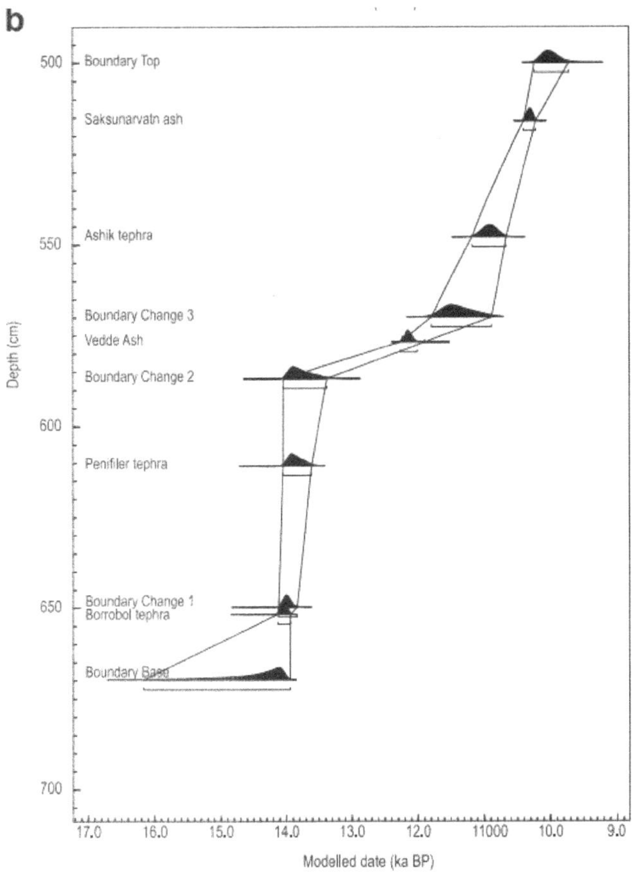

Figure 2: The *P_Sequence* age models used in this study: (a) Abernethy Forest; (b) Loch Ashik. The models were generated using OxCal v. 4.1 (Bronk Ramsey, 2009) and the IntCal09 calibration set (Reimer et al., 2009) (full details are reported in main text). The model displays highest probability density functions for the radiocarbon dates, major lithostratigraphic boundaries, and tephra age-estimates. Boundaries were placed where the major shifts in lithostratigraphy were identified (Bronk Ramsey, 2008). Both models have relatively small uncertainties within the early Lateglacial interstadial period. These chronological uncertainties are larger in the latter part of the Lateglacial interstadial and Younger Dryas, and remain so at Abernethy Forest during the early-Holocene. However, the presence of the Saksunarvatn ash at Loch Ashik means that the age model is better constrained during the early-Holocene.

Loch Ashik

No radiocarbon dates were obtained here, so an age-depth model was developed using the five ash layers (Fig. 2b) detected through the sequence. This is the first published age-depth model for the Lateglacial period in Northern Europe derived entirely from tephra dating. The Borrobol and Penefiler tephras were detected in Lateglacial Interstadial sediments at 652 cm and 610 cm, respectively, while the Vedde Ash was recognised as a visible tephra layer at 577 cm. In the early-Holocene sediments, tephra layers were detected at 548 cm and 515 cm and these were correlated with the Ashik and Saksunarvatn tephras. The ages of the Borrobol (14140–13950 cal yr BP) and Penifiler (14090–13650 cal yr BP) tephras were based on the estimated ages for these layers at Abernethy Forest (Matthews et al., 2011). The ages used for the Vedde Ash (12171 ± 114 GICC05 yr BP) and Saksunarvatn Ash (10347 ± 89 GICC05 yr BP), are those from the NGRIP ice-core (Rasmussen et al., 2007). The age range used for the Ashik tephra is relatively broad as no precise age estimate is currently available for this tephra. Its stratigraphical position, just above a change in sediment composition that has been correlated with the Pre-boreal Oscillation, suggests it post-dates this event and thus a relatively broad age of 11200–10700 calendar years BP was used here (Pyne-O›Donnell, 2007). This age determination could be significantly improved through radiometric dates on organic sediments associated with this layer elsewhere in Scotland. The ages of the tephra layers were used to construct an age model using the same approach as at Abernethy Forest. The *P_Sequence* age model includes a *k*-factor of 2 and produces an overall agreement index of 81.1%. The model produces generally well-constrained estimates through the first part of the Lateglacial Interstadial where the uncertainties are similar to Abernethy Forest and the total NGRIP counting errors (Lowe et al., 2008). The model is least well-constrained through the later parts of the Lateglacial Interstadial and in the YD, mostly due to the lack of dates.

LOI and Lithology

The detailed sediment lithologies are shown in Tables 1and 2 and summarised graphically in Fig. 3 and Fig. 4. LOI values are generally lower at Loch Ashik than Abernethy Forest (Fig. 7 and Fig. 8). In both sequences LOI is around 1% during the deglacial period, gradually rises throughout the Lateglacial Interstadial, and peaks at the end of the Lateglacial Interstadial at about 15% at Loch Ashik and 25% at Abernethy Forest. At both sites LOI abruptly falls at the beginning of the Younger Dryas (YD) and falls to below 5% at Loch Ashik in the late-YD or remains at between 5 and 10% throughout the YD and into the earliest Holocene at Abernethy Forest. In the early-Holocene there is a rapid rise in LOI to 25% at Loch Ashik and 35% at Abernethy Forest, and then more gradual rises, reaching 40% at Loch Ashik and 60% at Abernethy Forest by the top of each sequence.

Table 2: Lithostratigraphical description of the sediment sequence recovered from Abernethy Forest

Depth (cm from mire surface)	Composition
500–541	Coarse detritus gyttja with monocoyledonous rhizomes
541–543	Transition between silty and sandy gyttja and coarse detritus gyttja
543–544	Coarse silty and sandy gyttja
544–544.5	Sand layer
544–549	Coarse silty-sandy gyttja
549–549.5	Sand layer
549.5–570.5	Firm silty gyttja with thin fine-sand layers at 555, 556,558, 567, 568, 570.5 cm
570.5–600	Firm silty gyttja with indistinct darker bands of plant material. Lower boundary distinct
600–601	Pale silty band
601–608	Dark brown firm gyttja with silt
608–615	Medium brown firm gyttja
615–617.5	Dark brown gyttja
617.5–619.5	Paler gyttja
619.5–621	Dark brown gyttja
621–623	Firm brown gyttja with plant remains

623–623.5	Thin (0.5 cm) sand layer
623–632.5	Dark brown firm silty gyttja with plant remains
632.5–646	Medium brown firm silty gyttja
646–652.5	Dark brown silt gyttja
652.5–675	Complex of thin grey and cream silt, grey sand, brown gyttja layers
675–691	Grey silt

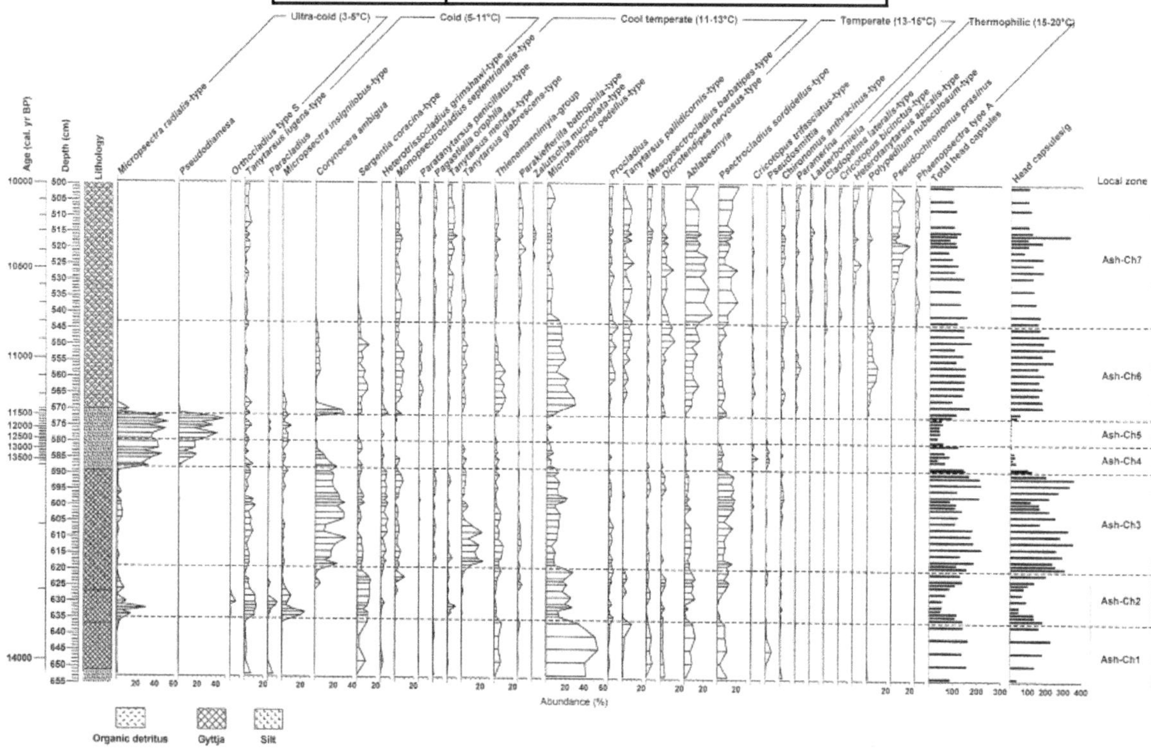

Figure 3: Relative abundances (percentages of the total) of the dominant chironomid taxa present during the Lateglacial and early-Holocene at Loch Ashik. Taxa are plotted left to right in order of their temperature optima in the modern Norwegian calibration set, coldest at left. The range of temperature optima that characterise each taxon group is shown.

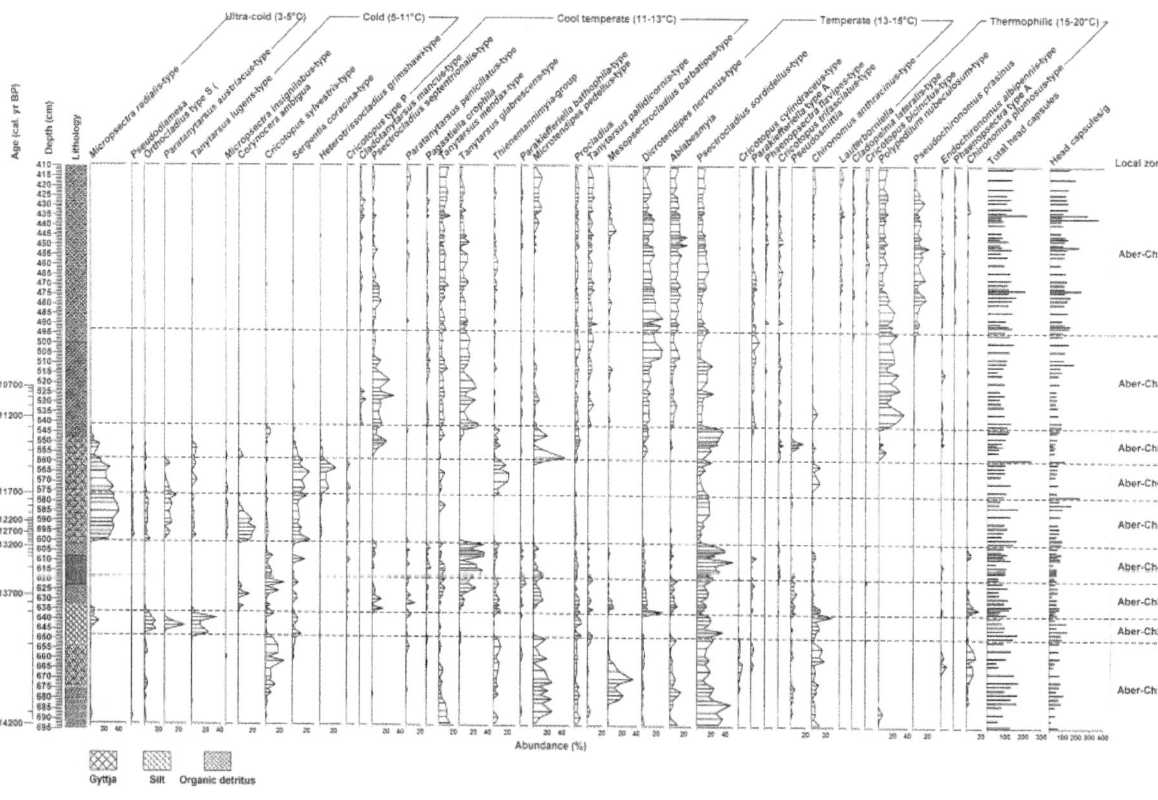

Figure 4: Relative abundances (percentages of the total) of the dominant chironomid taxa present during the Lateglacial and ear-

ly-Holocene at Abernethy Forest. Taxa are plotted left to right in order of their temperature optima in the modern Norwegian calibration set, coldest at left. The range of temperature optima that characterise each taxon group is shown.

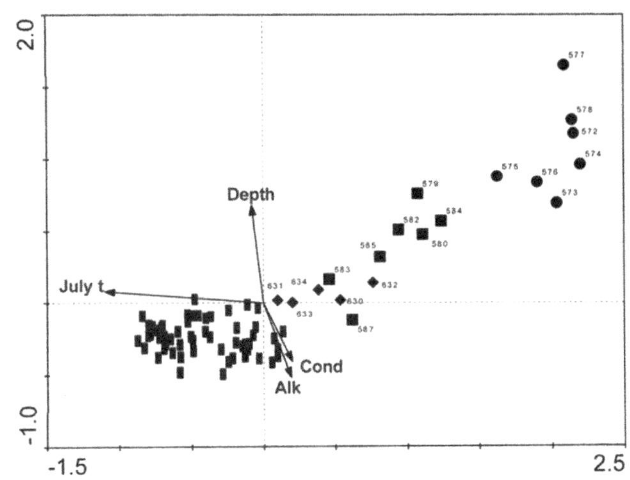

Figure 5: Fossil samples from Loch Ashik plotted passively in canonical correspondence analysis (CCA) space of the modern Norwegian calibration dataset. The four significant environmental variables explaining variance in the modern samples are shown as biplot arrows: July T (mean July air temperature), Depth (maximum lake water depth), Alk (alkalinity), Cond (conductivity). The number next to each plot denotes the sample depth (cm); diamonds represent samples during GI-1d, squares represent samples during early-YD, circles represent samples during late-YD, rectangles represent all other samples.

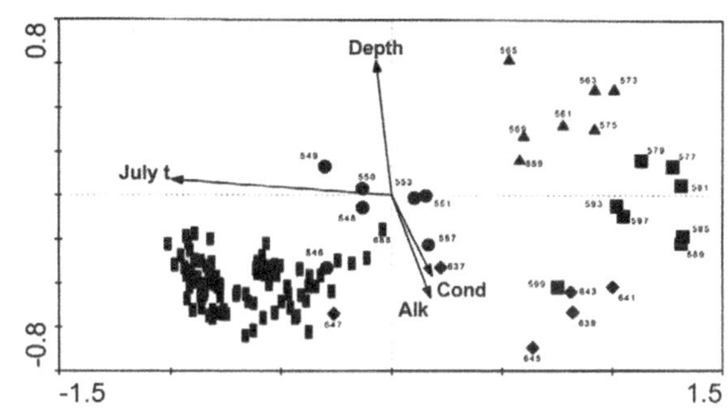

Figure 6: Fossil samples from Abernethy Forest plotted passively in canonical correspondence analysis (CCA) space of the modern Norwegian calibration dataset. The four significant environmental variables explaining variance in the modern samples are shown as biplot arrows: July T (mean July air temperature), Depth (maximum lake water depth), Alk (alkalinity), Cond (conductivity). The number next to each plot denotes the sample depth (cm); diamonds represent samples during GI-1d, squares represent samples during early-YD, triangles represent samples during the late-YD, circles represent samples during YD – Holocene transition, rectangles represent all other samples.

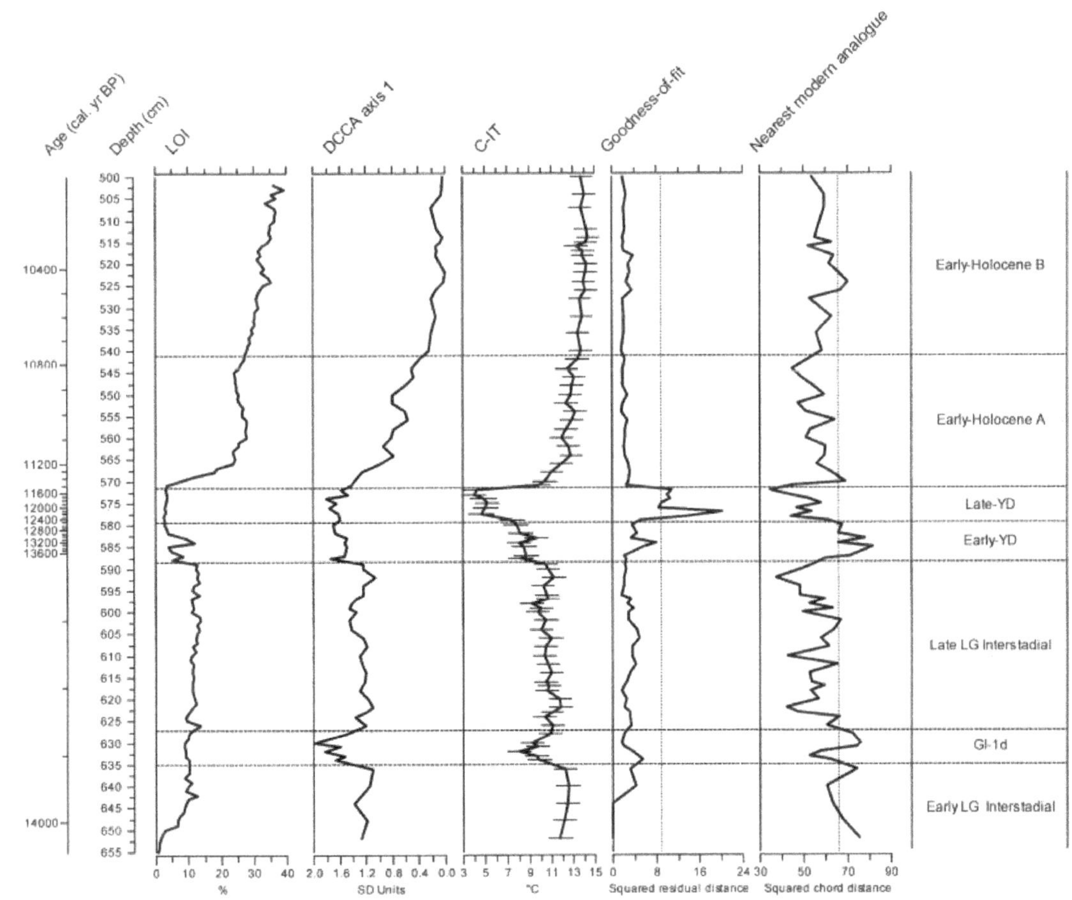

Figure 7: From left to right, Loch Ashik: organic matter content (as % loss-on-ignition (LOI) at 550 °C); detrended canonical correspondence analysis (DCCA) sample scores in which depth is the only constraining variable; chironomid-inferred temperature estimates (C-IT) with sample specific error bars; goodness-of-fit of the fossil assemblages to temperature, vertical dotted line indicates the 90th percentile of squared residual distances of modern samples to the first axis in a CCA, samples to the right of the line have a poor fit-to-temperature; nearest modern analogue analysis, vertical dotted line indicates the 5th percentile of squared chord distances of the fossil sample to samples in the modern calibration dataset, samples to right of line have no good modern analogue.

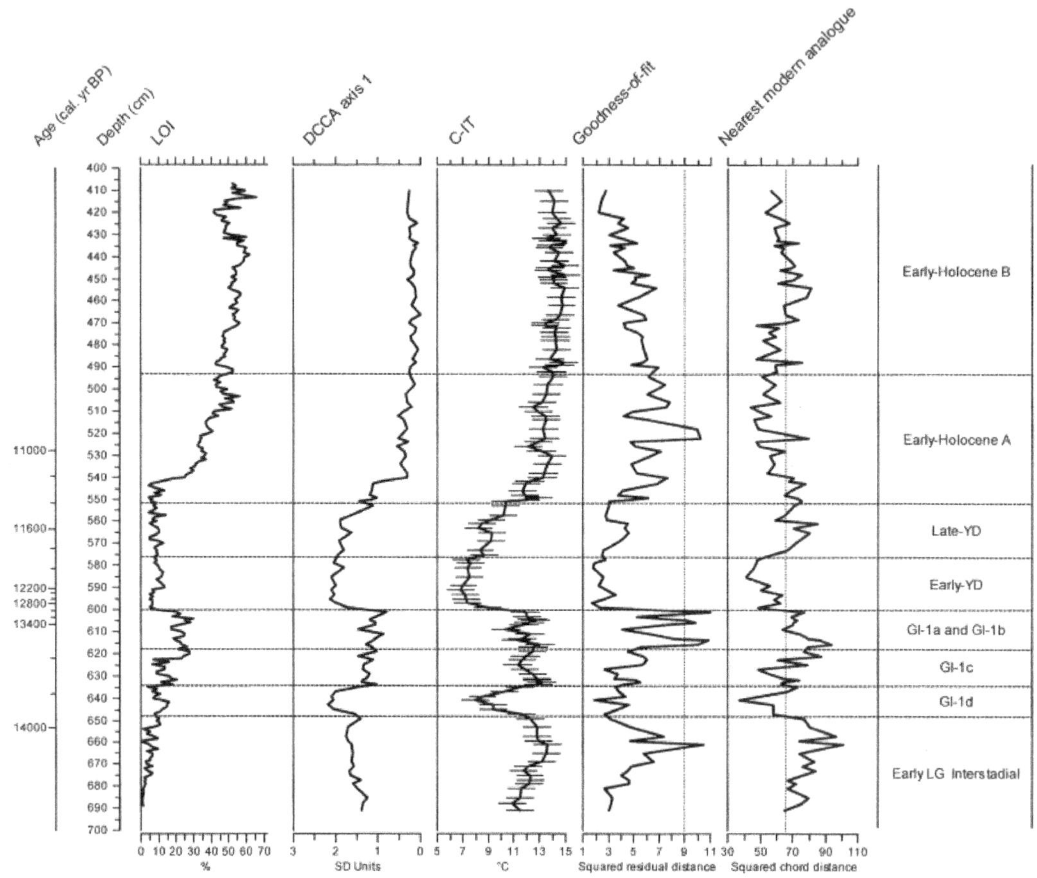

Figure 8: From left to right, Abernethy Forest: organic matter content (as % loss-on-ignition (LOI) at 550 °C); detrended canonical correspondence analysis (DCCA) sample scores in which depth is the only constraining variable; chironomid-inferred temperature estimates (C-IT) with sample specific error bars; goodness-of-fit of the fossil assemblages to temperature, vertical dotted line indicates the 90th percentile of squared residual distances of modern samples to the first axis in a CCA, samples to the right of the line have a poor fit-to-temperature; nearest modern analogue analysis, vertical dotted line indicates the 5th percentile of squared chord distances of the fossil sample to samples in the modern calibration dataset, samples to right of line have no good modern analogue.

Chironomid Assemblages

Loch Ashik

Altogether 83 chironomid taxa were identified to genus or species-morphotype. Seven significant assemblage zones were identified (Fig. 3).

The early Lateglacial Interstadial (zone Ash-Ch1) is dominated by cool-temperate taxa, in particular *Microtendipes pedellus*-type, which suggests that relatively warm, mesotrophic conditions prevailed even in the earliest samples. The presence of *Dicrotendipes nervosus*-type suggests there may have been aquatic macrophytes in the lake (Moller Pillot and Buskens, 1990; Brodersen et al., 2001). The appearance of several cold stenothermic taxa and the ultra-cold stenotherm *Micropsectra radialis*-type at around 13940 ± 250 cal yr BP (635 cm) (zone Ash-Ch2) suggests a response to climate climatic cooling at this time and the fall in head capsule concentrations suggests falling lake productivity, as chironomid biomass is linked with food availability (e.g.Tokeshi, 1986). This cold oscillation lasted about 90 years. By 13850 ± 380 cal yr BP (616 cm) the assemblage had become more diverse and included a wide range of cold and temperate taxa (zone Ash-Ch3). The high abundance of *Corynocera ambigua*suggests there may have been charophytes growing in the lake (Brodersen and Lindegaard, 1999).

In the early-YD (zone Ash-Ch4) there are large increases in ultra-cold stenotherms and declines in several temperate taxa, together with a fall in head capsule concentration, indicating rapidly cooling temperatures and decreasing lake productivity. The appearance of *Pseudosmittia*, which is typically found in terrestrial or semi-terrestrial habitats or in the splash-zone of lakes (Strenzke, 1950; Cranston, 1982), may be in response to a lowering of lake level and suggests a period of greater aridity. In the late-YD (zone Ash-Ch5) the assemblage is dominated by a few ultra-cold stenotherms and the head capsule concentration falls to its lowest in the entire sequence, suggesting extremely cold temperatures and low lake productivity.

In the earliest Holocene (zone Ash-Ch6 from 11400 ± 773 cal yr BP, 571 cm) rapidly rising temperature is suggested by a sudden change in the assemblage, which becomes more diverse and is dominated by cool-temperate taxa plus a few thermophilic taxa. The abundance of *D. nervosus* type and *Polypedilum nubeculosum*-type suggests aquatic macrophytes may have been present. Further declines in cold stenothermic and cool-temperate taxa, and the rise in temperate and thermophilic taxa after 10800 ± 475 cal yr BP (543 cm) (zone Ash-Ch7), suggests further gradual climate warming in the early-Holocene.

Abernethy Forest

Altogether 76 chironomid taxa were identified to genus or species-morphotype. Nine significant assemblage zones were identified (Fig. 4).

The early Lateglacial Interstadial (zone Aber-Ch1) is dominated by temperate and cool-temperate taxa. At about 13980 ± 120 cal yr BP (647.5 cm) several cold and ultra-cold stenotherms appear in the assemblage indicating a cold period (zone Aber-Ch2; GI-1d), which lasts until about 13870 ± 170 cal yr BP (635 cm) (110 years). Subsequently, a diverse assemblage of cool-temperate and temperate taxa dominate for the rest of the Lateglacial Interstadial (zones Aber-Ch3 and Aber-Ch4). The peak in the abundance of the cold stenotherm *Sergentia coracina*-type and the decline in the cool-temperate taxon *Tanytarsus glabrescens*-type at about 13550 cal yr BP ±200 (609 cm) may reflect a short period of cooler summers.

A response to rapid cooling at the start of the YD is evident at about 13000 ± 610 cal yr BP (600 cm) when ultra-cold and cold stenotherms dominate the assemblage (zone Aber-Ch5). In the late-YD, from about 11700 ± 270 cal yr BP (575 cm), rising temperatures are indicated by the increasing diversity of cold stenotherms (zone Aber-Ch6). Further warming during the YD – Holocene transition, after about 11550 ± 225 cal yr BP (557.5 cm), is indicated by declines in cold stenotherms and an increase in cool-temperate and temperate taxa (zone Aber-Ch7). The presence of *D. nervosus*-type and *P. nubeculosum*-type suggests that aquatic macrophytes were present in the lake. Plant macrofossil and pollen analyses (Birks and Mathewes, 1978) show that *Nitella* was abundant, *Ranunculus trichophyllus*-type was rare, and *Myriophyllum alterniflorum* and *Chara* were extremely rare at this time. These taxa, plus *Potamogeton filiformis*, were also present in the Allerød, so they did not become established in the YD. Although *D. nervosus*-type was present during the Allerød, *P. nubeculosum*-type was absent.

The complete replacement of cold stenothermic taxa by cool-temperate and temperate taxa, together with the sharp rise in the thermophilic taxon *P. nubeculosum*-type, occurs in the early-Holocene at about 11250 ± 400 cal yr BP (541 cm) (zone Aber-Ch8). Continuing rising temperatures in the early-Holocene are indicated by the appearance of several thermophilic taxa above core depth 493 cm, in particular *Pseudochironomus prasinus*.

Gradient Analysis of Fossil Chironomid Assemblages

At Loch Ashik the trajectory in the fossil sample scores is primarily along CCA axis 1 (Fig. 5) indicating the fossil assemblage was mostly responding to temperature change. However, during the late-YD, in particular, but also in the early-YD, there are shifts along CCA axis 2. Lake depth, conductivity and alkalinity are the major gradients on axis 2, together explaining 28% of the variation in the modern species data ($p = 0.002$). This suggests that changes in the composition of the chironomid fauna during the later part of the YD may in part be driven by an increase in lake depth or decreases in conductivity and alkalinity. DCCA (Fig. 7) shows the highest compositional turnover occurred in the GI-1d cold oscillation (635–619 cm: 0.89 standard deviation (SD) units) and during the earliest Holocene (571–541 cm: 1.02 SD units). The total taxon turnover through the whole sequence is 1.99 SD units.

The trajectory of assemblage change at Abernethy Forest is mostly along the CCA axis 1 temperature gradient but there are also some shifts along CCA axis 2 during the late-YD (Fig. 6). The overall compositional turnover (2.18 SD units) is greater than at Loch Ashik. The periods of highest assemblage turnover occurred in the GI-1d cold oscillation from 648 to 635 cm (1.15 SD units), during the transition between the Lateglacial Interstadial and start of the YD (601–595 cm: 1.35 SD units), and during the period of rising temperatures during the YD – Holocene transition (559–540 cm: 1.58 SD units) (Fig. 8).

Temperature Reconstructions

The LGIT chironomid-inferred mean July air temperature (C-IT) reconstructions for Loch Ashik and Abernethy Forest are shown in Fig. 7 and Fig. 8, respectively, and are compared with the NGRIP record in Fig. 9. At both sites, C-IT is around 12 °C at the base of the sequence, which is higher than expected and suggests that the basal sediments from the deglacial period were not found, or perhaps that the lakes were not open for colonisation before the rise in temperatures. In other published British and Irish records (Brooks and Birks, 2001; Lang et al., 2010; Watson et al., 2010; van Asch et al., 2012), temperatures similar to the YD have been inferred from chironomid assemblages early in the deglaciation phase. At both sites, early Lateglacial Interstadial C-IT increase to reach maxima of about 13.6 °C at Abernethy Forest and 12.5 °C at Loch Ashik. Both records then indicate a substantial cold oscillation as C-IT declines by 4.4 °C at Loch Ashik and 5.6 °C at Abernethy Forest. The age-depth model for Abernethy Forest gives a date of 13960 ± 120 cal yr BP (641 cm) for the coldest point in this oscillation, while at Loch Ashik this point occurs at 13930 ± 130 cal yr BP (633 cm). The ranges of both of these estimates fall within the range of the date of 14080 ± 169 GICC05 yr BP given for the GI-1d oscillation in the GRIP and NGRIP records (Björck et al., 1998; Lowe et al., 2008). This oscillation has been correlated with the Aegelsee Oscillation (Lotter et al., 1992), which has a reported age in the Hijkermeer (Netherlands) C-IT record of 14100–14000 cal yr BP (Heiri et al., 2007). After this cold oscillation, C-IT rises again to 13.2 °C at Abernethy Forest and to 11.9 °C at Loch Ashik.

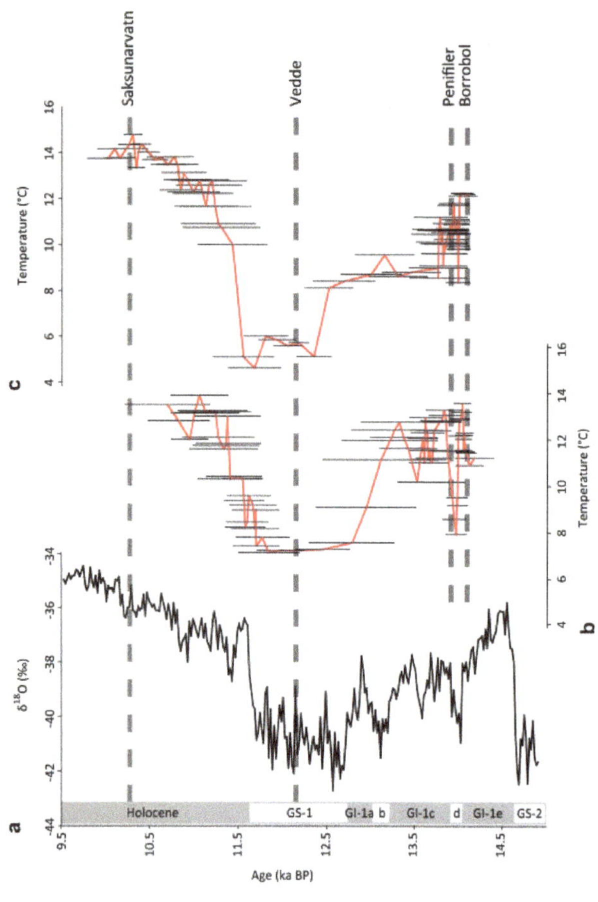

Figure 9: The Lateglacial and early-Holocene chironomid-inferred mean July air temperature reconstruction (°C) for (b) Loch Ashik and (c) Abernethy Forest compared with (a) NGRIP oxygen-isotope data (from Rasmussen et al., 2006). The chironomid-inferred temperature reconstruction is presented as mid-point ages with corresponding 95.4% ranges presented as error bars, while the NGRIP time-scale is in GICC05 ice-core years BP. The occurrence of the Vedde, Saksunarvatn, Borrobol, and Penifiler tephras are marked by grey dashed lines. The INTIMATE Greenland Event stratigraphy is listed below the NGRIP oxygen-isotope data (Lowe et al., 2008). The general GI-1, GS-1, and Holocene trend can be observed in all records as can a cold oscillation coincident with GI-1d.

The NGRIP record shows a cold oscillation of intermediate amplitude at 13640 ± 160 GICC05 yr BP within the phase that has been referred to as GI-1c. This oscillation may be reflected in the Abernethy Forest record by a cold oscillation of about 1.9 °C dated to 13680 ± 190 cal yr BP (626 cm). The NGRIP record indicates a final Lateglacial Interstadial cold oscillation, GI-1b, which reaches minimum temperatures at 13160 ± 150 GICC05 yr BP. However, the age model for Abernethy Forest is less well-constrained towards the end of the Lateglacial Interstadial, and a final Lateglacial Interstadial cold oscillation of about 2.6 °C, which is stratigraphically in the position of GI-1b, has a date of 13520 ± 220 cal yr BP (610 cm). At Loch Ashik a cold oscillation of about 1.7 °C may be indicated at 13760 ± 290 cal yr BP (598 cm), but due to age-depth model uncertainties it is difficult to determine which oscillation this may represent.

At the beginning of the YD, C-IT at Abernethy Forest falls by 5.8 °C to 6.8 °C and remains relatively stable in the early-YD. At Loch Ashik, C-IT initially falls by 3.0 °C, to a low of 8.2 °C, before falling again in the late-YD to a minimum of 4.2 °C at 11640 ± 300 cal yr BP (573 cm). The YD – Holocene transition is marked by rapidly rising temperatures at both sites. At Abernethy Forest, C-IT increases by 5.6 °C over a core length of 28 cm (about 330 years or 60 years per 1 °C rise) and at Loch Ashik a rise of 8.8 °C occurs over a core length of 9 cm (about 480 years or 55 years per 1 °C rise). At both sites the YD – Holocene temperature increase precedes the rise in LOI values by about 300 years. C-IT reaches around 14.0 °C near the core top at both sites. Minor cold oscillations may be indicated during the early-Holocene at both sites but poor chronological control in the age-depth models in this period prevents correlations with the NGRIP record.

At Loch Ashik all samples have a good fit-to-temperature except those in the late-YD (Fig. 7). This suggests that variables other than temperature may have been influencing the response of the chironomid assemblages at this time and concurs with the CCA results which suggest that the chironomid assemblage may have been responding to a change in lake depth. Most of the fossil samples also have good analogues in the modern calibration set (Fig. 7), although most of those in the early-YD and some in GI-1d have poor modern analogues. At Abernethy Forest most samples have a good fit-to-temperature, except for several in the later part of the Lateglacial Interstadial (Fig. 8). Many fossil samples have poor analogues in the modern calibration dataset, especially those in the early Lateglacial Interstadial, during GI-1b and GI-1a, the late-YD and early-Holocene, and towards the top of the sequence. Nevertheless, although these results imply that inferred temperatures in these poor-analogue parts of the sequence should be treated with more caution than those parts of the sequence with good modern analogues, WA-PLS inference models can perform surprisingly well in no-analogue situations (Birks et al., 2010).

DISCUSSION

Comparison of Temperature Records

Although general temperature trends during the Lateglacial and early-Holocene in north-west Europe are well-known from pollen, macrofossil and beetle records (e.g. Lowe et al., 1994; Coope et al., 1998; Birks and Ammann, 2000; Walker et al., 2003; Birks and Birks, 2008; Bakke et al., 2009) there are relatively few high temporal resolution, quantitative records available. The high resolution Greenland ice-core records (e.g. Rasmussen et al., 2006) provide an indication of mean annual temperature trends. However, they may not exactly match the local climatic trends in north-west Europe. High resolution, quantitative chironomid-inferred Lateglacial and early-Holocene mean July temperature reconstructions are available from Scotland (Brooks and Birks, 2000b), north-west England (Bedford et al., 2004; Lang et al., 2010) and Ireland (Watson et al., 2010; van Asch et al., 2012). Other north-west European records have been published from Norway (Brooks and Birks, 2001) and The Netherlands (Heiri et al., 2007).

Brooks and Birks (2000b) showed that the C-IT trends from Whitrig Bog, in south-east Scotland, closely followed the oxygen-isotope curve of the GRIP ice-core record (Johnsen et al., 1992; Dansgaard et al., 1993) (Fig. 10), although there were differences in the relative magnitude of the Lateglacial Interstadial cold oscillations and in the gradient of the cooling trend through the Lateglacial Interstadial. Unfortunately, the independent timing of events could not be addressed at this site due to restricted chronological information. The timing of events at Whitrig Bog and direct comparisons between Whitrig Bog, Loch Ashik and Abernethy Forest can now be partly addressed by using tephra layers common to all three sites (Fig. 10). Whitrig Bog contains both the Vedde and the Borrobol tephras at core depths 114 cm and 239 cm, respectively (Turney et al., 1997). In the three Scottish sites, the cold oscillation aligned with GI-1d was colder than the cold oscillation during GI-1b, whereas variations in ^{18}O in the ice-core records suggest the reverse was likely in Greenland (Fig. 9 and Fig. 10). In addition, the GRIP and NGRIP records indicate that the cooling trend through the Lateglacial Interstadial was steeper in Greenland than suggested by the chironomid records from Scotland (Fig. 10). Similar trends are also apparent in the C-IT records from sites in the English Lake District (Bedford et al., 2004; Lang et al., 2010), Lough Nadourcan in north-west Ireland (Watson et al., 2010) and Fiddaun in western Ireland (van Asch et al., 2012). Lang et al. (2010) explored the trends in Lateglacial C-IT at five adjacent lakes in the English Lake District and found the major climate fluctuations at all the sites. However, between-site variance was related to differences in altitude, lake area, and depth. Large, deep lakes at high altitude were found to have the greatest magnitude C-IT changes. Trends similar to these C-IT records are also apparent in the Hawes Water oxygen-isotope record (Bedford et al., 2004). This suggests that the consistent differences between the British chironomid records and the Greenland ice-core records may not necessarily be explained by the latter reflecting annual climate and the chironomid records reflecting summer temperature but that there may be differences in climate development between the regions. The other chironomid records, however, cannot be independently correlated with GI-1d due to limited or imprecise chronological data either through low numbers of radiocarbon determinations, date-sample selection problems (mineral carbon error), small amounts of dateable organic material, or chronologies derived through wiggle-matching regional bio-events (pollen variations).

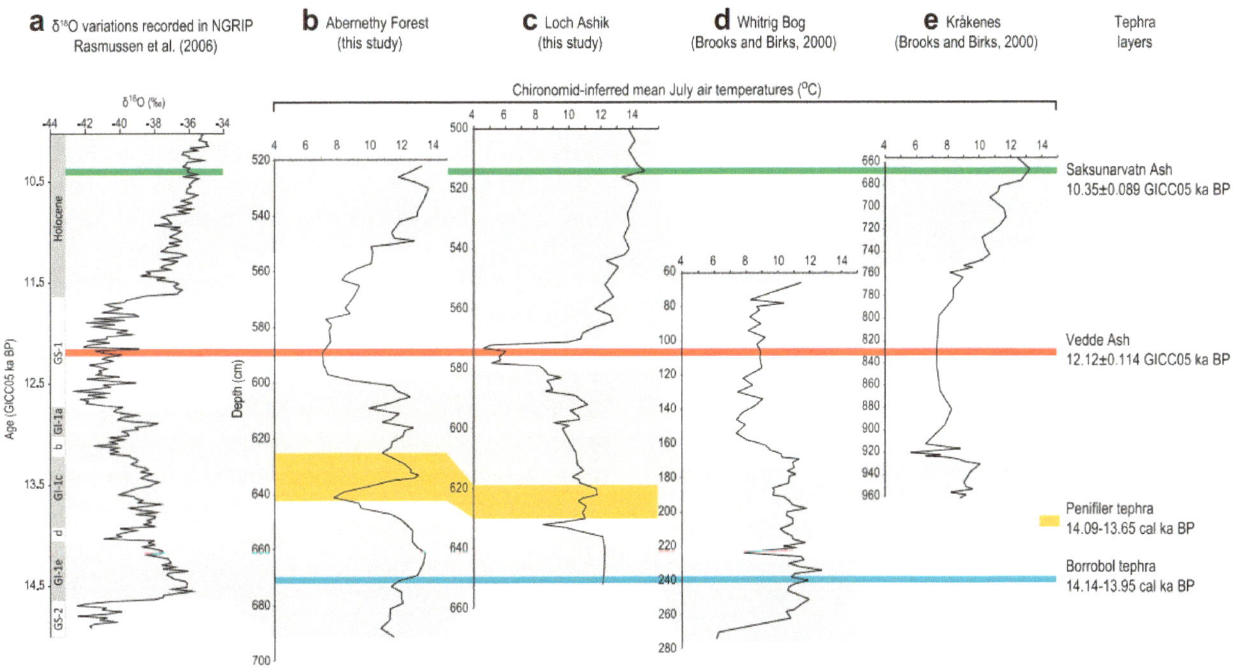

Figure 10: Chironomid-inferred mean July air temperature (C-IT) reconstructions (°C) across Northern Europe from sites reported to contain common tephra layers compared with the NGRIP oxygen-isotope data (a) (from Rasmussen et al., 2006). C-IT data are presented against depth (cm) as the Whitrig Bog record (d) has insufficient chronological information to construct a robust age model. The *y*-axis of each site (depth cm) has been scaled so that the tephra layers align against their chronological position in the NGRIP record. Grey shading is used to mark the main chronostratigraphical boundaries (based on the INTIMATE Event stratigraphic scheme of Lowe et al., 2008) in each record. The oscillations observed within the later part of GI-1 appear to be represented in each record; however, the age models developed for these sites are not yet sufficiently precise to link the events directly.

The unique cold C-IT at Loch Ashik during the final approximately 650 years of the YD contrasts with records from other sites in the region (cited above), which tend to show stable or slightly increasing trends (Fig. 10). In most sites the YD is typically only a little colder than the coldest temperatures attained during GI-1d, whereas at Loch Ashik the later part of the YD is about 4.0 °C colder than GI-1d. The samples from Loch Ashik during this period have a poor fit-to-temperature, which suggests that July air temperature may not have been the main influence on chironomid distribution and abundance at this time. Nevertheless, these samples do have good analogues in the modern calibration dataset so the temperature reconstructions are acceptable. At this time the chironomid fauna was dominated by just two taxa, *M. radialis*-type and *Pseudodiamesa*, together with *Paracladius* and *M. insignilobus*-type at lower abundance. Almost all other taxa were eliminated and head capsule concentrations are extremely low. This suggests that temperature and lake productivity were also extremely low. The closest modern analogues to these samples in the modern Norwegian calibration set all come from cold, relatively deep lakes, at high altitude or latitude, and having low conductivity and alkalinity (Brooks and Birks, 2000a, 2001; 2004; unpublished data). When the Loch Ashik samples are plotted passively in CCA space of the modern Norwegian calibration dataset (Fig. 5) the composition of the chironomid assemblage in the late-YD samples is clearly different from the other samples in the sequence, possibly reflecting the extreme environmental conditions at this time.

An ice-field occupied the Skye mountains during the YD which Ballantyne (1989)suggested covered 155 km^2. Golledge (2010) shows that the eastern perimeter of this ice-field was about 10 km from Loch Ashik. Therefore, the close proximity of this ice-field and the influence of catabatic winds may have depressed air temperature (Levesque et al., 1997) supposing that the ice-field expanded due to increased winter precipitation during the last 600 years of the YD. Although the timing of maximum glacial extent cannot be confirmed on Skye, at Loch Lomond and Lochaber, in the Scottish Highlands, the YD ice-field reached its greatest extent late in the YD around 12000 cal yr BP due to increased winter precipitation and cold temperatures (MacLeod et al., 2011). Increase in winter precipitation may also have resulted in the expansion of long-lasting snow beds in the catchment of Loch Ashik resulting in cold melt-water entering the lake. Conversely, YD ice-fields in the Cairngorms were relatively small and distant from Abernethy Forest (Golledge, 2010). Benn and Ballantyne (2005) calculated precipitation values in eastern and western Scotland at the end of the YD and suggest that precipitation values in the eastern Highlands were little different from the present day, whereas in north-west Scotland precipitation was 500–1600 mm yr^{-1} greater than present. Interestingly, CCA results suggest that the lake level at both Loch Ashik and Abernethy Forest may have risen during the late-YD (Fig. 5 and Fig. 6). However, this seems less likely at Abernethy Forest as the high percentages of *Artemisia* pollen in the Cairngorms during the Younger Dryas suggest the prevalence of a relatively arid climate (Birks and Mathewes, 1978). *Artemisia* pollen was considerably less abundant on Skye at this time (Birks, 1973; Walker and Lowe, 1991). The increase in precipitation levels in the west and the steeper west-east precipitation gradient during the YD probably resulted from atmospheric chilling by the West Highland Icefield and consequent reduction of snow carried by the westerly air masses to the Cairngorms (Benn and Ballantyne, 2005).

The rapid warming temperature trends in the early-Holocene inferred at Loch Ashik and Abernethy Forest are followed by more gradual warming until temperatures exceed those inferred in the Lateglacial Interstadial. This is consistent with the Greenland ice-core records and other British chironomid records (cited above). The absolute C-IT values inferred at Loch Ashik and Abernethy Forest are also similar to those inferred at the other British sites in most parts of each sequence.

Comparison of the British C-IT records with the two published NW European LGIT chironomid records reveals broad agreement but also some differences (Brooks and Birks, 2000a; Heiri et al., 2007). At Kråkenes, western Norway, (Brooks and Birks, 2000a) the Lateglacial Interstadial is truncated at the base due the late withdrawal of ice and so highest temperatures are not reached until the end of the Lateglacial Interstadial and the GI-1d and GI-1b cold oscillations are not apparent (see also Birks et al., 1994; Lowe et al., 1994). In the YD, C-IT is about 3 °C cooler than in the Lateglacial Interstadial, a warming trend is apparent in the YD, and early-Holocene temperatures climb steeply to reach values about 3 °C higher than peak temperatures in the Lateglacial Interstadial. Table 3 shows C-IT from Abernethy Forest, Loch Ashik, Whitrig Bog, and Kråkenes at the levels of tephra deposits and the temperature trends at these sites are compared in Fig. 10, which shows that climate development was similar at all these sites during the LGIT. Holocene and Lateglacial Interstadial temperatures at Kråkenes are cooler than the Scottish sites, probably reflecting the more northerly location and extremely oceanic climate of the site (present mean July air temperature at Kråkenes is 12.6 °C, about 2 °C cooler than present at Loch Ashik). The temperatures inferred at Loch Ashik during the late-YD are about 0.5 °C lower than those at Kråkenes, which was receiving melt waters from a cirque glacier in its catchment (Birks et al., 2000), may be due to the proximity of Loch Ashik to the large Skye ice-field.

Table 3: Comparison of chironomid-inferred mean July air temperature at Abernethy Forest, Loch Ashik, Whitrig Bog and Kråkenes at times of tephra deposits

Tephras	Sites			
	Abernethy Forest	Loch Ashik	Whitrig Bog	Krakenes
Saksunarvatn (10347 ± 89 GICC05 yr BP)	x	14.0 °C	x	11.9 °C
Ashik tephra (precise age unknown)	x	12.8 °C	x	x
Vedde (12,171 ± 114 GICC05 yr BP)	7.8 °C	5.8 °C	8.9 °C	6.4 °C
Penifiler (14.09–13.65 cal ka BP)[a]	7.8–11.4 °C	8.3–11.8 °C	x	x
Borrobol (14.14–13.95 cal ka BP)	12.2 °C	12.1 °C	11.9 °C	x

[a]The Penifiler tephra has a relatively large vertical distribution in Abernethy Forest and Loch Ashik. Matthews et al. (2011) demonstrated this tephra layer is closely associated with upper boundary of a cold oscillation correlated with GI-1d. At Loch Ashik the Penifiler tephra slightly post-dates a similar oscillation and thus it is suggested the Penifiler tephra just post-dates this event. The temperatures present here reflect this slight uncertainty and are presented as a range of temperatures from the coldest point of the oscillation and the return to warmer temperatures.

In contrast to the British records, C-IT records from Hijkermeer, The Netherlands (Heiri et al., 2007), indicate that Lateglacial Interstadial peak temperatures were not attained until after GI-1d. In addition, cooling correlated with GI-1b was stronger than cooling associated with GI-1d. These trends are consistent with other palaeoclimate records from central Europe (Heiri and Millet, 2005; Magny et al., 2006; Ilyashuk et al., 2009; Larocque-Tobler et al., 2010) and suggest regional differences in the influence of atmospheric and oceanic circulation on European climate, probably reflecting the stronger influence of the North Atlantic Current on Scottish and Norwegian sites.

Lake Ecology and Faunistics

In the Abernethy Forest sequence there is an absence of certain cold stenotherms which are present at Loch Ashik, such as *Paracladius*, and a low abundance of others, such as *Pseudodiamesa*, which are much more abundant at Loch Ashik. This may account for the lower compositional turnover in response to the YD at Loch Ashik, as revealed by DCCA, since the fossil assemblage already included a prevalence of cold-adapted taxa during the Lateglacial Interstadial, which continued into the YD. At Abernethy Forest the cold stenotherms colonized the lake during the cold GI-1d oscillation and YD but were absent during warmer intervals in the Lateglacial Interstadial.

Throughout most of the Lateglacial and early-Holocene, C-IT is higher at Abernethy Forest than Loch Ashik, which may be a consequence of its more continental climate resulting in warmer summers. LOI is also generally higher at Abernethy Forest, perhaps reflecting higher organic productivity or less mineral inwash or autogenic silica production (Birks and Birks, 2006). The head capsule concentrations were generally lower and more constant at Abernethy Forest, except in the YD when they were extremely low at Loch Ashik, probably reflecting the extremely cold, unproductive environment and possibly long-lasting winter ice-cover. Despite the cooler conditions at Loch Ashik, chironomid diversity was higher throughout the sequence than at Abernethy Forest, although many taxa occur in both lakes.

In both sequences, most changes in sediment lithology do not coincide with significant changes in chironomid assemblage composition (Fig. 3 and Fig. 4). This suggests that chironomids were not responding to changes in the lake-sediment but that an external driver (i.e. changes in temperature) was causing these changes in the chironomid assemblages. The difference in timing between sedimentological change and chironomid assemblage change is most apparent at the end of the YD at Abernethy Forest where the C-IT begins to rise at a core depth of 575 cm, about 30 cm before there is a change in the sediment lithology and LOI.

The increase in abundance of the thermophilic taxon *Heterotanytarsus apicalis*-type at both sites towards the tops of the sequences is probably a response to Holocene warming. However, this species is also indicative of dissolved organic carbon in the water and its higher abundance at Loch Ashik may reflect a greater development of peat surrounding the loch. Only later did the Abernethy site become overgrown by bog.

Comparison with Pollen Records

Both Loch Ashik and Abernethy Forest have been designated as key reference sites for the LGIT on the basis of their highly resolved pollen records. Therefore, the C-IT records now available for these sites are invaluable for placing these pollen records in an improved climatic context. The Loch Ashik pollen record (Walker and Lowe, 1991) reflects the rapid warming at the start of the Lateglacial Interstadial, followed by gradual climatic cooling, the return of arctic-like vegetation during the YD, and the rapid warming in the early-Holocene. In addition, during a period correlated with GI-1d, declines in *Juniperus* and *Empetrum*, and increases in *Rumex* and *Salix* were probably in response to the 4 °C summer cooling inferred from the chironomid assemblages. Although there are some similarities between the pollen sequences at Loch Ashik and Abernethy Forest, the percentages of *Artemisia* pollen in the Younger Dryas are higher at Abernethy Forest suggesting that the climate may have been more arid (Birks and Mathewes, 1978). A vegetation response to the GI-1d cooling is not apparent in the pollen record at Abernethy Forest possibly because an ecological threshold was not passed, or the sample resolution was too low to detect short-term changes.

Comparison with Beetle Records

Coleoptera have been used to infer the mean temperature of the warmest and coldest months during the Lateglacial and early-Holocene at several sites in north-west Europe, including some in Scotland (Lowe et al., 1994; Coope et al., 1998). The temperature trends are similar to those inferred from chironomids, although the beetle records lack the high temporal resolution of our chironomid stratigraphies. Coleoptera-inferred early-Lateglacial Interstadial temperature estimates for southern Scotland peak at 16–17 °C, which is about 3–4 °C higher than maximum chironomid-inferred temperatures, and falls outside the range of C-IT prediction errors. Later in the Lateglacial Interstadial Coleoptera-inferred temperatures range from 10 to 14 °C, which is encompassed by the range of those derived from chironomids. During the YD, Coleoptera-inferred temperature estimates from two sites in southern Scotland are 8.5 °C, about 1.0–1.5 °C warmer than the chironomid-inferred estimates from Whitrig Bog (Brooks and Birks, 2000b) and the early-YD at Abernethy Forest, although within the range of C-IT error estimates, but they are similar to the Loch Ashik estimates in the first part of the YD.

CONCLUSIONS

The high resolution, low uncertainty, chironomid-inferred July air-temperature reconstructions from Loch Ashik and Abernethy Forest provide important, independently dated, records of climate change during the Lateglacial and early-Holocene in Scotland, and provide a context for the previously reported vegetational changes at these two key reference sites. The age-depth models developed for Abernethy Forest and Loch Ashik allow independent comparisons to be made with the Greenland ice-core records. It is most likely that cold oscillations in the Scottish Lateglacial Interstadial were synchronous with oscillations GI-1d and GI-1c in the NGRIP record. However, poor dating control in both Scottish records precludes comparison of the relative timing of oscillation GI-1b. For similar reasons there are large discrepancies in the dates of the YD – Holocene boundary. The general similarities in most of the trends between these two Scottish sites, Norwegian records, and ice-core records from Greenland emphasise the widespread nature of the temperature changes in the eastern North Atlantic region. The most important driver of these changes is probably fluctuations in the North Atlantic thermohaline circulation (Bakke et al., 2009). Nevertheless, consistent differences between British and Greenland records, and also those from The Netherlands and central-southern Europe (Heiri and Millet, 2005; Ilyashuk et al., 2009), indicate that there are variations in the responses to climate drivers at the local level and highlight the importance of high resolution and precise chronologies to make detailed temporal comparisons of events in space. The key findings identified within our new Scottish records are:

- A strong cold oscillation during the Lateglacial Interstadial has been identified in both records, which appears to be synchronous with GI-1d in the GICC05 ice-core records. This is the highest amplitude oscillation detected during the Lateglacial Interstadial in both sequences. Comparison of the magnitude of this event in other British and European proxy records suggests regional climate variability. Although this oscillation appears to be synchronous with GI-1d, there are still chronological uncertainties regarding the timing of its onset and absolute duration in our records. These uncertainties cannot be resolved here and require further chronological refinement at sites that offer higher (decadal or annual) resolution. However, our results currently provide the best independent timing for this oscillation in Britain and strongly suggest that it correlates with Greenland GI-1d.
- The differences in C-IT observed between Loch Ashik and Abernethy Forest in the late part of the Younger Dryas may indicate differences in local climate drivers. These local differences are important in understanding terrestrial environmental responses to regional climatic forcing as shown by Lang et al. (2010).
- The reconstructed C-IT for the basal sediments at Loch Ashik and Abernethy Forest are between 12 °C and 13 °C indicating that the lakes may not have been open for colonisation before the rise in temperatures in the early deglacial phase. This might reflect the timing of the withdrawal of ice from the regions, placing it after the initial Lateglacial Interstadial warming but prior to the deposition of the Borrobol tephra (14.14–13.95 cal ka BP).
- Making robust comparisons between C-IT reconstructions across Europe is made more complicated by chronological uncertainties associated with each record. The identification of tephras in sequences can provide isochronous markers which facilitate direct chronological comparisons. At sites where this has been achieved, similarities in the timing of temperature trends are apparent in Greenland, Scotland and Norway.

ACKNOWLEDGEMENTS

Thanks are due to Pat Haynes who prepared the chironomid samples and to Lene S. Halvorsen for measuring the loss-on-ignition. Some of the material for the radiocarbon dates at Abernethy Forest was picked out by Lene S. Halvorsen. Doug Benn and Ewan Shilland assisted with fieldwork. Alison MacLeod processed and analysed the tephra samples at Loch Ashik. Tephra samples from Abernethy Forest were processed and analysed by Anna J. Bourne. We are grateful to all these people for their assistance. Funding for the project was provided by NERC (NER/B/S/2001/00839). The radiocarbon dates were funded by the Bjerknes Centre for Climate Research. This is publication number A392 from the Bjerknes Centre for Climate Research.

REFERENCES

1. Bakke, J., Lie, O., Heegaard, E., Dokken, T., Haug, G.H., Birks, H.H., Dulski, P., Nilsen, T., 2009. Rapid oceanic and atmospheric changes during the Younger Dryas cold period. Nature Geoscience 2, 202e205.
2. Ballantyne, C.K., 1989. The Loch Lomond Readvance on the Isle of Skye, Scotland: glacier reconstruction and palaeoclimatic implications. Journal of Quaternary Science 4, 95e108.
3. Bedford, A., Jones, R.T., Lang, B., Brooks, S.J., 2004. A Late-glacial chironomid record from Hawes Water, N.W. England. Journal of Quaternary Science 19, 281e290.
4. Benn, D.I., Ballantyne, C.K., 2005. Palaeoclimatic reconstruction from Loch Lomond Readvance glaciers in the West Drumochter Hills, Scotland. Journal of Quaternary Science 20, 577e592.
5. Bennett, K.D., 1996. Determination of the number of zones in a biostratigraphical sequence. New Phytologist 132, 155e170.
6. Birks, H.H., Ammann, B., 2000. Two terrestrial records of rapid climatic change during the glacialeHolocene transition (14,000e9,000 calendar years B.P.) from Europe. Proceedings of the National Academy of Sciences of the United States of America 97, 1390e1394.
7. Birks, H.H., Birks, H.J.B., 2006. Multi-proxy studies in palaeolimnology. Vegetation History and Archaeobotany 15, 235e251.
8. Birks, H.J.B., Birks, H.H., 2008. Biological responses to rapid climate change at the Younger Dryas - Holocene transition at Kråkenes, western Norway. The Holocene 18, 19e30.
9. Birks, H.H., Mathewes, R.W., 1978. Studies in the vegetational history of Scotland. V. Late Devensian and early Flandrian pollen and macrofossil stratigraphy at Abernethy Forest, Inverness-shire. New Phytologist 80, 455e484.
10. Birks, H.J.B., Juggins, S., Line, J.M., 1990. Lake Water Chemistry Reconstruction. In: Mason, B.J. (Ed.), The Surface Waters Acidification Programme. Cambridge University Press, Cambridge, pp. 301e313.
11. Birks, H.H., Paus, A., Svendsen, J.I., Alm, T., Mangerud, J., Landvik, J.Y., 1994. Late Weichselian environmental change in Norway, including Svalbard. Journal of Quaternary Science 9, 133e145.
12. Birks, H.H., Battarbee, R.W., Birks, H.J.B., Bradshaw, E.G., Brooks, S.J., Duigan, C.A., Jones, V.V., Lemdahl, G., Peglar, S.M., Solem, J.O., Solhøy, I.W., Solhøy, T., Stalsberg, M.K., 2000. The development of the aquatic

ecosystem at Kråkenes Lake, western Norway, during the late-glacial and early Holocene e a synthesis. Journal of Paleolimnology 23, 91e114.

13. Birks, H.J.B., Heiri, O., Seppä, H., Bjune, A.E., 2010. Strengths and weaknesses of quantitative climate reconstructions based on late-Quaternary biological proxies. The Open Ecology Journal 3, 68e110.

14. Birks, H.H., 1970. Studies in the vegetational history of Scotland. I. A pollen diagram from Abernethy Forest, Inverness-shire. Journal of Ecology 58, 827e846.

15. Birks, H.J.B., 1973. Past and Present Vegetation of the Isle of Skye e a Palaeoecological Study. Cambridge University Press, London.

16. Birks, H.J.B., 1995. Quantitative palaeoenvironmental reconstructions. In: Maddy, D., Brew, J.S. (Eds.), Statistical Modelling of Quaternary Science Data. Technical Guide 5. Quaternary Research Association, Cambridge, pp. 161e254.

17. Birks, H.J.B., 1998. Numerical tools in palaeolimnology-progress, potentialities, and problems. Journal of Paleolimnology 20, 307e332.

18. Birks, H.J.B., 2007. Estimating the amount of compositional change in lateQuaternary pollen-stratigraphical data. Vegetation History and Archaeobotany 16, 197e202.

19. Björck, S., Walker, M.J.C., Cwynar, L.C., Johnsen, S., Knudsen, K.L., Lowe, J.J., Wohlfarth, B., 1998. An event stratigraphy for the Last Termination in the north Atlantic region based on the Greenland ice-core record: a proposal by the INTIMATE group. Journal of Quaternary Science 13, 283e292.

20. Blockley, S.P.E., Pyne-O'Donnell, S.D.F., Lowe, J.J., Matthews, I.P., Stone, A., Pollard, A.M., Turney, C.S.M., Molyneux, E.G., 2005. A new and less destructive laboratory procedure for the physical separation of distal glass tephra shards from sediments. Quaternary Science Reviews 24, 1952e1960.

21. Brodersen, K.P., Lindegaard, C., 1999. Mass occurrence and sporadic distribution of Corynocera ambigua Zetterstedt (Diptera, Chironomidae) in Danish lakes. Neoand palaeolimnological records. Journal of Paleolimnology 22, 41e52.

22. Brodersen, K.P., Odgaard, B.V., Vestergaard, O., Anderson, N.J., 2001. Chironomid stratigraphy in the shallow and eutrophic Lake Søbygaard, Denmark: chironomid-macrophyte co-occurrence. Freshwater Biology 46, 253e267.

23. Bronk Ramsey, C., 2008. Deposition models for chronological records. Quaternary Science Reviews 27, 42e60.

24. Bronk Ramsey, C., 2009. Bayesian analysis of radiocarbon dates. Radiocarbon 51, 337e360.

25. Brooks, S.J., Birks, H.J.B., 2000a. Chironomid-inferred Lateglacial - early Holocene mean July air temperatures for Kråkenes Lake, western Norway. Journal of Palaeolimnology 23, 77e89.

26. Brooks, S.J., Birks, H.J.B., 2000b. Chironomid-inferred late-glacial air temperatures at Whitrig Bog, south-east Scotland. Journal of Quaternary Science 15, 759e764.

27. Brooks, S.J., Birks, H.J.B., 2001. Chironomid-inferred air temperatures from lateglacial and Holocene sites in north-west Europe: progress and problems. Quaternary Science Reviews 20, 1723e1741.

28. Brooks, S.J., Birks, H.J.B., 2004. The dynamics of Chironomidae assemblages in response to environmental change during the past 300 years in Spitsbergen. Journal of Paleolimnology 31, 483e498.

29. Brooks, S.J., Langdon, P.G., Heiri, O., 2007. The Identification and Use of Palaearctic Chironomidae Larvae in Palaeoecology. Technical Guide No. 10. Quaternary Research Association, London.

30. Coope, G.R., Lemdahl, G., Lowe, J.J.,Walkling, A.,1998. Temperature gradients in northern Europe during the last glacial-Holocene transition (14e9 14C kyr BP) interpreted from coleopteran assemblages. Journal of Quaternary Science 13, 419e433.

31. Cranston, P.S., 1982. A Key to the Larvae of the British Orthocladiinae (Chironomidae). In: Scientific Publication No. 45. Freshwater Biological Association, Ambleside.

32. Dansgaard, W., Johnsen, S.J., Clausen, H.B., Dahl-Jensen, D., Gundestrup, N.S., Hammer, C.U., Hvidberg, C.S., Steffensen, J.-P., Sveinbjörnsdottir, A.E., Jouzel, J., Bond, G., 1993. Evidence for general instability of past climate from a 250-kyr ice-core record. Nature 364, 218e220.

33. Davies, S.M., Wohlfarth, B., Wastegård, S., Andersson, M., Blockley, S., Possnert, G., 2004. Were there two Borrobol Tephras during the early Late-glacial period: implications for tephrochronology? Quaternary Science Reviews 23, 581e589.

34. Golledge, N.R., 2010. Glaciation of Scotland during the Younger Dryas stadial: a review. Journal of Quaternary Science 25, 550e566.

35. Grimm, E.C., 2004. Tilia and Tilia*Graph software. Illinois State Museum. Heiri, O., Millet, L., 2005. Reconstruction of Lateglacial summer temperatures from chironomid assemblages in Lac Lautrey (Jura, France). Journal of Quaternary Science 20, 33e44.

36. Heiri, O., Lotter, A.F., Lemcke, G., 2001. Loss on ignition as a method for estimating organic and carbonate content in sediments: reproducibility and comparability of results. Journal of Paleolimnology 25, 101e110.

37. Heiri, O., Cremer, H., Engels, S., Hoek, W.Z., Peeters, W., Lotter, A.F., 2007. Lateglacial summer temperatures in the Northwest European lowlands: a chironomid record from Hijkermeer, the Netherlands. Quaternary Science Reviews 26, 2420e2437.

38. Ilyashuk, B., Gobet, E., Heiri, O., Lotter, A.F., van Leeuwen, J.F.N., van der Knaap, W.O., Ilyashuk, E., Oberli, F., Ammann, B., 2009. Lateglacial environmental and climatic changes at the Maloja Pass, Central Swiss Alps, as recorded by chironomids and pollen. Quaternary Science Reviews 28, 1340e1353.

39. Johnsen, S.J., Hammer, C.U., Iversen, P., Jouzel, J., Stauffer, B., Steffensen, J.-P., Clausen, H.B., Dansgaard,

W., Fuhrer, K., Gundestrup, N., 1992. Irregular glacial interstadials recorded in a new Greenland ice core. Nature 359, 311e313.

40. Juggins, S., ter Braak, C.J.F., 2001. WAPLS v. 1.5 software. University of Newcastle upon Tyne.

41. Juggins, S., 1991. ZONE Software. University of Newcastle upon Tyne.

42. Lang, B., Brooks, S.J., Bedford, A., Jones, R.T., Birks, H.J.B., Marshall, J., 2010. Regional consistency in Lateglacial chironomid-inferred temperatures from five sites in north-west England. Quaternary Science Reviews 29, 1528e1538.

43. Larocque-Tobler, I., Heiri, O., Wehrli, M., 2010. Lateglacial and Holocene temperature changes at Egelsee, Switzerland, reconstructed using subfossil chironomids. Journal of Paleolimnology 43, 649e666.

44. Levesque, A.J., Cwynar, L.C., Walker, I.R., 1997. Exceptionally steep north-south gradients in lake temperatures during the last deglaciation. Nature 385, 423e426.

45. Lotter, A.F., Birks, H.J.B., Eicher, U., Siegenthaler, U., 1992. Lateglacial climatic oscillations as recorded in Swiss lake sediments. Journal of Quaternary Science 7, 187e204.

46. Lowe, J.J., Ammann, B., Birks, H.H., Björck, S., Coope, G.R., Cwynar, L.C., De Beaulieu, J.-L., Mott, R.J., Peteet, D.M., Walker, M.J.C., 1994. Climatic changes in areas adjacent to the North Atlantic during the last glacial-interglacial transition (14e9 ka BP): a contribution to IGCP-253. Journal of Quaternary Science 9, 185e198.

47. Lowe, J.J., Rasmussen, S.O., Björck, S., Hoek, W.Z., Steffensen, J.P., Walker, M.J.C., Yu, Z.C., the INTIMATE Group, 2008. Synchronisation of palaeoenvironmental events in the North Atlantic region during the Last Termination: a revised protocol recommended by the INTIMATE group. Quaternary Science Reviews 27, 6e17.

48. MacLeod, A., Palmer, A., Lowe, J., Rose, J., Bryant, C., Merritt, J., 2011. Timing of glacier response to Younger Dryas climatic cooling in Scotland. Global and Planetary Change 79, 264e274.

49. Magny, M., Aalbersberg, G., Begeot, C., Benoit-Ruffaldi, P., Bossuet, G., Disnar, J.-R., Heiri, O., Laggoun-Defarge, F., Mazier, F., Millet, L., Peyron, O., Vannière, B., Walter-Simonnet, A.-V., 2006. Environmental and climatic changes in the Jura mountains (eastern France) during the Lateglacial-Holocene transition: a multiproxy record from Lake Lautrey. Quaternary Science Reviews 25, 414e445.

50. Mangerud, J., Andersen, S.T., Berglund, B.E., Donner, J.J., 1974. Quaternary stratigraphy of Norden, a proposal for terminology and classification. Boreas 3, 109e128.

51. Matthews, I.P., Birks, H.H., Bourne, A., Brooks, S.J., Lowe, J.J., MacLeod, A., PyneO'Donnell, S.D.F., 2011. New age estimates and climatostratigraphic correlations for the Borrobol and Penifiler tephras: evidence from Abernethy Forest, Scotland. Journal of Quaternary Science 26, 247e252.

52. Moller Pillot, H.K.M., Buskens, R.F.M., 1990. De larven der Nederlandse Chironomidae. Autoecologie en verspreiding. In: Nederlandse Faunistische Mededelingen 1c. 1e87pp.

53. Pyne-O'Donnell, S.D.F., 2007. Three new distal tephras in sediments spanning the Last Glacial-Interglacial Transition in Scotland. Journal of Quaternary Science 22, 559e570.

54. Pyne-O'Donnell, S.D.F., 2011. The taphonomy of Last GlacialeInterglacial Transition (LGIT) distal volcanic ash in small Scottish lakes. Boreas 40, 131e145.

55. Rasmussen, S.O., Andersen, K.K., Svensson, A.M., Steffensen, J.P., Vinther, B.M., Clausen, H.B., Siggaard-Andersen, M.-L., Johnsen, S.J., Larsen, L.B., DahlJensen, D., Bigler, M., Röthlisberger, R., Fischer, H., Goto-Azuma, K., Hansson, M.E., Ruth, U., 2006. A new Greenland ice core chronology for the last glacial termination. Journal of Geophysical Research-Atmospheres 111, D06102. doi:10.1029/2005JD006079.

56. Rasmussen, S.O., Vinther, B.M., Clausen, H.B., Andersen, K.K., 2007. Early Holocene climate oscillations recorded in three Greenland ice cores. Quaternary Science Reviews 26, 1907e1914.

57. Reimer, P.J., Baillie, M.G.L., Bard, E., Bayliss, A., Beck, J.W., Blackwell, P.G., Bronk Ramsey, C., Buck, C.E., Burr, G.S., Edwards, R.L., Friedrich, M., Grootes, P.M., Guilderson, T.P., Hajdas, I., Heaton, T.J., Hogg, A.G., Hughen, K.A., Kaiser, K.F., Kromer, B., McCormac, F.G., Manning, S.W., Reimer, R.W., Richards, D.A., Southon, J.R., Talamo, S., Turney, C.S.M., van der Plicht, J., Weyhenmeyer, C.E., 2009. IntCal09 and Marine09 radiocarbon age calibration curves, 0e50,000 years cal BP. Radiocarbon 51, 1111e1150.

58. Rieradevall, M., Brooks, S.J., 2001. An identification guide to subfossil Tanypodinae larvae (Insecta: Diptera: Chironomidae) based on cephalic setation. Journal of Paleolimnology 23, 81e99.

59. Strenzke, K., 1950. Wassertiere als Bewohner der Waldsimse. In: Die Heimat 57. 14e16.

60. Svensson, A., Andersen, K.K., Bigler, M., Clausen, H.B., Dahl-Jensen, D., Davies, S.M., Johnsen, S.J., Muscheler, R., Parrenin, F., Rasmussen, S.O., Röthlisberger, R., Seierstad, I., Steffensen, J.P., Vinther, B.M., 2008. A 60,000 year Greenland stratigraphic ice core chronology. Climate of the Past 4, 47e57. ter Braak, C.J.F.,

61. Smilauer, P., 2002. CANOCO Reference Manual and User's Guide to Canoco for Windows: Software for Canonical Community Ordination (version 4.5). Microcomputer Power, Ithaca, NY, USA, 352 pp.

62. Tokeshi, M., 1986. Resource utilization, overlap and temporal community dynamics: a null model analysis of an epiphytic chironomid community. Journal of Animal Ecology 55, 491e506.

63. Turney, C.S.M., Harkness, D.D., Lowe, J.J., 1997. The use of microtephra horizons to correlate Lateglacial lake sediment successions in Scotland. Journal of Quaternary Science 12, 525e531.

64. van Asch, N., Lutz, A.F., Duijkers, M.C.H., Heiri, O., Brooks, S.J., Hoek, W.Z., 2012. Rapid climate

change during the Weichselian Lateglacial in Ireland: a multiproxy record from Fiddaun, Co. Galway. Palaeogeography, Palaeoclimatology, Palaeoecology 315e316, 1e11.

65. Velle, G., Brooks, S.J., Birks, H.J.B., Willassen, E., 2005. Chironomids as a tool for inferring Holocene climate: an assessment based on six sites in southern Scandinavia. Quaternary Science Reviews 24, 1429e1462.

66. Walker, M.J.C., Lowe, J.J., 1990. Reconstruction of the environmental history of the last glacial-interglacial transition: evidence from the Isle of Skye, Inner Hebrides, Scotland. Quaternary Science Reviews 9, 15e49.

67. Walker, M.J.C., Lowe, J.J., 1991. Vegetational history of the Isle of Skye: I. The Late Devensian Lateglacial period (13-10 cal. ka BP). In: Ballantyne, C.K., Benn, D.I., Lowe, J.J., Walker, M.J.C. (Eds.), The Quaternary of the Isle of Skye. Field Guide. Quaternary Research Association, Cambridge, pp. 98e118.

68. Walker, M.J.C., Bohncke, S.J.P., Coope, G.R., O'Connell, M., Usinger, H., Verbruggen, C., 1994. The Devensian/Weichselian Lateglacial in northwest Europe (Ireland, Britain, north Belgium, The Netherlands, northwest Germany). Journal of Quaternary Science 9, 109e118.

69. Walker, M.J.C., Coope, G.R., Sheldrick, C., Turney, C.S.M., Lowe, J.J., Blockley, S.P.E., Harkness, D.D., 2003. Devensian lateglacial environmental changes in Britain: a multi-proxy environmental record from Llanilid, South Wales, UK. Quaternary Science Reviews 22, 475e520.

70. Walker, M.J.C., 1993. Loch Ashik (lateglacial profile). In: Gordon, J.E., Sutherland, D.G. (Eds.), Quaternary of Scotland. Geological Conservation Review Series, No. 6. Chapman and Hall, London, pp. 1e5.

71. Watson, J., Brooks, S.J., Whitehouse, N.J., Reimer, P.J., Birks, H.J.B., Turney, C., 2010. Chironomid-inferred Lateglacial summer air temperatures from Lough Nadourcan, Co. Donegal, Ireland. Journal of Quaternary Science 25, 1200e1210.

72. Wiederholm, T. (Ed.), 1983. Chironomidae of the Holarctic Region. Keys and Diagnoses. Part 1. Larvae. Entomologica Scandinavica Supplement, vol. 19, pp. 1e457.

73. Wright Jr., H.E., 1967. A square-rod piston sampler for lake sediments. Journal of Sedimentary Research 37, 975e976.

CHAPTER 5 MICROBIAL COMMUNITY COMPOSITION SHAPES ENZYME PATTERNS IN TOPSOIL AND SUBSOIL

Jörg Schnecker, Birgit Wild, Mounir Takriti, Ricardo J. Eloy Alves, Norman Gentsch, Antje Gittel, Angelika Hofer, Karoline Klaus, Anna Knoltsch, Nikolay Lashchinskiy, Robert Mikutta, Andreas Richter

Keywords: Extracellular enzymes, PLFA, Tundra, Boreal forests, Steppe, Permafrost

INTRODUCTION

Extracellular enzymes break down soil organic matter (SOM) at every depth of the soil profile. Nonetheless most studies on enzyme activities focused on topsoil horizons in the upper 20 cm of the soil profile (e.g. Sinsabaugh et al., 2008, Wallenstein et al., 2009 and Kaiser et al., 2010) although up to 60% of the carbon stored in soils are located below 30 cm (Jobbágy and Jackson, 2000). These subsoil horizons differ from well studied topsoil horizons in a number of physical and chemical conditions that might influence enzyme activities and decomposition in general (Rumpel and Kögel-Knabner, 2011): Temperature decreases from topsoils to subsoils whereas soil moisture increases with depth, either improving conditions for decomposition in arid systems (Rovira and Vallejo, 2002), or impairing them in systems where water logging occurs and O_2availability is low (Kleber, 2010 and Davidson et al., 2012). Soil pH, one of the factors often associated with enzyme activities (Sinsabaugh et al., 2008), also changes with depth (Eilers et al., 2012). In addition to these direct influences on enzyme activities, the availability of substrate for enzymatic breakdown decreases with depth. First, SOM is less abundant in subsoils, which leads to a high probability of a spatial disconnection of enzyme and substrate (Holden and Fierer, 2005). Second, a high proportion of SOM in subsoils is bound to minerals, stabilized by metal ions, or occluded in aggregates and therefore access for microorganism is limited (von Lützow et al., 2006). In addition to physical hurdles for decomposition, SOM in subsoils is chemically different from topsoil SOM. While the main proportion of SOM in topsoils is plant derived material, SOM in subsoils is microbially transformed (Wallander et al., 2003). During this microbial transformation of SOM, carbon is lost, mainly as CO_2, whereas most of the nitrogen (N) is recycled and remains in the system, resulting in lower C/N ratios of subsoil SOM (Rumpel and Kögel-Knabner, 2011).

To fulfill the microbial demand for energy and nutrients, microorganisms need to adapt to the chemical composition of SOM and to the C/N ratio of the available substrate by adjusting their enzyme production (Sinsabaugh et al., 2008). Changes in enzyme production might either be physiological (Stone et al., 2014) or they might result from a shift in microbial community composition (Kaiser et al., 2014). Although the influence of microbial community composition on major microbial processes, such as C mineralization and N mineralization, has been recently challenged (Colman and Schimel, 2013), its influence on enzyme activities has been demonstrated repeatedly (e.g. Strickland et al., 2009, McGuire and Treseder, 2010 and Schnecker et al., 2014). Microbial community composition, as another potential control on enzyme activities, has already been shown to change more strongly with soil

Citation: Jörg Schnecker, Birgit Wild, Mounir Takriti, Ricardo J. Eloy Alves, Norman Gentsch, Antje Gittel, Angelika Hofer, Karoline Klaus, Anna Knoltsch, Nikolay Lashchinskiy, Robert Mikutta, Andreas Richter, Microbial community composition shapes enzyme patterns in topsoil and subsoil, Soil Biology and Biochemistry, Volume 83, April 2015, Pages 106-115, ISSN 0038-0717, http://dx.doi.org/10.1016/j.soilbio.2015.01.016.

depth, within ecosystems, than between topsoils of different ecosystems (Eilers et al., 2012 and Gittel et al., 2014).

Relations of enzyme activities to key factors such as pH, moisture, SOM content (Keeler et al., 2009), chemical composition of SOM (Grandy et al., 2009 and Sinsabaugh and Follstad Shah, 2010), and microbial community composition (Waldrop and Firestone, 2006 and Talbot et al., 2013) are well established in topsoil horizons. Whether enzyme activities in the subsoils are related to these key factors is still largely unknown since few studies have addressed changes of enzyme activities and their potential controls with soil depth so far (e.g.: Brockett et al., 2012 and Kramer et al., 2013; Turner et al., 2014, Schnecker et al., 2014 and Stone et al., 2014).

In this study we investigated enzyme patterns in different soil horizons, including mineral subsoils, from a wide range of ecosystems to identify potential drivers for these enzyme patterns. We measured potential activities of six extracellular enzymes in organic topsoil horizons, mineral topsoil horizons and mineral subsoil horizons in seven ecosystems along a 1500 km-long north–south transect in Western Siberia. In addition to enzyme activities, we analyzed microbial community composition (using phospholipid fatty acid analysis) as well as abiotic soil parameters and related these factors to the enzyme patterns.

We hypothesized: (1) enzyme patterns in topsoil and subsoil horizons are both related to the same key parameters, such as SOM content, pH and microbial community composition. Microbial community composition has been shown to differ more strongly between topsoils and subsoils than between topsoils of different ecosystems (Meyer et al., 2006 and Eilers et al., 2012). Since enzyme activities and enzyme patterns are often related to microbial community composition; (2) enzyme activities and enzyme patterns change with depth and differ more strongly between horizons than between ecosystems. The ecosystems along the transect showed large differences in vegetation and presumably in the chemical composition of litter entering the soil; (3) enzyme patterns would be more variable and show greater differences between ecosystems in the topsoil horizons, where the main constituents of SOM are plant-derived, than in mineral subsoil horizons.

MATERIAL AND METHODS

Sampling Sites

Soil samples were taken from seven ecosystems along a 1500 km latitudinal transect in Western Siberia, in August and September 2012. The ecosystems included tundra, northern taiga, middle taiga, and southern taiga, forest steppe (one forest site and one meadow site), and steppe. All soils were sampled from the active layer in an unfrozen state. Basic soil and climate parameters are provided in Table 1 and Table S1. Climate data are derived from Stolbovoi and McCallum (2002), soil classification follows the World Reference Base for Soil Resources (IUSS Working Group WRB, 2006).

Table 1: Basic soil and site characterization of sites along the latitudinal transect. MAT is mean annual temperature; MAP is mean annual precipitation. Aridity index has a threshold for drylands at 0.65 (Maestre et al., 2012)

	Coordinates	MAT °C	MAP mm	Aridity index	Soil type	Organic topsoils		Mineral topsoils		Mineral subsoils	
						Horizon	Depth cm	Horizon	Depth cm	Horizon	Depth cm
Tundra	67°16′N 78°50′E	−7.6	391	1.30	Turbic Cryosol	O	0–6	A	2–13	Bg, BCg	6–57
Northern taiga	63°17′N 74°32′E	−4.6	430	1.06	Histic Podzol	Oi, Oe	0–22	AE, EA	8–30	Bg	14–47
Middle taiga	60°09′N 71°43′E	−2.2	438	0.89	Endogleyic Regosol	Oi	0–6	A, AE, EA	6–14	E, EA	12–55
Southern taiga	58°18′N 68°35′E	−0.5	396	0.71	Albic Podzol	Oi	0–7	A, AE	4–18	E, EA	15–59
Forest steppe: Forest	56°14′N 70°43′E	0.7	340	0.53	Haplic Phaeozem	O, Oa	0–10	A	7–46	B	57–109
Forest steppe: Meadow	56°14′N 70°43′E	0.7	340	0.53	Luvic Phaeozem	Oa	0–7	A	4–35	Bt	26–84
Steppe	54°41′N 71°38′E	1.0	309	0.44	Calcic Kastanozem	OA	0–12	Ak	8–37	Bk	27–109

At all sites, we sampled the three dominant soil horizons of five replicate soil pits. We categorized the three horizon types as organic topsoil horizon (uppermost horizon, O), mineral topsoil horizon (second horizon, A), and mineral subsoil horizon (third horizon, M). We removed living plant roots from the samples and sieved them to <2 mm. We did this for samples from all sites, except for the tundra, where samples were manually homogenized because they were too moist for sieving. Before further analyses, soil water content was adjusted to a minimum of 60% for organic topsoils (except steppe), to 15% for mineral topsoils and steppe organic topsoils, and to 10% for mineral subsoils, respectively.

Soil Parameters

Soil pH was determined in 1 M KCl extracts. Samples for determination of organic C, total N content, and $\delta^{13}C$ were dried at 60 °C and ground with a ball mill. Ground samples were analyzed with EA-IRMS (CE Instrument EA 1110 elemental analyzer, coupled to a Finnigan MAT DeltaPlus IRMS with a Finnigan MAT ConFlo II Interface, Thermo Fisher Scientific, Waltham, MA, USA). Mineral topsoils and subsoils at both forest steppe sites, as well as all horizons of the steppe site, contained traces of carbonate. Carbonate was removed from these samples by acidification with HCl before EA-IRMS analysis. Water holding capacity (WHC) was determined as the amount of water that remained in saturated soil, from which water could be lost by drainage but not by evaporation after two days (Reynolds and Topp, 2007).

Microbial C and N were estimated using chloroform-fumigation–extraction (Kaiser et al., 2010 modified after Brookes et al., 1985): Soil samples, fumigated with chloroform, as well as unfumigated samples were extracted with 0.5 M K_2SO_4. Dissolved organic C and total dissolved N were determined in both sets of extracts with a DOC/TN analyzer (Shimadzu TOC-VCPH/CPN/TNM-1, Vienna, Austria). Microbial C and N were calculated as the difference between fumigated and non-fumigated samples, without correction for extraction efficiency. C/N ratios of SOM and microbial biomass were calculated on a mass basis.

Potential Extracellular Enzyme Activities

We measured potential enzyme activities fluorimetrically and photometrically using a microplate assay (Kaiser et al., 2010). For the fluorimetric assay, we used MUF (4-methylumbelliferyl) labeled substrates: β-d-cellobioside for cellobiohydrolase (CBH), triacetylchitotrioside for chititriosidase (CHT), N-acetyl-β-d-glucosaminide for N-acetyl-glucosaminidase (NAG) and phosphate for phosphatase (PHO). l-leucine-7-amido-4-methyl coumarin was used as substrate for leucine-amino-peptidase (LAP). Phenoloxidase (POX) activities were measured using L-3,4-dihydroxyphenylalanine (DOPA) as substrate in a photometric assay. Assays for CBH, CHT, NAG, PHO and LAP were incubated for 140 min at room temperature in a sodium acetate buffer (pH 5.5) and activity was measured fluorimetrically (excitation 365 nm and emission 450 nm). Plates for POX activity were measured photometrically (absorbance 450 nm) at the beginning and after incubation for 20 h at room temperature. POX activity was than calculated as the increase in color during the incubation time.

Phospholipid Fatty Acid (PLFA) Analysis

Extraction and measurement of PLFAs followed the procedure described by Frostegård et al. (1991) with the modifications by Kaiser et al. (2010). PLFAs were extracted from 1 g soil with chloroform/methanol/citric acid buffer and purified on silica columns (LC-Si SPE, Supleco, Bellefonte, PA, USA) using chloroform, acetone, and methanol. After addition of the internal standard (methyl-nonadecanoate), PLFAs were converted to fatty acid methyl esters (FAMEs) by alkaline methanolysis. Samples were analyzed on a Thermo Trace GC with FID detection (Thermo Fisher Scientific, Waltham, MA, USA), using a DB-23 column (Agilent, Vienna, Austria). FAMEs were identified using qualitative standard mixes (37 Components FAME Mix and Bacterial Acid Methyl Esters CP Mix, Supelco) and quantified using the internal standard. We categorized the fatty acids according to Kaiser et al. (2010). The markers 18:1ω9, 18:2ω6,9, and 18:3ω3,6,9 were used as markers for fungi; i15:0, a15:0, i16:0, i17:0, a17:0, cy17:0 (9/10), cy19:0 (9/10), 16:1ω7, 16:1ω9, 18:1ω7, 15:0, and 17:0 as bacterial markers. We used the above mentioned markers together with 14:0, i14:0, 16:0, 18:0, 20:0, 22:0, 16:1ω11, and 19:1ω8 for the calculation of total PLFA content (Schnecker et al., 2012).

Statistics

We calculated enzyme patterns to identify differences between horizons and between sites. To account for the different methods of measuring enzyme activities and the inherent differences in enzyme activities of different horizons, the individual enzyme activities per gram dry soil were log transformed and standardized by calculating the proportion of each enzyme to the sum of all enzymes. With these values, we calculated Euclidean distance matrixes. We used these matrixes to create Nonmetric Multidimensional Scaling (NMDS) plots. To identify differences between sites and horizons we used Permutational Multivariate Analysis of Variance Using Distance Matrices (ADONIS). This analysis is implemented in the R-package vegan (Oksanen et al., 2013). Additionally we used Mantel tests based on Spearman correlations of the calculated enzyme distance matrices with soil parameters and with microbial community composition (represented as a distance matrix based on relative abundances of individual PLFA biomarkers). We performed these analyses for the whole data set, as well as for data sets of the three horizon classes individually.

To evaluate whether differences between horizons or sites were stronger, we used two-way-ANOVA. To find differences within sites or within horizons, we used one-way ANOVA and Tukey HSD as post-hoc test. We did this for soil parameters, enzyme activities, and fungi:bacteria ratios, as well as for distances between different horizons and within horizons (variability). Before analysis, data were log-transformed or rank-normalized to meet the assumptions for ANOVA. Differences and correlations were assumed to be significant at $p < 0.05$. Statistics were performed in R 3.0.2 (R Development Core Team, 2013) using the vegan package (Oksanen et al., 2013).

RESULTS

Enzyme Activities

All measured hydrolytic enzyme activities – calculated per gram dry soil – differed more strongly between horizons than between sites (Table 2). Hydrolytic enzyme activities were highest in organic topsoil horizons followed by mineral topsoils and mineral subsoil horizons, with the exception of LAP in steppe where the highest values were found in organic topsoil horizons and mineral subsoil horizons and lower values were found in mineral topsoil horizons (Fig. 1). The oxidative enzyme POX did not follow this pattern and showed greater differences between sites than between horizons (Table 2, Fig. 1). When hydrolytic enzyme activities were calculated on a microbial C basis, differences between horizons were often smaller than differences between sites (Table 2, Fig. 2). POX activity, on a microbial C basis, was highest in the mineral subsoils followed by mineral topsoils (except northern taiga). The lowest rates were found in organic topsoil horizons at all sites. Differences between sites were ambiguous and POX activities, on a microbial C basis, in mineral subsoil horizons was the only enzyme activity that showed a clear north–south trend and decreased from the tundra site in the north to the steppe site in the South (Fig. 2).

Table 2: Two-way ANOVA R^2 for enzyme activities, based on dry soil and on microbial C basis; only significant differences are shown. Bold letters indicate whether horizon or site have the stronger influence on enzyme activities (higher R^2). Asterisks indicate significance (**mean $p < 0.01$; ***mean $p < 0.001$).

	Activities per g DM			Activities per g Cmic		
	Horizon	Site	Interaction	Horizon	Site	Interaction
cellobiohydrolase (CBH)	**0.80**	0.05	***	0.36	**0.52**	***
chititriosidase (CHT)	**0.82**	0.06	***	0.24	**0.32**	***
N-acetyl-glucosaminidase (NAG)	**0.84**	0.06	***	**0.43**	0.16	**
phosphatase (PHO)	**0.79**	0.16	***	0.03	**0.56**	***
leucine-amino-peptidase (LAP)	**0.83**	0.01	***	0.05	**0.28**	***
phenoloxidase (POX)	0.03	**0.17**	***	**0.73**	0.10	***

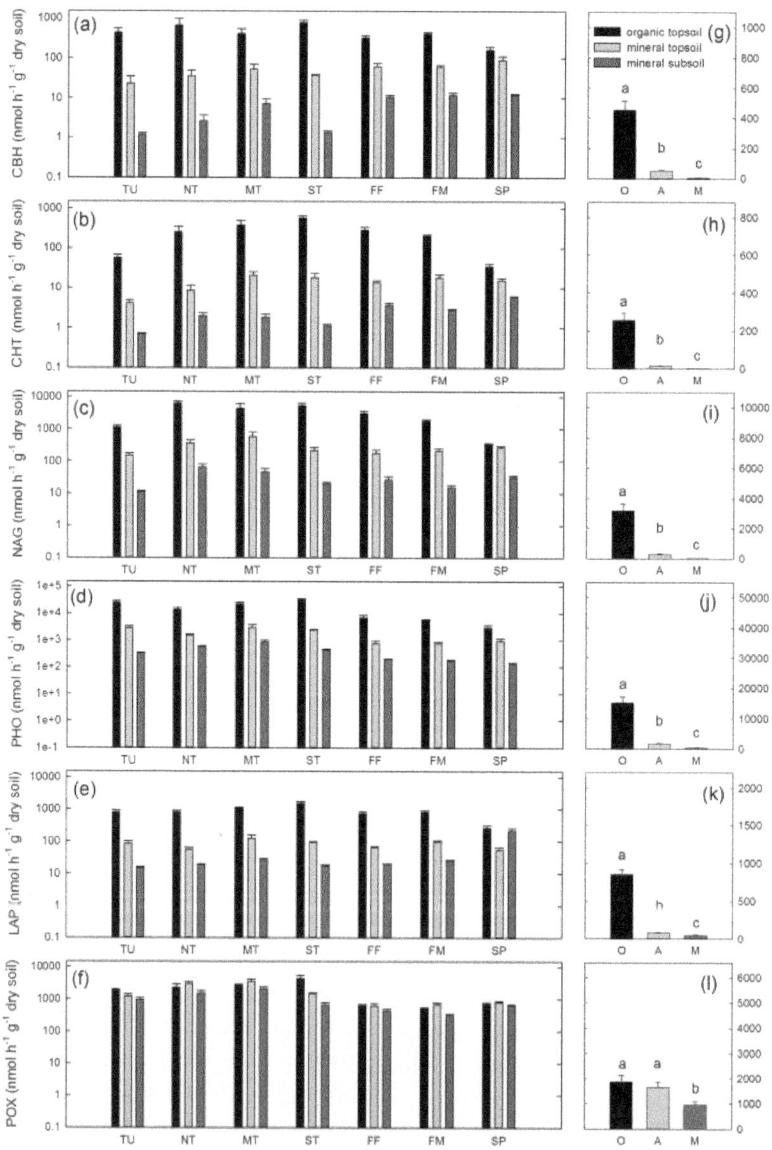

Figure 1: Extracellular enzyme activities per gram dry soil. Left panels (a–f) show activities on a log-scale for each horizon (O are organic top-

soils, black; A are mineral topsoils, light gray; M are mineral subsoils, dark gray) at each site (TU = Tundra; NT = northern taiga; MT = middle taiga; ST = southern taiga; FF = forest steppe forest; FM = forest steppe meadow; SP = steppe) individually. Right panels (g–l) show the mean of the individual horizons over all sites. Significant differences for horizon means are derived from ANOVA and Tukey HSD tests and are indicated by small letters ($p > 0.05$). Results from two-way ANOVAs are given in Table 2.

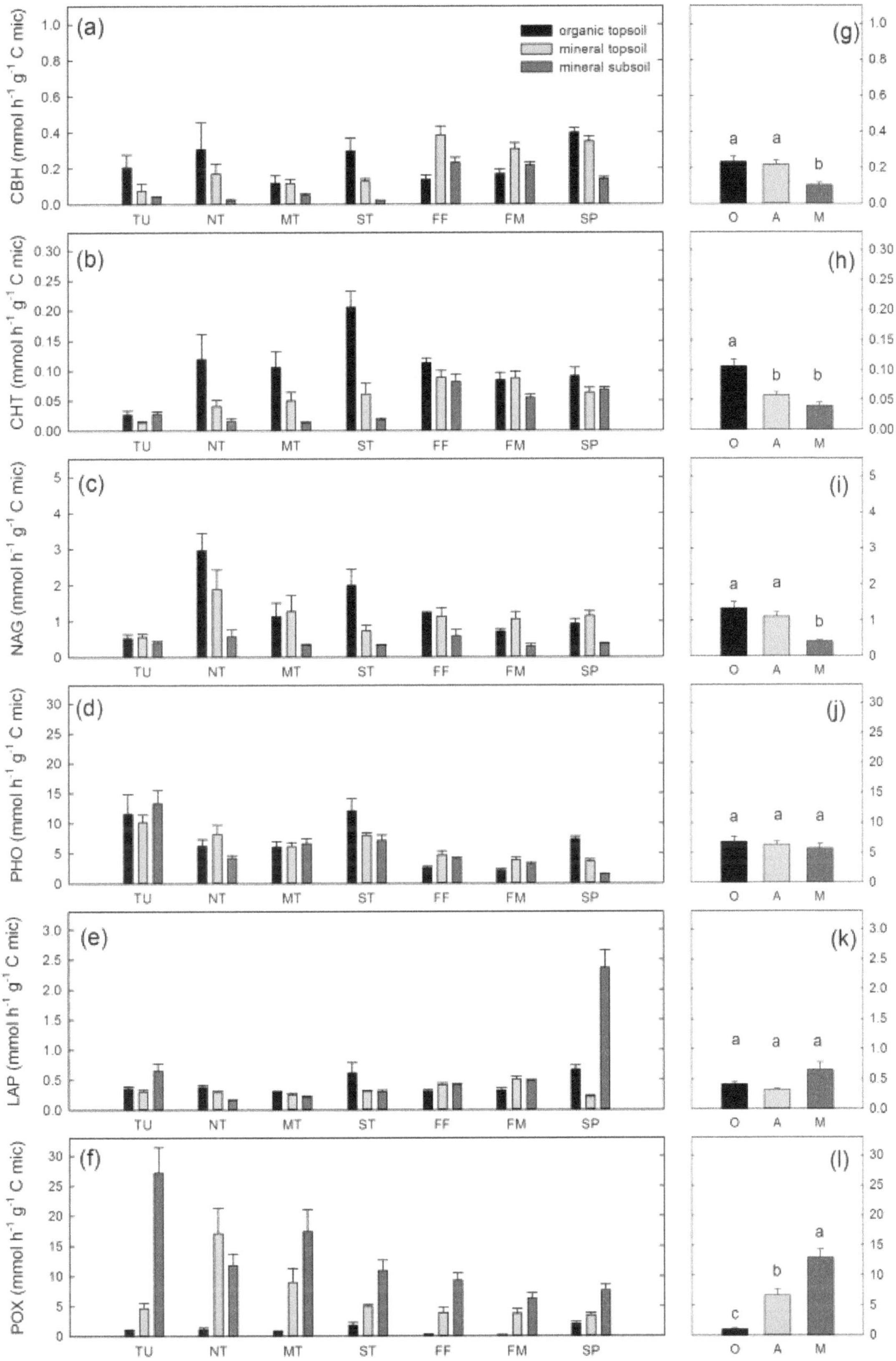

Figure 2: Extracellular enzyme activities per gram microbial C. Left panels (a–f) show activities for each horizon (O are organic topsoils, black; A are mineral topsoils, light gray; M are mineral subsoils, dark gray) at each site (TU = Tundra; NT = northern taiga; MT = middle taiga; ST = southern taiga; FF = forest steppe forest; FM = forest steppe meadow; SP = steppe) individually. Right panels (g–l) show the mean of the individual horizons over all sites. Significant differences for horizon means are derived from ANOVA and Tukey HSD tests and are indicated by small letters ($p > 0.05$). Results from two-way ANOVAs are given in Table 2.

Enzyme Patterns

To evaluate differences in the way microbes decompose SOM in different horizons, we used log-transformed and standardized enzyme activities to calculate distance matrixes. The resulting enzyme patterns differed more strongly between horizons than between sites (ADONIS: horizons $R^2 = 0.66$, sites $R^2 = 0.14$; Fig. 3). The mean distances and thus the variability of enzyme patterns were greatest in the mineral subsoil horizons, followed by mineral topsoil horizons, and organic topsoil horizons (Fig. 3b). Enzyme patterns of different horizons in the South clustered closer together than enzyme patterns of different horizons in the North in the NMDS plot (Fig. 3). This trend was more pronounced for differences (mean distances) between organic topsoils and mineral subsoils, which were significantly higher in the northern sites (tundra and taiga) than in the southern sites (forest-steppe and steppe; Fig. 5a). This trend was weaker but could also be seen for differences between organic topsoils and mineral topsoils (Fig. 5b). The differences between mineral topsoils and mineral subsoils did not show a decrease from North to South (Fig. 5c).

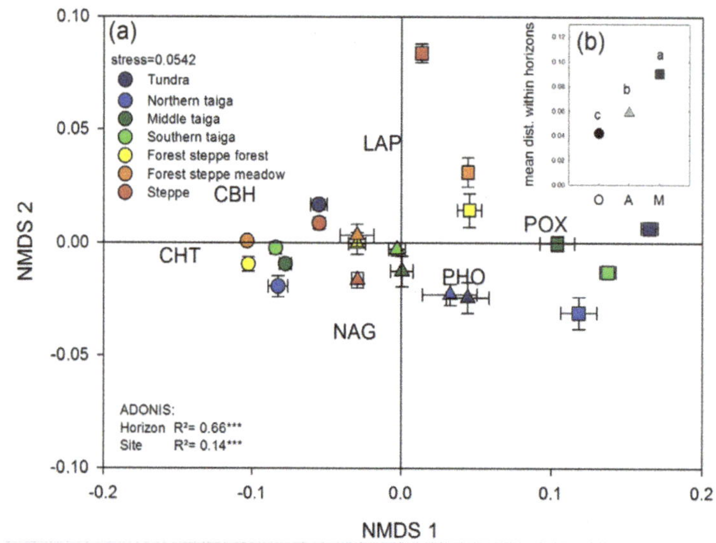

Figure 3: Enzyme patterns. NMDS plot of enzyme patterns calculated from a distance matrix of standardized enzyme activities (a). Symbols are the mean values of the replicated individual horizons of each site. Error bars are SE. Sites are indicated by color (Tundra is purple; northern taiga is blue; middle taiga is dark green; southern taiga light green; forest steppe forest is yellow; forest steppe meadow is orange; steppe is red). Horizons are indicated by different symbols (circles are organic topsoils; triangles are mineral topsoils; squares are mineral subsoils). The insert in the upper right corner (b) shows the distances within the horizons. The results of the ADONIS analysis show that horizon has a stronger effect than site. Asterisks indicate significance (*** mean $p > 0.001$). (For interpretation of the references to color in this figure legend, the reader is referred to the web version of this article.)

Correlations (Mantel tests) of enzyme patterns with biotic and abiotic factors varied between horizons (Table 3). The most striking difference between organic topsoils and mineral subsoils is the absence of a correlation between enzyme patterns and SOM properties (C, N, C/N ratio) in the mineral subsoils.

Table 3: Results of Mantel tests of enzyme patterns (distance matrix with standardized enzyme activities) and microbial community composition (distance matrix of relative amounts of PLFA) with abiotic and biotic variables. Only significant relations are shown ($p < 0.05$). Values are R

	Enzyme patterns			Microbial community composition		
	Organic topsoils	Mineral topsoils	Mineral subsoils	Organic topsoils	Mineral topsoils	Mineral subsoils
C content	0.06			0.29	0.03	
N content	0.22			0.30		
SOM C/N	0.05	0.04		0.27	0.04	
SOM $\delta^{13}C$	0.27			0.29	0.12	0.03
Microbial C				0.11	0.03	
Microbial N	0.03		0.02	0.18		
Microbial C/N		0.05		0.24	0.02	
pH	0.02	0.03	0.29	0.05	0.14	0.09

Water holding capacity	0.03		0.20	0.25	0.05	
Fungi:bacteria ratio	0.10	0.03	0.24	0.49	0.08	0.20
Mic. community comp.	0.16	0.04	0.17	–	–	–
Enzyme patterns	–	–	–	0.16	0.04	0.17

Microbial Community Composition

Microbial community composition differed significantly between horizons and between sites with both factors exerting a similar influence (site $R^2 = 0.26$ and horizon $R^2 = 0.25$). Differences between horizons were mainly caused by a decrease in fungal markers (18:3ω3 and 18:2ω6) with soil depth (Fig. 4). The differences in microbial community composition between horizons were further reflected in the fungi:bacteria ratios which decreased from organic topsoils, to mineral topsoils and mineral subsoils (Fig. 6). A clear north–south trend is also seen in the fungi:bacteria ratios in the topsoil and therefore differences between horizons were lower in the south located ecosystems. Although differences between horizons were not as pronounced for microbial community composition as for enzyme patterns, the variability within the horizons was also highest in mineral subsoils followed by mineral topsoils and organic topsoils (Fig. 4b). Microbial community composition in organic topsoils and mineral subsoils differed more in the North than in the South (Fig. 5d). With the exception of tundra, differences between organic topsoils and mineral topsoils also showed a decrease from North to South (Fig. 5e). Correlations of microbial community composition with biotic and abiotic factors were similar to those observed for enzyme patterns (Table 3). The correlations of microbial community composition and SOM parameters (C, N, CN ratio, $\delta^{13}C$) were strongest in the organic topsoils, but decreased to mineral topsoils and further to mineral subsoils (Table 3).

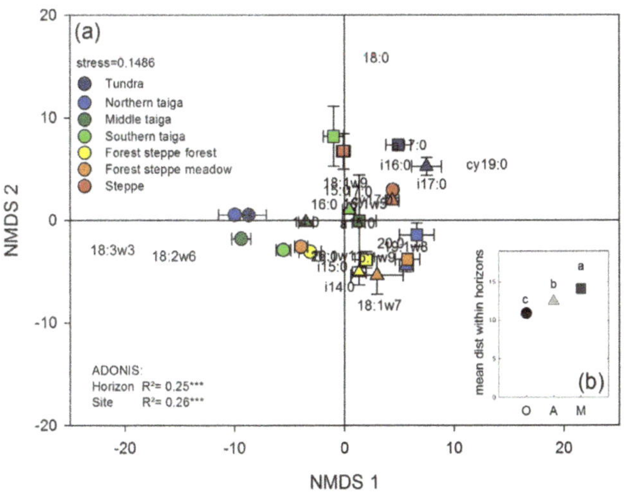

Figure 4: Microbial community composition. NMDS plot of microbial community composition calculated as distance matrix of the relative abundances of all individual PLFA markers (a). Symbols are the mean values of the replicated individual horizons of each site. Error bars are SE. Sites are indicated by color (Tundra is purple; northern taiga is blue; middle taiga is dark green; southern taiga light green; forest steppe forest is yellow; forest steppe meadow is orange; steppe is red). Horizons are indicated by different symbols (circles are organic topsoils; triangles are mineral topsoils; squares are mineral subsoils). The insert in the lower right corner (b) shows the distances within the horizons. The results of the ADONIS analysis show that horizon and site have equal influence. Asterisks indicate significance (*** mean p > 0.001). (For interpretation of the references to color in this figure legend, the reader is referred to the web version of this article.).

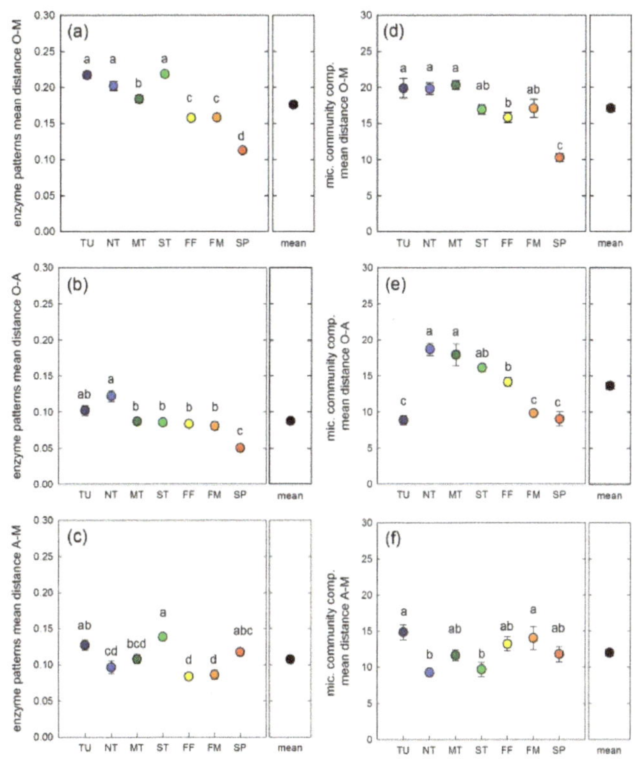

Figure 5: Differences between horizons in enzyme pattern (a–c) and microbial community composition (d–f). Values are the mean distances between the respective shown horizons. Sites are indicated by color (Tundra is purple; northern taiga is blue; middle taiga is dark green; southern taiga light green; forest steppe forest is yellow; forest steppe meadow is orange; steppe is red). Black dots are the mean values over all sites. (For interpretation of the references to color in this figure legend, the reader is referred to the web version of this article.).

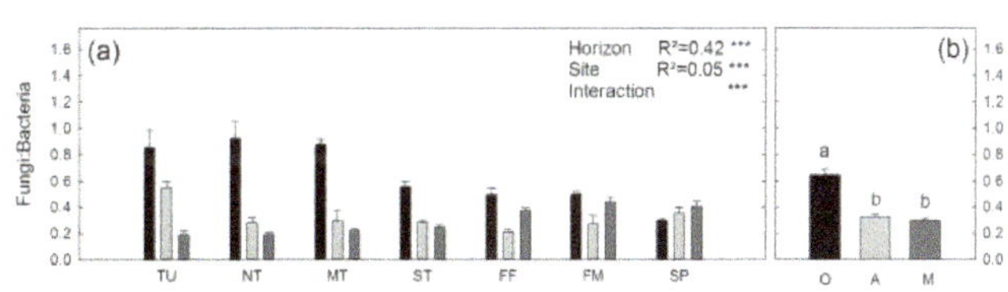

Figure 6: Fungi:bacteria ratios. Left panel (a) shows each horizon (O are organic topsoils, black; A are mineral topsoils, light gray; M are mineral subsoils, dark gray) at each site (TU = Tundra; NT = northern taiga; MT = middle taiga; ST = southern taiga; FF = forest steppe forest; FM = forest steppe meadow; SP = steppe) individually. Right panel (b) shows the mean of the individual horizons over all sites. Results for two-way ANOVA are shown in the left panel. Asterisks indicate significance (*** mean p > 0.001). In the right panel significant differences for horizon means are derived from ANOVA and Tukey HSD tests and are indicated by small letters (p > 0.05).

DISCUSSION

Enzyme Activities and Enzyme Patterns

Subsoil horizons differ in a range of physical and chemical parameters from topsoil horizons (Fierer et al., 2003, Salome et al., 2010 and Rumpel and Kögel-Knabner, 2011). They also exhibit a reduced influence of plants and a higher proportion of the present SOM is associated with minerals (Rumpel and Kögel-Knabner, 2011). These factors are most likely responsible for the clear separation of enzyme patterns according to horizons (R^2 = 0.66), which was stronger than the differences in enzyme patterns between sites (R^2 = 0.14; Fig. 3), in our study. A similar picture in enzyme patterns, with R^2 for horizon of 0.48 and R^2 for site of 0.23 could be found when POX was not included in the analysis (data not shown). The higher R^2 for horizon when POX was included, indicate that the different change of oxidative and hydrolytic enzyme activities with depth was a major factor responsible for the distinct enzyme patterns in topsoils and subsoils. While hydrolytic enzyme activities generally decreased with depth, POX activities did not change from topsoil to subsoil (Fig. 1). An explanation for this might be that the activity of hydrolytic enzymes is often directly related to SOM content (Sinsabaugh et al., 2008 and Schnecker et al., 2014). Since SOM content and the amount of regular polymers, that can be broken down hydrolytically, decrease with depth (Rumpel and Kögel-Knabner, 2011), hydrolytic enzyme activities can be expected to decrease accordingly. Oxidative enzymes, in contrast, are unspecific and are often not produced to directly acquire nutrients (Sinsabaugh, 2010). Instead, oxidative enzymes can degrade humic complexes and thereby free substrates for other enzymes (Hobbie and Horton, 2007 and Talbot et al., 2008) or degrade toxic substances such as phenols (Sinsabaugh, 2010). The production of oxidative enzymes might thus be related to the amount of irregular polymers or the amount of toxins, which are both independent from SOM content. In addition, and in contrast to hydrolytic enzymes, oxidative enzymes are preferentially stabilized on mineral surfaces and might thus prevail longer in mineral subsoils (Kramer et al., 2013). Overall, the contrasting behavior of hydrolytic and oxidative enzymes presumably led to the more pronounced differences in enzyme pattern between horizons than between sites.

Differences between sites and a latitudinal trend could be found for differences between enzyme patterns (expressed as mean distances), in organic topsoil and mineral subsoil horizons (Fig. 3 and Fig. 5). We found that enzyme patterns in topsoils and subsoils were most similar at the southernmost site, although organic topsoils and mineral subsoils in this steppe ecosystem were up to one meter apart from each other (Table 1). In contrast, at the tundra site enzyme patterns showed the greatest differences of all sites between organic topsoils and mineral subsoils, which are less than 50 cm apart. This trend from North to South was not caused by the variability of enzyme patterns in topsoils, but by the large variability in mineral subsoils (Fig. 3). This is in contrast to our hypothesis that greater differences between ecosystems would occur in organic topsoil horizons due to the diverse litter inputs in different ecosystems. Again, these differences between ecosystems in mineral horizons could be found with and without POX, but POX activity, on a microbial C basis (Fig. 2) was the only individual enzyme activity that decreased from North to South. Also in this case enzyme activities might have been controlled by physical parameters, which vary especially in subsoils of different ecosystems. Anoxia and water saturation for instance are common features of subsoil horizons of high latitude ecosystems, whereas they can be neglected as important factors in arid steppe subsoils. Fluctuating oxygen conditions might lead to abiotic oxidation of organic material in the presence of iron or manganese and mimic oxidative enzyme activities (Bach et al., 2013 and Hall and Silver, 2013). This might explain the high oxidative activities found in the northern ecosystems. Lack of oxygen influences oxidative enzyme gravely and although it has been proposed that anoxia does not directly affect hydrolytic enzyme activities (Hall et al., 2014), individual enzymes or their substrates might be differently stabilized on mineral surfaces (Turner et al., 2014). These stabilization mechanisms can be influenced by physical and chemical factors such as O_2 availability and pH or by the mineralogical composition of soils (von Lützow et al., 2006). Along the transect parent material changed from marine deposits at the tundra site to fluvi-glacial deposits in the northern and middle taiga to eolian deposits south of the middle taiga (Archipov et al., 1970). These differences in parent material might be reflected in the mineralogical composition of the soils at these sites.

Enzyme Patterns and Microbial Community Composition

Physical factors in subsoil horizons might also have indirectly influenced enzyme patterns by affecting microbial community composition (Schnecker et al., 2014). In all three soil horizon types, we found significant correlations between enzyme patterns and microbial community composition (Table 3). Microbial community composition itself was significantly different between horizons along this 1500 km long Siberian transect (ADONIS in Fig. 4). The differences between horizons even outranked differences between ecosystems in topsoil horizons, with a mean distance between organic topsoils and mineral subsoils of 17.1 ± 0.4 (Fig. 5) and mean distance within organic topsoil horizons of 10.9 ± 0.2 (Fig. 4b). Similar trends were found for North-American forest systems where the variability of the microbial community composition was greater within soil profiles than between 54 topsoil horizons collected from a wide range of ecosystems (Eilers et al., 2012). This clear picture of the stronger influence of depth than of geographical distance on microbial community composition becomes however blurred when subsoil horizons were also considered. When we included mineral topsoils and mineral subsoils in the statistical analyses, we found significant differences in microbial community composition between ecosystems (R^2 = 0.26), which were as strong as differences between horizons (R^2 = 0.25).

These findings might have two implications: First, the consistent correlations of enzyme patterns and microbial community composition, and the different enzyme activities on a microbial C basis (Fig. 2), point to distinct functional capacities of individual microbial communities in different soil horizons. This also indicates that enzyme patterns are an estimate for a functional community composition. Second, the differences between horizons were more pronounced in enzyme patterns than in the microbial community composition (Fig. 5), which might indicate an additional physiological adaptation of the microbial communities on top of community shifts from topsoils to subsoils.

Enzyme Patterns and Soil Organic Matter

So far, we have shown relations of enzyme patterns and microbial community composition and argued the potential controls of physical parameters over both. Enzyme patterns may however also reflect the availability of different substrates, as well as microbial energy and nutrient demand, and are often related to C/N ratios of the microbial biomass or of the SOM (Sinsabaugh et al., 2008). While, in this present study, enzyme patterns and microbial community composition were related to C and N content in the organic topsoil horizons, these relations were not found in mineral topsoils and mineral subsoils. In contrast to SOM quantity, SOM chemistry might be more important in subsoil horizons. Here, the distinct microbial communities that are presumably adapted to the different environments along the transect could have led to a diverging chemical composition of SOM, similar to a proposed diverging of litter chemistry with ongoing decomposition (Wickings et al., 2012). This divergence has been found in a litter decomposition study and has not been shown for SOM. Enzyme patterns might, nonetheless, reflect a diverged SOM chemistry, and therefore show greater variability in subsoils, where a high proportion of SOM is microbially transformed, than in topsoils, where a higher amount of plant components is present.

In summary our findings show that topsoil horizons and subsoil horizons harbor different microbial communities, which support distinct ways to decompose the available SOM. In accordance to our first hypothesis (i.e., that enzyme patterns in topsoil and subsoil horizons are both related to the same key parameters), we found that differences in enzyme patterns between horizons outranked the differences between ecosystems. However, in contradiction to our third hypothesis (i.e., that enzyme patterns would be more variable and show greater differences between ecosystems in topsoil than in subsoil horizons), we found a higher variability of enzyme patterns in subsoil horizons, which might have been caused by an interplay of physical conditions, microbial community composition and chemical composition of SOM. Although we were not able to identify and describe the mechanisms that shape the microbial community and control enzyme patterns in subsoil horizons in detail, we found that enzyme patterns and thus the strategy of the microbial community to decompose SOM, were not related to SOM content and C/N ratios, which contradicts our second hypothesis. In subsoil horizons, the microbial community, with its functional abilities, might be responsible for the way in which SOM is decomposed. In addition to SOM content or C/N ratios, the microbial community composition might therefore constitute an important factor controlling decomposition rates, especially in subsoil horizons.

ACKNOWLEDGMENTS

This work was funded by the Austrian Science Fund as part of the International Program CryoCARB – Long-term Carbon Storage in Cryoturbated Arctic Soils (FWF – I370-B17).

REFERENCES

1. Archipov, S.A., Vdovon, V.V., Mizerov, B.V., Nikolaev, V.A., 1970. Zapadno-Sibirskaja ravnina. Moskva Nauka, 278 p. (in Russian).
2. Bach, C.E., Warnock, D.D., Van Horn, D.J., Weintraub, M.N., Sinsabaugh, R.L., Allison, S.D., German, D.P., 2013. Measuring phenol oxidase and peroxidase activities with pyrogallol, l-DOPA, and ABTS: effect of assay conditions and soil type. Soil Biology & Biochemistry 67, 183e191.
3. Brockett, B.F., Prescott, C.E., Grayston, S.J., 2012. Soil moisture is the major factor influencing microbial community structure and enzyme activities across seven biogeoclimatic zones in western Canada. Soil Biology & Biochemistry 44, 9e20.
4. Brookes, P.C., Landman, A., Pruden, G., Jenkinson, D.S., 1985. Chloroform fumigation and the release of soil nitrogen: a rapid direct extraction method to measure microbial biomass nitrogen in soil. Soil Biology & Biochemistry 17, 837e842.
5. Colman, B.P., Schimel, J.P., 2013. Drivers of microbial respiration and net N mineralization at the continental scale. Soil Biology & Biochemistry 60, 65e76.
6. Davidson, E.A., Samanta, S., Caramori, S.S., Savage, K., 2012. The Dual Arrhenius and MichaeliseMenten kinetics model for decomposition of soil organic matter at hourly to seasonal time scales. Global Change Biology 18, 371e384.
7. Eilers, K.G., Debenport, S., Anderson, S., Fierer, N., 2012. Digging deeper to find unique microbial communities: the strong effect of depth on the structure of bacterial and archaeal communities in soil. Soil Biology & Biochemistry 50, 58e65.
8. Fierer, N., Allen, A.S., Schimel, J.P., Holden, P.A., 2003. Controls on microbial CO2 production: a comparison of surface and subsurface soil horizons. Global Change Biology 9, 1322e1332.
9. Frostegård, Å., Tunlid, A., Bååth, E., 1991. Microbial biomass measured as total lipid phosphate in soils of different organic content. Journal of Microbiological Methods 14, 151e163.
10. Gittel, A., B arta, J., Kohoutov a, I., Mikutta, R., Owens, S., Gilbert, J., Schnecker, J., Wild, B., Hannisdal, B., Maerz, J., Lashchinskiy, N., Capek, P., Santruckova, H., Gentsch, N., Shibistova, O., Guggenberger, G., Richter, A., Torsvik, V.L., Schleper, C., Urich, T., 2014. Distinct microbial communities associated with buried soils in the Siberian tundra. ISME Journal 841e853.
11. Grandy, A., Strickland, M., Lauber, C., Bradford, M., Fierer, N., 2009. The influence of microbial communities, management, and soil texture on soil organic matter chemistry. Geoderma 150, 278e286.
12. Hall, S.J., Silver, W.L., 2013. Iron oxidation stimulates organic matter decomposition in humid tropical forest soils. Global Change Biology 19, 2804e2813.
13. Hall, S.J., Treffkorn, J., Silver, W.L., 2014. Breaking the enzymatic latch: impacts of reducing conditions on hydrolytic enzyme activity in tropical forest soils.

Ecology. http://dx.doi.org/10.1890/13-2151.1.

14. Hobbie, E.A., Horton, T.R., 2007. Evidence that saprotrophic fungi mobilise carbon and mycorrhizal fungi mobilise nitrogen during litter decomposition. New Phytologist 173, 447e449.

15. Holden, P.A., Fierer, N., 2005. Microbial processes in the Vadose Zone. Vadose Zone Journal 4, 1.

16. IUSS Working Group WRB, 2006. World Reference Base for Soil Resources 2006-A Framework for International Classification, Correlation and Communication. Rome. Jobb agy, E.G., Jackson, R.B., 2000. The vertical distribution of soil organic carbon and its relation to climate and vegetation. Ecological Applications 10, 423e436.

17. Kaiser, C., Koranda, M., Kitzler, B., Fuchslueger, L., Schnecker, J., Schweiger, P., Rasche, F., Zechmeister-Boltenstern, S., Sessitsch, A., Richter, A., 2010. Belowground carbon allocation by trees drives seasonal patterns of extracellular enzyme activities by altering microbial community composition in a beech forest soil. New Phytologist 187, 843e858.

18. Kaiser, C., Franklin, O., Dieckmann, U., Richter, A., 2014. Microbial community dynamics alleviate stoichiometric constraints during litter decay. Ecology Letters 17, 680e690.

19. Keeler, B.L., Hobbie, S.E., Kellogg, L.E., 2009. Effects of long-term nitrogen addition on microbial enzyme activity in eight forested and grassland sites: implications for litter and soil organic matter decomposition. Ecosystems 1e15.

20. Kleber, M., 2010. What is recalcitrant soil organic matter? Environmental Chemistry 7, 320.

21. Kramer, S., Marhan, S., Haslwimmer, H., Ruess, L., Kandeler, E., 2013. Temporal variation in surface and subsoil abundance and function of the soil microbial community in an arable soil. Soil Biology & Biochemistry 61, 76e85.

22. Maestre, F.T., Salguero-Gomez, R., Quero, J.L., 2012. It is getting hotter in here: determining and projecting the impacts of global environmental change on drylands. Philosophical Transactions of the Royal Society B: Biological Sciences 367, 3062e3075.

23. McGuire, K.L., Treseder, K.K., 2010. Microbial communities and their relevance for ecosystem models: decomposition as a case study. Soil Biology & Biochemistry 42, 529e535.

24. Meyer, H., Kaiser, C., Biasi, C., Hammerle, R., Rusalimova, O., Lashchinsky, N., € Baranyi, C., Daims, H., Barsukov, P., Richter, A., 2006. Soil carbon and nitrogen 114 Jdynamics along a latitudinal transect in Western Siberia, Russia. Biogeochemistry 81, 239e252.

25. Oksanen, J., Blanchet, F.G., Kindt, R., Legendre, P., Minchin, P.R., O›Hara, R.B., Simpson, G.L., Solymos, P., Steven, M.H.H., Wagner, H., 2013. Vegan: Community Ecology Package.

26. R Development Core Team, 2013. R: a Language and Environment for Statistical Computing.

27. Reynolds, W.D., Topp, G.C., 2007. Soil water analyses. In: Carter, M.R., Gregorich, E.G. (Eds.), Soil Sampling and Methods of Analysis, second ed. CRC Press.

28. Rovira, P., Vallejo, V.R., 2002. Mineralization of carbon and nitrogen from plant debris, as affected by debris size and depth of burial. Soil Biology & Biochemistry 34, 327e339.

29. Rumpel, C., Kogel-Knabner, I., 2011. Deep soil organic matter € da key but poorly understood component of terrestrial C cycle. Plant and Soil 338, 143e158.

30. Salome, C., Nunan, N., Pouteau, V., Lerch, T.Z., Chenu, C., 2010. Carbon dynamics in topsoil and in subsoil may be controlled by different regulatory mechanisms. Global Change Biology 16, 416e426.

31. Schnecker, J., Wild, B., Fuchslueger, L., Richter, A., 2012. A field method to store samples from temperate mountain grassland soils for analysis of phospholipid fatty acids. Soil Biology & Biochemistry 51, 81e83.

32. Schnecker, J., Wild, B., Hofhansl, F., Eloy Alves, R.J., Barta, J., Capek, P., Fuchslueger, L., Gentsch, N., Gittel, A., Guggenberger, G., Hofer, A., Kienzl, S., Knoltsch, A., Lashchinskiy, N., Mikutta, R., Santruckova, H., Shibistova, O., Takriti, M., Urich, T., Weltin, G., Richter, A., 2014. Effects of soil organic matter properties and microbial community composition on enzyme activities in cryoturbated arctic soils. PLoS One 9, e94076.

33. Sinsabaugh, R.L., 2010. Phenol oxidase, peroxidase and organic matter dynamics of soil. Soil Biology & Biochemistry 42, 391e404.

34. Sinsabaugh, R.L., Follstad Shah, J.J., 2010. Ecoenzymatic stoichiometry of recalcitrant organic matter decomposition: the growth rate hypothesis in reverse. Biogeochemistry 102, 31e43.

35. Sinsabaugh, R.L., Lauber, C.L., Weintraub, M.N., Ahmed, B., Allison, S.D., Crenshaw, C., Contosta, A.R., Cusack, D., Frey, S., Gallo, M.E., Gartner, T.B., Hobbie, S.E., Holland, K., Keeler, B.L., Powers, J.S., Stursova, M., Takacs-Vesbach, C., Waldrop, M.P., Wallenstein, M.D., Zak, D.R., Zeglin, L.H., 2008. Stoichiometry of soil enzyme activity at global scale. Ecology Letters 11, 1252e1264.

36. Stolbovoi, V., McCallum, I., 2002. Land Resources of Russia (CD). International Institute for Applied Systems Analysis and the Russian Academy of Science, Laxenburg, Austria.

37. Stone, M.M., DeForest, J.L., Plante, A.F., 2014. Changes in extracellular enzyme activity and microbial community structure with soil depth at the Luquillo Critical Zone Observatory. Soil Biology & Biochemistry 75, 237e247.

38. Strickland, M.S., Lauber, C., Fierer, N., Bradford, M.A., 2009. Testing the functional significance of microbial community composition. Ecology 90, 441e451.

39. Talbot, J.M., Allison, S.D., Treseder, K.K., 2008. Decomposers in disguise: mycorrhizal fungi as regulators of soil C dynamics in ecosystems under global change. Functional Ecology 22, 955e963.

40. Talbot, J.M., Bruns, T.D., Smith, D.P., Branco, S., Glassman, S.I., Erlandson, S., Vilgalys, R., Peay,

K.G., 2013. Independent roles of ectomycorrhizal and saprotrophic communities in soil organic matter decomposition. Soil Biology & Biochemistry 57, 282e291.

41. Turner, S., Schippers, A., Meyer-Stüve, S., Guggenberger, G., Gentsch, N., Dohrmann, R., Condron, L.M., Eger, A., Almond, P.C., Peltzer, D.A., Richardson, S.J., Mikutta, R., 2014. Mineralogical impact on long-term patterns of soil nitrogen and phosphorus enzyme activities. Soil Biology & Biochemistry 68, 31e43.

42. Von Lützow, M., Kogel-Knabner, I., Ekschmitt, K., Matzner, E., Guggenberger, G., Marschner, B., Flessa, H., 2006. Stabilization of organic matter in temperate soils: mechanisms and their relevance under different soil conditions - a review. European Journal of Soil Science 57, 426e445.

43. Waldrop, M.P., Firestone, M.K., 2006. Response of microbial community composition and function to soil climate change. Microbial Ecology 52, 716e724.

44. Wallander, H., Nilsson, L.O., Hagerberg, D., Rosengren, U., 2003. Direct estimates of C: N ratios of ectomycorrhizal mycelia collected from Norway spruce forest soils. Soil Biology & Biochemistry 35, 997e999.

45. Wallenstein, M.D., McMahon, S.K., Schimel, J.P., 2009. Seasonal variation in enzyme activities and temperature sensitivities in Arctic tundra soils. Global Change Biology 15, 1631e1639.

46. Wickings, K., Grandy, A.S., Reed, S.C., Cleveland, C.C., 2012. The origin of litter chemical complexity during decomposition. Ecology Letters 15, 1180e1188.

CHAPTER 6

INPUT OF EASILY AVAILABLE ORGANIC C AND N STIMULATES MICROBIAL DECOMPOSITION OF SOIL ORGANIC MATTER IN ARCTIC PERMAFROST SOIL

Birgit Wild, Jörg Schnecker, Ricardo J. Eloy Alves, Pavel Barsukov, Jiří Bárta, Petr Čapek, Norman Gentsch, Antje Gittel, Georg Guggenberger, Nikolay Lashchinskiy, Robert Mikutta, Olga Rusalimova, Hana Šantrůčková, Olga Shibistova, Tim Urich, Margarete Watzka, Galina Zrazhevskaya, Andreas Richter

Keywords: Priming, Organic matter decomposition, Phospholipid fatty acid (PLFA), Tundra, Permafrost

INTRODUCTION

Soil organic matter (SOM) decomposition rates in permafrost soils are expected to increase with rising temperatures in the Arctic (Hartley et al., 2008 and Conant et al., 2011). In addition to the direct temperature effect, warming might also indirectly affect SOM decomposition, mediated by an increase in plant net primary production. Higher plant productivity is accompanied by an increased input of plant-derived C into the soil (as root litter or root exudates), and can thus increase soil C stocks, as observed for a mineral subsoil in the Alaskan tundra (Sistla et al., 2013). In contrast, higher productivity was found to reduce soil C stocks in a sub-arctic system, offsetting the increase in above- and belowground plant biomass, and, consequently, leading to a net loss of C from the ecosystem (Hartley et al., 2012). Plants supply the soil microbial community with a range of organic compounds that can be either immediately taken up by microorganisms (e.g., sugars, amino acids and organic acids from root exudation), or that can be easily decomposed (e.g., cellulose and protein from root litter). These organic compounds can stimulate the soil microbial community to decompose more SOM ("priming effect"; Bingeman et al., 1953), (i) by promoting microbial groups that target complex compounds of SOM (Fontaine et al., 2003), (ii) by providing the energy to break down these compounds (Blagodatskaya and Kuzyakov, 2008), or (iii) by providing C for microbial growth, thus increasing microbial N demand and facilitating N mining, i.e., the microbial breakdown of SOM to get access to N (Craine et al., 2007 and Dijkstra et al., 2013).

The latter two mechanisms might be of particular importance in arctic soils. Microbial activity in arctic soils is considered N limited (Sistla et al., 2012), suggesting that an increased allocation of plant C to the soil might strongly stimulate N mining. Additionally, microbial activity in subsoil horizons in general is considered energy limited (Fontaine et al., 2007), as the subsoil is poorly rooted, and supply of plant-derived compounds from root exudation and root litter is scarce. With 80% of arctic SOM located below 30 cm (Tarnocai et al., 2009), a large amount of SOM might be protected from decomposition by energy limitation of microbial decomposers, and might thus be particularly susceptible to an increased input of plant-derived compounds. So far, it is unknown how SOM decomposition in different horizons of arctic permafrost soils will respond to an increased input of plant-derived organic compounds, and what mechanisms might be involved. We here report on the susceptibility of different soil horizons from a tundra ecosystem in the central Siberian Arctic to an increased

Citation: Birgit Wild, Jörg Schnecker, Ricardo J. Eloy Alves, Pavel Barsukov, Jiří Bárta, Petr Čapek, Norman Gentsch, Antje Gittel, Georg Guggenberger, Nikolay Lashchinskiy, Robert Mikutta, Olga Rusalimova, Hana Šantrůčková, Olga Shibistova, Tim Urich, Margarete Watzka, Galina Zrazhevskaya, Andreas Richter, Input of easily available organic C and N stimulates microbial decomposition of soil organic matter in arctic permafrost soil, Soil Biology and Biochemistry, Volume 75, August 2014, Pages 143-151, ISSN 0038-0717, http://dx.doi.org/10.1016/j.soilbio.2014.04.014.

availability of organic compounds. In a priming experiment, we compared organic topsoil and mineral subsoil material, as well as cryoturbated material, i.e., topsoil material that was buried in the subsoil by freeze–thaw processes (Bockheim, 2007 and Tarnocai et al., 2009). Cryoturbated organic matter is common in arctic soils, accounting for approximately 400 Gt of C (Harden et al., 2012). Although it is chemically similar to topsoil organic matter (Xu et al., 2009), it shows retarded decomposition as indicated by low respiration rates (Kaiser et al., 2007) and relatively old radiocarbon ages (Kaiser et al., 2007, Xu et al., 2009 and Hugelius et al., 2010). We hypothesized that an increased availability of organic compounds would stimulate SOM decomposition in the subsoil (i.e., in mineral subsoil and cryoturbated horizons) by providing energy for microbial decomposers, but less so in the topsoil, where energy is not limiting. Additionally, we tested if priming of SOM decomposition was connected to N mining, by comparing the effect of organic substrates with and without N. We expected that N-containing substrates would result in a weaker priming effect than substrates without N, by reducing the dependence of the microbial community on SOM as an N source. Finally, we investigated if priming of SOM decomposition was connected to a shift in microbial community composition.

To that end, we analyzed SOM-derived respiration and microbial community composition in soil samples amended with ^{13}C-labeled glucose, cellulose, amino acids, or protein, in comparison with unamended controls. We thus compared substrates containing N to substrates without N, as well as monomeric substrates to polymeric substrates. Since monomeric substrates are immediately available for microorganisms, whereas polymeric substrates need to be broken down by extracellular enzymes before microbial uptake, they might differ in their effect on microbial community composition and function, and thus on SOM decomposition (Fontaine et al., 2003).

MATERIAL & METHODS

Soil Sampling

Soils were sampled on the Taymyr peninsula in the central Siberian Arctic (72° 29.57′ N, 101° 38.62′ E), from a shrubby moss tundra (bioclimatic subzone D; CAVM Team, 2003) dominated by *Cassiope tetragona*, *Carex arctisibirica*, *Tomentypnum nitens* and *Aulacomnium turgidum*. The soil was described as a Turbic Cryosol according to the World Reference Base for Soil Resources (IUSS Working Group WRB, 2007) or Typic Aquiturbel according to the US Soil Taxonomy (Soil Survey Staff, 1999), with fine to coarse loamy texture and an active layer depth of around 80 cm at the time of sampling in August 2011. We took samples from three soil horizons in the active layer: We sampled the OA horizon (topsoil material), as well as a buried Ajj (cryoturbated material) and the adjacent BCg horizon (mineral subsoil material), the latter two from a depth of 50–70 cm. Soils were sampled in a 2 m-wide soil profile by pooling samples taken horizontally with a metal soil corer within each horizon. We took care that the mineral subsoil material sampled did not include buried organic material and vice versa. Living roots were carefully removed and samples homogenized by hand directly after sampling. Carbon and nitrogen contents of the individual horizons were 9.4% C and 0.5% N for topsoil material, 4.3% C and 0.2% N for cryoturbated material, and 0.6% C, 0.1% N for mineral subsoil material (determined with a Perkin Elmer 2400 Series II CHNS/O analyzer).

Incubation Experiment

To investigate the effect of increased C availability on SOM decomposition, we amended the soils with ^{13}C-labeled glucose, amino acids, cellulose, or protein. ^{13}C-labeled glucose was purchased from Sigma–Aldrich (U-^{13}C, 99 at%), ^{13}C-labeled amino acids from Cambridge Isotope Laboratories (algal amino acid mixture, U-^{13}C, 97–99 at%), ^{13}C-labeled cellulose from Isolife (low degree of polymerization *Cichorium intybus*, U-^{13}C, >97 at%), and ^{13}C-labeled protein from Sigma–Aldrich (algal crude protein extract, U-^{13}C, 98 at%). All substrates were mixed with the respective unlabeled compounds to 10 at% ^{13}C before application, and all substrates were applied in dry form.

Aliquots of fresh soil were amended with glucose, amino acids, cellulose or protein of 10 at% ^{13}C, in five replicates of 25 g per treatment, or left unamended as controls (three sets of five 25 g replicates). We adjusted the amount of substrate to the approximate C content of each horizon by adding 554 µg C g^{-1} to topsoil material, 138 µg C g^{-1} to cryoturbated material and 55 µg C g^{-1} to mineral subsoil material. One set of controls was immediately harvested to determine the initial state before the start of the incubation; the remaining samples were filled into microcosms (50 ml polypropylene tubes with aeration holes in the bottom, for a detailed description see Inselsbacher et al., 2009) that were then loosely plugged with synthetic wool, and incubated at 10 °C. Since we expected a faster response to monomeric (glucose, amino acids) than to polymeric substrates (cellulose, protein), we incubated samples for either six days (glucose, amino acids, and one set of controls) or ten days (cellulose, protein, and one set of controls).

Respiration

Respiration rates were determined in four replicate microcosms on days 0, 1, 2, 3, 4, 6 (glucose and amino acids, plus controls), or 0, 2, 4, 6, 8, 10 (cellulose and protein, plus controls) after the start of the incubation, with the first sampling immediately after substrate amendment. For respiration measurements, we removed the synthetic wool from the microcosms, attached polypropylene tubes to the microcosms (resulting in an overall headspaces of 133–141 ml) and tightened the systems with rubber seals as described elsewhere (Inselsbacher et al., 2009). We incubated microcosms at 10 °C, took samples of 15 ml from the headspace after 10, 25 and 40 min (5, 10, 20 and 30 min for the glucose treatment), and injected the gas samples into pre-evacuated gas vials. After each sampling, we replaced the removed gas volume with air from a gas bag, filled with ambient air of known CO_2 concentration and isotopic composition (CO_2concentrations of 473–632 ppm depending on the sampling date, 1.09 atom% ^{13}C). Concentration and isotopic composition of CO_2 were measured with a GasBench II system coupled to a Delta V Advantage IRMS (Thermo Scientific), and corrected for the replaced gas volume. We then calculated the contribution of SOM-derived C to respiration using the equation:

$$R_{SOM}=R_{total}\cdot at\%_{Rtotal}-at\%_{sub}/at\%_{SOM}-at\%_{sub}, \quad (4)$$

where R_{total} and R_{SOM} are total and SOM-derived respiration, and $at\%_{Rtotal}$, $at\%_{SOM}$, and $at\%_{sub}$ are C isotope compositions (in atom% ^{13}C) of total respiration, SOM and the added substrate, respectively.

Microbial Biomass and Community Composition

We estimated microbial biomass as concentration of phospholipid fatty acids (PLFAs), described microbial community composition using PLFAs as biomarkers for different microbial groups, and assessed microbial substrate preferences by tracing C derived from the added substrates into individual PLFAs. Aliquots of 1 g fresh soil were stored in RNAlater (Schnecker et al., 2012), and processed following Frostegård et al. (1991), with the modifications described by Kaiser et al. (2010b). Briefly, PLFAs were extracted from the samples with chloroform/methanol/citric acid buffer, purified on silica columns (LC-Si SPE, Supelco) with chloroform, acetone, and methanol, amended with methyl-nonadecanoate as internal standard, and converted to fatty acid methyl esters (FAMEs) by alkaline methanolysis. We determined concentration and isotopic composition of individual FAMEs on a GC-IRMS system consisting of a Trace GC coupled to a Delta V Advantage IRMS over a GC Isolink interface (Thermo Scientific). Samples were injected in splitless mode at 300 °C, and individual FAMEs were separated on a DB-23 column (Agilent; GC-method: 70 °C for 1.5 min, ramp of 30 °C min^{-1} to 150 °C, 150 °C for 1 min, ramp of 4 °C min^{-1} to 230 °C, 230 °C for 20 min) with 1.5 ml helium min^{-1} as carrier gas. We used qualitative standard mixes (37 Comp. FAME Mix and Bacterial Acid Methyl Esters CP Mix, Sigma–Aldrich) for peak identification, and the internal standard methyl-nonadecanoate for quantification. Concentration and isotopic composition of each PLFA were corrected for C added during derivatization. We further considered only PLFAs that were detectable in all horizons, using i15:0, a15:0, i16:0, i17:0 and a17:0 as markers for Gram-positive bacteria, 16:1ω7, 18:1ω7 and cy17:0(9/10) as markers for Gram-negative bacteria, 18:1ω9 and 18:2ω6 as markers for fungi, and 16:0, 17:0, 18:0, 20:0, 16:1ω5, 16:1ω11 and 19:1ω8 as non-specific PLFAs (Kaiser et al., 2010a and Kaiser et al., 2010b). We expressed all PLFA values on the basis of C in PLFAs: Total concentrations are presented as total C in all PLFAs, the relative contributions of individual PLFAs as percentages thereof. To compare the microbial utilization of different substrates across soil horizons, we calculated substrate use efficiency by comparing the amounts of substrate-derived C respired and incorporated into PLFAs, using the equations

$$CR_{sub}=CR_{total}\cdot at\%_{CRtotal}-at\%_{SOM}/at\%_{sub}-at\%_{SOM}, \quad (2)$$

$$PLFA_{sub}=PLFA_{total}\cdot at\%_{PLFAtotal}-at\%_{SOM}/at\%_{sub}-at\%_{SOM}, \quad (3)$$

and

$$\text{Substrate use efficiency} = (PLFA_{sub})/(PLFA_{sub} + CR_{sub}). \quad (4)$$

CR_{total} and CR_{sub} represent total and substrate-derived C in cumulative respiration, $PLFA_{total}$ and $PLFA_{sub}$ total and substrate-derived C in PLFAs, and $at\%_{CRtotal}$, $at\%_{PLFAtotal}$, $at\%_{SOM}$, and $at\%_{sub}$ are C isotope compositions (in atom% ^{13}C) of total cumulative respiration, total PLFAs, SOM and the added substrate, respectively. Substrate use efficiency values thus reflect the partitioning of C between respiration and microbial growth (incorporation into PLFAs). Since we did not apply correction factors to convert PLFA concentrations into total microbial C, substrate use efficiency values should only be used for comparison between horizons and treatments.

We additionally determined microbial biomass using chloroform-fumigation-extraction (Kaiser et al., 2011). Samples fumigated with chloroform, as well as unfumigated samples, were extracted with 0.5 M K_2SO_4 and analyzed for C concentration with an HPLC-IRMS system in direct injection mode against sucrose standards (for description of the system see Wild et al., 2010). Microbial C was calculated as the difference between fumigated and non-fumigated samples.

Statistical Analyses

All statistics were performed with R 2.15.0 (R Development Core Team, 2012), with packages vegan (Oksanen et al., 2012) and ecodist (Goslee and Urban, 2007). We preferentially used parametric methods, and log-transformed data, if necessary, to meet conditions of normal distribution and homoscedasticity. If conditions could not be met, we applied non-parametric methods. We performed one-way ANOVAs with Tukey's HSD tests between treatments and controls for each soil horizon to test for effects of substrate addition on cumulative SOM-derived respiration, and two-way ANOVA to test for differences in substrate use efficiency across all horizons and substrates. Since PLFA data did not meet the conditions for parametric methods, we applied Mann–Whitney-U tests to assess differences in absolute amounts of total and substrate-derived C, and the relative contributions of individual PLFAs to both. We further performed non-metric multidimensional scaling (NMDS) to describe the effects of substrate addition on microbial community composition (% of total C in PLFAs; scaled data), and to describe the distribution of substrate derived C across the community (% of total substrate-derived C in PLFAs; scaled data). Differences in PLFA patterns were assessed using cluster analysis (Ward hierarchical clustering) and analysis of similarities (ANOSIM). Throughout this text, we use the term "significant" only when referring to statistical results (with differences considered significant at $p < 0.05$).

RESULTS

Basal Respiration and Priming Effect

Respiration rates of unamended control samples differed between soil horizons by more than one order of magnitude, with higher rates in topsoil (0.379 ± 0.030 µmol CO_2 g^{-1} dry soil h^{-1}, mean ± standard error) than in cryoturbated and mineral subsoil horizons (0.024 ± 0.002 and 0.019 ± 0.002 µmol CO_2 g^{-1} dry soil h^{-1}, respectively). Considering the differences in C content between soil horizons, respiration rates were in the same range for topsoil and mineral subsoil (4.050 ± 0.317 and 3.227 ± 0.351 µmol CO_2 g^{-1} soil C h^{-1}), but accounted for only 14% of the topsoil rates in the cryoturbated horizon (0.547 ± 0.043 µmol CO_2 g^{-1} soil C h^{-1}), indicating a slower decomposition of cryoturbated organic matter (Fig. 1, Supplementary Table 1).

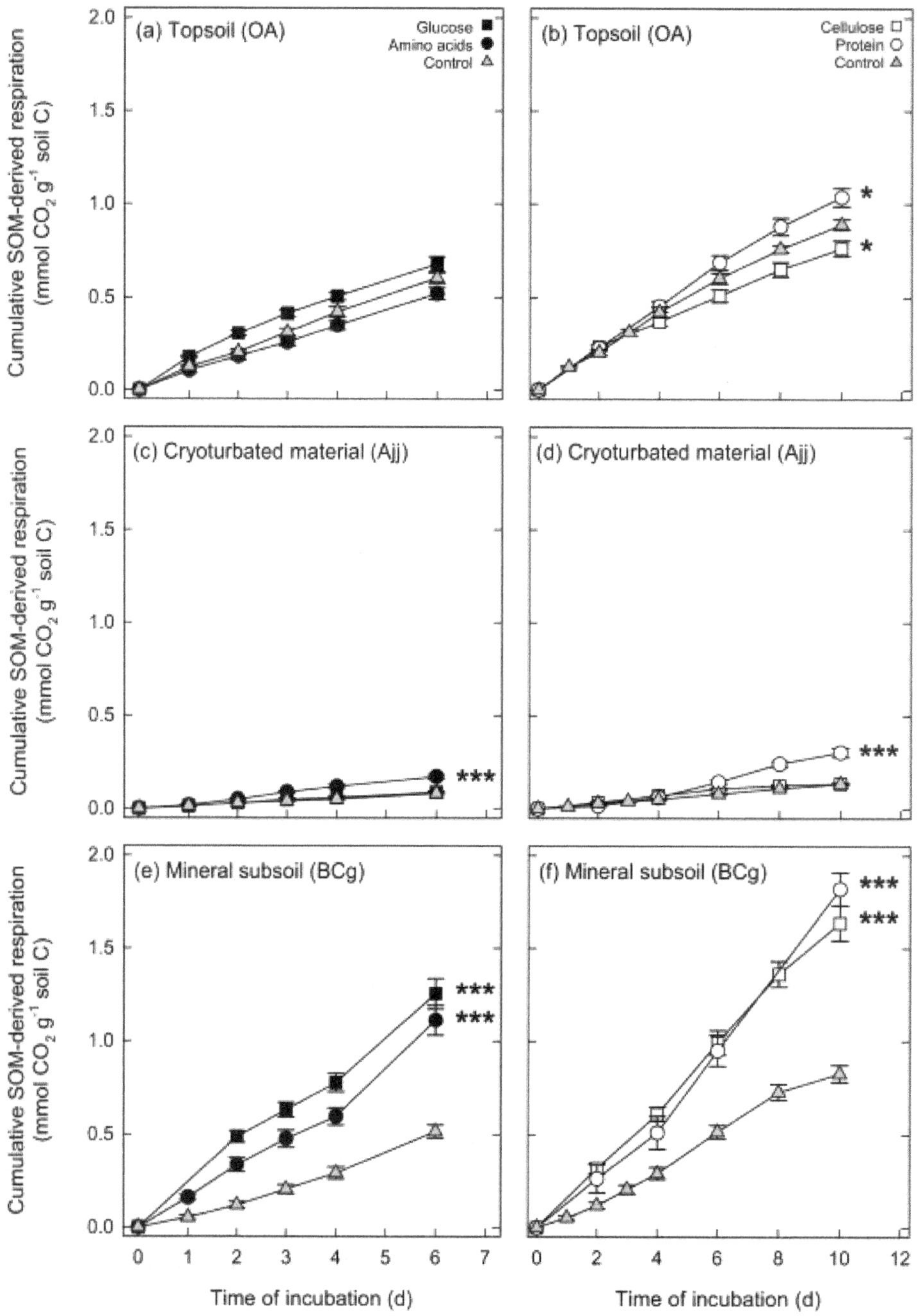

Figure 1: Cumulative SOM-derived respiration of topsoil, cryoturbated, and mineral subsoil horizons from a tundra ecosystem in the central Siberian Arctic. Samples were amended with glucose, amino acids, cellulose or protein (added at levels corresponding to approximately 1% of soil organic C), or left unamended as controls. Points represent means ± standard errors of four replicates. Significant differences between amended samples and controls at the end of the incubation are indicated (*, $p < 0.05$; **, $p < 0.01$; ***, $p < 0.001$).

The addition of organic compounds significantly affected SOM-derived respiration in all horizons. The response of SOM-derived respiration in the topsoil was weak, and significant differences were only observed for cellulose (−14%) and protein treatments (+17%; Fig. 2). In the mineral subsoil, however, all amendments (glucose, amino acids, cellulose, and protein) increased SOM-derived respiration two- to threefold (Fig. 2), demonstrating a strong priming of SOM decomposition by all substrates, which ultimately lead to a higher cumulative respiration than in the topsoil (per soil C; Fig. 1). For cryoturbated material, additions of amino acids and protein doubled SOM-derived respiration, whereas additions of glucose and cellulose had no significant effect (Fig. 2). This suggests that N availability was a major control on SOM decomposition in the cryoturbated horizon. The addition of organic substrates can increase the turnover of microbial C, resulting in higher respiration rates without changing SOM decomposition ("apparent priming", Blagodatskaya and Kuzyakov, 2008). In our study, the priming effect in both mineral subsoil (all substrates) and cryoturbated material (amino acids and protein) was in all cases much larger than the initial microbial biomass (Supplementary Table 2), eliminating apparent priming as an explanation for the observed increase in SOM-derived respiration.

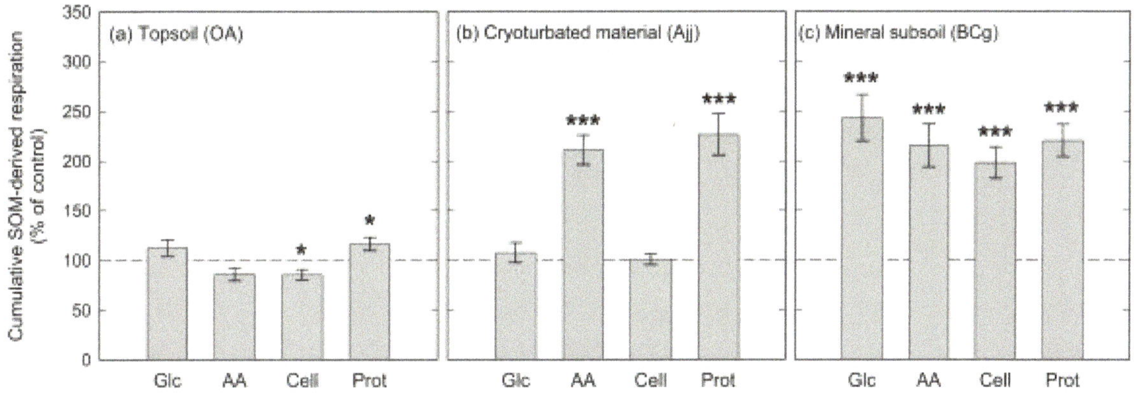

Figure 2: Cumulative SOM-derived respiration of topsoil, cryoturbated, and mineral subsoil horizons from a tundra ecosystem in the central Siberian Arctic, after incubation with glucose (Glc), amino acids (AA), cellulose (Cell) or protein (Prot). Bars represent means ± standard errors of four replicates, expressed as percent of unamended controls which are depicted by the dashed line. Significant differences between amended samples and controls are indicated (*, $p < 0.05$; **, $p < 0.01$; ***, $p < 0.001$).

Microbial Communities

We found different microbial communities in topsoil, cryoturbated, and mineral subsoil horizons, with the latter two harboring the most similar communities. Communities in cryoturbated and mineral subsoil horizons were distinguished from topsoil communities by lower abundances of fungi (18:1ω9, 18:2ω6) and Gram-negative bacteria (18:1ω7, cy17:0), and higher abundances of the non-specific PLFAs 16:0, 18:0 and 20:0 (Supplementary Fig. 1, Supplementary Table 3).

The addition of monomeric substrates (glucose, amino acids) caused a significant increase in the total amount of PLFAs in topsoil and cryoturbated horizons, and a decrease in the mineral subsoil (significant only for glucose; Supplementary Table 3). Microbial community composition in the topsoil was hardly affected by substrate additions, whereas microbial community composition in cryoturbated and mineral subsoil material changed significantly after addition of monomeric substrates (Fig. 3). Specifically, relative abundances of the non-specific PLFAs 18:0 and 20:0 decreased, and abundances of fungi (18:1ω9) and Gram-negative bacteria (cy17:0, 16:1ω7, 18:1ω7) increased (Supplementary Table 3). The addition of monomeric substrates to cryoturbated and mineral subsoil material thus shifted the microbial community composition closer to the topsoil. In contrast, the addition of polymeric substrates (cellulose, protein) had little effect on microbial community composition (Fig. 3, Supplementary Table 3). The community changes in cryoturbated and mineral subsoil material thus suggest a specific stimulation of fast-growing microbial groups by substrates that can be immediately taken up by microorganisms (i.e., glucose and amino acids).

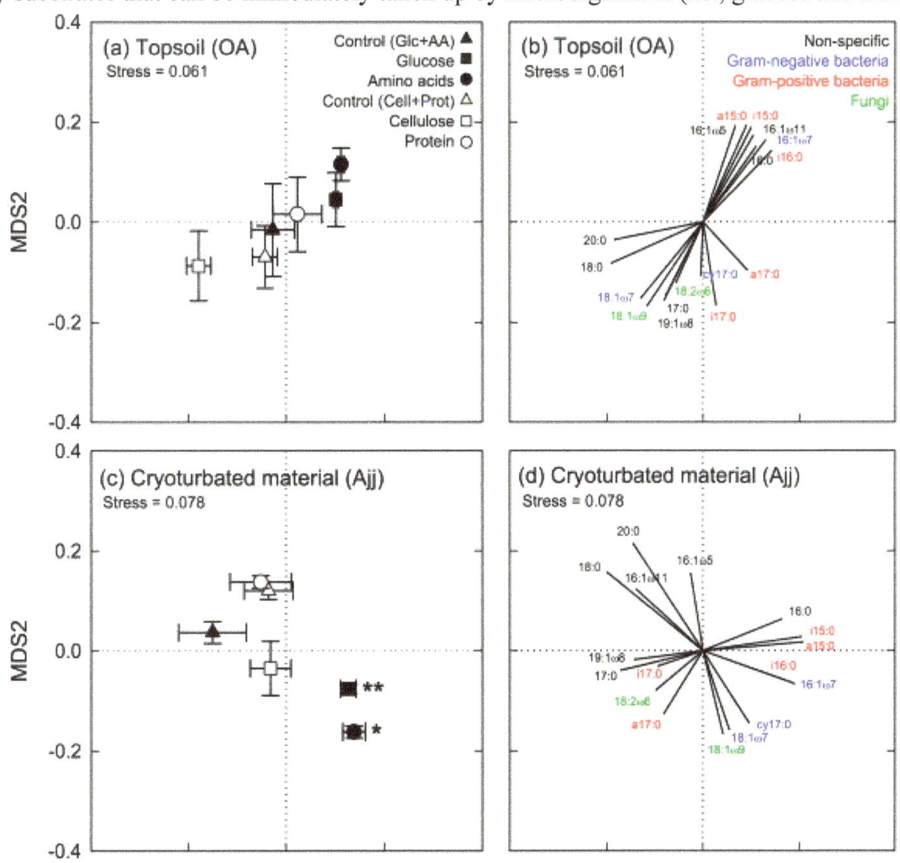

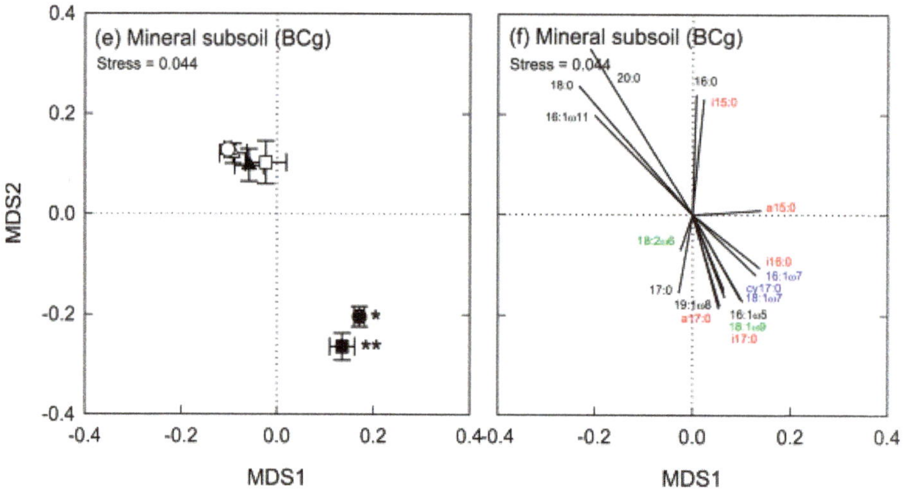

Figure 3: Phospholipid fatty acid (PLFA) patterns in topsoil, cryoturbated, and mineral subsoil horizons from a tundra ecosystem in the central Siberian Arctic. Samples were amended with glucose, amino acids, cellulose or protein, or left unamended as controls. Patterns are displayed by non-metric multidimensional scaling (NMDS) of PLFA abundances (% of total C in PLFAs). Points represent means ± standard errors of five replicates; significant differences to the respective controls are indicated (*, $p < 0.05$; **, $p < 0.01$; ***, $p < 0.001$). Data of PLFA abundances are presented in Supplementary Table 3.

The response of microbial biomass and microbial community composition to substrate additions, did, however, not reflect the response of SOM decomposition. Changes in biomass and community composition were connected to substrate complexity, whereas priming of SOM decomposition was unspecific (mineral subsoil) or connected to N (cryoturbated material).

Substrate Use

Substrate use efficiency was significantly lower in the mineral subsoil than in topsoil and cryoturbated horizons, i.e., a higher proportion of substrate-derived C was respired and a smaller proportion incorporated into PLFAs (Fig. 4). Comparing different substrates, C from amino acids was more efficiently incorporated into PLFAs than C from glucose, cellulose, and protein.

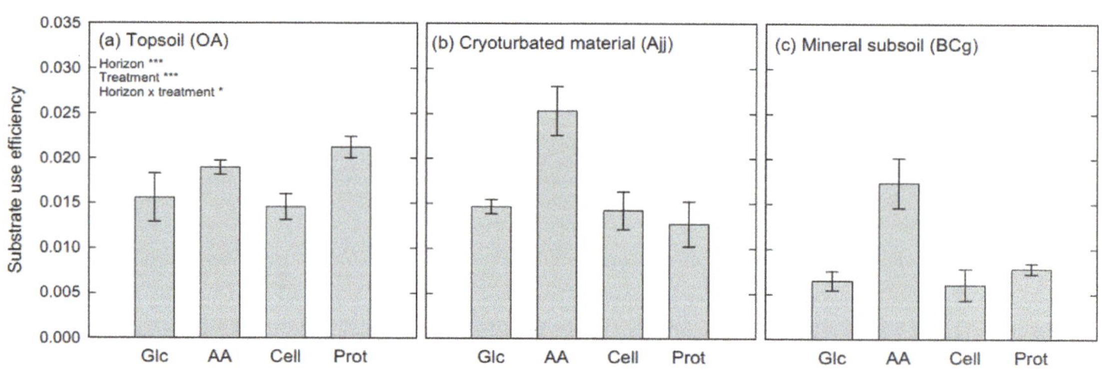

Figure 4: Substrate use efficiency of topsoil, cryoturbated, and mineral subsoil horizons from a tundra ecosystem in the central Siberian Arctic, after incubation with ^{13}C-labeled glucose (Glc), amino acids (AA), cellulose (Cell) or protein (Prot). Substrate use efficiency was estimated as the proportion of substrate-derived C in PLFAs over substrate-derived C in both PLFAs and respiration. Bars represent means ± standard errors of four replicates.

The incorporation of C from different substrates into PLFAs suggests specific substrate preferences of different microbial groups. In cryoturbated material, the incorporation pattern of substrate-derived C depended on the presence of N in the substrate (Fig. 5; separation displayed along MDS 1), and on substrate complexity (separation along MDS 2). In particular, C from N-containing substrates was preferentially incorporated into the Gram-negative biomarkers 18:1ω7, cy17:0, and into the fungal biomarker 18:1ω9, which together contained 47% and 39% of C from amino acids and protein incorporated into PLFAs, but only 25% and 22% of C from glucose and cellulose (Supplementary Table 4). Carbon from monomeric substrates was preferentially incorporated into a15:0 (Gram-positive biomarker) and 16:1ω7 (Gram-negative biomarker), and C from polymeric substrates into 18:2ω6 (fungal biomarker), as well as the non-specific PLFAs 16:0, 17:0, 18:0 and 20:0. A separation due to substrate N content and substrate complexity was not found for other horizons. In topsoil and mineral subsoil, glucose, amino acid and protein treatments had more similar incorporation patterns, whereas cellulose was distinct due to a preferential incorporation into a set of Gram-positive biomarkers (i17:0, a17:0) and non-specific PLFAs (16:0, 16:1ω5, and 16:1ω11) in the topsoil, and into the non-specific PLFA 16:0 in the mineral subsoil (Supplementary Table 4).

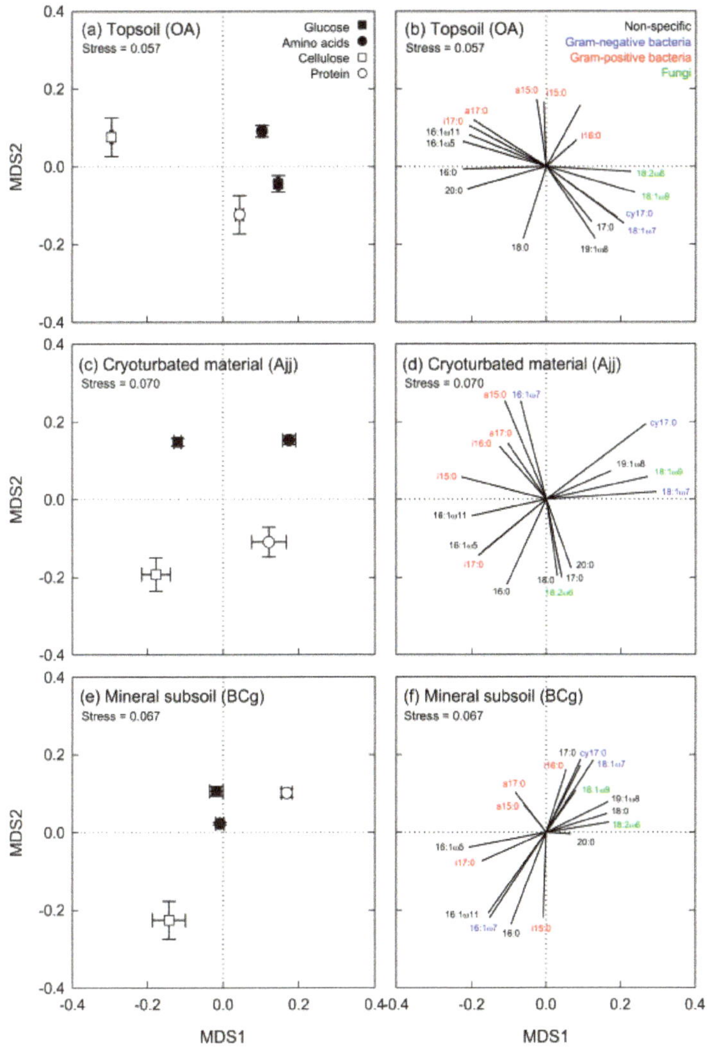

Figure 5: Incorporation patterns of C from glucose, amino acids, cellulose or protein into phospholipid fatty acids (PLFAs), in topsoil, cryoturbated, and mineral subsoil horizons from a tundra ecosystem in the central Siberian Arctic. Patterns are displayed by non-metric multidimensional scaling (NMDS). Points represent means ± standard errors of five replicates; in all horizons, patterns for all substrates were significantly different from each other. Data on the incorporation of substrate-derived C into PLFAs are presented in Supplementary Table 4.

DISCUSSION

Arctic permafrost soils contain large amounts of organic C, with recent estimates of approximately 400 Gt C in mineral subsoil and cryoturbated subsoil horizons each, and 250 Gt C in organic topsoil horizons (Harden et al., 2012). This organic C might be vulnerable to priming, caused by an increasing availability of plant-derived organic compounds with rising temperatures.

We found that organic topsoil, mineral subsoil and cryoturbated horizons responded differently to an increased availability of organic compounds. Substrate additions hardly affected rates of SOM-derived respiration in the organic topsoil, whereas rates more than doubled in the mineral subsoil, irrespective of the substrate added (Fig. 2). Substrate use efficiency (the fraction of substrate taken up invested into PLFAs) was lower in the mineral subsoil than in the other horizons (Fig. 4), indicating that a higher proportion of the available substrate was needed to generate energy. These findings thus support our hypothesis of energy limitation in the mineral subsoil in arctic permafrost soil. Energy limitation of SOM decomposition in mineral subsoil horizons has also been shown for a temperate grassland, where the addition of cellulose induced the mineralization of millennia-old soil organic C (Fontaine et al., 2007).

In our study, priming of SOM decomposition in the mineral subsoil was not connected to an increase in microbial biomass or a change in community composition, since significant community changes were only observed for glucose and amino acid amendments, but not for cellulose or protein (Fig. 3, Supplementary Table 3). Priming was also not connected to N mining of microbial decomposers, since substrates with and without N similarly stimulated SOM decomposition (Fig. 2).

In contrast to the mineral subsoil, SOM decomposition in cryoturbated material was not altered by additions of glucose and cellulose, but doubled when amino acids or protein were added (Fig. 2). This specific response to the N-containing substrates, as well as the higher substrate use efficiency than in the mineral subsoil (Fig. 4), suggest that microbial decomposers in the cryoturbated horizon were not limited in energy, but in N. Cryoturbated material is usually similar in N content and C/N ratio to organic or mineral topsoil horizons, but has lower N transformation rates (Kaiser et al., 2007 and Wild et al., 2013), suggesting that N bound in cryoturbated organic material is poorly available for microorganisms (e.g., due to binding to phenolic compounds; Olk et al., 2006). In our study, the C/N ratio of cryoturbated material (21.4) even slightly exceeded the C/N ratio of the organic topsoil (18.5); this might have further promoted N limitation in cryoturbated material.

Carbon from N-containing substrates (amino acids and protein) was preferentially incorporated into fungal (18:1ω9) and Gram-negative bacterial biomarkers (18:1ω7, cy17:0; Fig. 5). Each of these markers was of significantly lower relative abundance in the cryoturbated than in the topsoil horizon, accounting for a total of 13% (cryoturbated) and 40% (topsoil) of C in PLFAs (means of controls; Supplementary Fig. 1). Differences in microbial community composition between topsoil and cryoturbated horizons have previously been suggested to contribute to the slower decomposition of cryoturbated material (Schnecker et al., 2014), in particular the lower abundance of fungi in cryoturbated horizons (Wild et al., 2013 and Gittel et al., 2014). Fungi are pivotal for the decomposition of organic material in most terrestrial systems (de Boer et al., 2005 and Strickland and Rousk, 2010), and can benefit from higher N availability (Rousk and Bååth, 2007, Grman and Robinson, 2013 and Koranda et al., 2014). However, although the preference of specific microbial groups for N-containing compounds is striking, this does not necessarily identify these groups as responsible for priming.

Nitrogen addition often decreases SOM decomposition rates, by reducing N mining of microbial decomposers (e.g., Craine et al., 2007). On the other hand, in systems of strong N limitation, where low N availability limits the production of SOM-degrading enzymes, N addition can stimulate decomposition by facilitating enzyme production by fungi (Allison et al., 2009). Since in our study the amendment of amino acids and protein to cryoturbated material did not systematically increase microbial biomass or the abundance of specific microbial groups (Fig. 3, Supplementary Table 3), we suggest that microorganisms invested the additional N in the production of extracellular enzymes.

Within a relatively short period, SOM decomposition in the subsoil, but not in the topsoil, was strongly stimulated by increased availability of organic C (mineral subsoil) or organic N (cryoturbated material). Rising temperatures in the Arctic (IPCC, 2007) are expected to promote soil C availability by increasing plant primary productivity (Bhatt et al., 2010 and Xu et al., 2013) and thus C allocation from plants to the soil. Nitrogen availability might also benefit from higher plant productivity, if the additional input of N-containing compounds by plants (e.g., proteins) exceeds the increase in plant N uptake (Weintraub and Schimel, 2005). In addition, N availability might further increase if SOM decomposition rates increase (Nadelhoffer et al., 1991, Hobbie, 1996, Schimel et al., 2004 and Schaeffer et al., 2013).

In our study, mineral subsoil material was particularly susceptible to priming, with rates of SOM-derived respiration exceeding topsoil rates (as related to soil C) after substrate addition. Our findings are thus in contrast to observations of increased C storage in mineral subsoil horizons with warming (Sistla et al., 2013), but support predictions of high C losses from arctic soils if plant productivity increases (Hartley et al., 2012). These contrasting findings on the impact of higher plant productivity on soil C stocks point to a strong context-dependency of priming effects in arctic soils. Our study provides key insights into the mechanisms behind priming effects in permafrost soils; this will allow possible feedbacks from belowground plant–microbe interactions to be incorporated into larger-scale models of arctic C balances in a future climate.

ACKNOWLEDGMENTS

This study was funded by the Austrian Science Fund (FWF) as part of the International Program CryoCARB (Long-term Carbon Storage in Cryoturbated Arctic Soils; FWF – I370-B17).

REFERENCES

1. Allison, S.D., LeBauer, D.S., Ofrecio, M.R., Reyes, R., Ta, A., Tran, T.M., 2009. Low levels of nitrogen addition stimulate decomposition by boreal forest fungi. Soil Biology & Biochemistry 41, 293e302.
2. Bhatt, U.S., Walker, D.A., Raynolds, M.K., Comiso, J.C., Epstein, H.E., Jia, G., Gens, R., Pinzon, J.E., Tucker, C.J., Tweedie, C.E., Webber, P.J., 2010. Circumpolar Arctic tundra vegetation change is linked to sea ice decline. Earth Interactions 14, 1e20.
3. Bingeman, C.W., Varner, J.E., Martin, W.P., 1953. The effect of the addition of organic materials on the decomposition of an organic soil. Soil Science Society of America Journal 17, 34e38.
4. Blagodatskaya, E., Kuzyakov, Y., 2008. Mechanisms of real and apparent priming effects and their dependence on soil microbial biomass and community structure: critical review. Biology and Fertility of Soils 45, 115e131.
5. Bockheim, J.G., 2007. Importance of cryoturbation in redistributing organic carbon in permafrost-affected soils. Soil Science Society of America Journal 71, 1335.
6. CAVM Team, 2003. Circumpolar Arctic Vegetation Map. (1:7,500,000 scale), Conservation of Arctic Flora and Fauna (CAFF) Map No. 1. U.S. Fish and Wildlife Service, Anchorage, Alaska.
7. Conant, R.T., Ryan, M.G., Ågren, G.I., Birge, H.E., Davidson, E.A., Eliasson, P.E., Evans, S.E., Frey, S.D., Giardina, C.P., Hopkins, F.M., Hyvönen, R., Kirschbaum, M.U., Lavallee, J.M., Leifeld, J., Parton, W.J., Steinweg, J.M., Wallenstein, M.D., Wetterstedt, J.Å., Bradford, M.A., 2011. Temperature and soil organic matter decomposition rates e synthesis of current knowledge and a way forward. Global Change Biology 17, 3392e3404.
8. Craine, J., Morrow, C., Fierer, N., 2007. Microbial nitrogen limitation increases decomposition. Ecology 88, 2105e2113.
9. de Boer, W., Folman, L.B., Summerbell, R.C., Boddy, L., 2005. Living in a fungal world: impact of fungi on soil bacterial niche development. FEMS Microbiology Reviews 29, 795e811.
10. Dijkstra, F.A., Carrillo, Y., Pendall, E., Morgan, J.A., 2013. Rhizosphere priming: a nutrient perspective. Frontiers in Microbiology 4, 1e8.
11. Fontaine, S., Barot, S., Barré, P., Bdioui, N., Mary, B.,

Rumpel, C., 2007. Stability of organic carbon in deep soil layers controlled by fresh carbon supply. Nature 450, 277e280.

12. Fontaine, S., Mariotti, A., Abbadie, L., 2003. The priming effect of organic matter: a question of microbial competition? Soil Biology & Biochemistry 35, 837e843.

13. Frostegård, Å., Tunlid, A., Bååth, E., 1991. Microbial biomass measured as total lipid phosphate in soils of different organic content. Journal of Microbiological Methods 14, 151e163.

14. Gittel, A., Bárta, J., Kohoutová, I., Mikutta, R., Owens, S., Gilbert, J., Schnecker, J., Wild, B., Hannisdal, B., Maerz, J., Lashchinskiy, N., Capek, P., Santrucková, H., Gentsch, N., Shibistova, O., Guggenberger, G., Richter, A., Torsvik, V.L., Schleper, C., Urich, T., 2014. Distinct microbial communities associated with buried soils in the Siberian tundra. The ISME Journal 8, 841e853.

15. Goslee, S.C., Urban, D.L., 2007. The ecodist package for dissimilarity-based analysis of ecological data. Journal of Statistical Software 22, 1e19.

16. Grman, E., Robinson, T.M., 2013. Resource availability and imbalance affect plantmycorrhizal interactions: a field test of three hypotheses. Ecology 94, 62e67.

17. Harden, J.W., Koven, C.D., Ping, C., Hugelius, G., McGuire, A.D., Camill, P., Jorgenson, T., Kuhry, P., Michaelson, G.J., O'Donnell, J.A., Schuur, E.A.G., Taronocai, C., Johnson, K., Grosse, G., 2012. Field information links permafrost carbon to physical vulnerabilities of thawing. Geophysical Research Letters 39, L15704.

18. Hartley, I.P., Garnett, M.H., Sommerkorn, M., Hopkins, D.W., Fletcher, B.J., Sloan, V.L., Phoenix, G.K., Wookey, P.A., 2012. A potential loss of carbon associated with greater plant growth in the European Arctic. Nature Climate Change 2, 875e879.

19. Hartley, I.P., Hopkins, D.W., Garnett, M.H., Sommerkorn, M., Wookey, P.A., 2008. Soil microbial respiration in arctic soil does not acclimate to temperature. Ecology Letters 11, 1092e1100.

20. Hobbie, S.E., 1996. Temperature and plant species control over litter decomposition in Alaskan tundra. Ecological Monographs 66, 503e522.

21. Hugelius, G., Kuhry, P., Tarnocai, C., Virtanen, T., 2010. Soil organic carbon pools in a periglacial landscape: a case study from the central Canadian Arctic. Permafrost and Periglacial Processes 21, 16e29.

22. IPCC, 2007. Climate Change 2007: Impacts, Adaptation and Vulnerability. Contribution of Working Group II to the Fourth Assessment Report of the Intergovernmental Panel on Climate Change. Cambridge University Press, Cambridge, UK.

23. Inselsbacher, E., Ripka, K., Klaubauf, S., Fedosoyenko, D., Hackl, E., Gorfer, M., HoodNovotny, R., von Wirén, N., Sessitsch, A., Zechmeister-Boltenstern, S., Wanek, W., Strauss, J., 2009. A cost-effective high-throughput microcosm system for studying nitrogen dynamics at the plant-microbe-soil interface. Plant and Soil 317, 293e307.

24. IUSS Working Group WRB, 2007. World Reference Base for Soil Resources 2006, First Update 2007. Rome.

25. Kaiser, C., Frank, A., Wild, B., Koranda, M., Richter, A., 2010a. Negligible contribution from roots to soil-borne phospholipid fatty acid fungal biomarkers 18:2u6,9 and 18:1u9. Soil Biology & Biochemistry 42, 1650e1652.

26. Kaiser, C., Fuchslueger, L., Koranda, M., Gorfer, M., Stange, C.F., Kitzler, B., Rasche, F., Strauss, J., Sessitsch, A., Zechmeister-Boltenstern, S., Richter, A., 2011. Plants control the seasonal dynamics of microbial N cycling in a beech forest soil by belowground C allocation. Ecology 92, 1036e1051.

27. Kaiser, C., Koranda, M., Kitzler, B., Fuchslueger, L., Schnecker, J., Schweiger, P., Rasche, F., Zechmeister-Boltenstern, S., Sessitsch, A., Richter, A., 2010b. Belowground carbon allocation by trees drives seasonal patterns of extracellular enzyme activities by altering microbial community composition in a beech forest soil. New Phytologist 187, 843e858.

28. Kaiser, C., Meyer, H., Biasi, C., Rusalimova, O., Barsukov, P., Richter, A., 2007. Conservation of soil organic matter through cryoturbation in arctic soils in Siberia. Journal of Geophysical Research 112, G02017.

29. Koranda, M., Kaiser, C., Fuchslueger, L., Kitzler, B., Sessitsch, A., ZechmeisterBoltenstern, S., Richter, A., 2014. Fungal and bacterial utilization of organic substrates depends on substrate complexity and N availability. FEMS Microbiology Ecology 87, 142e152.

30. Nadelhoffer, K.J., Giblin, A.E., Shaver, G.R., Laundre, J.A., 1991. Effects of temperature and substrate quality on element mineralization in six arctic soils. Ecology 72, 242e253.

31. Oksanen, J., Blanchet, F.G., Kindt, R., Legendre, P., Minchin, P.R., O'Hara, R.B., Simpson, G.L., Solymos, P., Stevens, M.H., Wagner, H., 2012. vegan: Community Ecology Package. R Package Version 2.0-3.

32. Olk, D.C., Cassman, K.G., Schmidt-Rohr, K., Anders, M.M., Mao, J.D., Deenik, J.L., 2006. Chemical stabilization of soil organic nitrogen by phenolic lignin residues in anaerobic agroecosystems. Soil Biology & Biochemistry 38, 3303e3312.

33. R Development Core Team, 2012. R: a Language and Environment for Statistical Computing. R Foundation for Statistical Computing.

34. Rousk, J., Bååth, E., 2007. Fungal and bacterial growth in soil with plant materials of different C/N ratios. FEMS Microbiology Ecology 62, 258e267.

35. Schaeffer, S.M., Sharp, E., Schimel, J.P., Welker, J.M., 2013. Soileplant N processes in a High Arctic ecosystem, NW Greenland are altered by longterm experimental warming and higher rainfall. Global Change Biology 19, 3529e3539.

36. Schimel, J.P., Bilbrough, C., Welker, J.M., 2004. Increased snow depth affects microbial activity and nitrogen mineralization in two Arctic tundra communities. Soil Biology & Biochemistry 36, 217e227.

37. Schnecker, J., Wild, B., Fuchslueger, L., Richter, A., 2012. A field method to store samples from temperate

mountain grassland soils for analysis of phospholipid fatty acids. Soil Biology & Biochemistry 51, 81e83.

38. Schnecker, J., Wild, B., Hofhansl, F., Alves, R.J.E., Bárta, J., Capek, P., Fuchslueger, L., Gentsch, N., Gittel, A., Guggenberger, G., Hofer, A., Kienzl, S., Knoltsch, A., Lashchinskiy, N., Mikutta, R., Santrucková, H., Shibistova, O., Takriti, M., Urich, T., Weltin, G., Richter, A., 2014. Effects of soil organic matter properties and microbial community composition on enzyme activities in cryoturbated arctic soils. PLoS One 9, e94076.

39. Sistla, S.A., Asao, S., Schimel, J.P., 2012. Detecting microbial N-limitation in tussock tundra soil: implications for Arctic soil organic carbon cycling. Soil Biology & Biochemistry 55, 78e84.

40. Sistla, S.A., Moore, J.C., Simpson, R.T., Gough, L., Shaver, G.R., Schimel, J.P., 2013. Long-term warming restructures Arctic tundra without changing net soil carbon storage. Nature 497, 615e618.

41. Soil Survey Staff, 1999. Soil Taxonomy: a Basic System of Soil Classification for Making and Interpreting Soil Surveys, second ed. U. S. Department of Agriculture Handbook 436.

42. Strickland, M.S., Rousk, J., 2010. Considering fungal:bacterial dominance in soils e methods, controls, and ecosystem implications. Soil Biology & Biochemistry 42, 1385e1395.

43. Tarnocai, C., Canadell, J.G., Schuur, E.A., Kuhry, P., Mazhitova, G., Zimov, S., 2009. Soil organic carbon pools in the northern circumpolar permafrost region. Global Biogeochemical Cycles 23, GB2023.

44. Weintraub, M.N., Schimel, J.P., 2005. Seasonal protein dynamics in Alaskan arctic tundra soils. Soil Biology & Biochemistry 37, 1469e1475.

45. Wild, B., Schnecker, J., Bárta, J., Capek, P., Guggenberger, G., Hofhansl, F., Kaiser, C., Lashchinsky, N., Mikutta, R., Mooshammer, M., Santrucková, H., Shibistova, O., Urich, T., Zimov, S.A., Richter, A., 2013. Nitrogen dynamics inTurbic Cryosols from Siberia and Greenland. Soil Biology & Biochemistry 67, 85e93.

46. Wild, B., Wanek, W., Postl, W., Richter, A., 2010. Contribution of carbon fixed by Rubisco and PEPC to phloem export in the Crassulacean acid metabolism plant Kalanchoë daigremontiana. Journal of Experimental Botany 61, 1375e1383.

47. Xu, C., Guo, L., Ping, C., White, D.M., 2009. Chemical and isotopic characterization of size-fractionated organic matter from cryoturbated tundra soils, northern Alaska. Journal of Geophysical Research 114, G03002.

48. Xu, L., Myneni, R.B., Chapin III, F.S., Callaghan, T.V., Pinzon, J.E., Tucker, C.J., Zhu, Z., Bi, J., Ciais, P., Tømmervik, H., Euskirchen, E.S., Forbes, B.C., Piao, S.L., Anderson, B.T., Ganguly, S., Nemani, R.R., Goetz, S.J., Beck, P.S., Bunn, A.G., Cao, C., Stroeve, J.C., 2013. Temperature and vegetation seasonality diminishment over northern lands. Nature Climate Change 3, 581e586. B. Wild et.

CHAPTER 7

LAND SYSTEM SCIENCE AND SUSTAINABLE DEVELOPMENT OF THE EARTH SYSTEM

Peter H. Verburg, Neville Crossman, Erle C. Ellis, Andreas Heinimann, Patrick Hostert, Ole Mertz, Harini Nagendra, Thomas Sikor, Karl-Heinz Erb, Nancy Golubiewski, Ricardo Grau, Morgan Grove, Souleymane Konaté, Patrick Meyfroidt, Dawn C. Parker, Rinku Roy Chowdhury, Hideaki Shibata, Allison Thomson, Lin Zhen

Keywords: Land use, Land cover, Land systems, Ecosystem, Human-environment systems, Global Land Project

INTRODUCTION

Land systems constitute the terrestrial component of the Earth system and encompass all processes and activities related to the human use of land, including socioeconomic, technological and organizational investments and arrangements, as well as the benefits gained from land and the unintended social and ecological outcomes of societal activities (Verburg et al., 2013). Changes in land systems have large consequences for the local environment and human well-being and are at the same time pervasive factors of global environmental change. Land provides vital resources to society, such as food, fuel, fibres and many other ecosystem services that support production functions, regulate risks of natural hazards, or provide cultural and spiritual services. By using the land, society alters and modifies the quantity and quality of the provision of these services.

Land system changes are the direct result of human decision making at multiple scales ranging from local land owners decisions to national scale land use planning and global trade agreements. The aggregate impact of many local land system changes has far reaching consequences for the Earth System that feedback on ecosystem services, human well-being and decision making (Crossman et al., 2013). As a consequence, land system change is both a cause and consequence of socio-ecological processes.

The land system science community is organised through the Global Land Project (GLP). The orientation of land system science at the interface of social, physical and ecological systems is reflected by the origin of GLP, which, until the transition to Future Earth, was a core project of both the International Geosphere Biosphere Programme (IGBP) and the International Human Dimensions Programme on Global Environmental Change (IHDP) commissioned by the International Council for Science (ICSU) and the International Social Science Council (ISSC). The Global Land Project started in 2006 after publishing its science plan in 2005 (GLP, 2005). GLP is a successor of the previous Land Use and Land Cover Change project (LUCC; 1994–2005) and the Global Change and Terrestrial Ecosystems project (GCTE; 1992–2003). GLP aims at synthesis and integration of insights, knowledge and methodologies in research across the land system science community. A core task of GLP is the identification of scientific priorities and agenda setting through synthesis of existing knowledge, meta-analysis of land-based research and targeted workshops. In addition, GLP provides a platform for the land system science community through networking activities, such as the organization of workshops and conferences. This paper aims to describe the development of land system science and its main achievements and provides a perspective of land system science within global environmental change research. Finally, the paper provides an outlook on emerging issues in land system science and identifies the challenges for the land system science community to translate explorative research into solution-oriented research, one of the priorities under the Future Earth initiative.

Citation: Peter H. Verburg, Neville Crossman, Erle C. Ellis, Andreas Heinimann, Patrick Hostert, Ole Mertz, Harini Nagendra, Thomas Sikor, Karl-Heinz Erb, Nancy Golubiewski, Ricardo Grau, Morgan Grove, Souleymane Konaté, Patrick Meyfroidt, Dawn C. Parker, Rinku Roy Chowdhury, Hideaki Shibata, Allison Thomson, Lin Zhen, Land system science and sustainable development of the earth system, Volume 12, December 2015, Pages 29-41, ISSN 2213-3054, http://dx.doi.org/10.1016/j.ancene.2015.09.004.

LAND SYSTEM SCIENCE ACHIEVEMENTS

The Development of Land System Science

Land system science has developed over the past twenty years. The study of land use and land cover change (coordinated through the LUCC project) was initially dominated by monitoring and modelling of the ecological impacts of major land cover changes on the natural system (Turner et al., 1993, Lambin et al., 2000 and Lambin and Geist, 2006). An orientation towards sustainability always has been a constitutive element of land system science (Turner et al., 2007), and therefore key processes studied in case-study synthesis and land use modelling were deforestation and desertification. Achievements were made in terms of observing land cover changes by remote sensing for single case studies as well as in global datasets (Walsh and Crews-Meyer, 2002). As part of LUCC activities, Belward (1996) developed definitions of land cover classes. The classification employed was developed to meet the needs of other IGBP projects. However, later it was recognised that specific projects and study objectives required different legends. The development of a common land cover legend became less of a priority for the LUCC community.

One of the main achievements of the early LUCC work was the synthesis of case studies to identify common driving factors of change and causation patterns (Geist and Lambin, 2002 and Geist and Lambin, 2004). At the same time, land use models were developed that allowed the exploration of future scenarios of land use change (Verburg et al., 1999 and Pontius et al., 2001).

Besides the LUCC project the Global Change and Terrestrial Ecosystems (GCTE) project contributed by the research on terrestrial ecosystem changes under local, regional and global environmental changes such as increasing concentrations of greenhouse gases, changes in global and regional climate, habitat destruction, and increases in number and impacts of exotic invasive species (Pitelka et al., 2007). The overarching goal of the GCTE project was to project the effect of changes in climate, atmospheric composition, and land use on terrestrial ecosystem and to determine how these effects lead to feedbacks on the atmosphere and physical climate system. GCTE took the lead in analyzing the nature of nonlinear changes in Earth System functioning. This work played a central role in the emergence of abrupt change, surprises and extreme events as unifying themes in the second phase of IGBP research.

Gradually, the research field of land use and land cover change matured and became more integrative, focusing on both the drivers and impacts of land change and including a wider range of interacting processes of land use change. This development was stimulated by the strong exchanges with other communities working on related issues. The growing group of researchers engaged in this field led to the emergence of 'Land Change Science' as a separate, interdisciplinary, research field engaging scientists across the social, economic, geographical and natural sciences (Rindfuss et al., 2004 and Turner et al., 2007). The increasing attention for the feedbacks between drivers and impacts (Verburg, 2006), the interactions between social and ecological systems and telecoupling between world regions (Lambin and Meyfroidt, 2011 and Liu et al., 2013) and between cities and their rural hinterlands (Seto et al., 2012) have motivated an integrated socio-ecological systems perspective. In this integrated concept, land systems are conceptualised as the result of dynamic interactions within the socio-ecological system that operate across spatial and temporal scales. This perspective has also moved land system science from a focus on the most dramatic land cover changes to greater attention for subtle changes of human interactions with the natural surroundings, including land management (Erb et al., 2013 and Kuemmerle et al., 2013) and the provisioning of a wide range of ecosystem services (Crossman et al., 2013). The following sections describe a number of major achievements of land system science over the past decade through illustrations of the different types of research conducted in this field.

Meta-Analysis of Land use Drivers and Impacts

Place-based research in small case-studies has always formed a key component of land system science. In order to better contextualise individual case studies and identify generic patterns across case-studies, meta-analyses were conducted, focusing on the causes of key land change processes or consequences of land system change (Geist and Lambin, 2002, Geist and Lambin, 2004, Keys and McConnell, 2005, Don et al., 2011, Gibson et al., 2011, Seto et al., 2011, van Vliet et al., 2012, Ziegler et al., 2012 and Van Asselen et al., 2013). Methodological development of meta-analysis for the specific purpose of global environmental change and land system science has also been carried out, e.g. on the utility of qualitative comparative analysis for comparing highly diverse case studies from different regions and the possibilities and caveats of using statistical analysis when case studies are not designed in a similar way (Rudel, 2008). Two recent studies provide an overview of synthesis and meta-analytical studies in land system science since 1995 (Magliocca et al., 2014 and van Vliet et al., 2015). From the total of 181 studies analysed it became clear that while the number of meta-studies is increasing rapidly, more interaction between researchers focusing on meta-studies and those conducting case-study is needed to ensure that meta-analyses capture the essence of case-studies and that case-studies target the research gaps identified in meta-analysis. A tool developed in GLP context to further such integration is the GLOBE project. GLOBE (global collaboration engine; http://globe.umbc.edu) is an online collaborative environment designed for real-time assessment of the global context and relevance of local case studies, both individually and in collections, as used for meta-studies. To make this possible, GLOBE leverages an advanced geocomputation and visualization engine that couples local case-study locations with more than 100 global datasets characterizing social and environmental patterns across the Earth. Using GLOBE, site-based researchers quantify, map and publish the global relevance of their studies while identifying additional studies and researchers working in globally similar environments around the world. Researchers can engage in cross-site synthesis to assess geographic biases and global knowledge gaps across their case study collections (Fig. 1) and can correct them by adding more studies in underrepresented areas and by reweighting studies to better reflect their global abundance. With this type of tool, researchers can collaborate globally across sites, environments and social contexts with the ultimate aim of nurturing a robust culture of globally relevant knowledge creation and sharing in land system science.

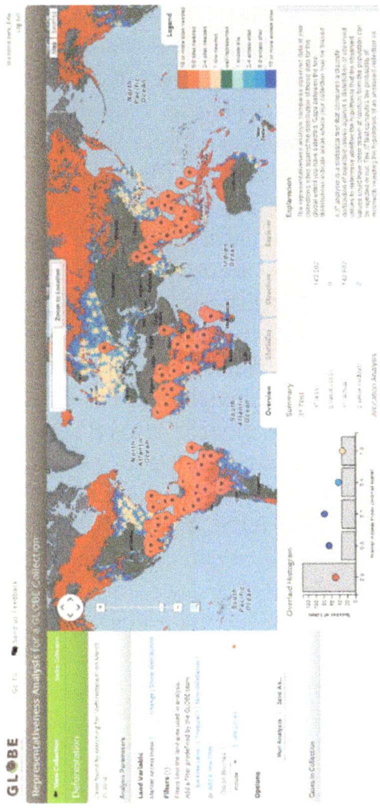

Figure 1: Global representativeness of 273 deforestation case studies identified by searching in GLOBE (of 875 total; published collection data (Ellis, 2014)). Colours indicate the coverage of case-studies across different classes of market access across Earth's woodlands, with red areas indicating significant gaps in study coverage in more remote regions.

Long-term Histories of Land use Change

Land system science studies the past, current and projected state and dynamics of land use. Human use of land is a major component of the long-term anthropogenic global changes in the Earth system that have underpinned the call for the Anthropocene as a new epoch of geologic time (Zalasiewicz et al., 2012). A collaborative effort was made to review and compare the latest global datasets on human populations and land use over the course of the past 8000 years to evaluate human use of land as a global force transforming the terrestrial biosphere (Ellis et al., 2013). Different historic reconstructions of global land use relying on different models of land use intensification produced radically different assessments of the emergence, history, and future of land use as a process transforming the Earth System (Fig. 2; (Klein Goldewijk et al., 2011, Kaplan et al., 2012 and Ellis et al., 2013). By combining these historical reconstructions with evidence on land use systems and their intensification from archaeology, paleoecology and environmental history, it was possible to explain why relatively small human populations could have caused widespread and profound ecological changes more than 3000 years ago, while in recent times the largest and wealthiest human populations in history are using less arable land per person every decade. While human population as well as biomass consumption by humans has grown fourfold and economic output 17-fold in the last decade, the global human appropriation of net primary production (HANPP) has only doubled in this period of time (Krausmann et al., 2013), mainly due to surges in food production per unit of land and the use of external (e.g., energy) inputs in agriculture. Reconstructions of past developments that take such adaptive changes in land-use systems over time into account, including land-use intensification, offered a more spatially detailed and plausible assessment of our planet›s history, with a biosphere and perhaps even climate, long ago affected by humans (Fig. 2). Based on this assessment, further work to empirically validate a date for the emergence of land use as a global force transforming the biosphere and initiating the Anthropocene has been proposed (Ellis et al., 2013). While land-use processes are now diverging rapidly from historical patterns into new patterns and novel land systems, integrative global land-use models that incorporate dynamic adaptations in human-environment relationships help to advance our understanding of both past and future land-use changes, including their sustainability and potential global effects.

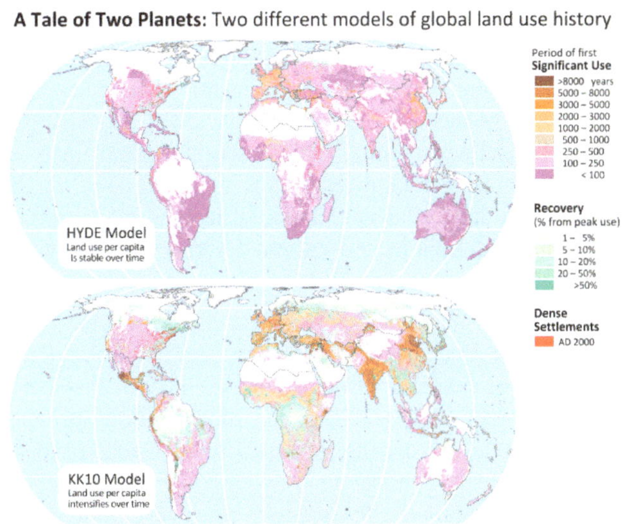

Figure 2: Different land use intensification models produce different views of land use as a global force transforming the terrestrial biosphere. Maps show time period of first significant land use and recovery from peak land use, 6000 B.C.–A.D. 2000, based on two different historical land-use reconstructions (upper plate, HYDE, and lower plate, KK10 model that accounts for changes in land productivity per person).

Reproduced with permission from Ellis et al. (2013).

Observing Land Change

Progress in land system science depends on both conceptual progress and availability of accurate land change data. Accelerating global change and increasing global connections forces land system scientists to strive for more timely and accurate data. At the same time, our increasing knowledge of land change and the more sophisticated questions arising from that knowledge enforce the need for increased semantic depth in land use data. Data for land use change analyses are derived largely from remote sensing analyses or census information, or from finer scale field surveys, cadastres or participatory mapping. While much progress has been made in harmonizing such heterogeneous data, there are multiple challenges associated with spatially, temporally and thematically heterogeneous data sources and their inherent uncertainties (Verburg et al., 2011). During the last decade two major developments have transformed land change data generation and analysis. Firstly, there has been a move from land cover to land use information as well as increased attention for the collection of qualitative information. This development is largely driven by the verdict that better information on land use intensification is needed to improve our understanding of global land changes which will in the future relate to land use intensification rather than land use conversions sensu stricto (Kuemmerle et al., 2013).

Secondly, higher resolution and more accurate global remote sensing based datasets have been developed since the early 2000s. While most of those global land observation products are based on data with a spatial resolution of hundreds of meters to one km (Achard et al., 2007 and Fritz et al., 2013), new remote sensing capabilities can create higher resolution land change products at landscape scale (Hansen et al., 2013 and Roy et al., 2014).

One prominent, but largely under-researched land change process is agricultural land abandonment. Studies relying on imagery-derived data help quantifying land abandonment and improving our understanding concerning this under-researched land change process (Kuemmerle et al., 2011 and Griffiths et al., 2013). For example, remote sensing based analyses created much of our knowledge how the breakdown of the Soviet Union and collapsing markets triggered massive land abandonment in central and eastern Europe (Hostert et al., 2011 and Prishchepov et al., 2012).

Another major challenge is differentiating similar land cover classes that have significantly different ecological properties. For example, in Latin America, natural grasslands (high diversity, high soil carbon, moderate/low primary productivity), are hardly discriminated from planted pastures (low biodiversity, moderate/low carbon, high primary productivity). Similarly, it is often also very difficult to discriminate between tree plantations and natural forests, and of course, between crops with very different ecological, economic and socioeconomic implications.

Modelling Land System Change

Land use modelling takes place using simulation tools covering a wide range of different model concepts originating from different disciplines. Models used in land system science range from global scale, coarse resolution models assessing global demand and supply of commodities produced by land systems (e.g. Computational General Equilibrium models), to local multi-agent models that simulate individual land use decisions at the level of individual actors. This wide range of models and modelling approaches can be explained by the different scales that land system science addresses, the different research and policy questions to which land use models are applied and the different disciplinary traditions from which land system science is originating. The range of different modelling approaches available is synthesised in various review papers (Verburg et al., 2004, Matthews et al., 2007, Priess and Schaldach, 2008 and Brown et al., 2013). In a recent review for the National Research Council of the United States models were classified according to the role they play within the process of supporting policy decisions (Brown et al., 2013 and National Research Council, 2013). While some models are targeted as learning-tools to test alternative conceptualisations of land system dynamics, other models are specifically designed to evaluate alternative policy proposals to support decision making by assessing effectiveness and identifying potential trade-offs resulting from the direct and indirect impacts of the intended policies on land use. Other models are targeted at scenario analysis, either to inform large scale assessments of climate change or biodiversity impacts (Lotze-Campen et al., 2010 and Pereira et al., 2010) or to detect specific land use trajectories foreseen under alternative scenarios (Verburg et al., 2010 and Sleeter et al., 2012) in order to raise important policy issues and provide an early warning signal that policies can anticipate. While such use is increasing, land use models are not yet standard in major assessments such as those for IPCC and IPBES, even though land use change is a major driver of climate change, biodiversity and ecosystem services. Each model type and approach has its own specific niche of applications, and it is not possible to classify one approach as superior to other approaches. While pixel-based approaches using spatial data of site conditions as major determinants of the transition probability of land use are extremely popular due to their good fit with available data, many have made pleas for a wider adaptation of so called multi-agent models that can explicitly address variation in decision making and interactions amongst decision makers (Brown et al., 2013, Filatova et al., 2013 and Rounsevell et al., 2014). Development of such models still faces some outstanding challenges, including modelling agents' behaviour, sensitivity analysis, verification, and validation; coupling socio-demographic, ecological, and biophysical models, and spatial representation. However, forward progress is being made on these challenges (Filatova et al., 2013). At the same time, in spite of the many developments of multi-agent models in land system science at the local scale it has proven difficult to operationalise and parameterise such models at regional and coarser scales, largely due to data limitations and lack of empirical knowledge of decision mechanisms (Smajgl et al., 2011).

Apart from using different land use model types for different applications and at different scales, the linking of models is becoming more popular. Especially given the requirement to establish an empirical grounding for land-change models, many scholars are moving towards the development of hybrid models that combine participatory, experimental, statistical, and simulation methods (National Research Council, 2013). To better represent cross-scale feedbacks and relationships, pixel-based models are used to downscale results of macro-economic models and nested approaches in which local dynamics feedback to higher-scale dynamics are being explored, making optimal use of the complementarities of different model types (Verburg et al., 2008, Rounsevell et al., 2012 and Herrero et al., 2014). However, often model coupling is still following a top-down approach in which higher scale models constrain more detailed models that translate regional dynamics to land use change patterns. In such approaches feedbacks from adaptive behaviour and changes in decision making at the local scale to higher level system dynamics are ignored. Both empirical data and knowledge and modelling concepts are lacking to implement such feedbacks (Verburg et al., 2015).

While land use models have strongly evolved over the past decade, further development is needed to enhance their use in both academia and in interaction with stakeholders. Such development should focus on strengthening the spatio-temporal representation of land use decision making, cross-scale dynamics, and a stronger link to empirical data collection for parameterization and validation.

Telecoupling of Land Systems

Globalised markets, decisions by distant governments, and global agenda setting (e.g. in light of global change) influence local land use decisions to an ever increasing degree. Land system science has taken up the challenge to both enhance understanding of the interaction of coupled human-environmental system in a telecoupled world as well as provide new frameworks and concepts to inform evidence based decision and policy making.

For example, the recent reforestation in Vietnam is to a large degree based on displacement of logging through import of wood products from other countries (of which 50% was illegally logged) (Meyfroidt and Lambin, 2009). On a global level 52% of the reforestation (2003–2007) of seven countries that have recently undergone forest transition is based on such displacement effects (Meyfroidt et al., 2010). Economic globalisation therefore facilitates forest transition in one country through displacement effects and leads to an export of negative externalities (frequently to countries with weak land governance systems). In another example, forest exports and imports of China in 2010 illustrates the degree to which land is now a globalised good (Fig. 3). Between 1997 and 2010 China exported forest products to over 160 countries and imported forest products from over 170 countries (Liu, 2014). Such empirical evidence is crucial to enhance respective global policies, such as those related to global deforestation.

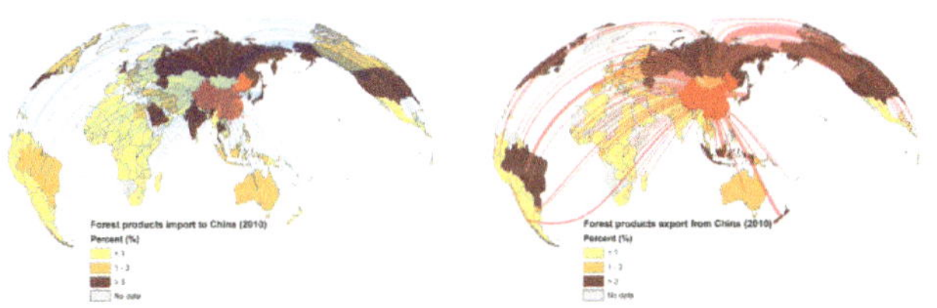

Figure 3: Map of imports and exports of forest products to and from China. (The countries are classified cased on the percentage of imported or exported value.).

(Source: Liu 2014, p. 7; reproduced with kind permission of the author).

Recently, there have been different attempts to quantify these displacement effects of resource use, and land use specifically based on trade and consumption data (e.g. (Qiang et al., 2013, Weinzettel et al., 2013 and Yu et al., 2013)). While the results are partly contradictory, they all show the high degree to which land and the resulting respective externalities have been globalised. These results motivate the clear need for further methodological advances in this field (Kastner et al., 2014) so that the land system science community can provide critical information and tools to support national, regional and global policy decision-making.

In addition to displacement effects, teleconnections of land systems involves other socioecological processes such as human and species migrations, large land acquisition by foreign owners (Messerli et al., 2014), or distant "transfers" of ecosystem services (e.g. through watershed conservation). Land system science has reacted to the conceptual challenge of these increased interactions of distant places and has already proposed different frameworks to address them (e.g. (Seto et al., 2012 and Liu et al., 2013). Liu et al. (2013) discussed an integrated concept of telecoupling that encompasses both socioeconomic and environmental interactions among coupled human environmental system over wide distances. Their concept centres on differentiating between sending, receiving and spill over systems, the flow between these systems, and the agents, causes and effects within these systems. This move from place based to process based conceptualisation offers much potential to enhance the system understanding of a telecoupled world and contributes to more effective policies and action towards sustainable development. While the Liu et al. (2013) framework emphasises geographical distance, other works (Eakin et al., 2014) emphasised the importance of institutional distance, or governance disconnect, in telecoupling: distant systems can be strongly influencing nearby systems while institutions are local and unable to govern these flows. This view also emphasises the role of feedbacks arising from these mismatches to modify the governance systems.

Land System Science as a Crucial Component of Earth System Science

Changes in the land system have many interactions with other components of the earth system. Aquatic and atmospheric systems are influenced by the land system in manifold ways, e.g. through emissions of greenhouse gases, the water consumption of terrestrial vegetation and sediment run-off through rivers. Changes and configurations of land systems have a substantial impact on terrestrial biodiversity through changes in habitat conditions by fragmenting and modifying natural ecosystems. In all these interactions, the land system operates at the interface of human and biophysical systems, in which the environment is modified by humans responding to the opportunities offered by land resources and adapting to environmental change. In this section we outline a number of major interactions of land systems with other components of global environmental change that are often studied independently.

Land-Atmosphere Processes

Land-use and land-cover changes have contributed substantially to climate change and are expected to continue to do so in the future (Le Quéré et al., 2009, Pitman et al., 2009 and Houghton et al., 2012) due to the release of large quantities of carbon when natural ecosystems (mostly forests) are converted into croplands or pastures, or due to changes in management or land-use intensity (Erb et al., 2013). Intensification of crop management leads to release of additional greenhouse gases, like N_2O from fertiliser application (Zaehle et al., 2011), or CH_4 from cattle and rice production (Verburg and Denier van der Gon, 2001 and Steinfeld et al., 2006). Changes that follow land use change in the surface reflectivity and the way absorbed energy is distributed towards evapotranspiration or heating at the near-surface affect regional climate substantially (Pitman et al., 2009 and Pongratz et al., 2009). Changes in land systems can also result in increased carbon sequestration, due to e.g. the land-sparing effects of intensification, if not overcompensated by rebound effects (Lambin and Meyfroidt, 2011) or due to management changes in forests that do not result in changes in land cover, such as forest grazing or litter raking. Although much emphasis has been put on carbon fluxes related to land cover changes, the complex suite of biogeophysical effects of land use and cover change (Mahmood et al., 2014), as well as the equally important effects of changes in land management and intensity within land use classes (Luyssaert et al., 2014) are now increasingly recognised. To adequately describe the functioning of the Earth system, land use and management (change) need to be accounted for along other components including climate and natural ecosystems. Nevertheless, the recent 5th Assessment Report by the IPCC WG1 is the first where land use change has been explicitly, although rudimentarily, accounted for in projections of climate change. Recent, model-based studies, have shown that land use change has an important impact on the radiative forcing calculation (Jones et al., 2012) while, often ignored, spatio-temporal dynamics in land systems are not accounted for, thus underestimating the land system change impacts on climate change (Stocker et al., 2014 and Fuchs et al., 2015Fuchs et al., 2015). From a human system›s perspective, climate and climate change also contribute to land use change (Mertz et al., 2010). Climate determines the types of crops that can be grown (Easterling et al., 2007 and Gornall et al., 2010). Harvest failures following floods, heat-waves or droughts can lead to food-shortages, and increases in local and global grain prices (Beddington, 2010). Indirect climate-effects like fires or insect-outbreaks can also affect the yield of forests and crops. At the same time, in other regions climate change also favours higher agriculture yields in areas currently limited by temperature and rainfall. As an example, the agriculture production of Argentina (one of the major global food producers) has been favoured by climate change (via rainfall increase) in recent decades. While such feedbacks through biophysical processes have been captured in integrated assessment models, the human feedbacks through adaptive behaviour and decision making have been largely ignored and require both empirical research and representation in models (Rounsevell et al., 2014).

An emerging challenge is, therefore, to quantify impacts and feedbacks among land systems, the societies managing these systems, and the climate system, and to take into consideration regional to global scales and short- and longer-term time perspectives. A global scale perspective is needed because climate systems operate on global scales, and land systems are increasingly influenced by drivers at these scales, as illustrated in section 2.6. The debate on indirect land use change arising from land-based bioenergy production is a prominent example of global-scale dynamics (Fargione et al., 2008), but equally important is the possibility to supply food to regions that suffer from, for example, a climate-induced crisis. Issues of time arise from legacy effects of land use change in the climate system (Houghton et al., 2012) or from time-lags between introduction of a climate policy and its actual take-up by local farmers (Alexander et al., 2013). The climate, environmental and socio-economic research communities are confronted with providing the required understanding of the fundamental processes that operate at the interface of the climate and land systems, and their manifold interactions across local, regional, and global scales. A key challenge is to find ways to bridge, philosophically and methodologically, the diverse scientific communities. Finding ways to synthesise available data and knowledge in these communities will allow further development of the mechanisms represented in models, advance our capacity to evaluate model performance, and yield information to support policy development and societies towards successful adaptation and mitigation strategies (Hibbard et al., 2010 and Rounsevell et al., 2014).

Urban Rural Teleconnections

Over half of the world's population already lives in cities, with a massive increase in urban population projected by 2050, with an anticipated increase in urban land cover about twice the increase in urban population (Seto et al., 2011). Rates of urbanization are spatially uneven, with Asia and Africa projected to account for as much as 86% of projected urban growth (United Nations, 2010). Alongside rapid growth, urban shrinkage is also taking place in many parts of Europe and North America (Haase et al., 2012). Conversion of land to built-up is often considered one of the most problematic trajectories of land change, due to its perceived irreversibility and severe consequences for climate, biodiversity, ecosystem quality and ecosystem services, which are difficult to mitigate and manage (Elmqvist et al., 2013). Urban green spaces, wetlands and water bodies provide critical biodiversity and ecosystem services for all city dweller, and are especially important for poor and vulnerable populations, such as urban slum residents (Nagendra et al., 2013c). Thus, protecting and restoring urban ecosystems is an important issue for sustainable urbanization.

Economic and demographic factors appear to be particularly strong drivers of urban land change in China, India and Africa (Seto et al., 2011). Macroeconomic factors have increased in their influence, while the drivers are often spatially removed from the locations of change, and stakeholders increasingly globalised African urbanization further constitutes a particular information gap, with studies of land change in African cities being relatively sparse compared to other parts of the world, as a recent meta-analysis indicates (Seto et al., 2011). Peri-urban regions constitute areas of particularly rapid change that are especially vulnerable to land acquisitions and tenure changes with potentially disrupting socioeconomic effects and ecosystem degradation (Seto et al., 2012), further increasing the social vulnerability of the urban poor, migrants and people practising traditional rural livelihoods. In order to understand, model and manage the multi-level drivers of urbanization in cities and their peri-urban surroundings, recent efforts are beginning to focus on developing a better understanding of urban-rural teleconnections (Seto et al., 2012). As an example, Aide and Grau (2004) show that by reducing rural population pressure in areas marginal for agriculture production, rural to urban migration may facilitate the recovery of natural ecosystems while at the same time, the quality of life for the rural-to-urban migrants improves. Land system scientists suggest a more nuanced, continuous representation of urbanity and rurality across multiple dimensions of livelihood, lifestyle and teleconnectivity, as an advance beyond traditional conceptualizations of an rural-urban gradient or divide (Nagendra et al., 2013b; Boone et al., 2014). Achieving sustainable land uses in an urban era requires moving beyond the vision of cities as centres of consumption that externalise land use impacts on rural hinterlands, and acknowledging that cities and urban lifestyles and institutions can also contribute to solutions (Seto and Reenberg, 2014).

Land Governance

Land systems are increasingly affected by changes in global governance and wider revalorizations of land, and in turn influence the wider transformations in governance and value (see Fig. 4). For example, global demand for food and biofuels drives one of the most visible revalorizations of land over the past decade, giving rise to large-scale land acquisitions by states, transnational corporations and financial investors (Anseeuw et al., 2012 and White et al., 2012). The land acquisitions have been enabled by larger governance changes at the international, national and local levels, such as the ascendance of the World Trade Organization, national policies on food, agriculture and trade, and the rolling out of commercial land markets (Peluso and Lund, 2011, McMichael, 2012 and Margulis et al., 2013). Simultaneously, indigenous peoples have mobilised in the pursuit of political and cultural goals, highlighting the value of land as a place of belonging, territory for self-determination and religious practice (Sikor and Stahl, 2011). Changes in land systems drive global revalorizations of land and wider transformations of governance, including multilateral agreements, as illustrated by the inclusion of forests in global climate mitigation efforts due to concerns over land-related carbon emissions. Thousands of small-scale reforestation projects in the so-called voluntary sector and the United Nations initiative of Reducing Emissions from Deforestation and Forest Degradation (REDD+) have brought increasing attention to carbon storage and changes in land governance that accommodate the challenge of climate change (Bumpus and Liverman, 2008 and Corbera and Schroeder, 2011), but land systems science has also highlighted the complexities of governing climate change with often unpredictable and complex changes that occur in land systems (Ankersen et al., 2015).

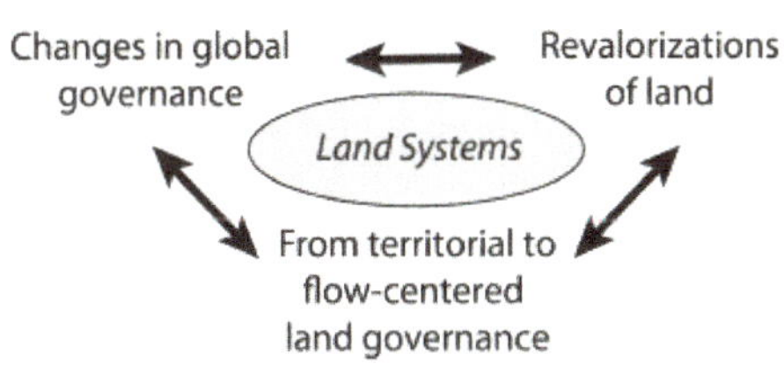

Figure 4: Dynamics of land governance and value.

In reaction to shifts in governance, value and land systems, land governance is partly shifting from 'territorial' toward 'flow-centered' arrangements (Sikor et al., 2013). Flow-centered governance targets particular flows of resources or goods, such as certification of agricultural or wood products or voluntary regulation in the mining sector (Auld, 2014). For example, concentration in global agri-food supply chains has enabled industry to introduce production and sustainability standards (Bailis and Baka, 2011). Initially NGOs, but later also governments have promoted certification schemes for food at global or regional levels (Auld, 2010). The European Union and USA are now seeking to regulate global timber production through trade-related measures. These forms of flow-centered governance complement classic territorial forms of land governance, such as the designation of protected areas, regulation of land use, and land use planning (Sikor and Müller, 2009). They also add to the new instruments used in territorial governance, in particular financial mechanisms such as Payments for Ecosystem Services (Muradian et al., 2010).

Land Change and Ecosystem Services/Biodiversity

Ecosystem services are supplied directly by natural capital, i.e. the major components of the land system: soil, water, vegetation and biota. Land use and land use change decisions therefore have direct bearing on natural capital and ecosystem service supply (Foley et al., 2005). The impacts of these decisions typically unfold at local to regional scales, affecting local livelihoods that depend on ecosystem services and biodiversity (Wu, 2013b). Trade-offs often arise between bundles of ecosystem services supplied by alternative land uses (Chisholm, 2010, Crossman et al., 2011 and Smith et al., 2013). For example, land use change from natural forest to annual cropping or grazing systems (Fig. 5) may enhance supply of the food production ecosystem service, but at the cost of a number of other services, such as water purification, local climate regulation, carbon sequestration, biodiversity and habitat, and important cultural and spiritual values (Smith et al., 2013). Robust quantification of the many services supplied by natural capital is critical for understanding the trade-offs between services under alternative land use scenarios. Land system science plays a pivotal role in quantifying the bio-physical processes underpinning ecosystem services. In some instances the processes are well understood (e.g. sediment transport processes under alternative land use arrangements), whereas in other cases the processes have yet to be fully described (e.g. the distribution of biodiversity and supply of habitat-related ecosystem services (Nagendra et al., 2013a).

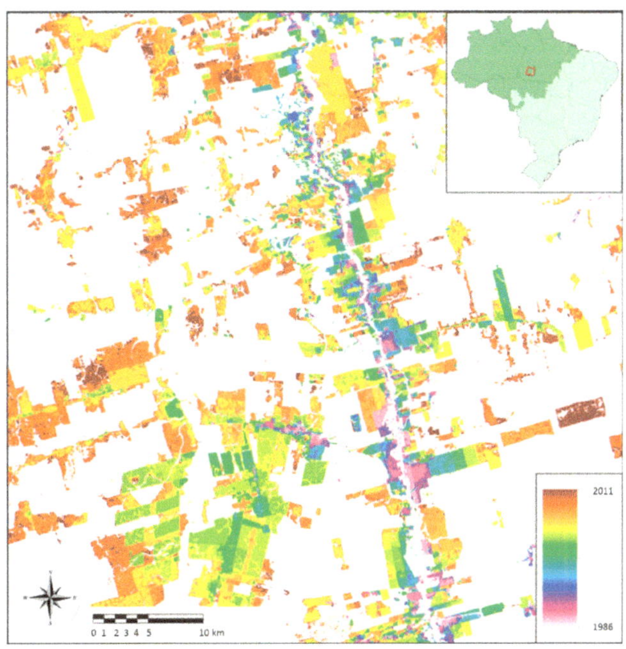

Figure 5: Yearly map of primary forests converted to rangelands in the Brazilian Amazon along the BR 163 highway. Improved food production services on the expense of carbon sequestration, biodiversity and habitat.

There is an important social dimension to natural capital and ecosystem services that land change science can help better elucidate. Arguably, ecosystem services are co-produced by human-environment interactions, and do not exist if there are no beneficiaries (Wolff et al., 2015). There is a clear need to better understand the spatial and temporal dimensions of beneficiaries, and whether they are spatially and temporally connected or disconnected from where ecosystem services are supplied (Serna-Chavez et al., 2014). Furthermore, land system science also aims to understand the impact of human decisions on natural capital and ecosystem service supply. Yet this is an area of greatest uncertainty necessitating deeper collaborations between land system science and the social sciences (Crossman et al., 2013).

LAND SYSTEM SCIENCE: FROM UNDERSTANDING TO SUSTAINABILITY SOLUTIONS

Land System Science Contributions to Sustainability Solutions

In 2013, a new initiative gathering all previous global environmental change programmes was established and named 'Future Earth', in response to a visioning process on Earth System Science and Global Sustainability undertaken by ICSU (Reid et al., 2010). This Global Land Project transitioned to Future Earth in 2014. Future Earth prioritises a stronger focus on interdisciplinarity and science that supports sustainability transitions through co-design and co-production of research together with stakeholders. While interdisciplinarity is central to land system science, land system science could benefit from a stronger engagement of stakeholders to develop sustainability solutions. Although the orientation towards sustainability has a long tradition in land system science, a lot of research is focused on understanding the drivers and impacts of land system changes (see Section 2.1), where large knowledge gaps still remain. However, there is also a need to further engage in using the acquired knowledge to develop and prototype sustainable land management practices and policies. Land system science is closely related to the fields of land use planning and land use policy. However, with much land being owned and managed by private land owners that are, for various reasons, not always directly responding to planning and policy, new ways of linking science and practice need to be developed to effectively create scientific findings that contribute to sustainability solutions.

Important ways forward in this perspective include the evaluation and design of alternative ways to govern land resources (Deininger et al., 2011, Bourgoin et al., 2012 and Sikor et al., 2013). Studies in Laos exemplify how trans-disciplinary research can contribute to this. In Laos there is an immense commercial pressure on land and authorities and private operators seek to replace the traditional shifting cultivation practices with more commercially oriented agriculture. Development-oriented research endeavors have been co-designed with local and governmental stakeholders leading to more effective approaches to local land use planning (Bourgoin et al., 2012). These approaches capture and contextualise for the first time the immense magnitude of land investment in the country (Schönweger et al., 2012) as well as traditional smallholder claims on land through shifting cultivation (van Vliet et al., 2012 and Heinimann et al., 2013). This co-production of knowledge led to the recognition by officials of the need for more efficient and participatory strategies for land management and improved access to land for smallholder communities. It also contributed to the declaration of a land concession moratorium (for rubber, eucalyptus and mining) by the prime minister office of Laos, and to increased recognition of the importance of shifting cultivation in the new Upland Strategy of the government. Whether this is then translated into locally appropriate development has yet to be seen and land system science has an important role to document such policy changes with case studies at local level in Laos and elsewhere.

Land system science can also make major contributions to the search for solutions to meet the multiple, conflicting, demands on the land system. In particular the challenge to produce more food for a growing and more demanding population while avoiding the detriments of many output-enhancing land-use practices, such as biodiversity loss and the degradation of ecosystem service provision, has given rise to debate about alternative strategies of land management (Rosegrant and Cline, 2003, Godfray et al., 2010, Ingram, 2011 and Lambin and Meyfroidt, 2011). This debate has been influenced by strong opinions in favour of either intensification of agricultural production on a relatively limited area, so-called land sparing, or multi-functional agriculture areas, referred to as land sharing (Phalan et al., 2011, Butsic et al., 2012 and Tscharntke et al., 2012). Whereas intensive agriculture has been characterised by negative environmental and social externalities, multi-functional agriculture has often been characterised by lower yields, thus requiring larger areas to produce the same quantities of agricultural product. However, in reality these tradeoffs are very dependent on the biophysical and socio-economic context. In sensitive environments, organic or other extensive, multifunctional, forms of agriculture can be suitable options. At the same time, intensive agriculture can, with appropriate technologies and management, produce large quantities of commodities fulfilling the food demand of many people on a relatively small land area. Matching land use systems with the abilities and willingness of the land managers, the local environmental conditions and the demand for ecosystem services is important to achieve sustainable land management. In a globalised world locally optimal solutions need to be contextualised as choices that always have wider implications: the choice for relatively unproductive systems has trade-offs due to displacement of production to other places (Lambin and Meyfroidt, 2011). It is incorrect to assume that intensification will always spare land that can be used for conservation purposes. Increased production can trigger increased consumption as a result of lower prices and improved agricultural opportunities may attract new activities on 'spared' land (Matson and Vitousek, 2006, Lambin and Meyfroidt, 2011, Phalan et al., 2011, Grau et al., 2013 and Hertel et al., 2014). Land system science can provide insights in both the local and global tradeoffs. The quantification of the global scale tradeoffs of locally sustainable solutions can inform land management and conservation decisions, and bring nuance into discussions about land sparing and sharing paradigms by accounting for local context and environmental heterogeneity (Bryan and Crossman, 2013).

The conceptual and methodological studies to evaluate tradeoffs can support a land system architecture, which draws on principles of landscape ecology about spatial structure and pattern (Wu, 2013a), to design novel land systems that more optimally account for spatial and temporal interactions and landscape configuration (Bateman et al., 2013, Crossman et al., 2013, Seppelt et al., 2013 and Turner et al., 2013). By modifying spatial landscape structure and allocating land use activities to the most optimal places in the landscape, it is possible to enhance the production of multiple services and the resilience of the land system (Bryan et al., 2011). Such a design-based approach needs a comprehensive knowledge base, thus requiring engaging with stakeholders through co-design and transdisciplinary approaches. In this way the designed systems may better fit the local interests and ecosystem service demand, be sustainable from both local and global perspectives and fit to the local socio-economic and land governance systems (Bryan et al., 2010). Such research requires the use of participatory techniques but also the involvement of practitioners to prototype the newly designed land systems and test their suitability in reality. The approach aims to use the capacity of land systems and the architecture of these systems to simultaneously respond to changing societal demand and support mitigation and adaption to environmental change. That way, land system science would not only play a role at the interface of the social and natural sciences, but also at the interface between science and practice.

Land System Science as a Platform for Interdisciplinary and Transdisciplinary Collaboration in Global Change Research

The position of land system science at the interface of human and environmental systems makes it central to studies across different scales and disciplinary perspectives, thus creating a platform to bring the different communities and disciplines together, exchange and compare approaches, and create a mutual understanding of the challenges and knowledge gaps. International global change programmes projects LUCC, GCTE and GLP have been successful in creating such platforms. In addition to creating an awareness of different perspectives and methods, the interactions in these projects have facilitated, at various stages, knowledge synthesis and evolution of research agendas and priorities. The knowledge gaps related to global environmental change cannot be addressed by a single core project or within one research community, re-iterating the grand challenge for more interdisciplinarity in global change research identified during the process towards the establishment of Future Earth (Reid et al., 2010). Land systems at the interface of human and environmental systems provide a unique platform for interaction amongst the different global change communities.

CONCLUSIONS

Land use has been a central element for society throughout human history and remains even more so today. However, land systems, their characteristics, dynamics, constraints and impacts, have traditionally been studied as part of different disciplines, including economics, ecology and geography. The increasing pressures on land resources and the key role of land systems within the Earth system dynamics have given rise to the development of land system science as an interdisciplinary field that acts as a platform to integrate different perspectives and dimensions of global change research. The field has matured during recent decades giving rise to novel approaches to bridge the different scales of analysis as well as facilitating the establishment of links between science and practice.

Land systems are both a cause of global environmental change and a possible powerful means of mitigation of and adaptation to global environmental change. In order to exploit this potential, the community needs to move from a dominant focus on exploratory research towards understanding the functioning of the land system and its dynamics to approaches that use this knowledge together with stakeholders to mitigate and adapt to the changing environmental and socio-economic context. The concept of land governance opens up many points of connection with stakeholders, including both formal organizations various local, national and global fora where people make collective decisions about land. The competition for land resources and the services provided by the land are increasingly dynamic in response to the high demands of a growing population under a changing climate. In this paper we have identified priority areas that require more attention during the coming years to be able to effectively address these challenges. The knowledge and commitment embedded in land system science can contribute to co-designing land system solutions to these global change challenges.

ACKNOWLEDGEMENTS

Peter Verburg acknowledges funding provided by the European Commission through ERC project GLOLAND and K.-H. Erb through the ERC project LUISE. Neville Crossman acknowledges the support of the CSIRO Water for a Healthy Country National Research Flagship. Erle Ellis acknowledges the US National Science Foundation (CNS 1125210) for funding the GLOBE project as presented in this paper. We are grateful to Jianguo Liu for the kind permission to reproduce the figure from Liu (2014).

REFERENCES

1. Achard, F., DeFries, R., Eva, H., Hansen, M., Mayaux, P., Stibig, H.J., 2007. Pan-tropical monitoring of deforestation. Environ. Res. Lett. 2, 045022.
2. Aide, T.M., Grau, H.R., 2004. Globalization, migration, and Latin American ecosystems. Science 305, 1915–1916.
3. Alexander, P., Moran, D., Rounsevell, M.D.A., Smith, P., 2013. Modelling the perennial energy crop market: the role of spatial diffusion. J. R. Soc. Interface 10, 20130656.
4. Ankersen, J., Grogan, K., Mertz, O., Fensholt, R., Castella, J.-C., Lestrelin, G., Nguyen, D., Danielsen, F., Brofeldt, S., Rasmussen, K., 2015. Vietnam's forest transition in retrospect: demonstrating weaknesses in business-as-usual scenarios for REDD +. Environ. Manage. 55, 1080–1092.
5. Anseeuw, W., Boche, M., Breu, T., Giger, M., Lay, J., Messerli, P., Nolte, K., 2012. Transnational Land Deals for Agriculture in the Global South. Analytical Report based on the Land Matrix Database. CDE/CIRAD/GIGA, Bern, Montpellier, Hamburg.
6. Auld, G., 2010. Assessing certification as governance: effects and broader consequences for coffee. J. Environ. Dev. 19, 215–241.
7. Auld, G., 2014. Private market-based regulation:What they are, and what they mean for land-use governance. In: Seto, K.C., Reenberg, A. (Eds.), Rethinking Global Land Use in an Urban Era. MIT Press.

8. Bailis, R., Baka, J., 2011. Constructing sustainable biofuels: governance of the emerging biofuel economy. Ann. Assoc. Am. Geogr. 101, 827–838.
9. Bateman, I.J., Harwood, A.R., Mace, G.M., Watson, R.T., Abson, D.J., Andrews, B., Binner, A., Crowe, A., Day, B.H., Dugdale, S., Fezzi, C., Foden, J., Hadley, D., HainesYoung, R., Hulme, M., Kontoleon, A., Lovett, A.A., Munday, P., Pascual, U., Paterson, J., Perino, G., Sen, A., Siriwardena, G., van Soest, D., Termansen, M., 2013. Bringing ecosystem services into economic decision-making: land use in the United Kingdom. Science 341, 45–50.
10. Beddington, J., 2010. Global food and farming futures. Philos. Trans. R. Soc. B 365, 2767.
11. Belward, A.S., 1996. The IGBP-DIS Global 1 km Land Cover Data Set DISCover: Proposal and Implementation Plans: Report of the Land Recover Working Group of IGBP-DIS. IGBP-DIS.
12. Boone, C.G., Redman, C.L., Blanco, H., Haase, D., Koch, J., Lwasa, S., Nagendra, H., Pauleit, S., Pickett, S.T.A., Seto, K.C., Yokohari, M., 2014. Reconceptualizing land for sustainable urbanity. In: Seto, K.C., Reenberg, A. (Eds.), Rethinking Global Land Use in an Urban Era. MIT Press, Cambridge, MA.
13. Bourgoin, J., Castella, J.-C., Pullar, D., Lestrelin, G., Bouahom, B., 2012. Toward a land zoning negotiation support platform: Tips and tricks for participatory land use planning in Laos. Landscape Urban Plan. 104, 270–278.
14. Brown, D.G., Verburg, P.H., Pontius Jr., R.G., Lange, M.D., 2013. Opportunities to improve impact, integration, and evaluation of land change models. Curr. Opin. Environ. Sustain. 5, 452–457.
15. Bryan, B.A., Crossman, N.D., 2013. Impact of multiple interacting financial incentives on land use change and the supply of ecosystem services. Ecosyst. Serv. 4, 60–72.
16. Bryan, B.A., Crossman, N.D., King, D., Meyer, W.S., 2011. Landscape futures analysis: assessing the impacts of environmental targets under alternative spatial policy options and future scenarios. Environ. Model. Software 26, 83–91.
17. Bryan, B.A., Raymond, C.M., Crossman, N.D., Macdonald, D.H., 2010. Targeting the management of ecosystem services based on social values: where, what, and how? Landscape Urban Plan. 97, 111–122.
18. Bumpus, A.G., Liverman, D.M., 2008. Accumulation by decarbonization and the governance of carbon offsets. Econ. Geogr. 84, 127–155.
19. Butsic, V., Radeloff, V.C., Kuemmerle, T., Pidgeon, A.M., 2012. Analytical solutions to trade-offs between size of protected areas and land-use intensity. Conserv. Biol. 26, 883–893.
20. Chisholm, R.A., 2010. Trade-offs between ecosystem services: Water and carbon in a biodiversity hotspot. Ecol. Econ. 69, 1973–1987.
21. Corbera, E., Schroeder, H., 2011. Governing and implementing REDD+. Environ. Sci. Policy 14, 89–99.
22. Crossman, N.D., Bryan, B.A., de Groot, R.S., Lin, Y.-P., Minang, P.A., 2013. Land science contributions to ecosystem services. Curr. Opin. Environ. Sustain. 5, 509–514.
23. Crossman, N.D., Bryan, B.A., Summers, D.M., 2011. Carbon payments and low-cost conservation. Conserv. Biol. 25, 835–845.
24. Deininger, K., Selod, H., Burns, T., 2011. The Land Governance Assessment Framework: Identifying and Monitoring Good Practice in the Land Sector. World Bank, Washington.
25. Don, A., Schumacher, J., Freibauer, A., 2011. Impact of tropical land-use change on soil organic carbon stocks—a meta-analysis. Global Change Biol. 17, 1658–1670.
26. Eakin, H., Defries, R., Kerr, S., Lambin, E.F., Liu, J., Marcotullio, P.J., Messerli, P., Reenberg, A., Rueda, X., Swaffield, S.R., 2014. Significance of telecoupling for exploration of land-use change. In: Seto, K.C., Reenberg, A. (Eds.), Rethinking Global Land Use in an Urban Era. MIT Press, Cambridge, MA.
27. Easterling, W.E., Aggarwal, P.K., Batima, P., Brander, K.M., Erda, L., Howden, S.M., Kirilenko, A., Morton, J., Soussana, J.-F., Schmidhuber, J., Tubiello, F.N., 2007. Food, fibre and forest products. Climate Change 2007: impacts, adaptation and vulnerability. In: Parry, M.L., Canziani, O.F., Palutikof, J.P., van der Linden, P.J., Hanson, C.E. (Eds.), Climate Change 2007: Impacts, Adaptation and Vulnerability. Cambridge University Press, Cambridge, UK, pp. 273–313.
28. Ellis, E. 2014. Deforestation. GLOBE Collection of Georeferenced Case Studies. phttp://dx.doi.org/10.7933/K100001H.
29. Ellis, E.C., Kaplan, J.O., Fuller, D.Q., Vavrus, S., Klein Goldewijk, K., Verburg, P.H., 2013. Used planet: a global history. Proc. Natl. Acad. Sci. U. S. A. 110, 7978–7985.
30. Elmqvist, T., Fragkias, M., Goodness, J., Güneralp, B., Marcotullio, P., McDonald, R., Parnell, S., Schewenius, M., Sendstad, M., Seto, K., Wilkinson, C., Alberti, M., Folke, C., Frantzeskaki, N., Haase, D., Katti, M., Nagendra, H., Niemelä, J., Pickett, S.A., Redman, C., Tidball, K., 2013. Stewardship of the biosphere in the urban era. In: Elmqvist, T., Fragkias, M., Goodness, J., Güneralp, B., Marcotullio, P.J., McDonald, R.I., Parnell, S., Schewenius, M., Sendstad, M., Seto, K.C., Wilkinson, C.
31. (Eds.), Urbanization, Biodiversity and Ecosystem Services: Challenges and Opportunities. Springer, Netherlands, pp. 719–746.
32. Erb, K.-H., Haberl, H., Jepsen, M.R., Kuemmerle, T., Lindner, M., Müller, D., Verburg, P. H., Reenberg, A., 2013. A conceptual framework for analysing and measuring land-use intensity. Curr. Opin. Environ. Sustain. 5, 464–470.
33. Fargione, J., Hill, J., Tilman, D., Polasky, S., Hawthorne, P., 2008. Land clearing and the biofuel carbon debt. Science 319, 1235–1238.
34. Filatova, T., Verburg, P.H., Parker, D.C., Stannard, C.A., 2013. Spatial agent-based models for socio-ecological systems: Challenges and prospects. Environ. Model.

Software 45, 1–7.

35. Foley, J.A., DeFries, R., Asner, G.P., Barford, C., Bonan, G., Carpenter, S.R., Chapin, F.S., Coe, M.T., Daily, G.C., Gibbs, H.K., Helkowski, J.H., Holloway, T., Howard, E.A., Kucharik, C.J., Monfreda, C., Patz, J.A., Prentice, I.C., Ramankutty, N., Snyder, P.K., 2005. Global Consequences of land use. Science 309, 570–574.

36. Fritz, S., See, L., You, L., Justice, C., Becker-Reshef, I., Bydekerke, L., Cumani, R., Defourny, P., Erb, K., Foley, J., Gilliams, S., Gong, P., Hansen, M., Hertel, T., Herold, M., Herrero, M., Kayitakire, F., Latham, J., Leo, O., McCallum, I., Obersteiner, M., Ramankutty, N., Rocha, J., Tang, H., Thornton, P., Vancutsem, C., van, der Velde M, Wood, S., Woodcock, C., 2013. The need for improved maps of global cropland. Eos, Transactions American Geophysical Union 94, 31–32.

37. Fuchs, R., Herold, M., Verburg, P.H., Clevers, J.G.P.W., Eberle, J., 2015. Gross changes in reconstructions of historic land cover/use for Europe between 1900 and 2010. Global Change Biol. 21, 299–313.

38. Geist, H.J., Lambin, E.F., 2002. Proximate causes and underlying driving forces of tropical deforestation. Bioscience 52, 143–150.

39. Geist, H.J., Lambin, E.F., 2004. Dynamic causal patterns of desertification. Bioscience 54, 817–829.

40. Gibson, L., Lee, T.M., Koh, L.P., Brook, B.W., Gardner, T.A., Barlow, J., Peres, C.A., Bradshaw, C.J.A., Laurance, W.F., Lovejoy, T.E., Sodhi, N.S., 2011. Primary forests are irreplaceable for sustaining tropical biodiversity. Nature 478, 378–381.

41. GLP, 2005. Science Plan and Implementation Strategy Global Land Project. IGBP Report No. 53/IHDP Report No. 19. IGBP Secretariat, Stockholm.

42. Godfray, H.C., Beddington, J.R., Crute, I.R., Haddad, L., Lawrence, D., Muir, J.F., Pretty, J., Robinson, S., Thomas, S.M., Toulmin, C., 2010. Food security: the challenge of feeding 9 Billion people. Science 327, 812–818.

43. Gornall, J., Betts, R., Burke, E., Clark, R., Camp, J., Willett, K., Wiltshire, A., 2010. Implications of climate change for agricultural productivity in the early twenty-first century. Philos. Trans. R. Soc. B 365, 2973–2989. doi:http://dx.doi.org/ 10.1098/rstb.2010.0158.

44. Grau, R., Kuemmerle, T., Macchi, L., 2013. Beyond 'land sparing versus land sharing': environmental heterogeneity, globalization and the balance between agricultural production and nature conservation. Curr. Opin. Environ. Sustain. 5, 477–483.

45. Griffiths, P., Müller, D., Kuemmerle, T., Hostert, P., 2013. Agricultural land change in the Carpathian ecoregion after the breakdown of socialism and expansion of the European Union. Environ. Res. Lett. 8, 045024.

46. Haase, D., Haase, A., Kabisch, N., Kabisch, S., Rink, D., 2012. Actors and factors in land-use simulation: the challenge of urban shrinkage. Environ. Model. Software 35, 92–103.

47. Hansen, M.C., Potapov, P.V., Moore, R., Hancher, M., Turubanova, S.A., Tyukavina, A., Thau, D., Stehman, S.V., Goetz, S.J., Loveland, T.R., Kommareddy, A., Egorov, A., Chini, L., Justice, C.O., Townshend, J.R.G., 2013. High-resolution global maps of 21st-century forest cover change. Science 342, 850–853.

48. Heinimann, A., Hett, C., Hurni, K., Messerli, P., Epprecht, M., Jørgensen, L., Breu, T., 2013. Socio-economic perspectives on shifting cultivation landscapes in Northern Laos. Hum. Ecol. 41, 51–62.

49. Herrero, M., Thornton, P.K., Bernués, A., Baltenweck, I., Vervoort, J., van de Steeg, J., Makokha, S., van Wijk, M.T., Karanja, S., Rufino, M.C., Staal, S.J., 2014. Exploring future changes in smallholder farming systems by linking socio-economic scenarios with regional and household models. Global Environ. Change 24, 165–182.

50. Hertel, T.W., Ramankutty, N., Baldos, U.L.C., 2014. Global market integration increases likelihood that a future African Green Revolution could increase crop land use and CO2 emissions. Proc. Natl. Acad. Sci. U.S. A. 111, 13799–13804.

51. Hibbard, K., Janetos, A., van Vuuren, D.P., Pongratz, J., Rose, S.K., Betts, R., Herold, M., Feddema, J.J., 2010. Research priorities in land use and land-cover change for the Earth system and integrated assessment modelling. Int. J. Climatol. 30, 2118–2128.

52. Hostert, P., Kuemmerle, T., Prishchepov, A., Sieber, A., Lambin, E.F., Radeloff, V.C., 2011. Rapid land use change after socio-economic disturbances: the collapse of the Soviet Union versus Chernobyl. Environ. Res. Lett. 6, 045201.

53. Houghton, R.A., House, J.I., Pongratz, J., van der Werf, G.R., DeFries, R.S., Hansen, M.C., Le Quéré, C., Ramankutty, N., 2012. Carbon emissions from land use and landcover change. Biogeosciences 9, 5125–5142.

54. Ingram, J., 2011. A food systems approach to researching food security and its interactions with global environmental change. Food Secur. 3, 417–431.

55. Jones, A.D., Collins, W.D., Edmonds, J., Torn, M.S., Janetos, A., Calvin, K.V., Thomson, A., Chini, L.P., Mao, J., Shi, X., Thornton, P., Hurtt, G.C., Wise, M., 2012. Greenhouse gas policy influences climate via direct effects of land-use change. J. Clim. 26, 3657–3670.

56. Kaplan, J.O., Krumhardt, K.M., Zimmermann, N.E., 2012. The effects of land use and climate change on the carbon cycle of Europe over the past 500 years. Global Change Biol. 18, 902–914.

57. Kastner, T., Schaffartzik, A., Eisenmenger, N., Erb, K.-H., Haberl, H., Krausmann, F., 2014. Cropland area embodied in internationaltrade: contradictory results from different approaches. Ecol. Econ. 104, 140–144.

58. Keys, E., McConnell, W.J., 2005. Global change and the intensification of agriculture in the tropics. Global Environ. Change 15, 320–337.

59. Klein Goldewijk, K., Beusen, A., van Drecht, G., de Vos, M., 2011. The HYDE 3.1 spatially explicit database of human-induced global land-use change over the past 12,000 years. Global Ecol. Biogeogr. 20, 73–86.

60. Krausmann, F., Erb, K.-H., Gingrich, S., Haberl,

H., Bondeau, A., Gaube, V., Lauk, C., Plutzar, C., Searchinger, T.D., 2013. Global human appropriation of net primary production doubled in the 20th century. Proc. Natl. Acad. Sci. U. S. A. 110, 10324–10329.

61. Kuemmerle, T., Erb, K., Meyfroidt, P., Müller, D., Verburg, P.H., Estel, S., Haberl, H., Hostert, P., Jepsen, M.R., Kastner, T., Levers, C., Lindner, M., Plutzar, C., Verkerk, P. J., van der Zanden, E.H., Reenberg, A., 2013. Challenges and opportunities in mapping land use intensity globally. Curr. Opin. Environ. Sustain. 5, 484–493.

62. Kuemmerle, T., Olofsson, P., Chaskovskyy, O., Baumann, M., Ostapowicz, K., Woodcock, C.E., Houghton, R.A., Hostert, P., Keeton, W.S., Radeloff, V.C., 2011. Post-Soviet farmland abandonment, forest recovery, and carbon sequestration in western Ukraine. Global Change Biol. 17, 1335–1349.

63. Lambin EF, Baulies X, Bockstael NE, Fischer G, Krug T, Leemans R, Moran EF, Rindfuss RR, Sato Y, Skole D, Turner II BL, Vogel C. 2000. Land-Use and Land-Cover Change (LUCC), Implementation Strategy. In. Stockholm, Bonn.

64. Land-Use and Land-Cover Change—Local Processes and Global Impacts. In: Lambin, E.F., Geist, H. (Eds.), Springer, Berlin.

65. Lambin, E.F., Meyfroidt, P., 2011. Global land use change, economic globalization, and the looming land scarcity. Proc. Natl. Acad. Sci. U. S. A. 108, 3465–3472.

66. Le Quéré, C., Raupach, M.R., Canadell, J.G., Marland, G., Bopp, L., Ciais, P., Conway, T.J., Doney, S.C., Feely, R.A., Foster, P., 2009. Trends in the sources and sinks of carbon dioxide. Nat. Geosci. 2, 831–836.

67. Liu, J., 2014. Forest sustainability in china and implications for a telecoupled world. Asia Pacific Policy Stud. 1, 230–250.

68. Liu, J.Q., Hull, V., Batistella, M., DeFries, R., Dietz, T., Fu, F., Hertel, T.W., Izaurralde, R. C., Lambin, E.F., Li, S., Martinelli, L.A., McConnell, W., Moran, E.F., Naylor, R., Ouyang, Z., Polenske, K.R., Reenberg, A., de Miranda Rocha, G., Simmons, C.S., Verburg, P.H., Vitousek, P.M., Zhang, F., Zhu, C., 2013. Framing sustainability in a telecoupled world. Ecol. Soc. 186 (2).

69. Lotze-Campen, H., Popp, A., Beringer, T., Müller, C., Bondeau, A., Rost, S., Lucht, W., 2010. Scenarios of global bioenergy production: The trade-offs between agricultural expansion, intensification and trade. Ecol. Model. 221, 2188–2196.

70. Luyssaert, S., Jammet, M., Stoy, P.C., Estel, S., Pongratz, J., Ceschia, E., Churkina, G., Don, A., Erb, K., Ferlicoq, M., 2014. Land management and land-cover change have impacts of similar magnitude on surface temperature. Nat. Clim. Change 4, 389–393.

71. Magliocca, N., Rudel, T., Verburg, P., McConnell, W., Mertz, O., Gerstner, K., Heinimann, A., Ellis, E., 2014. Synthesis in land change science: methodological patterns, challenges, and guidelines. Reg. Environ. Change 15, 211–226.

72. Mahmood, R., Pielke, R.A., Hubbard, K.G., Niyogi, D., Dirmeyer, P.A., McAlpine, C., Carleton, A.M., Hale, R., Gameda, S., Beltrán-Przekurat, A., 2014. Land cover changes and their biogeophysical effects on climate. Int. J. Climatol. 34, 929–953.

73. Margulis, M.E., McKeon, N., Borras, S.M., 2013. Land grabbing and global governance: critical perspectives. Globalizations 10, 1–23.

74. Matson, P.A., Vitousek, P.M., 2006. Agricultural intensification: will land spared from farming be land spared for nature? Conserv. Biol. 20, 709–710.

75. Matthews, R., Gilbert, N., Roach, A., Polhill, J., Gotts, N., 2007. Agent-based land-use models: a review of applications. Landscape Ecol. 22, 1447–1459.

76. McMichael, P., 2012. The land grab and corporate food regime restructuring. The Journal of Peasant Studies 39, 681–701.

77. Mertz, O., Mbow, C., Nielsen, J.Ø., Maiga, A., Diallo, D., Reenberg, A., Diouf, A., Barbier, B., Moussa, I.B., Zorom, M., Ouattara, I., Dabi, D., 2010. Climate factors play a limited role for past adaptation strategies in West Africa. Ecol. Soc. 15.

78. Messerli, P., Giger, M., Dwyer, M.B., Breu, T., Eckert, S., 2014. The geography of largescale land acquisitions: analysing socio-ecological patterns of target contexts in the global South. Appl. Geogr. 53, 449–459.

79. Meyfroidt, P., Lambin, E.F., 2009. Forest transition in Vietnam and displacement of deforestation abroad. Proc. Natl. Acad. Sci. U. S. A. 106, 16139–16144.

80. Meyfroidt, P., Rudel, T.K., Lambin, E.F., 2010. Forest transitions, trade, and the global displacement of land use. Proc. Natl. Acad. Sci. U. S. A. 107, 20917–20922.

81. Muradian, R., Corbera, E., Pascual, U., Kosoy, N., May, P.H., 2010. Reconciling theory and practice: An alternative conceptual framework for understanding payments for environmental services. Ecol. Econ. 69, 1202–1208.

82. Nagendra, H., Reyers, B., Lavorel, S., 2013a. Impacts of land change on biodiversity: making the link to ecosystem services. Curr. Opin. Environ. Sustain. 5, 503–508.

83. Nagendra, H., Sudhira, H.S., Katti, M., Schewenius, M., 2013b. Sub-regional assessment of India: effects of urbanization on land use, biodiversity and ecosystem services. In: Elmqvist, T., Fragkias, M., Goodness, J., Güneralp, B., Marcotullio, P.J., McDonald, R.I., Parnell, S., Schewenius, M., Sendstad, M., Seto, K.C., Wilkinson, C. (Eds.), Urbanization, Biodiversity and Ecosystem Services: Challenges and Opportunities. Springer, Netherlands, pp. 65–74.

84. Nagendra, H., Unnikrishnan, H., Sen, S., 2013c. Villages in the city: spatial and temporal heterogeneity in rurality and urbanity in Bangalore, India. Land 3, 1–18.

85. National Research Council, 2013. Advancing Land Change Modeling: Opportunities and Research Requirements (committee members: Brown, D., Band, L.E., Green, K.O., Irwin, E.G., Jain, A., Lambin, E.F., Pontius, R.G., Seto, K.C., Turner, II B.L., Verburg, P.H.). The National Academies Press.

86. Peluso, N.L., Lund, C., 2011. New frontiers of land

control: introduction. J. Peasant Stud. 38, 667–681.

87. Pereira, H.M., Leadley, P.W., Proenca, V., Alkemade, R., Scharlemann, J.P.W., Fernandez-Manjarres, J.F., Araujo, M.B., Balvanera, P., Biggs, R., Cheung, W.W.L., Chini, L., Cooper, H.D., Gilman, E.L., Guenette, S., Hurtt, G.C., Huntington, H.P., Mace, G.M., Oberdorff, T., Revenga, C., Rodrigues, P., Scholes, R.J., Sumaila, U.R., Walpole, M., 2010. Scenarios for global biodiversity in the 21st century. Science 330, 1496–1501.
88. Phalan, B., Onial, M., Balmford, A., Green, R.E., 2011. Reconciling food production and biodiversity conservation: land sharing and land sparing compared. Science 333, 1289–1291.
89. Pitelka, L., Canadell, J., Pataki, D., 2007. Global Ecology, Networks, and Research Synthesis. In: Canadell, J., Pataki, D., Pitelka, L. (Eds.), Terrestrial Ecosystems in a Changing World. Springer, Berlin, Heidelberg, pp. 1–5.
90. Pitman, A.J., de Noblet-Ducoudré, N., Cruz, F.T., Davin, E.L., Bonan, G.B., Brovkin, V., Claussen, M., Delire, C., Ganzeveld, L., Gayler, V., van den Hurk, B.J.J.M., Lawrence, P.J., van der Molen, M.K., Müller, C., Reick, C.H., Seneviratne, S.I., Strengers, B.J., Voldoire, A., 2009. Uncertainties in climate responses to past land cover change: First results from the LUCID intercomparison study. Geophys. Res. Lett. 36.
91. Pongratz, J., Raddatz, T., Reick, C.H., Esch, M., Claussen, M., 2009. Radiative forcing from anthropogenic land cover change since AD 800. Geophys. Res. Lett. 36. Pontius, R.G., Cornell, J.D., Hall, C.A.S., 2001. Modeling the spatial pattern of land-use change with GEOMOD2: application and validation for Costa Rica. Agric. Ecosyst. Environ. 85, 191–203.
92. Priess, J.A., Schaldach, R., 2008. Integrated models of the land system: a review of modelling approaches on the regional to global scale. Living Rev. Landscape Res. 2.
93. Prishchepov, A.V., Radeloff, V.C., Baumann, M., Kuemmerle, T., Müller, D., 2012. Effects of institutional changes on land use: agricultural land abandonment during the transition from state-command to market-driven economies in postSoviet Eastern Europe. Environ. Res. Lett. 7, 024021.
94. Qiang, W., Liu, A., Cheng, S., Kastner, T., Xie, G., 2013. Agricultural trade and virtual land use: the case of China's crop trade. Land Use Policy 33, 141–150.
95. Reid, W.V., Chen, D., Goldfarb, L., Hackmann, H., Lee, Y.T., Mokhele, K., Ostrom, E., Raivio, K., Rockström, J., Schellnhuber, H.J., Whyte, A., 2010. Earth system science for global sustainability: grand challenges. Science 330, 916–917.
96. Rindfuss, R.R., Walsh, S.J., Turner, B.L., Fox II, J., Mishra, V., 2004. Developing a science of land change: challenges and methodological issues. Proc. Natl. Acad. Sci. U. S. A. 101, 13976–13981.
97. Rosegrant, M.W., Cline, S.A., 2003. Global Food Security: Challenges and Policies. Science 302, 1917–1919.
98. Rounsevell, M.D.A., Arneth, A., Alexander, P., Brown, D.G., de Noblet-Ducoudré, N., Ellis, E., Finnigan, J., Galvin, K., Grigg, N., Harman, I., Lennox, J., Magliocca, N., Parker, D., O'Neill, B.C., Verburg, P.H., Young, O., 2014. Towards decision-based global land use models for improved understanding of the Earth system. Earth Syst. Dynam. 5, 117–137.
99. Rounsevell, M.D.A., Robinson, D.T., Murray-Rust, D., 2012. From actors to agents in socio-ecological systems models. Philos. Trans. R. Soc. B 367, 259–269.
100. Roy, D.P., Wulder, M.A., Loveland, T.R.C.E., W, Allen R.G., Anderson, M.C., Helder, D., Irons, J.R., Johnson, D.M., Kennedy, R., Scambos, T.A., Schaaf, C.B., Schott, J.R., Sheng, Y., Vermote, E.F., Belward, A.S., Bindschadler, R., Cohen, W.B., Gao, F., Hipple, J.D., Hostert, P., Huntington, J., Justice, C.O., Kilic, A., Kovalskyy, V., Lee, Z. P., Lymburner, L., Masek, J.G., McCorkel, J., Shuai, Y., Trezza, R., Vogelmann, J., Wynne, R.H., Zhu, Z., 2014. Landsat-8: science and product vision for terrestrial global change research. Remote Sens. Environ. 145, 154–172.
101. Rudel, T.K., 2008. Meta-analyses of case studies: a method for studying regional and global environmental change. Global Environ. Change 18, 18–25.
102. Schönweger, O., Heinimann, A., Epprecht, M., Lu, J., Thalongsengchanh, P., 2012. Concessions and leases in the Lao PDR: taking stock of land investments. Geographica Bernensia. Bern and Vientiane: Centre for Development and Environment (CDE). University of Bern, pp. 98.
103. Seppelt, R., Lautenbach, S., Volk, M., 2013. Identifying trade-offs between ecosystem services, land use, and biodiversity: a plea for combining scenario analysis and optimization on different spatial scales. Curr. Opin. Environ. Sustain. 5, 458–463.
104. Serna-Chavez, H.M., Schulp, C.J.E., van Bodegom, P.M., Bouten, W., Verburg, P.H., Davidson, M.D., 2014. A quantitative framework for assessing spatial flows of ecosystem services. Ecol. Indic. 39, 24–33.
105. Seto, K.C., Fragkias, M., Güneralp, B., Reilly, M.K., 2011. A meta-analysis of global urban land expansion. PLoS One 6, e23777.
106. Seto, K.C., Reenberg, A. (Eds.), 2014. Rethinking Global Land Use in an Urban Era. MIT Press, Cambridge, MA.
107. Seto, K.C., Reenberg, A., Boone, C.G., Fragkias, M., Haase, D., Langanke, T., Marcotullio, P., Munroe, D.K., Olah, B., Simon, D., 2012. Urban land teleconnections and sustainability. Proc. Natl. Acad. Sci. U. S. A..
108. Sikor, T., Auld, G., Bebbington, A.J., Benjaminsen, T.A., Gentry, B.S., Hunsberger, C., Izac, A.-M., Margulis, M.E., Plieninger, T., Schroeder, H., Upton, C., 2013. Global land governance: from territory to flow? Curr. Opin. Environ. Sustain. 5, 522–527.
109. Sikor, T., Müller, D., 2009. The limits of state-led land reform: an introduction. World Dev. 37, 1307–1316.
110. Sikor, T., Stahl, J., 2011. Introduction: the rights agenda in international forestry. In: Sikor, T., Stahl, J. (Eds.), Forests and Peoples: Property, Governance and Human Rights. Earthscan, London.
111. Sleeter, B.M., Sohl, T.L., Bouchard, M.A., Reker, R.R.,

Soulard, C.E., Acevedo, W., Griffith, G.E., Sleeter, R.R., Auch, R.F., Sayler, K.L., Prisley, S., Zhu, Z., 2012. Scenarios of land use and land cover change in the conterminous United States: utilizing the special report on emission scenarios at ecoregional scales. Global Environ. Change 22, 896–914.

112. Smajgl, A., Brown, D.G., Valbuena, D., Huigen, M.G.A., 2011. Empirical characterisation of agent behaviours in socio-ecological systems. Environ. Model. Software 26, 837–844.

113. Smith, P., Ashmore, M.R., Black, H.I.J., Burgess, P.J., Evans, C.D., Quine, T.A., Thomson, A.M., Hicks, K., Orr, H.G., 2013. REVIEW: the role of ecosystems and their management in regulating climate, and soil, water and air quality. J. Appl. Ecol. 50, 812–829.

114. Steinfeld, H., Gerber, P., Wassenaar, T.D., Castel, V., Food, Nations AOotU, Livestock, E., Development, de Haan, C., 2006. Livestock's Long Shadow: Environmental Issues and Options. Food and Agriculture Organization of the United Nations.

115. Stocker, B.D., Feissli, F., Strassmann, K.M., Spahni, R., Joos, F., 2014. Past and future carbon fluxes from land use change, shifting cultivation and wood harvest. Tellus 66.

116. Tscharntke, T., Clough, Y., Wanger, T.C., Jackson, L., Motzke, I., Perfecto, I., Vandermeer, J., Whitbread, A., 2012. Global food security, biodiversity conservation and the future of agricultural intensification. Biol. Conserv. 151, 53–59.

117. Turner, B.L., Lambin, E.F., Reenberg, A., 2007. The emergence of land change science for global environmental change and sustainability. Proc. Natl. Acad. Sci. U.S. A. 104, 20666–20671.

118. Turner II, B.L., Janetos, A.C., Verburg, P.H., Murray, A.T., 2013. Land system architecture: using land systems to adapt and mitigate global environmental change. Global Environ. Change 23, 395–397.

119. Turner II, B.L., Ross, R.H., Skole, D.L., 1993. Relating land use and global land-cover change: a proposal for an IGBP-HDP core project. Report from the IGBP-HDP working group on land-use/land-cover change. Joint publication of the International Geosphere-Biosphere Programme (Report No. 24) and the Human Dimensions of Global Environmental Change Programme (Report No. 5). Royal Swedish Academy of Sciences, Stockholm.

120. United Nations, 2010. World Urbanization Prospects: The 2009 Revision (United Nations Department of Economic and Social Affairs/Population Division, New York). http://esa.un.org/unpd/wup/index.htm.

121. Van Asselen, S., Verburg, P.H., Vermaat, J.E., Janse, J., 2013. A global meta-analysis of wetland conversion. PLoS One 8, e81292.

122. van Vliet, J., Magliocca, N., Büchner, B., Cook, E., Rey Benayas, J., Ellis, E., Heinimann, A., Keys, E., Lee, T., Liu, J., Mertz, O., Meyfroidt, P., Moritz, M., Poeplau, C., Robinson, B., Seppelt, R., Seto, K., Verburg, P., 2015. Meta-studies in land use science: current coverage and prospects. Ambio 015–s13280 (in press).

123. van Vliet, N., Mertz, O., Heinimann, A., Langanke, T., Pascual, U., Schmook, B., Adams, C., Schmidt-Vogt, D., Messerli, P., Leisz, S., Castella, J.C., Jörgensen, L., BirchThomsen, T., Hett, C., Bech-Bruun, T., Ickowitz, A., Vu, K.C., Yasuyuki, K., Fox, J., Padoch, C., Dressler, W., Ziegler, A.D., 2012. Trends, drivers and impacts of changes in swidden cultivation in tropical forest-agriculture frontiers: a global assessment. Global Environ. Change 22, 418–429.

124. Verburg, P., van Berkel, D., van Doorn, A., van Eupen, M., van den Heiligenberg, H., 2010. Trajectories of land use change in Europe: a model-based exploration of rural futures. Landscape Ecol. 25, 217–232.

125. Verburg, P.H., 2006. Simulating feedbacks in land use and land cover change models. Landscape Ecol. 21, 1171–1183.

126. Verburg, P.H., Dearing, J., Dyke, J., Leeuw, S.V.D., Seitzinger, Steffen, W., Syvitski, J., 2015. Methods and approaches to modelling the anthropocene. Global Environ. Change doi:http://dx.doi.org/10.1016/j.gloenvcha.2015.08.007.

127. Verburg, P.H., Denier van der Gon, H.A.C., 2001. Spatial and temporal dynamics of methane emissions from agricultural sources in China. Global Change Biol. 7, 31–47.

128. Verburg, P.H., Eickhout, B., van Meijl, H., 2008. A multi-scale, multi-model approach for analyzing the future dynamics of European land use. Ann. Reg. Sci. 42, 57–77.

129. Verburg, P.H., Erb, K.-H., Mertz, O., Espindola, G., 2013. Land System Science: between global challenges and local realities. Curr. Opin. Environ. Sustain. 5, 433–437.

130. Verburg, P.H., Neumann, K., Nol, L., 2011. Challenges in using land use and land cover data for global change studies. Global Change Biol. 17, 974–989.

131. Verburg, P.H., Schot, P., Dijst, M., Veldkamp, A., 2004. Land use change modelling: current practice and research priorities. GeoJournal 61, 309–324.

132. Verburg, P.H., Veldkamp, A., de Koning, G.H.J., Kok, K., Bouma, J., 1999. A spatial explicit allocation procedure for modelling the pattern of land use change based upon actual land use. Ecol. Model. 116, 45–61.

133. Walsh, S.J., Crews-Meyer, K.A., 2002. Linking People, Place, and Policy: A GIScience Approach. Kluwer Academic Publishers, Boston.

134. Weinzettel, J., Hertwich, E.G., Peters, G.P., Steen-Olsen, K., Galli, A., 2013. Affluence drives the global displacement of land use. Global Environ. Change 23, 433–438.

135. White, B., Borras, Jr SM, Hall, R., Scoones, I., Wolford, W., 2012. The new enclosures: critical perspectives on corporate land deals. J. Peasant Stud. 39, 619–647.

136. Wolff, S., Schulp, C.J.E., Verburg, P.H., 2015. Mapping ecosystem services demand: a review of current research and future perspectives. Ecol. Indic. 55, 159–171.

137. Wu, J., 2013a. Key concepts and research topics in landscape ecology revisited: 30 years after the Allerton Park workshop. Landscape Ecol. 28, 1–11.

138. Wu, J., 2013b. Landscape sustainability science: ecosystem services and human well-being in changing landscapes. Landscape Ecol. 28, 999–1023.

139. Yu, Y., Feng, K., Hubacek, K., 2013. Tele-connecting local consumption to global land use. Global Environ. Change 23, 1178–1186.

140. Zaehle, S., Ciais, P., Friend, A.D., Prieur, V., 2011. Carbon benefits of anthropogenic reactive nitrogen offset by nitrous oxide emissions. Nat. Geosci. 4, 601–605. doi: http://dx.doi.org/10.1038/ngeo1207.

141. Zalasiewicz, J., Crutzen, P.J., Steffen, W., 2012. Chapter 32—the anthropocene. In: Gradstein, F.M., Schmitz, J.G.O.D., Ogg, G.M. (Eds.), The Geologic Time Scale. Elsevier, Boston, pp. 1033–1040.

142. Ziegler, A.D., Phelps, J., Yuen, J.Q.I., Webb, E.L., Lawrence, D., Fox, J.M., Bruun, T.B., Leisz, S.J., Ryan, C.M., Dressler, W., Mertz, O., Pascual, U., Padoch, C., Koh, L.P., 2012. Carbon outcomes of major land-cover transitions in SE Asia: great uncertainties and REDD+ policy implications. Global Change Biol. 18, 3087–3099.

CHAPTER 8

ASSESSING ON-FARM PRODUCTIVITY OF MISCANTHUS CROPS BY COMBINING SOIL MAPPING, YIELD MODELLING AND REMOTE SENSING

Goetz M. Richter, Francesco Agostini, Alexandra Barker Delphine Costomirisa, Aiming Qi

Keywords: Agronomy, Crop establishment, Land use, Optical remote sensing, Soil maps, Yield gap

INTRODUCTION

Bioenergy from agricultural crops has come under severe criticism when sourced from 1st generation, starch/sugar providing feedstocks due to environmental concerns [1] and impacts on food security [2]. In contrast, 2nd generation, ligno-cellulosic feedstocks are more acceptable due to their superior greenhouse gas (GHG) benefit due to low fertiliser and management inputs [3]. For large parts of the United Kingdom, a previous comparative study on GHG emissions [4] was based on feedstock estimated and up-scaled with empirical yield models for two major agricultural biomass crops: Miscanthus and short-rotation coppice [5] and [6]. These feedstocks were rated according to their potential regional performances [7] based on experimental observations, but there was no information with regard to the actual on-farm performance or yield gap.

On-farm yield prediction is vital for the successful implementation of a sustainable energy supply chain based on biomass crops. Bottom-up verification of model-based supply estimates is essential to assess the wider impact on food and energy security, land use change, and GHG emission. In England, farmers have been encouraged to grow short rotation coppice and Miscanthus as biomass within the Energy Crops Scheme (ECS). Planting of Miscanthus has increased between 2006 and 2011 from 6 to 8 thousand ha[1][8]. According to expert knowledge one could produce an estimated total of 80–120 thousand tonnes of oven dry matter (odmt which corresponds to "Mg") per annum. Potential yield maps [9][2] were based on long-term temperature and radiation only and categorised land into three different yield classes (low/medium/high), suggesting an economic threshold of 12 Mg ha^{-1} in a well-established crop. In fact, the average Miscanthus yield was 12.8 Mg ha^{-1} in experimental observations [6] and the simulated, obtainable average yield was 12.5 Mg ha^{-1} if grown on agricultural land in England [10]. However, the national average of delivered on-farm oven dry matter biomass yield is about 9 Mg ha^{-1}. This would correspond to a yield gap of close to 30%. Questions arise about the reasons for this gap, whether soils differ in terms of yield realisation and variation, and whether a diagnostic tool can be developed to overcome the sub-optimal performance of these new crops. Considering the economic threshold of between 12.5 [9] and [11] and 9 Mg ha^{-1} for Miscanthus [10] many farmers would be well below the break-even point and, therefore, it is crucial to identify avenues to improve productivity.

No studies have been conducted to assess the extent to which farmers successfully grow these new crops, although this has been identified as a future research need in particular in relation to marginal lands [12]. It is not clear, whether there is a yield difference due to the effect of soil type in terms of water

Citation: Goetz M. Richter, Francesco Agostini, Alexandra Barker, Delphine Costomiris, Aiming Qi, Assessing on-farm productivity of Miscanthus crops by combining soil mapping, yield modelling and remote sensing, Biomass and Bioenergy, Volume 85, February 2016, Pages 252-261, ISSN 0961-9534, http://dx.doi.org/10.1016/j.biombioe.2015.12.024.

availability or texture class, and how achievable on-farm yields compare with model-based estimates of biologically obtainable production (i.e. accounting for bio-physical constraints). It would be valuable to examine whether available remote sensing (RS) data enable us to quantify patchiness which has been shown to impact on yield and economic feasibility [13], and trade-offs between two major ecosystem services, yield related soil carbon sequestration [13] and biodiversity [14].

The aim of this study was to compare model-derived productivity potentials, based on soil water availability and long-term climate data, with sustainable (i.e. achievable and consistently delivered) on-farm Miscanthus yields. The latter are defined as the reported average annual field or on-farm yields of well-established crops, i.e. ≥5 years after planting. In a second step we investigated how observed yields related to the variability of the crops by using geo-referenced optical satellite data to derive the Normalized Difference Vegetation Index (NDVI) per field from Miscanthus growing farms. Finally, this led us to discuss the potential of developing a decision support tool for farmers to improve crop cultivation (e.g. decide on gap remediation or total replanting), and for policy makers to monitor and improve the effectiveness of support mechanisms.

MATERIALS AND METHODS

This study takes a bottom-up approach and uses on-farm yield surveys at the field and farm scale to verify the outputs (yield maps) of productivity models established on the basis of experimental, plot-scale observations (Fig. 1). Google Earth was used in photo-interpretation for quality control and to verify the field locations given in the ECS database. Geolocation of fields was essential to create field- and crop-specific masks to overlay information layers, in particular the spatially explicit potential yield map and the satellite image-derived NDVI (see also Fig. 6). For a subset of fields, ground observations on crop density, height and leaf area (morphometric data) were also taken.

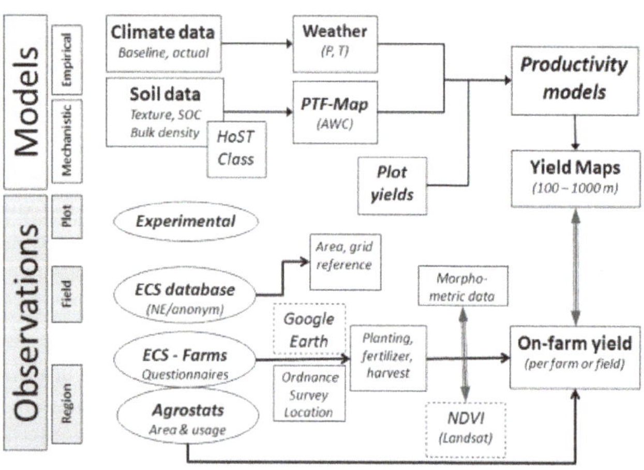

Figure 1: Data integration for on-farm yield evaluation, estimation and improvement.

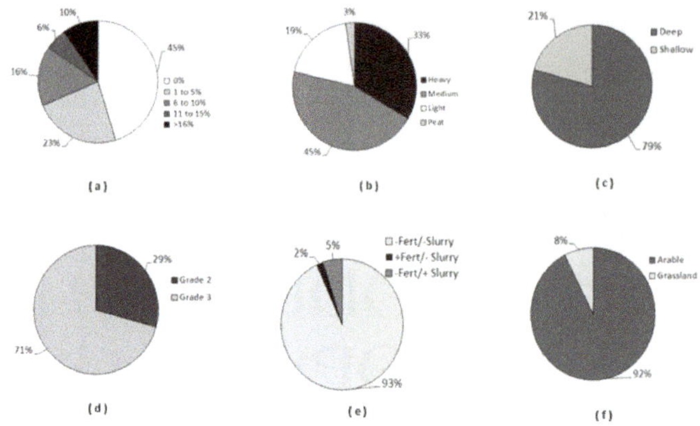

Figure 2: Qualitative information provided in questionnaires for 77 Miscanthus fields, field characterisation according to slope (a), soil texture (b), profile depth (c), grade according to Agricultural Land Classification (d), fertilizer/slurry application (e), and prior land use (f).

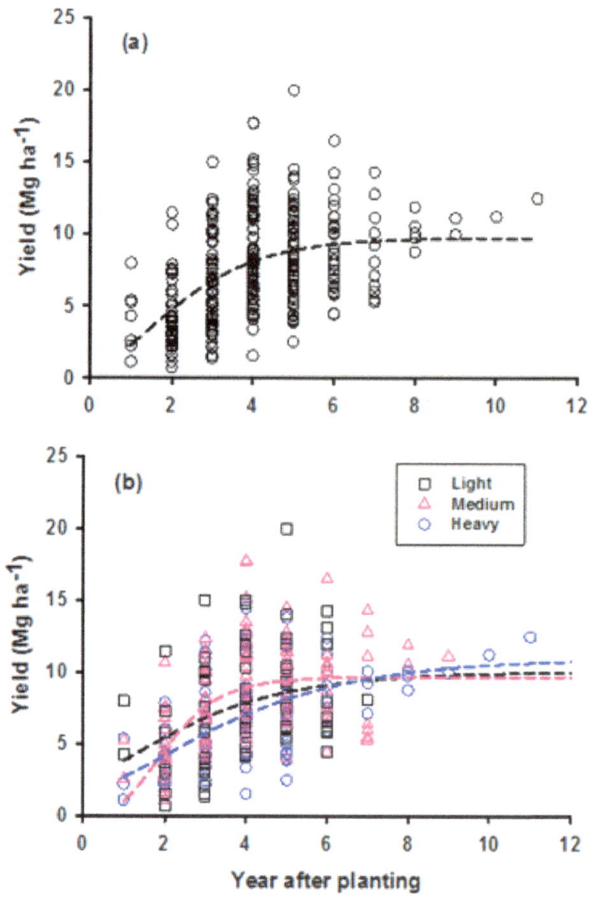

Figure 3: Fitted curves for progression in annual biomass yield of UK Miscanthus crops after planting, based on fields from all soils (a) and on fields grouped according to light, medium and heavy soil texture (b).

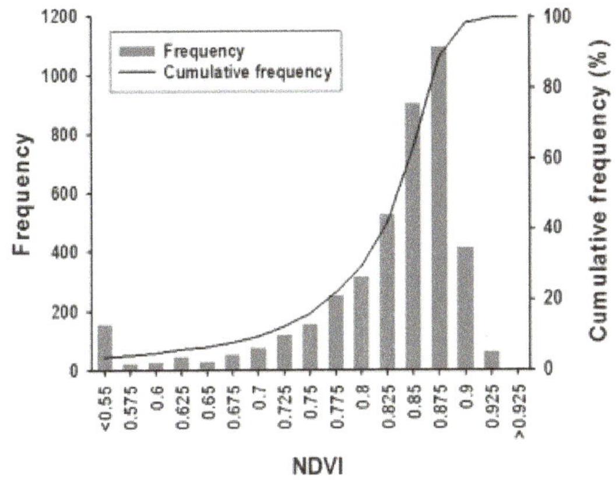

Figure 4: Frequency distribution of the normalised difference vegetation index (NDVI) across 98 Miscanthus fields sampled and covered by the Landsat-5 TM images for 2011 cropping season (50 separate field or farm observations).

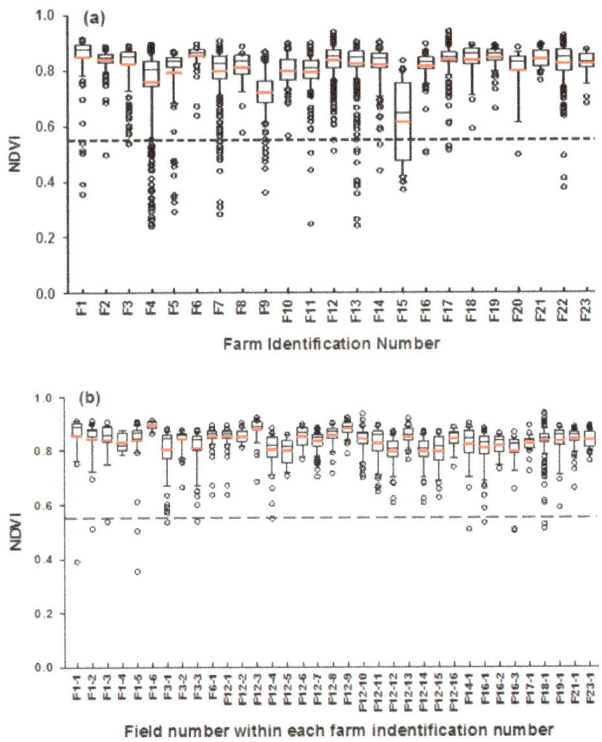

Figure 5: Box and whisker plots showing distribution of normalised difference vegetation index (NDVI) derived from images by Landsat-5 TM in September 2011 within farm-grouped fields on 23 different farms (a) and within 35 single fields on 11 different farms (b). Letters and numbers denote farmers and fields. For example, F3-1 means field 1 on farm 3. The lower boundary of the box indicates the 25th percentile while the upper boundary of the box indicates the 75th percentile. The lower and upper whiskers indicate the 5th and the 95th percentile while the black and red horizontal lines within the box indicate the median and the mean NDVI, respectively. The open circles both below and above the whiskers are outliers.

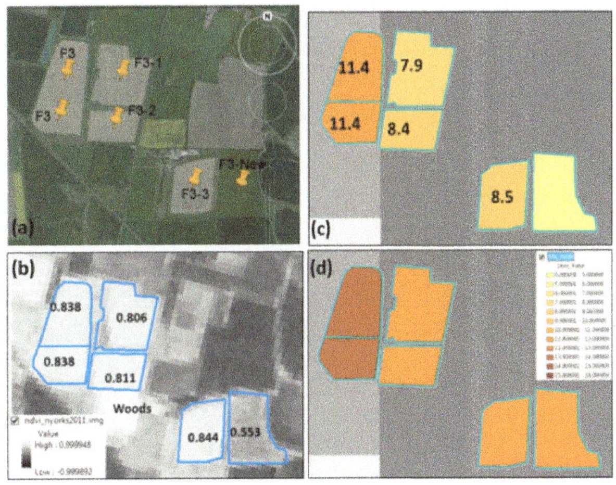

Figure 6: Field map of Google Earth identified Miscanthus fields (a) and NDVI derived from Landsat-5 TM in 2011 (b), management effects (year of planting: brown – 2006; orange – 2007; yellow - newly planted) on average field yield (c), and yield map (d).

On-Farm Survey

With the support of Natural England (NE) we conducted an anonymized field survey among farmers within the ECS. Three consecutive questionnaires were sent out in 2009, 2011 and 2013 to (1) establish a database of corresponding field locations with UK grid geographic references (Ordnance Survey map), soil type and hydrology, and soil texture class; (2) record data on land use history and management (Miscanthus planting date, fertiliser use, harvest dates); and (3) collate time series of biomass yield to estimate actual deliverable on-farm yield potentials.

From the coordinates (UK Ordnance Grid) or the postcode provided by some farmers, fields or farm sites were located using UK Streetmap software. Screenshots of the field sites using Google Earth time line and Ordnance survey maps (1:25,000) were instrumental to verify field locations initially starting from the coordinates recorded by NE in the ECS database.

The questionnaires were circulated to farmers to identify individual field or farm characteristics. Questionnaire 1 was centrally distributed by NE and designed to record site details in terms of soil water availability and fertility to understand site selection criteria and local yield formation. As feedback from farmers on questions regarding soil, climate and site hydrology was rare the simplified Questionnaire 2 was distributed by NE as well as being sent directly to farmers who had provided their addresses. Questionnaire 3 was further complemented with feedback reports to communicate results and prepare for personal interviews with background information. Its main purpose was to develop and verify a comprehensive yield database. Some farmers conveniently recorded yields as number of total bales per year across the whole farm, but some also distinguished field-specific yields. Conversion of bale number to weight was based on the assumption that each bale (size 0.9 × 1.2 × 2 m) consisted of 600 kg at 25% moisture content (i.e. an equivalence of 450 kg oven dry matter per bale).

Model-Estimated Obtainable Yields

Crop productivity depends strongly on variety, agronomy and environment but for commercial farms a single Miscanthus genotype is grown. The best yield achievable using current variety and recommended agronomic practices under any physical and biotic conditions is defined as obtainable yield. Obtainable Miscanthus yields were estimated using an empirical model (Y_{EM}) based on soil available water capacity (SAWC) and long-term monthly climate data (precipitation, temperature, radiation) for each field [6] and [10].

This empirical yield model is based on a multiple regression of experimental yields from 14 experiments across 10 different sites in the UK, with SAWC ranging from <100 mm–>300 mm within the rooting depth and observed local weather records [6]. A yield map had been generated using long-term weather data (1950–2000) [15] at a 1 km by 1 km grid scale across the UK and SAWC was estimated using the method described earlier [10]. In short, soil hydrological site properties combined water retention properties (derived from soil physical properties) [16] with depth to groundwater and water from underlying porous rock (i.e. hydrology of soil type (HoST) classes [17]). For the on-farm yield evaluation all relevant soil data were downscaled from NATMAP1000 to a 100 × 100 m^2 grid by combining the soil survey data and NATMAP vector maps [18] and selecting the respective soil properties and yield potential of the dominant soil series and association.

Calculating the Empirical on-Farm Yield Plateau

An empirical on-farm yield plateau (i.e. the likely peak or achievable yield of a given plantation), Y_{OF}, was estimated using the time series of site-specific commercially harvested oven dry matter yield reported in the survey. Unlike arable or fodder crops, maximum and peak productivity of a Miscanthus crop can take three to five years after planting [19] and [20]. In the case of long and complete time series (10–15 years) such as the plot size experiments at Rothamsted Research, a double exponential model [21] was found most appropriate to describe the first phase rise and the final phase decline of obtainable biomass yields (Y) over years after planting (X) as observed in many other experimental data [22]:

$$Y = Y_0 \exp\left(\mu_{min}(X-1) + \frac{\mu_0 - \mu_{min}}{\nu}(1 - \exp(-\nu \cdot (X-1))\right) \quad (1)$$

where Y0 is the potential biomass yield at the end of first establishing year, X the number of years after planting, μ_0 and μ_{min} are the initial and final annual net relative increase rate of biomass yield, respectively, and v is the parameter determining the rate of change from μ_0 to μ_{min}.

In the survey database, most farmers had annual biomass yields for fewer than seven years after planting. For practical and economic reasons only a small number of fields were harvested after the first year (ca. 9% of the recorded fields), this increased during the second and third year of growth to 50 and 90% of the recorded fields, respectively. A total of 79 fields represented about 1000 ha with annual yields for four or more years after planting, and only half of these extended beyond six years after planting. Therefore, Equation (1) was simplified to the Gompertz equation (Equation (2)) in order to estimate the maximum potential biomass yield at a pre-determined harvesting year (X = 5) beyond which biomass yield increased very little to reach an asymptote (i.e. the on-farm yield peak or plateau, Y_{OF}):

$$Y = Y_0 \exp(\mu(1 - \exp(-\nu \cdot (X-1))) \quad (2)$$

Equation (2) has only three parameters; as in Equation (1), Y0 is the potential biomass yield at the end of first year, μ equals μ_0/v since no decline of biomass yield was assumed (i.e. μ_{min} was at zero in Equation (1)) and v is rate of change relative increase of biomass yield.

Preliminary analyses showed that biomass yield reached more than 97% of the on-farm plateau yield on the 5th year after planting in all soil texture classes. Therefore, the 5th year after planting was used to estimate on-farm plateau biomass yield, Y_{OF}. This same pre-defined year was also used to calculate the average actual observed on-farm plateau of biomass yield using available yields from fields harvested on the 5th year after planting. This average on-farm biomass yield plateau was later used to calculate yield gaps. Curve-based yield plateaus and biomass yield by the end of the first year (Y0) were estimated for each soil texture category (heavy, middle and light; H/M/L) using the Gompertz model (Equation (2)).

Normalised difference Vegetation Index (NDVI)

Normalised Difference Vegetation Index (NDVI) was derived from Landsat-5 TM scenes (183 × 172.8 km) provided by the European Space Agency (ESA) [23] in the Universal Transverse Mercator (UTM) coordinate system with a 30 m pixel spacing. Scenes were selected for dates during the late growing season (September 2011) for comparison to the harvest data. Landsat-5 TM Level 1 System Corrected data had undergone pre-flight and in-flight radiometric calibration [24] and was geometrically accurate within 1 sigma [23] and [25]. The nearest neighbour (NN) resampling algorithm preserves the maximum original radiance values, and makes it free from distortions from the sensor, satellite or Earth [26] and [27]. The satellite imagery was pre-processed in ERDAS Imagine 10 to remove undesirable image characteristics, correct for geometric distortions and calculate NDVI.

Atmospheric and Topographic Correction

Dark Object Subtraction (DOS) is a radiative transfer approach applied to remove atmospheric distortions (within ERDAS Imagine software). This method corrects for the effect of atmospheric scattering (the dominant atmospheric affect for Landsat TM data) based on the darkest pixel in the image to produce apparent surface reflectance [28]. This was accomplished in four steps by: (1) selecting a clear water body as a dark object with low digital number (DN) values as previously suggested by Gordon [29]; (2) selecting a representative pixel from each reflective band from its DN frequency histogram where there was a sudden increase in the number of pixels; (3) implementing the conditional statement to subtract the per-band haze DN value from the respective spectral band of the whole scene; and (4) preventing the value of zero [30].

A 10 m spatial resolution Land Form Profile Digital Terrain Model (DTM) acquired from the Ordnance Survey [31] showed that terrain slopes varied from 2 to

15%. However, with the exception of a few fields in the 2011 crop survey, none of the fields in 2009 and 2013 crop survey exceeded the critical threshold of 7% slope. Only slopes of 7% or higher would be significant enough to impact on the crops' spectral signatures [32].

Calculating Radiance and Reflectance

Spectral radiance values (Lλ) at the sensor's aperture were calculated using Equation (3)from the total potential energy reaching the sensor (Qcal) normalised for solar irradiance using ratio of spectral radiance scaling factors LMAXλ and LMINλ (range 0–1; NASA, 2011) over the maximum and minimum energy ($Q_{cal\ max/}Q_{cal\ min}$) measured in each band pixel (values of DN range from 1 to 255 for 8 unsigned Level-1 Product Generation System products).

$$L_\lambda = (LMAX_\lambda - LMIN_\lambda)/(Q_{cal\ max} - Q_{cal\ min}) \times (Q_{cal} - Q_{cal\ min}) + LMIN_\lambda \quad (3)$$

Respective solar spectral irradiances ESUNλ 1533 and 1039 W/($m^2 \times \mu m$) for spectral band 3 (red) and 4 (near-infrared, NIR), respectively, were obtained from NASA (2011) to calculate the spectral reflectances, ρλ using Equation (4).

$$\rho_\lambda = \pi \times L_\lambda \times d^2 / ESUN_\lambda \times \cos\theta_s \quad (4)$$

where π is a mathematical constant (3.14159), Lλ is the solar spectral radiances for band 3 or 4, respectively, d2 the distance on the Julian day, ESUNλ is 1533 (for Landsat-5 TM band 3) or 1039 (for Landsat-5 TM band 4) and cosθs is the solar zenith.

Calculating the NDVI

The NDVI depends on the orientation of the surface with respect to the observer. As all selected sites were below 7% the NDVI was simply calculated using Equation (5) within ERDAS Imagine from spectral reflectance for band 3 and 4; ratios of red and NIR spectral bands determine the presence of vegetation [33]. Output numbers must be in float single data type.

$$NDVI = (\rho NIR - \rho RED)/(\rho NIR + \rho RED) \quad (5)$$

where ρRED and ρNIR are reflectance in band 3 and band 4, respectively

Building a GIS Data Base for Field Sites

A GIS database was created in ArcGIS (version 10) to overlay soil and obtainable yield data (both at a 100 m × 100 m grid resolution), derived from the empirical Miscanthus model which uses soil and long-term climatic inputs, and the NDVI for verified field locations. The analysis of variance was done with Genstat [34] for samples grouped according to soil texture group, which covered clayey (heavy-H), loamy (medium-M) and sandy (light-L) soil textures.

Box and whisker plots, using the Tukey method [35], were used to display the quartiles and outliers of NDVI from combined fields and within individual fields on various farms (SigmaPlot 13.0 software). When frequency histograms for NDVI and ground-based data deviated from the normal distribution non-parametric correlation (Spearman›s ranking) was used, which was more appropriate to handle data with outliers. Landsat image-derived NDVI were extracted from subplots to relate pixel properties to ground-based estimates, e.g. LAI, plant density, using a linear regression model [36].

RESULTS

Survey Results – Structure of the Sample

In 2009, farmers growing a collective 573 ha of Miscanthus responded to the first survey sent out by NE; in 2011, the response to direct contacts was >90% which in conjunction with additional contact through NE resulted in retrieval of information for a further 669 ha. Knowledge Transfer activities and farm visits within the NERC-funded Microclimates project increased the area with background information to >2100 ha. Of the approximate 2100 ha, 940 ha provided biomass yield data from crops planted >4 years ago, which could be analysed to estimate a yield plateau by fitting Equation (2) (see 3.2). The data set covered a similar number of sites with yield data in 2011, which could be examined in relation to the NDVI. Over all of the area, the average field size was about 6.7 ha.

The questionnaire characterised the topographic, pedological, management, land use history and other site properties (Fig. 2). In the 2011 crop survey samples, almost 70% of area was flat to slightly sloped (<5%) but 10% was above the suggested slope threshold of 15%, considered unsuitable for crop cultivation [10]. About half of the crop was grown on medium textured soils (Table 1), which usually have deep profiles and plenty of plant available water. With few exceptions, the farmers› classification of soil textures (H/M/L) agreed with the information on the soil map. In contrast to earlier scenarios [10] ~30% of the area was classified as good agricultural land (Grade 2) and no Grade 4 land was used. Less than 10% of the area surveyed in 2011 had grassland as the previous land use, and very few farmers (<10%) applied any nitrogen (N) or organic fertilizer to their Miscanthus crops.

Table 1: Summary survey data with mean actual on-farm plateau Miscanthus yields (Y_{OF}) in 5th year after planting (YaP) with standard error (SE) and coefficient of variation (CV) grouped according to soil texture group – light (L), medium (M) an heavy (H) in comparison to the long-term biologically obtainable potential yield map using an empirical model (Y_{EM}) and the calculated yield gap (YGap). n is the respective number of sampled fields

Soil group	Area	n	Y_{OF} (SE)	CV	Y_{EM}	N	YGap
	[ha]	[5th YaP]	[Mg ha^{-1}]	[%]	[Mg ha^{-1}]	[Y_{EM}]	[%]
L	242	15	9.51 (1.02)	41.4	10.4	25	8.6
M	471	29	9.48 (0.41)	23.3	10.7	48	11.5
H	227	22	7.85 (0.70)	41.9	12.4	27	36.7
All	940	66	8.94(0.38)	34.1	11.0	100	18.7

Actual on-Farm Yield (Y_{OF}) Plateaus

A total of 66 observations were available for estimating achievable on-farm plateau yields (Y_{OF}) harvested five years after planting. On average, Y_{OF} was 8.94 Mg ha^{-1} over all soils and varied between 7.85 and 9.51 Mg ha^{-1} depending on the soil texture group (Table 1). There was considerable variation within each soil texture group and the calculated coefficient of variation (CV) ranged from 23% to over 41% in medium and light/heavy textured soils, respectively.

The achievable on-farm yield plateau (Y_{OF}) was also estimated by fitting the three parameter Gompertz equation (Equation (2)). It was first fitted to all combined data points among all soil texture classes (Fig. 3a) and then separately to three subsets pooled according to distinctive soil texture groups (Fig. 3b). The estimated parameters are given in Table 2. The estimated biomass yield harvested in the first year was 2.36 Mg ha^{-1} over all soils but it varied appreciably from one soil texture group to another. There were few fields with 1st year harvests, which caused great uncertainty in these estimates, particularly in the loamy textured (M) soil group. The Gompertz equation-based estimate of achievable on-farm biomass yield plateau (i.e. in the 5th year harvest) was 8.90 Mg ha^{-1} over all soils while it varied between 8.13, 9.46 and 8.62 Mg ha^{-1} on H/M/L soils, respectively (Table 2). These estimates compared well with the corresponding achievable on-farm biomass potential yields calculated using observed yields on the 5th year after planting (see Table 1). The added benefit of this analysis was that it clearly showed that a longer timespan was required to reach the yield plateau in heavy clay soils (Fig. 3B).

Table 2: Parameter values and standard errors of the Gompertz model (Equation (2)) to estimate achievable on-farm plateau yields (Y_{OF}) for crops in the 5th year after planting on light(L), medium(M) and heavy(H) soil. Y_0 is the estimated yield harvested by the end of the first year, n is the number of data points.

Soil Group	Y_0 Mg ha^{-1}	μ	ν	p-value	Y_{OF} Mg ha^{-1}	n
L	3.86 (1.93)	0.962 (0.38)	0.453 (0.49)	<0.01	8.62	71
M	1.03 (0.89)	2.245 (0.84)	1.115 (0.32)	<0.01	9.46	147
H	2.71 (0.80)	1.400 (0.24)	0.383 (0.14)	<0.01	8.13	108
ALL	2.36 (0.68)	1.421 (0.26)	0.679 (0.16)	<0.01	8.90	327

Predictability of on-Farm Miscanthus Productivity – Yield Gap

The yield map was generated using the empirical relationship between peak yields of established crops (i.e. ≥5 years after planting) from experimental sites and biophysical control variables (precipitation, temperature and soil available water capacity) [6] and [10]. This yield represents the bio-physically obtainable biomass yield of the genotypes grown at the time. The downscaled vector map allowed readings at the hectare resolution, and a total of 100 map entries could be compared to 66 surveyed on-farm yield records (see Table 1). Site-specific biological yield potentials estimated using the empirical model (Y_{EM}) ranged from <5 to >15 Mg ha^{-1} with an average of 11 Mg ha^{-1}. The area-weighted observed average yield (8.94 Mg ha^{-1}) remains 18.7% below the site-specific average Y_{EM} (see Table 1) and 28.5% below the predicted national average (12.5 Mg ha^{-1}) [10]. The biggest yield gap, exceeding 35%, between the actual achievable on-farm yield (Y_{OF}) and the biologically obtainable yield estimated using the empirical model (Y_{EM}), was found for heavy clay soils. On-farm delivered yields on medium and light textured soils came closer to their biologically obtainable yields, with the yield gaps averaging about 10% (Table 1). The achievable on-farm plateau yields (Y_{OF}) estimated for >5 year old plantings using Equation (2) were very similar to those actual reported on-farm yields calculated using survey yields harvested from mature crops (i.e. on the 5th year after planting; see Table 1 and Table 2). Therefore, the yield gaps between the Y_{EM} and on-farm plateau yields estimated using the Gompertz equation were also similar to those between the Y_{EM} and the actual on-farm yields.

NDVI – a Measure for Patchiness of Miscanthus Fields?

A total of about 100 fields that covered ca. 700 ha were subjected to spatial analysis using the Landsat image-derived NDVI, which was related to field-specific yield measurements. A total of 23 farms were included in the analysis. For 36 individual fields (about 300 ha) on 11 of the 23 farms, field-specific yields could be compared to the corresponding field specific NDVI while for the other 12 farms, field-specific NDVI were aggregated to a farm average to be compared with yields reported at the farm scale. The NDVI ranged from 0.227 to 0.939 and its right skewed distribution (Fig. 4)indicates that the median was higher than the mean (0.836 vs. 0.805). About 200 values were smaller than 0.6, which corresponds to less than 5% of the total readings and of which 66 readings were associated with a recently established field (averaged at 0.553; see Fig. 6). Some of the low readings could be associated with shaded borders or other non-radiative bodies on the border, but could also be due to poor establishment in some of the fields, which was visible even on Google Earth images.

Considering this value of 0.553 as a reference threshold, the box and whisker plot (Fig. 5) reveals more NDVI outliers below this value where only farm-level aggregated (Fig. 5a) rather than field-specific yields were recorded (Fig. 5b). These pixels could represent areas with low stand density, where NDVI differed from the overall mean of field or farm beyond the inner and outer fences of their interquartile range. There were five samples in the farm-specific set (Fig. 5a) that show many outliers (Farm Nos. 4, 5, 7, 9 and 13), in particular the multiple field set of Farm No. 13. The suspected patchiness could be verified in all cases using Google Earth images. This box and whisker plot also identifies entire farms or fields which underperformed in terms of the mean (for example Farm No. 15) and revealed that patchiness further reduced the farm-specific NDVI (see Farm Nos. 7, 9 and 11). In the case of field-specific yield monitoring there were fewer outliers in total and hardly any below the NDVI threshold (Fig. 5b). Most of the outliers in these samples were observed in between the inner and outer interquartile range and above the threshold.

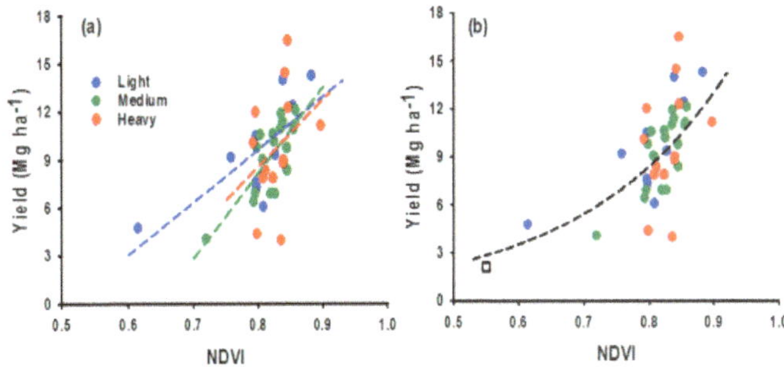

Figure 7: Field-specific Miscanthus biomass yield in simple linear relation to field-specific NDVI grouped according to soil texture group without the yield from the first year establishing crop (a), but in an exponential relation to field-specific NDVI with the yield from the first crop (open square) (b) in which Yield = $0.1687\exp^{4.8712*NDVI}$, $R^2 = 0.542$, n = 43.

Relationship between NDVI and Biomass Yield

NDVI derived from reflectance in the NIR and red spectrum has been used as a proxy for canopy leaf area index (LAI) and fraction of absorbed photosynthetically active radiation (fAPAR) and, consequently, can be an indirect measure of biomass productivity. As expected, the NDVI was a good overall indicator of increasing Miscanthus yield in mature stands, both at the farm level (Fig. 6) and in between different data sets across different farms, regions and soil properties (Fig. 7). At the farm scale, 4th year yield was about 25% below 5th year yields (Fig. 6) which was much more than the difference in NDVI (ca. −4%) derived for this set of fields. The newly planted crop on the same farm, for which a yield of about 2.4 Mg ha^{-1} could be expected, had already a substantial NDVI, which, however, was lower than other poorly performing Miscanthus crops (e.g. farms No 4 and 15). There are two implications from this example: (a) the difference in yield (biomass density) exceeds greatly that in NDVI (canopy density), and (b) there is a background reflectance which should be attributed to the undergrowth in spite of initial spray of herbicides.

There was one field from which the harvested yield was from the first year crop and therefore was not compatible with crops of more than four years after planting in other fields. Therefore, first, it was excluded in the simple linear regression analysis between biomass yield and NDVI. When a simple linear relation was fitted to data points among all soil texture groups, NDVI accounted for 32.4% (n = 42) of variation in the observed on-farm yield (see Table 3). When the simple linear relation was fitted to data points of separate soil texture classes, the respective best line is shown in Fig. 7a and the parameter estimates are shown in Table 3. Due to the narrow range in image-derived NDVI values, little variation in the biomass yield (~10%) can be explained by the NDVI in the heavy soil texture group (Table 3). However, the NDVI reflects more than 50% of the yield variation for both light and medium soils (Fig. 7 and Table 3). When the biomass yield from the first year crop was included, an exponential relation was better fitted to the data under all soils than the simple linear relation (Fig. 7b). With the fitted exponential curve, NDVI explained 54.2% of the yield variation.

Table 3: Relationship between mean NDVI and mean observed Miscanthus yield at single field scale under light (L), medium (M), heavy (H) and all combined soil groups (ALL). Yield = a *NDVI – b, p-value is the significant probability level, R^2 is proportion of variance in yield accounted for by NDVI, n is the total number of data pairs

Soil	Slope (a)	Intercept (b)	R^2	p-value	n	Mean yield [Mg ha^{-1}]
L	33.03 (10.43)	16.75(8.34)	0.556	<0.02	10	9.56
Ma	53.53 (10.95)	34.64(8.78)	0.584	<0.01	19	9.19
H	41.54 (35.48)	24.63(29.40)	0.111	>0.25	13	9.76
ALL	36.64 (8.37)	20.45(6.85)	0.324	<0.01	42	9.46

[a]One data point was omitted because the yield was from the first year establishing crop (see Fig. 7).

DISCUSSION

This work set out to obtain a comprehensive view of the actual on-farm productivity of Miscanthus as a novel crop in the UK to see how it compares to the model-derived biologically obtainable biomass yield (i.e. yield gap analysis), and to find out whether there are diagnostic tools and ways to identify measures to close the yield gap.

How Representative is Our Survey?

Surveyed on-farm Miscanthus yields reported here are not comprehensive but are still highly representative, being based on about 1000 ha of mature crops and representing about 20% of the whole Miscanthus crop area registered under the ECS until 2007 at the time. The present database is therefore an important resource to study biomass yield of Miscanthus in the UK and relevant to other regions where it is or will be grown for bioenergy. It can easily be expanded if the public funding were linked to clear reporting obligations at the source or user end. All data were provided on a voluntary basis and might bear some uncertainty due to reporting and conversion methods. For example, a biomass yield of 8 Mg ha^{-1} reported on a light soil for the first year after planting seems rather unrealistic. The harvested and delivered yields were extremely variable (2.4–20 Mg ha^{-1}). Although the overall average of about 9 Mg ha^{-1} stayed well below the predicted biophysically obtainable yield, as reported in the national yield map (see Table 1), it matches well the National Statistics. Compared to the biomass provided to the energy companies [8] and [9] the average on-farm peak yield plateau (Table 1) estimated from our database seems fairly representative. According to the latest statistics, biomass from Miscanthus used in power stations in the UK increased from 40 to 47 thousand tonnes [37] between 2010 and 2012, and was produced from a varying area of variable age crops. Assuming a fuel usage of about two thirds of the total production, and considering only mature stands, average Miscanthus yields from an increasing area (6400–7500 ha) would be ca. 9.5 Mg ha^{-1}, which is very close to the average reported here.

Although full establishments of Miscanthus crops can take as long as 7 years [22]depending on planting quality and density, the achievable on-farm biomass yield plateau (Y_{OF}) estimated using the Gompertz equation (Equation (2); X = 5 years) was in good agreement with the observed achievable on-farm biomass yield in the survey (see Table 1 and Table 2). Miscanthus yields can be limited by a host of biotic and abiotic factors including agronomic practices. Firstly, poor and/or delayed establishment led to a variable shoot density; e.g. on fields of Farm No 11, shoot density ranged from 14 to 60 m^{-2} (Barker, unpublished results). The lower end of the range is well below the critical value of 39 m^{-2} above which peak yield is independent of shoot density [38]. Secondly, farmers had applied either limited or no N fertilizer (Fig. 2), which could explain declining yields over time, even on experimental plots [22] and [39]. It was further likely that both limited agronomic research and growing experience with this new crop and the low cost and minimum risk strategy of its introduction through the ECS led to this variable and low yield reality. The in-depth analysis and classification of yields showed that crops can fail or flourish on any soil texture class but failure is more likely on heavy soils (Fig. 7).

Yield Maps and Yield Gaps

The yield maps generated by Lovett et al. [10] provided a long-term snapshot of biologically obtainable yield potential, derived from experimentally determined yields at the plot scale and their pedo-climatic inputs, largely explained by soil water availability [6]. Down-scaled to field-specific and farm-specific yields, it was assumed that the soil map of England and Wales of 1970s is still valid and precise both at the 1 km × 1 km grid (NATMAP1000) and its down-scaling using the vector-based field map (i.e. at hectare scale). The texture-based grouping of soils into light, medium and heavy soil categories by the farmers was correct for an overall 55% when compared to the NATMAP database, and more than 75% correctly classified for the heavy soil. The qualitative classification into light and medium textured soils was often interchangeably used by some farmers. We therefore, think that these pedo-climatic yield maps based on detailed soil information and soil hydrology [10] provided a good estimate of biological potential productivity of commercial Miscanthus. However, the probabilistic nature of the soil map (distribution of series within associations) introduces some uncertainty.

The individual farm- and field-specific yield plateaus varied greatly in comparison to the predicted yields (−71%–+262%), which indicates some limitation of the soil maps. Overall, the yield gap of quasi mature Miscanthus crops (i.e. 5 years after planting) averaged at 18.7% when compared to the mapped obtainable yields, whilst it was larger for heavy soils (>35%). The reasons for lower on-farm yields are many and well researched for arable crops [40] and [41] and well recognised in new biomass crops [12]. Our data however, are the first to compare actual, achievable yields to biologically obtainable potential yields. The reasons for the yield gap can be a lack of nutrients [39] and [42] and also crop establishment [13]. Miscanthus stands can have a large number of gaps, which may be beneficial for biodiversity [14] but severely reduce the actual achievable yields and consequently carbon sequestration in the soil [13]. Lesur-Dumoulin [38] found that yields on young commercial Miscanthus fields were about 20% lower than plot yields. Their analysis showed that shoot density <39 m^{-2} was the most limiting factor for yield formation, strongly related to weed cover occupying the gaps as a complementary indicator. There was no information on yield losses of Miscanthus crops while they were machine-harvested in the field. The uncertainty about such losses is large even under experimental conditions [43], [44] and [45] and this may be an important factor for assessing yield gaps. Harvest losses in the supply chain for combustion are assumed to be 30% of the standing biomass at the field scale [46], however, no on-farm data exist.

What is the Potential Benefit of Remote Sensing?

It is clear from this analysis that on-farm productivity for mature stands (≥5 years after planting) can be assessed from NDVI; however, its precision for biomass estimation could be improved by linking information from images to an underlying understanding of growth dynamics. The analysis of two subsamples of Miscanthus fields created by different operators showed a similar range of NDVI (sample 1, 0.227 to 0.913; sample 2, 0.240 to 0.939) over a similar range of yields. Landsat-5 TM data for the 2011 cropping season of interest were scarce, and availability and image quality with <10% cloud cover over the whole scene is a well-recognised limitation [47].

Although RS images from unmanned aerial vehicle (UAV) can be better suited in providing supplementary data for assessment of crop canopy health and development, satellite-derived data from new Earth Observation data could improve productivity and management of the crop, and increase the efficiency of the supply chain. New satellites provide high spatial, temporal, spectral and radiometric resolutions, e.g. Sentinel-1 and -2, SPOT5, QuickBird GeoEye, WorldView-1 and World-View-2. This would help to overcome the problems of frequent cloud cover in the higher latitudes, under the maritime climatic conditions in the UK. Landsat-5 scenes from which to derive in-season growth estimates were scarce. Images taken earlier in the season would allow groundcover to be added into process-based model analysis. Patchiness, shoot density and canopy height could be important adjustment factors for field scale predictions. More work is needed with improved products to define field-specific benchmark crop cover measured by NDVI as its variation can help growers to gauge how well the Miscanthus crops are being performing during the establishment phase. A well-established crop is the key to reach the peak yield early and to maintain both high and stable yield during the whole plantation lifetime. It is clear from the evidence on a specific farm (Fig. 6) that growth dynamics affected 4th year yields, most probably via lower crop height. New RS data from systems, such as synthetic-aperture radar (SAR) from Sentinel-1, need to be tested to see if they can provide information on canopy height to supplement optical data. Light detection and ranging (LiDAR) imagery data can be an important source in assessing variations in canopy height most relevant for biomass estimates in forestry[48] and [49] but also for managing agricultural crops [50]. As they are now being made freely available for England and Wales (UK Environment Agency[3]), LiDAR data could be used to improve management of novel crops like Miscanthus.

Strategic Planting of Perennial Crops

Our analysis showed that heavy soils can be potentially the most productive (with yields of >16 Mg ha^{-1}) but also the most variable in terms of stand density, most likely related to the soil being either too wet or too dry for planting. Limited contact in a dry, cloddy seedbed would expose the rhizomes to hazardous establishment conditions. However, too wet soils due to waterlogging can also cause large bare patches. The question is how to use the field-specific information on soil cover and its related patchiness to aid the decision-makers on termination or amelioration measures. Clay soils, that are difficult to work on an annual basis, should be the most likely to be selected for perennials. Annual sowing and harvest are likely to hit similar problems as the spring planted Miscanthus, which could be estimated using long-term weather scenarios. A lack of timeliness in planting but also harvesting due to limited machinery was a serious problem for the establishment phase of this crop and in need of serious investment.

CONCLUSION

1. On-farm yields of quasi mature Miscanthus stands (~5 years old) varied between 2.4 and 20 Mg ha^{-1} and currently average at a long-term equilibrium yield of around 10 Mg ha^{-1}.
2. Yields on medium and light textured soils are more predictable than on heavy, clay-textured soils, which have the highest yield potential but also the highest yield uncertainty.
3. Satellite imagery of plant cover was well correlated to yield, but, canopy cover (shoot density) should be augmented with information on canopy height which also determines total biomass.
4. Improved spatial and temporal resolution of new satellite data that combines estimates of crop cover and height would facilitate better early productivity forecasts.
5. To obtain continued improvement in crop productivity and efficiency of the biomass supply chain, it is crucial to develop diagnostic and predictive tools based on field-specific soil and satellite-derived information to support the reporting, and national statistics, of biomass yields.

ACKNOWLEDGEMENTS

This work was funded by the UK National Environment Research Council (NERC), through grant NE/H010386/1 (Microclimates[4]), and the Biotechnology and Biological Sciences Research Council (BBSRC) through its Institute Strategic Programme Grant "Cropping Carbon" (grant number BB/I014934/1) at Rothamsted Research. We would like to thank our colleagues at Natural England for their support and in particular many unnamed farmers who showed interest in our work and gave their time to provide yield and management data.

REFERENCES

1. T. Searchinger, R. Heimlich, A. Houghton, F. Dong, A. Elobeid, J. Fabiosa, et al., Use of U.S. Croplands for biofuels increases greenhouse gases through emissions from land-use change, Science 319 (2008) 1238.
2. D. Tilman, J. Hill, C. Lehman, Carbon-negative biofuels from low-input highdiversity grassland biomass, Science 314 (5805) (2006) 1598e1600.
3. K.A. Smith, T.D. Searchinger, Crop-based biofuels and associated environmental concerns, Glob. Change Biol. Bioenergy 4 (5) (2012) 479e484.
4. J. Hillier, C. Whittaker, A.G. Dailey, M. Aylott, E. Casella, G.M. Richter, et al., Greenhouse gas emissions from four bio-energy crops in England and Wales: Integrating spatial estimates of yield and soil C balance in life cycle analyses, Glob. change Biol. Bioenerg. 1 (4) (2009) 267e281.
5. M.J. Aylott, E. Casella, I. Tubby, N.R. Street, P. Smith, G. Taylor, Yield and spatial supply of bioenergy poplar and willow short-rotation coppice in the UK, New Phytol. 178 (2) (2008) 358e370.
6. G.M. Richter, A.B. Riche, A.G. Dailey, S.A. Gezan, D.S. Powlson, Is UK biofuel supply from Miscanthus water-limited? Soil Use Manag. 24 (3) (2008) 235e245.
7. A.W. Bauen, A.J. Dunnett, G.M. Richter, A.G. Dailey, M. Aylott, E. Casella, et al., Modelling supply and demand of bioenergy from short rotation coppice and Miscanthus in the UK, Bioresour. Technol. 101 (21) (2010) 8132e8143.
8. DEFRA, Area of Crops Grown for Bioenergy in England and the UK : 2008- 2011, Department of Environment FaRA, York, England, 2013, p. 32. Department for Environment, Food and Rural Affairs.
9. Defra, Opportunities and Optimum Sitings for Energy Crops, Natural England D, Worcester, UK, 2014.
10. A.A. Lovett, G.M. Sünnenberg, G.M. Richter, A.G. Dailey, A.B. Riche, A. Karp, Biomass production and land use trade-offs revealed by GIS constraint and yield mapping of Miscanthus in England, BioEnergy Res. 2 (1) (2009) 17e29.
11. L. Price, M. Bullard, H. Lyons, S. Anthony, P. Nixon,

Identifying the yield potential of Miscanthus x giganteus: an assessment of the spatial and temporal variability of M-x giganteus biomass productivity across England and Wales, Biomass Bioenerg. 26 (1) (2004) 3e13.

12. B. Gabrielle, L. Bamiere, N. Caldes, S. De Cara, G. Decocq, F. Ferchaud, et al., Paving the way for sustainable bioenergy in Europe: technological options and research avenues for large-scale biomass feedstock supply, Renew. Sustain. Energy Rev. 33 (2014) 11e25.
13. J. Zimmermann, D. Styles, A. Hastings, J. Dauber, M.B. Jones, Assessing the impact of within crop heterogeneity ('patchiness') in young Miscanthus x giganteus fields on economic feasibility and soil carbon sequestration, Glob. Change Biol. Bioenerg. 6 (5) (2014) 566e576.
14. J. Dauber, S. Cass, D. Gabriel, K. Harte, S. Astrom, E. O'Rourke, et al., Yieldbiodiversity trade-off in patchy fields of Miscanthus x giganteus, Glob. Change Biol. Bioenerg. 7 (3) (2015) 455e467.
15. R.J. Hijmans, S.E. Cameron, J.L. Parra, P.G. Jones, A. Jarvis, Very high resolution interpolated climate surfaces for global land areas, Int. J. Climatol. 25 (15) (2005) 1965e1978.
16. J.H.M. Wosten, A. Lilly, A. Nemes, C. Le Bas, Development and use of a database € of hydraulic properties of European soils, Geoderma 90 (3e4) (1999) 169e185.
17. D.B. Boorman, J.M. Hollis, A. Lilly, Hydrology of Soil Types: a Hydrologicallybased Classification of the Soils of the United Kingdom, Institute of Hydrolgy, Wallingford, 1995.
18. NSRI, The National Soil Map and Soil Classification, National Soil Resource Institute, 2004 (Soil map of England and Wales).
19. I. Lewandowski, J.C. Clifton-Brown, J.M.O. Scurlock, W. Huisman, Miscanthus: European experience with a novel energy crop, Biomass Bioenerg. 19 (4) (2000) 209e227.
20. F.E. Miguez, M.B. Villamil, S.P. Long, G.A. Bollero, Meta-analysis of the effects of management factors on Miscanthus x giganteus growth and biomass production, Agric. For. Meteorol. 148 (8e9) (2008) 1280e1292.
21. A.R. Werker, K.W. Jaggard, Modelling asymmetrical growth curves that rise and then fall: applications to foliage dynamics of sugar beet (Beta vulgaris L), Ann. Bot. 79 (6) (1997) 657e665.
22. C. Lesur, M.-H. Jeuffroy, D. Makowski, A.B. Riche, I. Shield, N. Yates, et al., Modeling long-term yield trends of Miscanthus x giganteus using experimental data from across Europe, Field Crops Res. 149 (2013) 252e260.
23. European Space Agency, ESA Earthnet Online. Browse Data Products, Available from:, 2012 www.earth.esa.int. Accessed 04/2012.
24. European Space Agency, Observing the Earth, Available from:, 2011 www.esa. int/esaEO/. Accessed 04/2012.
25. US Geological Survey, Landsat Missions. Landsat Processing Details, Available from:, 2012 http://landsat. usgs.gov.
26. ESA/ESRIN - GAEL Consultant, Landsat TM/ETMþ CEOS/ESA Products: Format Specification, Issue 3.1, available from: Documents Library, European Space Agency, 2005, https://earth.esa.int/.
27. G.A. Ippoliti-Ramilo, J.C.N. Epiphanio, Y.E. Shimabukuro, Landst-5 Thematic Mapper data for pre-planting crop area evaluation in tropical countries, Int. J. Remote Sens. 24 (2003) 1521e1534.
28. C. Song, C.E. Woodcock, K.C. Seto, M.P. Lenney, S.A. Macomber, Classification and change detection using Landsat TM data: when and how to correct atmospheric effect, Remote Sens. Environ. 75 (2001) 230e244.
29. H.R. Gordon, Removal of atmospheric effects from satellite imagery of the oceans, Appl. Opt. 17 (1978) 1631e1636.
30. D.G. Hadjimitsis, G. Papadavid, A. Agapiou, K. Themistocleous, M.G. Hadjimitsis, A. Retalis, et al., Atmospheric correction for satellite remotely sensed data intended for agricultural applications: impact on vegetation indices, Nat. Hazards Earth Syst. Sci. 10 (2010) 89e95.
31. Ordnance Survey, Digital Map Data from EDINA AgCensus, Digimap and UK Borders. 1:10 000 Scale Digital Terrain Model (Land-form Profile DTM), Ordnance Survey, UK, 2008.
32. Biomass Energy Centre, Sources of Biomass, Available from:, 2011 http:// www.biomassenergycentre.org.uk/.
33. J.W. Rouse, R.H. Haas, J.A. Schell, D. D.W, Monitoring vegetation systems in the Great Plains with ERTS, in: S.C. Freden, E.P. Mercanti, M.A. Becker (Eds.), Third Earth Resources Technology Satellite-1 Symposium NASA SP-351, NASA, Washington, D.C., 1973, pp. 309e317.
34. R.W. Payne, S.A. Harding, D.A. Murray, D.M. Soutar, D.B. Baird, A.I. Glaser, et al., The Guide to GenStat Release 14, VSN International, Hemel Hempstead, UK, 2011.
35. J.W. Tukey, Exploratory Data Analysis, Addison-Wesley, Reading, PA, 1977.
36. M. Launay, M. Guerif, Assimilating remote sensing data into a crop model to improve predictive performance for spatial applications, Agric. Ecosyst. Environ. 111 (1e4) (2005) 321e339.
37. DEFRA, Area of Crops Grown for Bioenergy in England and the UK: 2008-2013, Department of Environment FaRA, York, England, 2015, p. 32. Department for Environment, Food and Rural Affairs.
38. C. Lesur-Dumoulin, M. Lorin, M. Bazot, M.-H. Jeuffroy, C. Loyce, Analysis of young Miscanthus giganteus yield variability: a survey of farmers' fields in east central France, GCB Bioenergy (2015), http://dx.doi. org/10.1111/ gcbb.12247.
39. I.F. Shield, T.J.P. Barradough, A.B. Riche, N.E. Yates, The yield and quality response of the energy grass Miscanthus x giganteus to fertiliser applications of nitrogen, potassium and sulphur, Biomass Bioenerg. 68 (2014) 185e194.
40. R.A. Fischer, G.O. Edmeades, Breeding and cereal yield progress, Crop Sci. 50 (2) (2010) S85eS98.

41. D.B. Lobell, K.G. Cassman, C.B. Field, Crop yield gaps: their importance, magnitudes, and causes, Annu. Rev. Environ. Resour. 34 (2009) 179e204.

42. S. Cadoux, A.B. Riche, N.E. Yates, J.-M. Machet, Nutrient requirements of Miscanthus x giganteus: conclusions from a review of published studies, Biomass Bioenerg. 38 (2012) 14e22.

43. S. Beuch, B. Boelcke, L. Belau, Effect of the organic residues of Miscanthus x giganteus on the soil organic matter level of arable soils, J. Agron. Crop Sci. Zeitschrift Fur Acker Und Pflanzenbau 184 (2) (2000) 111e119.

44. E.M. Hansen, B.T. Christensen, L.S. Jensen, K. Kristensen, Carbon sequestration in soil beneath long-term Miscanthus plantations as determined by C-13 abundance, Biomass Bioenerg. 26 (2) (2004) 97e105.

45. P. Kahle, S. Beuch, B. Boelcke, P. Leinweber, H.R. Schulten, Cropping of Miscanthus in Central Europe: biomass production and influence on nutrients and soil organic matter, Eur. J. Agron. 15 (3) (2001) 171e184.

46. D. Styles, F. Thorne, M.B. Jones, Energy crops in Ireland: an economic comparison of willow and Miscanthus production with conventional farming systems, Biomass Bioenerg. 32 (5) (2008) 407e421.

47. X. Chen, L. Vierling, D. Deering, A simple and effective radiometric correction method to improve landscape change detection across sensors and across time, Remote Sens. Environ. 98 (1) (2005) 63e79.

48. C.J. Gleason, J. Im, A review of remote sensing of forest biomass and biofuel: options for small-area applications, Gisci. Remote Sens. 48 (2) (2011) 141e170.

49. S. Kaasalainen, M. Holopainen, M. Karjalainen, M. Vastaranta, V. Kankare, K. Karila, et al., Combining Lidar and synthetic aperture Radar data to estimate forest biomass: status and prospects, Forests 6 (1) (2015) 252e270.

50. Y. Zhang, X.H. Liu, S.L. Su, C.Z. Wang, Retrieving canopy height and density of paddy rice from Radarsat-2 images with a canopy scattering model, Int. J. Appl. Earth Obs. Geoinfor. 28 (2014) 170e180.

CHAPTER 9 HOLOCENE CULTURAL HISTORY OF RED JUNGLE FOWL (GALLUS GALLUS) AND ITS DOMESTIC DESCENDANT IN EAST ASIA

Joris Peters, Ophélie Lebrasseur, Hui Deng, Greger Larson

Keywords: Red jungle fowl (Gallus gallus), Domestic chicken, Holocene, East Asia, Palaeoclimate, Palaeozoogeography, Zooarchaeology, Ancient DNA analysis, Shang Dynasty

INTRODUCTION

Chickens are by far the most numerous domestic animal on earth and are of immense importance as a global food species. The domestication history of Red jungle fowl (*Gallus gallus*) and the spatio-temporal dispersal of chicken husbandry prior to the modern era, however, remain contentious. One hypothesis holds that middle Holocene [1]village economies in Northern China practiced poultry farming by at least ~8.0 ka calBP. Recently published mitogenetic signatures obtained by Xiang et al. (2014) from galliform bones collected at the early Neolithic site of Cishan that was inhabited ~8.0–7.7 ka calBP appear to substantiate a previous zooarchaeological study published nearly three decades ago (West and Zhou, 1988). Both of these studies also conclude that since poultry farming had been established in the middle Holocene Yellow River basin ~8.0 calBP, efforts to manage Red jungle fowl must have started during the early Holocene.

In terms of geography, both Ho (1977) and Crawford (1984) argued for an autochthonous episode of *Gallus* domestication in Northern China, linked to the favourable climatic conditions of the Holocene that would have allowed Red jungle fowl to extend its native distribution (currently restricted to south of 25°N) much further north. West and Zhou (1988) disagreed, however, and pointed out that the late Pleistocene and early Holocene palaeoclimatic signatures documented in palaeosols of the Loess Plateau pointed to semi-arid and not tropical climate and vegetation required by Red Jungle fowl. The lack of conditions suitable to Red Jungle fowl thus obviated the potential for chickens to be independently domesticated in the north. Instead, West and Zhou argued that flocks must have been domesticated in the south and then transported to the north where they required human-assisted sheltering to survive the low winter temperatures and lack of vegetative ground cover characteristic of Holocene Northern China. More recently, this diffusion model has been challenged by a study of ancient chicken DNA (Xiang et al., 2014) which presented evidence supporting both the presence of Red jungle fowl in the early Holocene North China Plain and local chicken domestication.

If true, the occurrence of Red jungle fowl would thus imply the existence of suitable habitat for this thermophilic bird shortly after the onset of the Holocene. Xiang et al. (2014) argued that abundant remains of tropical animal and plant species at early Holocene Nanzhuangtou, with main occupation likely dated between 11.5 and 11.0 ka calBP[2] (Yang et al., 2012, Table S1), and middle Holocene Cishan (CPAM, Hebei Province Relics Preservation Station et al., 1981) suggest warmer and more humid conditions and a much larger forest cover. These conditions would have allowed *Gallus*to be present in areas where Red Jungle fowl do not exist today, thus facilitating their domestication.

With regard to the domestication status of *Gallus,* West and Zhou (1988) considered the chickens from middle Holocene Cishan and contemporary early and middle Neolithic sites in Northern China to be domestic birds introduced by humans based on both environmental arguments and because osteometrical comparisons demonstrated that the bones were larger than those of the wild ancestor *Gallus gallus,* but

Citation: Joris Peters, Ophélie Lebrasseur, Hui Deng, Greger Larson, Holocene cultural history of Red jungle fowl (Gallus gallus) and its domestic descendant in East Asia, Quaternary Science Reviews, Volume 142, 15 June 2016, Pages 102-119, ISSN 0277-3791, http://dx.doi.org/10.1016/j.quascirev.2016.04.004.

smaller than those of modern chickens. A zooarchaeological explanation posits that size increase in early domestic flocks results from regular and sufficient feeding and protection against predators, thereby reducing necessity to fly in birds kept in anthropogenic environments whilst converting energy intake more efficiently into meat and fat (von den Driesch, 1987).

Though the genetic analyses presented by Xiang et al. (2014) did not allow for a classification of individual bones as either domestic or wild, they concluded the bone finds from early and middle Holocene contexts represent the remains of a population ancestral to at least some of modern chicken mtDNA diversity. They based this conclusion on four arguments. Firstly, the remains were found in sites representing transitional foraging (Nanzhuangtou) and developed agricultural (Cishan, Wangyin) societies. Secondly, the chicken bones are present across several thousand years in the archaeological record of Northern China. Thirdly, these findings predate archaeological chicken remains from any other region in East and Southeast Asia by millennia. And lastly, all major modern chicken haplogroups and also one of the most common haplotypes were represented in their ancient DNA sequences.

The presence of domestic fowl in middle Holocene Northern China also necessitated revisiting the species› pattern of dispersal across Asia to Europe. West and Zhou (1988)postulated that the translocation of chickens occurred along East-West trade routes across Central Asia, rather than travelling from the jungle fowl›s most western native distribution in South Asia, i.e. India across the Arabian Gulf to Mesopotamia and from there to Europe (Zeuner, 1963). At the time of publication, however, *Gallus* finds were remarkably absent in the assumed region of passage, which West and Zhou (1988) explained by Central Asia›s poor archaeological coverage.

Thus, taken at face value, the historical zooarchaeological and recent genetic analyses published thus far suggest that chicken husbandry was being practiced by the Neolithic inhabitants of the Yellow River basin ~8.0 ka calBP. Recent ancient DNA data also imply the presence of an autochthonous *Gallus* population in the North China Plain since the beginning of the Holocene (Xiang et al., 2014). More recent and on-going zooarchaeological work, however, does not support the conclusions of either West and Zhou (1988) or Xiang et al. (2014).

One major concern is the fact that morphological comparison of specimens identified as *Gallus* and figured in Chinese archaeological literature revealed that they are not derived from chickens, but from pheasants (Benecke, 1994, Peters, 1998, Deng et al., 2013, Peters et al., 2015, Li et al., 2015 and Eda et al., 2016). In addition, our current understanding of the early Holocene climate in Northern China suggests that *Gallus*could not survive in this region. This review, therefore, takes a closer look at final Pleistocene to middle Holocene climatic developments and associated vertebrate distribution patterns in Northwestern and Northern China, defined here as the region north of the Huai River – Qin Mountains line ranging between 33°03′N and 34°25′N. Against the background of Holocene climate and zoogeography, we reconsider the zooarchaeological record and related ancient DNA evidence alongside a discussion of methodological issues and cultural aspects relevant to the study of ancient poultry keeping in China. We close with an assessment of the Holocene cultural history of *Gallus*in northern China and adjacent areas.

LATE QUATERNARY CLIMATE, VEGETATION AND FAUNA IN NORTHERN CHINA

Palaeoclimate and Vegetation

The modern climate in Northern China is controlled by the East Asian Summer Monsoon (EASM), associated with climatic features of the northern high latitudes, low latitude ocean and the southern hemisphere. Seasonal migration of the monsoon front causes prominent seasonality in Northern China, winters being cold and dry since dominated by air masses from the interior of Asia, whilst the moist and warm summer months are influenced by south-easterly winds from the western Pacific (An, 2000). Obviously, fluctuations in the EASM regime influenced settlement in Late Quaternary Northern China, with rapid cooling events causing major declines in human population density, as is illustrated by a study linking demographic fluctuations with records captured in loess, lake, palaeo-ocean and cave deposits as well as ice cores and stalagmites (Wang et al., 2014).

Of course, differences in temporal resolution and sensitivity of responses to changes in climate complicate detailing the magnitude of regional influence of the EASM (Chen et al., 2015), not to mention the nature and scale of the vegetation changes subsequently induced by these (Zhao et al., 2007 and Stebich et al., 2009). Proxy data generated with different methods nonetheless agree that prior to the onset of the Holocene ~11.7 ka calBP, East Asia witnessed a severe climatic reversal compatible with the Younger Dryas event in the circum-Atlantic region. Signatures captured in for instance the stalagmites from Hulu Cave (32°05′N, 119°02′E) as well as peat deposits and lakes at the desert-loess transition belt in Northern China illustrate that this reversal brought about low temperatures, whilst humidity fluctuated (e.g., Zhou et al., 1998, Wang et al., 2001, Li et al., 2003, Xu et al., 2010 and Chen et al., 2015).

Turning to the Holocene, pollen-based quantitative analyses of sedimentary sequences from Gonghai Lake (38°54′N, 112°14′E; 1,860m ASL) located on the NE margin of the Chinese Loess Plateau (Chen et al., 2015) and Daihai Lake (40°29′-40°79′N, 112°33′-112°46′E) situated ~200 km to the north of Gonghai Lake in arid North China (Xu et al., 2010) seem particularly relevant here. Both sequences show increasing arboreal pollen percentages in post-Pleistocene sediments indicative of more tree cover and a warmer climate. Assemblages in the Gonghai Lake core are dominated by temperate-forest (*Pinus, Picea, Betula, Quercus, Ulmus*) and temperate-steppe (*Artemisia*, Chenopodiaceae and Poaceae) pollen types (Chen et al., 2015). However, the EASM was clearly weaker during the early Holocene than during the middle Holocene, with lowered precipitation ~9.5–8.5 ka calBP. Maximum monsoon intensity occurred in the middle Holocene from ~7.8–5.3 ka calBP, with average annual precipitation ~30% higher than the modern value (Chen et al., 2015). At Daihai Lake, climatic conditions were essentially cool and dry ~11.7–9.0 ka calBP, annual precipitation being 50–100 mm lower than the present. Conversely, precipitation was 100–200 mm higher than today during the subsequent middle Holocene thermal optimum. During the latter, temperatures averaged up to 3 °C higher, whereas the preceding early Holocene witnessed on average 1°-2 °C lower temperatures compared to today (Xu et al., 2010).

Similar climatic developments were noted east of Northern China across the Bohai Sea and Korea Bay. Until ~9.0 ka calBP winters remained cold preventing evergreen forests from prospering, higher mean annual temperatures compared to today (+2–3 °C) and warmer winters being evidenced during the middle Holocene in central South Korea, causing replacement of deciduous broadleaved forests by evergreen forests (Yi, 2011). In the Yangtze River delta (31°23′3″N, 121°58′59″E), middle Holocene temperatures were ~2–4 °C warmer than today (Yi et al., 2003).

Based on their results of arboreal pollen analyses, Yu et al. (2000) concluded that during the Holocene thermal optimum, the northern limit of the temperate deciduous forest, in which Northern China is presently situated (Liu, 1988), was located ca. 800 km north relative to today. This implies a northward shift of ca. 7° of the corresponding vegetation belts. However, even during the middle Holocene optimum climatic conditions were far from constant, with warm-humid periods alternating with cold-dry episodes and with corresponding effects on vegetation cover, fauna and human population density (e.g. Xu et al., 2010, Wang et al., 2014 and Chen et al., 2015).

Mammalian Zoogeography

Taxonomic composition and faunal distribution patterns place Northern China in the Palaearctic Region (e.g. Corbet and Hill, 1992, Hoffmann, 2001 and Smith and Xie, 2008). Corbet and Hill (1992: 4, map 2) position the boundary of the Palaearctic with the Indomalayan Region slightly south of the lower Yangtze River. Between these two Regions, the authors defined three Divisions, i.e. (1) the Central China Division with predominantly Palaearctic taxa situated ~30–35°N, (2) the Southern China Division, where Palaearctic taxa form a minority and located between ~25 and 30°N, and (3) the Indochinese Division, with a mammalian fauna overwhelmingly Indomalayan in character and located south of ~25°N. Hoffmann (2001) further detailed the boundaries of the transition zone (Fig. 1) and confirmed that its mid-point should be close to 30°N.

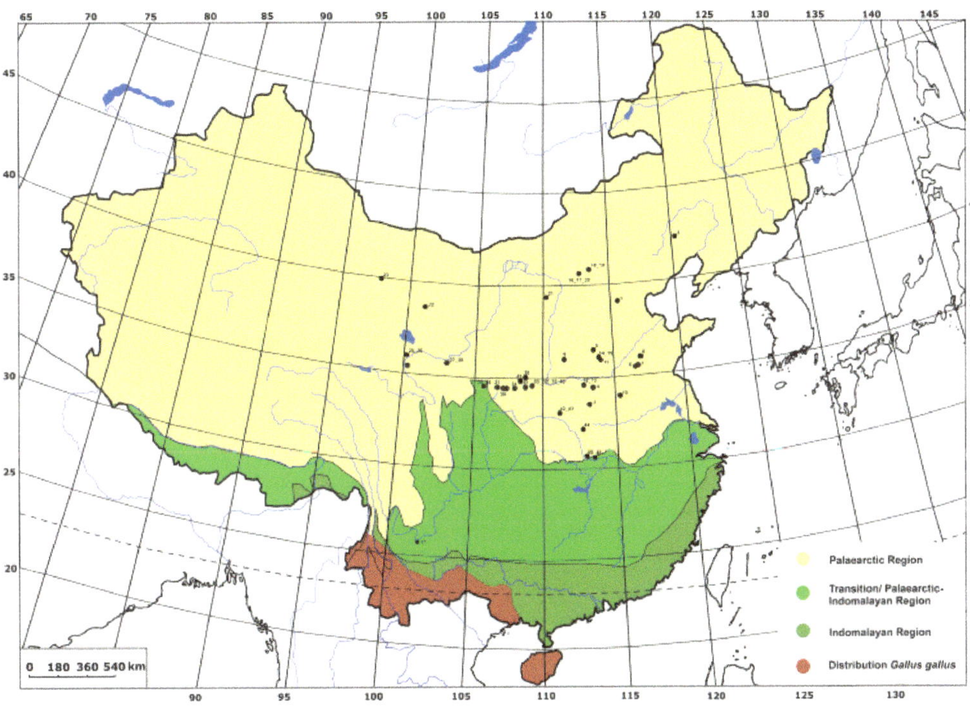

Figure 1: Map of China indicating the geographic range and boundaries of the Palaearctic (yellow) and Indomalayan Regions (dark green). In between is the transition zone as defined by Hoffmann (2001, Pl. 1). Superposed is the modern distribution of Red jungle fowl *Gallus gallus* in China based on Cheng (1987, 160, Map 114) and McGowan (1994, 530). Sites mentioned in the text and tables are: 1 Nanzhuangtou, 2 Jiahu, 3 Xinglongwa, 4 Peiligang, 5 Cishan, 6 Yuhuangding, 7 Wangyin, 8 Dawenkou, 9 Taosi, 10 Shantaisi, 11 Baiying, 12 Erlitou, 13 Yanshishangcheng, 14 Yinxu, 15 Dasikongcun, 16 Daihai Region, 17 Shihushan, 18 Miaozigou, 19 Dabagou, 20 Wangmushan poshang, 21 Zhukaigou, 22 Donghuishan, 23 Huoshaogou/Ganguyan, 24 Zongri, 25 Yangqu shierdang, 26 Xiangnagou, 27 Dahezhuang, 28 Qinweijia, 29 Xishanping, 30 Baijiacun, 31 Guantaoyuan, 32 Lingkoucun, 33 Beishouling, 34 Anban, 35 Jiangzhai, 36 Shizhaocun, 37 Banpo, 38 Wayaogou, 39 Fulinbao, 40 Kangjia, 41 Zaoshugounao, 42 Xiawanggang, 43 Shenmingpu, 44 Jiuliandun, 45 Qujialing, 46 Dengjiawan, 47 Dadunzi. (For interpretation of the references to colour in this figure legend, the reader is referred to the web version of this article.)

Table 1 lists the mammalian taxa evidenced in early to middle Holocene archaeological contexts in Northern China. From this, we notice the rareness of tropical Indomalayan and the dominance of Palaearctic genera notably adapted to marked daily and seasonal variation in temperature and cool winters in particular. Even during the Holocene thermal optimum, the ubiquitous presence of Wild boar, Red deer, Sika deer, Musk deer, Chinese water deer, Roe deer, European hare etc. characterizes landscape and vegetation in the study area essentially as temperate, not (sub-)tropical.

Table 1: Mammalian taxa evidenced in Holocene sites dating ~11.5–3.0 ka BP in Northwestern and Northern China, based on data published in Bo, 2011, Flad et al., 2007, Fu, 2000, Hu et al., 2007, Li and Han, 1963, Wang, 2011, Wu, 1993, Yuan and Li, 2010 and Zhong, 2010 and Zhou, 1981, Zhou, 1983a, Zhou, 1983b, Zhou, 1992, Zhou, 1994 and Zhou, 2000

Region	Northern China																Northwestern China	
Subregion	North China plain										Arid north China						Hexi Corr.	
Sites	Nanzhuangtou (1)	Jiahu (2)	Peiligang (4)	Cishan (5)	Yuhuangding (6)	Wangyin (7)	Dawenkou (8)	Taosi (9)	Baiying (11)	Yinxu (14)	Daihai Region (16)	Shihushan (17)	Miaozigou (18)	Dabagou (19)	Wangmushan poshang (20)	Zhukaigou (21)	Donghuishan (22)	Huoshaogou/ Ganguyan (23)
Dates in ka calBP (ca.)	12.5–11.0	9.0–7.5	9.0–7.0	8.1–7.7	6.5–6.0	6.5–5.5	6.3–4.5	4.3–3.9	4.1–3.7	3.3–3.0	6.7–5.0	6.5–6.4	5.8–5.0	5.8–5.0	5.5–5.0	5.0–3.4	4.0–3.5	4.0–3.0
Taxa	*Palaearctic mammalian taxa*																	
Sus	x	x	x	x	x	x	x	x	x	x	x	x	x	x	x	x	x	x
Bos		x	x	x	x	x		x	x				x	x		x		x
Bubalus	x	x				x					x				x			
Ovis (+Ovis/Capra)		x	x		x			x	x	x						x	x	x
Procapra												x	x	x				
Pseudois																		
Naemorhedus																x		
Gazella																		
Capricornis																		
Cervus																	x	
Cervus elaphus				x					x		x	x	x	x	x	x		
Elaphurus davidianus		x		x	x	x	x		x	x								
Cervus albirostris						x												
Cervus nippon	x	x		x	x	x	x					x		x				
Capreolus				x		x					x	x	x	x	x	x		
Moschus																	x	
Hydropotes		x	x	x	x	x	x		x									
Moschus/ Hydropotes	x																	
Muntiacus		x		x														
Camelus																x		
Equus									x	x				x				x
Canis	x	x	x	x	x	x		x	x	x		x		x	x	x	x	x
Nyctereutes		x				x						x						
Vulpes						x						x						
Cuon												x						
Meles		x	x	x	x	x						x				x		
Arctonyx																		
Mustela												x						
Martes		x																
Lutra						x												
Ursus/ Selenarctos						x						x		x		x		
Ailuropoda																		
Felis		x				x	x		x			x				x		
Panthera				x		x			x									
Marmota																		
Rhizomys																		
Myospalax				x	x							x			x			
Microtus								x										
Cricetulus																		
Sciurus																		
Castor																		
Spermophilus												x			x			
Ochotona												x						
Lepus	x	x		x	x			x				x						
Rhinopithecus																		
Taxa	*Indomalayan mammalian taxa*																	
Macaca				x														
Axis																		
Rusa								x										
Dicerorhinus																		
Paguma				x														
Hystrix							x											
Elephas										x								

Northwestern China																	Southern China	
Upper yellow river					Wei river												Dan river	Yunnan
Zongri (24)	Yangqu Shierdang (25)	Xiangnagou (26)	Dahezhuang (27)	Qinweijia (28)	Xishanping (29)	Baijiacun (30)	Guantaoyuan (31)	Lingkoucun (32)	Beishouling (33)	Anban (34)	Jiangzhai (35)	Shizhaocun (36)	Banpo (37)	Wayaogou (38)	Fulinbao (39)	Kangjia (40)	Xiawanggang (42)	Dadunzi (47)
5.6–4.0	5.6–4.0	5.6–4.0	4.1–3.9	4.1–3.9	8.0–3.9	7.5–6.3	7.3–6.9	7.0–4.8	7.0–4.5	7.0–4.3	6.8–4.0	6.8–3.9	6.7–5.6	6.5–6.0	5.5–5.0	4.5–4.0	5.0–2.7	3.2
x		x	x	x	x	x	x	x	x	x	x	x	x	x	x	x	x	x
x		x	x	x	x		x		x	x	x	x	x	x	x	x	x	x
						x	x									x	x	
			x	x				x				x	x	x		x		x
x	x	x				x					x							
x										x								
							x											
								x					x	x?				
																	x	
			x		x						x	x						
x	x	x			x	x			x					x	x			
							x	x										
							x	x		x	x		x	x	x	x	x	
x	x	x					x		x					x	x		x	

Northwestern China																	Southern China	
Upper yellow river					Wei river												Dan river	Yunnan
Zongri (24)	Yangqu Shierdang (25)	Xiangnagou (26)	Dahezhuang (27)	Qinweijia (28)	Xishanping (29)	Baijiacun (30)	Guantaoyuan (31)	Lingkoucun (32)	Beishouling (33)	Anban (34)	Jiangzhai (35)	Shizhaocun (36)	Banpo (37)	Wayaogou (38)	Fulinbao (39)	Kangjia (40)	Xiawanggang (42)	Dadunzi (47)
5.6–4.0	5.6–4.0	5.6–4.0	4.1–3.9	4.1–3.9	8.0–3.9	7.5–6.3	7.3–6.9	7.0–4.8	7.0–4.5	7.0–4.3	6.8–4.0	6.8–3.9	6.7–5.6	6.5–6.0	5.5–5.0	4.5–4.0	5.0–2.7	3.2
x	x	x			x		x	x	x		x	x		x	x		x	x
						x	x		x	x	x		x	x	x			
							x										x	x
			x		x		x					x		x				
x		x	x		x	x			x	x	x	x	x	x	x	x	x	x
						x		x	x	x	x			x		x	x	
							x		x				x	x		x		
											x					x		
								x	x		x		x				x	
							x											
														x		x		
														x				
																	x	
					x		x		x		x	x				x	x	x
																	x	
						x					x	x	x	x		x	x	
											x			x	x	x	x	
x	x	x												x				
					x	x	x		x		x	x	x	x		x		x
					x		x		x		x	x	x	x	x			
														x				
							x											
													x	x				
											x		x	x		x		x
							x											
					x				x		x						x	x
																	x	
							x										x	x
							x										x	
								x		x							x	x
																	x	

In accordance with the warm-humid conditions of the mid-Holocene optimum evidenced in all East Asian palaeoclimate archives, sites located in Northern China however yielded remains of mammals classified to-date as Indomalayan (Hoffmann, 2001, Appendix), including Rhesus macaque (*Macaca mulatta*), Masked palm civet (*Paguma larvata*), Reeves's muntjac (*Muntiacus reevesi*), and Sambar (*Rusa unicolor*) (Table 1). Since specimen counts are missing in the respective faunal reports, their quantitative importance cannot be properly determined.

A similar picture emerges for the contemporaneous Ordos Loop region west of the North China Plain (Flad et al., 2007, Table 1; Wang, 2011, 93). Even in the ecologically favourable landscapes bordering the Wei River (the most prominent tributary of the Yellow River flowing eastward along latitude ~34°-35°N), specimen counts illustrate that Indomalayan taxa, if present at all, account for less than 0.2% (see Supplementary Table S1). Most interestingly, for four Indomalayan taxa identified in Neolithic Northern China and the Ordos Loop, i.e. *Macaca, Muntiacus, Hystrix,* and *Rusa*, their actual native distributions extend to the northern boundary of the Central Chinese Division ~35°N or slightly further north (Corbet and Hill, 1992, maps 79, 80, 109, 111, 137, 172; Smith and Xie, 2008, maps 10, 528, 531). As to the Masked palm civet (*Paguma larvata*), this Indomalayan taxon seems climatically particularly versatile, since it has been recorded to-date close to Beijing at ~39°N.

According to Xiang et al., 2014 and Xiang et al., 2015b, indicator species of warm-humid, tropical conditions and much more extensive forest coverage during the early Holocene are the Chinese alligator (*Alligator sinensis*) and the extinct Short-horned water buffalo (*Bubalus mephistopheles*). Until recently, the Chinese alligator was much more widely distributed across China and in the past possibly even lived in Korea (Groppi, 2006), underscoring its eurytherm rather than thermophile properties. As to *B. mephistopheles,* its remains have been identified at early Holocene Nanzhuangtou in the North China Plain and from several middle Holocene sites located near major drainages in Northern and Northwestern China (Table 1). Throughout its former range, however, Short-horned water buffalo was found associated exclusively with Palaearctic artiodactyls frequenting cool temperate deciduous and coniferous forests, such as Wild boar, Sika deer, Red deer, Roe deer, Thorold's deer, Musk deer, Chinese water deer, and Muntjac. Rather than implying a (sub-)tropical environment, the synecological context clearly shows that Short-horned (Chinese) water buffalo must have been much better adapted to low temperatures and cold winters than its thermophile (sub-)tropical relatives, the Asian wild water buffalo *Bubalus arnee* and its domestic descendant *Bubalus bubalis*. This distinctness is reinforced by DNA analysis which failed to establish direct genetic links between the domestic water buffalo *B. bubalis* and *B. mephistopheles* (Yang et al., 2008). In sum, climatic, synecological and genetic data show that *B. mephistopheles*differed markedly from present-day water buffaloes. Adapted to the cool temperate landscape and vegetation characterising Holocene Northern China, the Short-horned water buffalo clearly represents a Palaearctic, not an Indomalayan taxon.

South of the dividing line separating Northern and Southern China, the River Dan takes its origin in the Qin Mountains flowing southeast to join the Han River, a major tributary of the Yangtze River. On its left bank lies the Neolithic settlement of Xiawanggang (33°01'N, 111°21'E). Here Indomalayan taxa like Sumatran Rhinoceros (*Dicerorhinus sumatrensis)* and Asian elephant (*Elephas maximus*) have been evidenced as well as Giant panda (*Ailuropoda melanoleuca*) (Jia et al., 1977), pointing to warmer climatic conditions and extensive woodlands and bamboo vegetation near the site. The latter can therefore be considered suitable habitat for Red jungle fowl.

Holocene Distribution of Red Jungle Fowl

The foregoing illustrates that only few Indomalayan taxa occurred in early and middle Holocene Northern China and that these are eurytherm rather than stenotherm. Conversely, not a single species typical of the Southern China (~25°-30°N) and Indochinese Divisions (south of ~25°N) has been identified in Northern Chinese transitional and Neolithic archaeofaunal assemblages. In its present-day tropical distribution range, Red jungle fowl are inhabitants of warm habitats at low and moderate altitudes and usually not exposed to any but slight changes in daily and seasonal temperatures. *Gallus gallus*, therefore, have no cause to make even limited seasonal migrations. However, wild fowl ranging up the valleys and southern slopes of the Himalayas will descend several thousand feet to warmer altitudes following the approach of cold weather (Beebe, 1921, 178).

Besides the availability of suitable habitat, another important but mostly overlooked aspect governing the natural dispersal of *Gallus* is the species' limited flight ability. Inhabiting forested habitats, ground-feeding jungle fowl needs only cover short distances, mainly to escape from immediate danger during daytime or to reach its roosting place a few metres above ground at sunset. However, *Gallus* dispersing to Northern China forcibly needed crossing vast East-West running water courses bordered by steep rock formations, extensive floodplains, swampy vegetation etc., raising the question how of the species coped with major ecogeographical barriers. Beebe (1921)and Delacour (1977) noted that jungle fowl avoid alluvial plains, turning many sections and particularly the lower reaches of the Yangtze and Yellow Rivers and their major tributaries into insuperable obstacles. Migrating jungle fowl along the Mekong River in Thailand illustrate this well: "*In the Chiengsen region the birds are apparently trying to cross the river eastwards, where it is about 700 yards* [=640 m] *across. In crossing, the birds fly up as high as they can go, and then attempt to glide across. A large number, however, fall in the river and are drowned*" (Giles, 1932). Besides wide water courses, the fact that during the middle Holocene optimum Northern China also witnessed frequent, heavy flooding and large inundations (Liu et al., 2013) certainly imposed additional constraints on *Gallus'* northward dispersal and acclimatisation. Finally, physiological (cold, snow) and biotic barriers (availability of food species, predators, disease) would have taken their toll as well.

In sum, despite marked influence of the warm-humid East Asian Summer Monsoon on climate development in Northern China, Holocene high-resolution archives do not warrant the conclusion that the study area witnessed (sub-)tropical conditions with moderate variation in seasonal temperatures allowing for the continuous presence of Red jungle fowl (*Gallus gallus*). Moreover, given the nearly exclusive Palaearctic species composition in early and middle Holocene archaeofaunas from Northern China (Table 1), it is difficult to imagine that even at the height of the thermal optimum, wild fowl would have found suitable habitat north of ~35°N that allowed for its long-term exploitation and domestication. If true, these observations suggest that despite evidence to the contrary, purported chicken bones in Northern Chinese archaeological sites dated to ~11.5–4.5 ka calBP are instead derived from a different bird species.

ZOOARCHAEOLOGY

Avifaunal specialists studying bird bones from prehistoric Asia are aware of the fact that pheasants in general and the ubiquitous Common or Ring-necked pheasant (*Phasianus colchicus*) in particular must have been a widely available food resource to Pleistocene and Holocene foragers and farmers inhabiting Central and East Asia. In China, for instance, 18 subspecies of *Phasianus colchicus* and an actual population counting tens of millions of birds have been reported (Cheng, 1987 and McGowan, 1994). Based on published archaeofaunal accounts, however, it would seem as if the Holocene inhabitants of Northern China were far better acquainted with *Gallus* than with *Phasianus,* given the much lower frequency of the latter taxon in the archaeofaunal record of that region. The high population density of this species and the fact that Common pheasants are known to frequent anthropogenic environments and cultivated plots in search of food renders this situation counterintuitive.

Osteomorphology

Contrary to statements in the literature, different genera and species of galliform birds can be distinguished based on morphological criteria (e.g., Erbersdobler, 1968, Tomek and Bocheński, 2009, Eda and Inoué, 2011 and Deng et al., 2013). Embedded in the tradition of vertebrate palaeontology, earlier zooarchaeological work in China figured single *'Gallus'* specimens, e.g. from middle Neolithic Banpo (Li and Han, 1963) and early Neolithic Cishan (Zhou, 1981). Morphological re-assessment of these bones showed that they represented exclusively pheasants (Peters, 1998, Deng et al., 2013 and Eda et al., 2016; our Table 2). Relative to the key site of Cishan, this conclusion was reinforced by direct analysis of specimens housed at the Institute of Archaeology, Chinese Academy of Social Sciences, Beijing. Subsequently, we re-evaluated additional specimens depicted in the literature as well as bone samples from other Holocene sites, with a focus on Northern China. The results of our study are presented in Table 2 that incorporates key data published by other researchers. The overwhelming majority of Neolithic and Bronze Age finds identified originally as *Gallus* cannot be confirmed. Eda et al. (2016, Table 1) reached the same conclusion based on their analysis of selected Phasianid specimens (femur, tibiotarsus, tarsometatarsus) from 18 Holocene sites.

Table 2: Overview of Chinese sites with remains attributed to '*Gallus*' in literature that could be (re-)evaluated either by study of the literature and re-assessment of the specimens figured in these faunal reports and/or direct analysis of specimens curated at the Institute of Archaeology, Chinese Academy of Social Sciences (IA, CASS), Beijing [a,b]. Dating of the archaeofaunas is in calibrated ka before Present and based on published dates in Flad et al., 2007 (Table 1) and Liu and Chen (2012) as well as ^{14}C-dates in excavation and other reports. Column 3 provides information about the Galliform bones in the original publications and the site's cultural sequence. Column 4 summarizes the state-of-the-art. Site numbers as in Fig 1

Northern China – North China plain			
Sites and their dating	Literature	Finds reported and stratigraphy of the sites	State-of-the-art
Nanzhuangtou (1)12.5–10.5 ka calBPHebei Province;site occupation mainly 11.5–11.0 ka calBP(Yang et al., 2012)	Baoding Prefectural Administration et al. (1992)Zhou (1992) Hebei Provincial Cultural Relics Institute et al. (2010)	Bone specimens reported: clavicula, humerus, radius, ulna, femur, tibiotarsus, tarsometatarsus.Identification: Wild fowl species, GallusStratigraphy: This is a multi-layered site with extensive historical occupation dating to the Warring States Period (476-221 BCE).	-Yuan and Li (2010) identified the bird remains from Nanzhuangtou only to the family level (Phasianidae). Contrary to earlier work, however, these faunal specialists did not confirm the presence of chickens at the site.-All specimens restudied by Eda et al. (2016, Table 1, Fig. 2C) were classified as non-chicken.
Jiahu (2)9.0–7.8 ka calBPHenan ProvincePeiligang Culture	Henan Institute of Cultural Relics and Archaeology (1999)	Bone specimens reported: 19 elements, including humerus, ulna, femur, tibiotarsus, tarsometatarsus.Identification: Phasianus colchicus; verifiable since specimens figured in the publication.Stratigraphy: Neolithic occupation covered with levels dating to the Han (202 BCE-220 CE) and Song (960–1279 CE) Dynasties.	-All specimens figured in literature are morphologically pheasants.-Analysis JP 2012: Phasianus colchicus: coracoid, femur-All femora, tibiotarsi and tarsometatarsi (n = 30) have been identified as non-chicken (Eda et al., 2016, Table 1).
Xinglongwa (3)8.0–7.0 ka calBPNeimenggu ProvinceXinglongwa Culture	Neimeng Archaeological Team, Institute of Archaeology, CASS (1997)	Bone specimens reported: none.Stratigraphy: The Xinglongwa occupation is superposed by levels dating to the Hongshan (6.5–5.0 ka calBP) and the Lower Xiajiadian (4.3–3.6 ka calBP) Cultures.	-Analysis JP 2012: Phasianus colchicus: coracoid, several tarsometatarsi-Eda et al. (2016, Table 1) identified one tibiotarsus and 43 tarsometatarsi as non-chicken, whilst candidate chicken bones were absent from the site.
Peiligang (4)9.0–7.0 ka calBPHenan ProvincePeiligang Culture	Henan Working Team, Institute of Archaeology, CASS (1984)West and Zhou (1988, Table 1)	Bone specimens reported: bones, tarsometatarsus.Identification: Gallus; not verifiable since specimens not figured.Stratigraphy: The Peiligang Culture occupation is located below top soil.	-In the original publication the remains were considered representing chicken because of their size. However, the use of size as distinctive criterion is problematic (Deng et al., 2013; see also text).
Cishan (5)8.0–7.7 ka calBPHebei ProvinceCishan-Beifudi Culture	CPAM, Hebei Provincial Relics Preservation Station et al. (1981)Zhou (1981) West and Zhou (1988, Table 1)	Bone specimens reported: clavicula, humerus, ulna, radius, femur, tibiotarsus, tarsometatarsus.Identification: Gallus gallus domesticus; verifiable in part because some humeri, femora, tibiotarsi and metatarsi have been figured.Stratigraphy: There is a Shang Dynasty (1600-1046 BCE) layer overlying the two Neolithic levels.	-Yuan (2010) and Deng et al. (2013) pointed out that the evidence presented by Zhou (1981) to claim the presence of Gallus is not tenable, necessitating revision of all Cishan finds.-Analysis JP 2012: Phasianus colchicus: humerus, tibiotarsus. In addition, we noted that the specimens figured on Plates 9 and 10 in Zhou (1981) exhibit the morphology of pheasants and are likely P. colchicus as well.-Five metatarsi displayed at the Handan City Museum, Hebei and labelled "the oldest domestic chicken in the world" were re-analysed by Eda et al. (2016, Table 1) and classified as non-chicken.

Yuhuangding (6)6.5–6.0 ka calBPShandong ProvinceBeixin and Dawenkou Cultures	Shandong Cultural Relics Institute (2010)Zhong (2010)	Bone specimens reported: humerus, femur, tarsometatarsus.Original identification: Gallus sp.; not verifiable since the specimens have not been figured.Stratigraphy: The bones all originate from the late Beixin and early Dawenkou Culture levels. However, Yuhuangding also exhibits extensive Han Dynasty (202 BCE-220 CE) occupation.	-The faunal report states that the remains have been identified as chicken because most of the specimens pertain to male birds. Not only is this argument invalid (Deng et al., 2013), it is also counterintuitive (see text).
Wangyin (7)6.5–5.5 ka calBPShandong Province-Late Beixin Culture and Dawenkou Culture	West and Zhou (1988, Table 1)Institute of Archaeology, CASS (2000)Zhou (2000)	Bone specimens reported: femur, tibiotarsus, tarsometatarsus.Identification: Gallus gallus domesticus; not verifiable since the specimens have not been figured.Stratigraphy: The Beixin and early Dawenkou Culture layers are located below an extensive, late Dawenkou graveyard.	-Zhou (2000) mentions the presence of domestic chicken at the site, but the evidence on which this is based is not presented (Deng et al., 2013). Prof. J. Yuan, Institute of Archaeology, CASS Beijing noted the presence of Phasianus at this site (pers. comm., 2015).-Analysis JP 2012: This site produced a large collection of pheasant bones (mainly of Phasianus colchicus) including 36 humeri, 27 coracoids, ten carpometacarpi, 17 femora and twelve tarsometatarsi.-All femora and tarsometatarsi analysed by Eda et al. (2016, Table 1) were classified as non-chicken.
Dawenkou (8)6.3–4.5 ka calBP Shandong ProvinceDawenkou Culture	Shandong Cultural Relics Institute (1974)Li (1974)West and Zhou (1988, Table 1)	Bone specimens reported: tarsometatarsus, bones.Identification: Gallus; verifiable in part because some specimens have been figured in the excavation report.Stratigraphy: The bones originate from a tomb.	-Li (1974) assigned the bones to Gallus but this identification must be rejected since the specimens figured in his contribution correspond to pheasants (Deng et al., 2013).
Taosi (9)4.3–3.9 ka calBP-Shanxi ProvinceLongshan Culture	Shanxi Archaeological Team (1980)West and Zhou (1988, Table 1)Bo (2011)	Bone specimens reported: femur, tarsometatarsus.Identification: Gallus; not verifiable since specimens were not figured.Stratigraphy: Neolithic occupation is located partly below top soil and partly below levels of Western Zhou (1046-771 BCE) and subsequent Dynasties.	-Bo (2011) refrained from assigning the Phasianid remains from Taosi to either chicken or pheasant since she considered it hard telling the difference.-Eda et al. (2016, Table 1) did not find any candidate chicken bones in this small assemblage.
Shantaisi (10)4.2–3.8 ka calBPHenan Province-mainly Longshan Culture	China-US Joint Archaeological Team (2010)	Bone specimens reported: none.Stratigraphy: This is a multi-layered site with Late Neolithic and historical occupations dating to the Shang Dynasty (1600-1046 BCE).	-There is no mention of chicken bones in the excavation report-Analysis JP 2012: Phasianus colchicus: coracoid, tarsometatarsus-Eda et al. (2016, Table 1) only observed remains of pheasants in this small assemblage.
Baiying (11)4.1–3.7 ka calBPHenan Province-mainly Longshan Culture	CPAM of Anyang Prefecture of Henan Province (1983)Zhou (1983a)West and Zhou (1988, Table 1)	Bone specimens reported: femur, tarsometatarsus.Identification: Gallus; not verifiable since specimens were not figured.Stratigraphy: There is a Western Zhou Dynasty layer (1046-771 BCE) overlying the Longshan Culture deposits.	-Zhou (1983a) considered the remains pertaining to chicken simply because domestic chicken had already been evidenced at Baiying before (Deng et al., 2013).
Erlitou (12)3.7–3.4 ka calBPHenan ProvinceErlitou-Erligang Cultures	Institute of Archaeology, CASS (1999) Yang (2008)	Bone specimens reported: 42 bones, including coracoid, humerus, femur, tarsometatarsus etc.Identification: Gallus; not verifiable since the specimens have not been all figured.Stratigraphy: The bones all originate from the Erlitou and Erligang Culture levels. However, besides Neolithic levels the site of Erlitou also exhibits a Han Dynasty (202 BCE-220 CE) occupation.	-Eda et al. (2016, Table 1) identified one tibiotarsus and ten tarsometarsi as non-chickens.

Yanshishangcheng (13)3.6–3.4 ka calBPHenan ProvinceEarly Shang Dynasty(1600-1400 BCE)	Luoyang Hanwei City Ruins Team, Institute of Archaeology, CASS (1984)Du and Wang (2004)	Bone specimens reported: none.Stratigraphy: This is a multi-layered site with extensive historical occupation dating to the Ming Dynasty (1368–1644 CE).	-There is no mentioning of chicken bones in the excavation report.-Analysis JP 2012: Phasianus colchicus: tarsometatarsus-Amongst the femora, tibiotarsi and tarsometatarsi analysed, Eda et al. (2016, Table 1) identified only non-chicken remains.
Yinxu (14)3.3–3.0 ka calBPHenan ProvinceLate Shang Dynasty(1320-1046 BCE)	Hou (1989)Institute of Archaeology, CASS (1994a)Chen (1994)	Bone specimens reported: cranium, tarsometatarsi.Identification: Gallus gallus domesticus.Stratigraphy: Later levels overly the late Shang Dynasty occupation.	-Yinxu was the capital of the Late Shang Dynasty (1320-1046 BCE).Deng et al. (2013) identified a skull fragment as Gallus based on the specimen's comparably small Condylus occipitalis as well as its deep and large Fossa occipitalis.-Eda et al. (2016, Table 1) noted the presence of pheasants as well.
Dasikongcun (15)3.3–3.0 ka calBPHenan Province-Late Shang Dynasty	Ma et al. (1955)	Specimens reported: Two chicken skeletons in different tombs.Identification: Gallus; not verifiable since the specimens have not been figured.	-The cemetery of Dasikongcun is considered the burial ground of the Late Shang Dynasty kings (1320-1046 BCE). See text for discussion of these finds.
Northwestern China – Wei River Valley			
Baijiacun (30)7.5–6.3 ka calBPShaanxi Province(Pre-)Yangshao Culture	Institute of Archaeology, CASS (1994b) Zhou (1994)	Bone specimens reported: mandible, vertebra, scapula, femur.Original identification: Gallus gallus; not verifiable since the specimens have not been figured.Stratigraphy: The bones all originate from Levels II and III. The pre-Yangshao Culture occupation phase is located below top soil.	-Zhou (1994) argued that the remains pertain to chicken simply because they look similar to the finds at that time already excavated at Banpo, Cishan and Beishouling (Deng et al., 2013). However, this conclusion must be rejected because all specimens reported from Cishan, Banpo and Beishouling are pheasants (Eda et al., 2016; this Table).
Guantaoyuan (31)7.3–6.9 ka cal BPShaanxi ProvincePre-Yangshao Culture	Shaanxi Provincial Institute of Archaeology et al. (2007)Hu et al. (2007)	Bone specimens reported: 12 bones + 47 bones of uncertain stratigraphic position. Original identification: Gallus sp.; not verifiable since the specimens have not been figured.Stratigraphy: This is a multi-layered site with extensive Neolithic inhabitation and historical occupation dating to the Ming Dynasty (1368–1644 CE). The 12 bones originate from Pre-Yangshao occupation Phases II and III.	-Hu et al. (2007) concluded that the remains pertain to chicken since they compare well with the figured specimens excavated at Banpo (Deng et al., 2013). However, the specimens excavated at Banpo represent pheasants (s. below).
Beishouling (33)7.0–4.5 ka calBPShaanxi ProvinceYangshao Culture	Institute of Archaeology, CASS (1983) Zhou (1983b)West and Zhou (1988, Table 1)	Specimens reported: Two skeletons in a pot.Identification: Gallus gallus domesticus; verifiable in part because some of the remains have been figured.Stratigraphy: An extensive Han Dynasty (202 BCE-220 CE) occupation superposes the Yangshao levels.	-The remains were assigned to chicken (Zhou, 1983b), but this identification must be rejected since the specimens figured in the publication correspond to pheasants (Deng et al., 2013).
Anban (34)7.0–4.3 ka calBPShaanxi ProvinceYangshao and Longshan Cultures	Dept. of Archaeology, Northwest Univ. (2000)Fu (2000)	Bone specimens reported: humerus, tarsometatarsus.Identification: Gallus domesticus; not verifiable since specimens were not figured.Stratigraphy: All Gallus finds would originate from Neolithic strata, but there are Zhou (1046-256 BCE) and Han Dynasty (202 BCE-220 CE) levels overlying the Neolithic occupation.	-The remains were considered representing chicken by Fu (2000) because the size of the Anban tarsometatarsus equalled that in similar finds from Cishan and Banpo (Deng et al., 2013).-Flad et al. (2007) had already expressed doubts about this identification since they addressed the finds as 'chicken (pheasant?)'.

Banpo (37)6.7–5.6 ka calBPShaanxi ProvinceYangshao Culture	Institute of Archaeology and Chinese Academy of Sciences (1963)Li and Han (1963)West and Zhou (1988, Table 1)	Bone specimens reported: femur, tibiotarsus, tarsometatarsus.Identification: tentatively as Gallus sp.; verifiable in part since two specimens have been figured by Li and Han (1963, Pl. 198).	-Banpo has an alluvial setting, which is hardly suitable for Gallus (see text). Although certainly not intended by the authors, their tentative identification likely paved the way for other Gallus finds in the Neolithic Yellow River basin.-The two specimens figured, i.e. one tarsometatarsus (Li and Han, 1963, Pl. 198, Figs. 8A, B) and one tibiotarsus (Figs. 9A, B) are morphologically pheasants, most likely Phasianus colchicus.
Fulinbao (39)5.5–5.0 ka calBPShaanxi ProvinceYangshao culture	Baoji Archaeological Team (1993), Wu (1993)	Bone specimens reported: bones.Identification: Phasianus colchicus; not verifiable, since specimens were not figured.Stratigraphy: There are historic levels overlying the Yangshao Culture occupation.	-Flad et al. (2007) reported the presence of P. colchicus at this site.
Zaoshugounao (41)3.2–3.0 ka calBPShaanxi ProvinceProto-Zhou Culture	School of Cultural Heritage, Northwest Univ. et al. (2013)	Bone specimens reported: noneStratigraphy: This is a multi-layered site with Neolithic and historical occupations dating to the Tang Dynasty (618–907 CE).	-Analysis by Eda et al. (2016, Table 1) of selected elements (femur, tibiotarsus, tarsometatarsus) showed that one femur and six tarsometatarsi represent non-chicken remains, whilst two tibiotarsi can be considered candidate chicken bones.
Southern China			
Xiawanggang (42)5.0–2.7 ka calBPHenan ProvinceYangshao Culture till Western Zhou period	Jia and Zhang (1977) Henan Provincial Institute of Cultural Relics et al. (1989)	Bone specimens reported: none.Stratigraphy: This is a multi-layered site with extensive Neolithic and historical occupations dating to the Western Zhou Period (1046-771 BCE).	-There is no mentioning of chicken bones in the original report.-Eda et al. (2016, Table 1) identified a single candidate chicken femur from this site. The specimen was found in the Longshan layer dating to between 5.0 and 4.0 ka calBP.
Shenmingpu (43)2.2–2.0 ka calBPHenan ProvinceEarly Han Dynasty202-141 BCE	Deng et al. (2013)	Bone specimens reported: Partial skeleton of chicken in a tomb.Identification: Gallus domesticus.	
Jiuliandun (44)2.5–2.2 ka calBPHubei ProvinceWarring States Period476-221 BCE	Hubei Provincial Museum (2007)	Bone specimens reported: several incomplete skeletons in Tomb 1.Identification: none.	-No report exists about the Jiuliandun fauna.-Analysis JP 2012: Gallus domesticus: coracoid, two humeri, femur and two tibiotarsi.
Quijaling (45)5.0–4.6 ka calBPHubei ProvinceQuijaling Culture	Institute of Archaeology, CAS (1965)West and Zhou (1988, Table 1)	Finds reported: pottery figurines of fowlIdentification: Gallus; only verifiable in part, since bone finds have not been figured.Stratigraphy: No post-Quijaling occupation.	-Plate LIV, Fig. 8 shows a bird figurine identified as Gallus. However, the absence of wattles and a comb (or any indication that these might have been broken off during excavation) makes this identification not very likely.
Dengjiawan (46)4.5–4.0 ka calBPHubei ProvinceShijiahe Culture	Wu (2001)	Specimens reported: Pottery figurines of fowl.Identification: Gallus domestica (Pl. 1, Fig. 12) and Phasianus colchicus torquatus (Pl. 1, Fig. 14).	-The assemblage of figurines depicted in Wu (2001, Plate 1) is rich in species and includes subtropical taxa including Gallus, Bambusicola and Pavo. Whether Gallus is domestic or not is difficult to decide based on the depiction (see text for discussion).

Dadunzi (47)3.2 ka calBPYunnan ProvinceDadunzi Culture	Yunnan Provincial Museum (1977), Zhang (1985), West and Zhou (1988, Table 1)	Bone specimens reported: tarsometatarsus, pottery figurines and a chicken-shaped pot. Identification: Gallus; the bone specimen is not verifiable since not figured.Stratigraphy: Two Late Neolithic occupation stages are located below top soil. A single14C-date comes from charcoal found in a post-hole of a dwelling associated with early occupation.	-Older publications considered domestic chicken to be present because of the chicken-shaped pot (Deng et al., 2013). However, Red jungle fowl (Gallus gallus) was distributed across Yunnan Province during that period as well. Thus, whether Gallus is wild or domestic birds is inconclusive. Detailed analysis of the Phasianid bones associated with this culture is crucial, since Yunnan Province could be a region of Gallus domestication.

[a]J. Peters compared bone specimens figured in the literature and housed at the Institute of Archaeology, Chinese Academy of Social Sciences (CASS) with reference skeletons from the Bavarian State Collection of Anthropology and Palaeoanatomy, Munich; H. Deng et al. (2013) verified identification of bone specimens with reference skeletons available at the Institute of Archaeology, CASS, Beijing.

[b]Apart from the sites listed here, 'chickens' have been reported from many other sites (e.g. Ren, 1996), but verification is impossible due to poor documentation (Deng et al., 2013). This applies for instance to the sites of Machengzi (Liaoning Institute of Cultural Relics and Archaeology, Benxi Museum, 1994), Beixin (Shandong Archaeological Team, Institute of Archaeology, CASS, 1984), Caiyuanzi and Mopandi (Yunnan Provincial Institute of Cultural Relics and Archaeology et al., 2003), Dongying (Hu, 2010), Shishanzi (Anhui Provincial Institute of Archaeology (1992)), Hougang (Anyang Archaeological Team, Institute of Archaeology, CASS, 1985), Kangjia (Liu et al., 2001), Longgangsi (Wu, 1990), Zaojiaoshu (Yuan, 2002), Zuojiashan (Chen, 1993), Qianbuxia (Kong, 2000), Xishanping (Zhou, 1999 and Flad et al., 2007), Dadiwan (Qi et al., 2006), Yinjiacheng (Lu et al., 1990) and Zengpiyan (Yuan, 2003).

Medium to large Phasianid species distributed across Northern and Northwestern China include: Blood Pheasant (*Ithaginis cruentus*), Koklass Pheasant (*Pucrasia macrolopha*), Golden Pheasant (*Chrysolophus pictus*), Temminck's Tragopan (*Tragopan temminckii*), Reeves's pheasant (*Syrmaticus reevesii*), Brown Eared-pheasant (*Crossoptilon mantchuricum*) and the ubiquitous Common or Ring-necked Pheasant (*Phasianus colchicus*) (Cheng, 1987). One key morphological feature distinguishing *Phasianus colchicus* from *Gallus* is the presence of a *Foramen pneumaticum* at the greater trochanter of the femur (e.g. Tomek and Bocheński, 2009 and Eda and Inoué, 2011). Except for a single specimen from Late Neolithic Xiawanggang (see below), all Neolithic and Bronze Age femora analysed by Eda et al. (2016) and those in this study exhibited this foramen.

A morphological analysis of modern comparatives, however, revealed that this foramen is missing in the genera *Pucrasia, Tragopan* and *Syrmaticus* as well, which excludes these taxa as possible identifications for Neolithic and Bronze Age femur specimens. Complementing this observation with other osteological criteria, for instance the morphology of the tarsometatarsus or the position, size and shape of the tarsometatarsal spur in male birds, we conclude that most archaeological specimens do not fit their homologues in the genera *Ithaginis*, *Chrysolophus* and *Crossoptilon*. In sum, a re-evaluation of early and middle Holocene specimens depicted in the literature and housed at the Institute of Archaeology, Chinese Academy of Social Sciences, Beijing (Table 2) excludes *Gallus* as possible identification while indicating that a high proportion of avifaunal assemblages from Northern and Northwestern China are made up of *Phasianus colchicus*.

In their study, Eda et al. (2016) noted a candidate chicken bone at Late Neolithic ~5.0–4.0 ka calBP Xiawanggang in the Yangtze River basin. These authors also mention two candidate chicken bones from the Proto-Zhou site of Zhaoshugounao (3.2–3.0 ka calBP) located in the Wei River valley. In the North China Plain, chicken appears present in a broadly contemporaneous context at Yinxu, the residential city of the late Shang Dynasty (1320-1046 BCE) (Deng et al., 2013).

Osteometry

Comparative osteometry can be useful for providing a secure morphological identification and confirming domestication status. Zhou (1981, Table 3) thus compared the Greatest Length of tarsometatarsi from Neolithic Cishan with metrical data from modern comparatives including Common pheasant (*Phasianus colchicus* subsp. *karpowi)*, Koklass pheasant (*Pucrasia macrolopha*) and Edward's pheasant (*Lophura edwardsii)*. The fact that the tarsometatarsi of modern *Phasianus colchicus* were significantly shorter than their homologues in modern *Gallus* reinforced the author's conclusion that the ancient finds from Cishan represented chickens.

Table 3: Metrical comparison of medium to large-sized Phasianid species and several subspecies of *P. colchicus*confined to Northern and Northwestern China. Tarsus Lengths of ♀♀ and ♂♂ pheasants and *P. colchicus*subspecies from China according to Cheng (1978) and for *P. colchicus* from Germany according to Glutz von Blotzheim et al. (1973). Greatest Length values of tarsometatarsi were obtained with modern specimens from the Bavarian State Collection of Anthropology and Palaeoanatomy, Munich, Smithsonian Institution, Washington, Louisiana State University Museum of Natural Science, Baton Rouge, Institute of Vertebrate Palaeontology and Palaeoanthropology, Beijing, and Institute of Archaeology, Chinese Academy of Social Sciences, Beijing. Table S2 provides more detail about the modern comparatives. M. Eda (Hokkaido University, Japan) kindly provided measurements of pheasant specimens and of *P. colchicus karpowi* in particular

Taxon		Tarsus length ♀	n	Tarsometarsus GL ♀	n	Tarsus length ♂	n	Tarsometatarsus GL ♂	n
Blood Pheasant	Ithaginis cruentus	58–63	13	63.8	1	61–68	28	64.8	1
Koklas Pheasant	Pucrasia macrolopha	46–60	15	–		57–69	10	69.1–70.8	2
Golden Pheasant	Chrysolophus pictus	68–76	5	66.9–84.4	5	73–75	5	68.0–77.9	7
Temminck's Tragopan	Tragopan temminckii	58–66	5	65.9–69.8	2	71–76	5	80.2–90.7	2
Reeves's Pheasant	Syrmaticus reevesii	66	1	66.2–69.0	6	73–83	10	79.9–86.2	6
Common Pheasant	Phasianus colchicus (Germany)	56–64	13	56.3–70.2	31	64.0–77.0	58	69.1–81.0	22
	P. colchicus karpowi	46–55	8	60.2–62.7	2	59.0–63.0	10	69.2–74.3	5
	P. colchicus kiangsuensis	57–58	2	69.0	1	61.0–65.0	2	–	–
	P. colchicus torquatus	53–62	8	66.5	1	66.0–72.0	10	68.0–80.0	3
Brown Eared-Pheasant	Crossoptilon mantchuricum	89–97	8	84.6–95.8	3	92.0–101.0	8	97.6–101.5	2

Our analysis of the Greatest Length (GL) of the tarsometatarsus (according to von den Driesch, 1976) in 22 adult male *P. colchicus* collected in Germany [3] produced values between 69.5 and 81.0 mm (Table 3). Comparison with values obtained for *P.c. karpowi*by Zhou (1981, Table 3; 59.0–63.0 mm), however, illustrates birds with distinctly shorter legs. One possible explanation for this discrepancy is variability in size due to the species' wide distribution range across Asia, but this is not consistent with other lines of evidence. For instance, body weight in male *P.c. karpowi* (1000–1312 g; Cheng, 1978) overlaps considerably with that in its European relative (1145–1565 g; Glutz von Blotzheim et al., 1973). A closer look at Zhou›s data revealed that his values were not obtained on macerated (dry) bone specimens but reproduced from the ornithological standard work edited by Cheng (1978). This compendium records the Tarsal Length (TL) in birds based on measurements of bird skins (Cheng, 1978, Pl. 6, no. 5). As such, TL values are often lower than GL measurements because small articular protrusions of the tarsometatarsus will not be included as they are 'hidden' in the joint.

Table 3 recapitulates the TL values published by Cheng (1978) for the major (sub-) species of pheasants confined to Northern and Northwestern China and compares this information with GL measurements from macerated specimens in modern collections. Our data confirm that TL and GL values can differ considerably and they illustrate that modern Chinese *P.c. karpowi, P.c. kiangsuensis, and P.c. torquatus* clearly overlap in size with European pheasants. Consequently, Zhou's assumption that the Cishan tarsometatarsi measuring between 70.0 and 86.5 mm fit male *Gallus* and not *Phasianus*must be rejected. As mentioned before, morphological analysis of the femora implied that the genera *Pucrasia, Tragopan* and *Syrmaticus* were absent from the assemblages studied. Metrical analysis of the tarsometatarsus allows extending this conclusion to the smaller and more slender-built genera *Ithaginis* and *Chrysolophus*. Because of its clearly larger size (Table 3), the genus *Crossoptilon* has been identified in the avifaunal record of the study area (Hou, 1989).

Finally, Zhou›s conclusion that the remains pertained to *Gallus gallus domesticus*because the Cishan specimens were intermediate in size between larger-sized modern domestic chickens and smaller-sized wild fowl can no longer be maintained in view of the exclusive presence of pheasants in this assemblage (Eda et al., 2016; our Table 2).

Flock Management and Health Status

Relative to Neolithic Cishan and Wangyin, *Gallus* has been considered domestic because agriculture was already practiced at considerable scale, including cultivation of millet (Lu et al., 2009) and pig husbandry (Larson et al., 2010 and Cucchi et al., 2011). However, the premise 'practicing agriculture' does not allow reaching the logically certain conclusion that poultry keeping was part of it. According to us, postulating chicken farming presumes human management and should be reflected in the demographic profiles and particularly in the male to female ratio. In traditional poultry husbandry intended for local consumption, adult hens (laying, brooding) clearly outnumber cocks of the same age, surplus males usually being eliminated from the flocks prior to reaching sexual maturity in order to reduce aggressiveness in the coop. In this respect, human management mimics the situation in free-living Red jungle fowl during reproduction, with a reported ratio of cocks to hens of 1 to 3–5 (Raethel, 1988, 574).

To assess the males to females ratio in Phasianids, zooarchaeologists can consider (*i*) the pronounced sexual dimorphism expressed in the length of long bones, (*ii*) the tarsometatarsal spur present in mature cocks and absent in hens except in old birds (Peters, 1997), and (*iii*) the presence of medullary bone in the marrow cavity of long bones. The latter serves as calcium repository in egg-laying hens and is recognized easily in fragmentary specimens. In complete tarsometatarsi, however, the first two criteria allow sexing specimens securely, whilst the morphology of the articulations as well as texture of the bone surface and shape and size of the spur enable distinguishing between infantile, juvenile, young-adult and mature birds.

Considering Early Neolithic Cishan once more, remains of adult male birds largely dominate the assemblage (Zhou, 1981, Table 2). This pattern has been observed in other Neolithic sites and interpreted as a conscious choice by flock keepers, with hens being raised for egg production and cocks for meat, whereby the latter may even have been used in religious rites (Ren, 1996). As mentioned above, keeping adult cocks in numbers is counterproductive, not only for behavioural reasons but also because feeding cocks predestined to meat consumption into adulthood is – economically speaking – not profitable. Moreover, meat production in early domestic *Gallus* (which was still close to the wild form) may not have been particularly rewarding, considering that brooding cycles and offspring per cycle number few. Furthermore, live weights in early breeds may still have been close to that in the wild form, ranging between 485–1050 g in females and 672–1450 g in males (McGowan, 1994). Contrary to Zhou, 1981 and Ren, 1996 and other authors and independently of the galliform species dealt with in Northern China, the dominance of adult male birds in Neolithic assemblages does not support their conclusion of human-manipulated flocks. Rather, we consider such pattern indicative of fowling, male birds being easier targets due to their larger size, colourful plumage, and conspicuous courtship behaviour, whilst hens and chicks benefit from protective mimicry and hiding behaviour. Moreover, Northern China witnessed millet cultivation since the early Holocene (Yang et al., 2012), a highly nutritional staple attractive to ground-feeding birds including *P. colchicus*. High numbers of pheasant remains in archaeological assemblages can therefore be best explained by protection of agricultural plots against crop raiders.

Another issue in poultry farming is the health status in populations managed by humans. Although the life span in domestic chickens is comparably short and most health problems not expressed in their skeleton, bone specimens exhibiting pathologies are regularly encountered in *Gallus* assemblages (e.g. von den Driesch and Pöllath, 2000). Conditions observed mainly concern lesions due to traumas related to captivity and handling by humans, such as fractures, but ancient chicken populations also suffered from conditions like arthritis and avian osteopetrosis (or thick leg disease) caused by the Avian Leukosis Virus (ALV) (Peters, 1997). Most striking, therefore, is the fact that the galliform remains analysed by us (Table 2) revealed no bone lesions regularly encountered in domestic populations.

Wild or Domestic: the Cultural Context

Like in most other species, difficulties arise in separating osseous remains of the wild from the domestic form if the finds originate from sites located in the native distribution of the wild ancestor (Dobney et al., 2013 and Peters et al., 2013). One phenomenon indicative of a species› domestic status in the natural range of the wild ancestor, however, is animals deposited in funerary contexts. Indeed, to accompany the deceased on their last journey, farming communities have since the Neolithic selected mostly barnyard species as mortuary offerings 'readily at hand'. The deposition of chickens *in toto* and sometimes even plucked in tombs and graves as well as the use of birds or parts thereof in offerings is a widespread custom as well, reported not only within the Southeast Asian native distribution of Red jungle fowl, e.g. at 3.0–2.0 ka calBP Thai Non Nok Tha (Higham, 1975) and Ban Non Wat (Higham and Kijngam, 2010 and Higham and Kijngam, 2012) or Cambodian Phum Snay (2.2–1.7 ka calBP; O›Reilly et al., 2006), but also well beyond this range, e.g. in prehistoric and Roman Europe (Lignereux et al., 1995 and Hochmuth et al., 2005).

Given the aforementioned absence of *Gallus gallus* in Northern China, birds intentionally deposited in funerary architecture and identified as chicken likely represent domesticates. This applies for instance to the finds recovered from the royal tombs at Jiuliandun and Shenmingpu dating respectively to the Warring States period and the Early Han Dynasty (Table 2). Of particular interest here is the fact that amongst the grave goods unearthed from the burial ground of the late Shang Dynasty Kings at Dasikongcun, two chicken skeletons dating ~3.2 ka calBP have been reported (Ma et al., 1955). Osteological details are lacking, but the archaeological context strongly suggests that we are dealing with *Gallus*, rather than *Phasianus*. Of course, considering the elite status of the persons interred at Dasikongcun, the possibility of Red jungle fowl being delivered to the royal household from (sub-) tropical China cannot be entirely ruled out, but this is not very likely given the distance. Moreover, other domestic animals including horses (with chariots) have been deposited in the royal tombs as well. Whether these birds were predestined as grave goods due to chickens at that time being particularly valued cannot be answered for the moment. However, the contemporaneity of these skeletons with isolated bone finds of chickens at the site of Yinxu allows postulating that Shang Dynasty farming communities played a role in the dispersal of *Gallus* in Northern China.

In summary, re-analysis of galliform bones depicted in publications and housed in archaeological collections at the Institute of Archaeology, Chinese Academy of Social Sciences, Beijing, suggest that all early and middle Holocene finds from Northern China hitherto classified as *Gallus* are instead derived from pheasants, and mainly *Phasianus colchicus*. Moreover, the predominance of adult males and the lack of bone pathologies usually associated with poultry farming in these assemblages contradict the hypothesis of chicken husbandry in Neolithic Northern China as well. Our re-evaluation suggests that the advent of poultry husbandry in this region postdates the middle Holocene thermal optimum and that chicken farming gained foothold in local food economies since the late Shang Dynasty (1320-1046 BCE).

CHICKENS IN ICONOGRAPHY AND EARLY WRITING

Apart from osseous remains in archaeological contexts, human interest in chickens is also expressed in iconography. Renowned early records comprise clay figurines in Late Neolithic contexts of the Shijiahe Culture at Dengjiawan (34°28′N, 109°43′E) on the middle Yangtze River. Found together with large amounts of slag and dating to ~4.5–4.0 ka calBP, these figurines are considered waste of local craft production (Liu and Chen, 2012, 245). Taxonomic richness in this assemblage is surprising (Wu, 2001, Plate 1). Besides domestic dog, pig, and sheep, Wu (2001) identified Daurian ground squirrel (*Citellus dauricus*), Siberian weasel (*Mustela sibirica*), Water deer (*Hydropotes inermis*), Rhesus macaque (*Macaca mulatta*), Asian elephant (*Elephas maximus*), Chinese soft-shell turtle (*Trionyx sinensis*), Chinese pond turtle (*Mauremys reevesi*), Chinese ring-necked pheasant (*Phasianus colchicus torquatus*), Chinese bamboo partridge (*Bambusicola thoracica thoracica*), green peafowl (*Pavo muticus imperator*), and chicken (*Gallus domesticus*). Classifying clay figurines to the subspecies level is questionable, though, since seldom detailed enough to separate closely related taxa. Even species identification is not always unequivocal, as is illustrated by a figurine classified either as Giant tapir (*Megatapirus augustus*) (Wu, 2001) or Asian elephant (Liu and Chen, 2012, Fig. 7.15).

If the habitus of dog, sheep and pig leaves little doubt about their domestic status, we consider the case of the chicken less clear in absence of features typical of domestic breeds, e.g. different body proportions. In this respect, chicken figurines are usually classified as representing domestic birds since found in (*i*) sites located outside the known native distribution of Red jungle fowl in southern China and (*ii*) archaeological contexts expressing ritual behaviour, for instance votive offerings. However, apart from *Gallus,* the species spectrum at Dengjiawan comprises other thermophile birds, such as bamboo partridge and green peafowl, as well as mammals confined to the (sub-)tropics, like the Asian elephant. It is worth noting that the faunal composition at the contemporaneous site of Xiawanggang mentioned afore possesses taxa adapted to warm-humid forests and bamboo vegetation as well, such as Asian elephant, Giant panda and (wild) fowl. From a spatio-temporal perspective, domestic chickens certainly arrived earlier in the Yangtze River than in the Yellow River Basin, but it remains to be seen if the candidate chicken found in levels dating ~5.0–4.0 ka calBP (Eda et al., 2016, Table 1) descended from a domestic flock.

Turning to the epigraphic record, the use of distinct characters for pheasant and chicken in early Chinese writing is noteworthy. At the late Shang Dynasty capital of Yinxu, an assemblage of oracle bones utilised in divination has been excavated. Usually written on flat bones such as plastrons of turtles and scapulae of mammals, these oracle texts mention pheasants and chickens, whereby the pictograph for chicken features a comb (Shen and Cao, 2001, No. 1921). In later writings such as The Book of Odes (*Shijing* 诗经), an anthology of 305 songs and odes collected during the Zhou Dynasty that provides insight into the life of ordinary people from the late 11th to the middle of the 6th century BCE (e.g., Ho, 1969), the crow of a rooster is mentioned (Yu Chong, pers. comm. 2016). Written during the subsequent Spring and Autumn period of the Eastern Zhou Dynasty (771-476 BCE), the narrative *Zuo zhuan (左传)* mentions six major domesticates: horse, cattle, sheep, pig, dog and chicken (Deng et al., 2013).

GENETICS OF MODERN AND ANCIENT CHICKENS

Molecular genetic studies of modern jungle fowl and domestic breeds have confirmed the hypothesis of Red jungle fowl (*Gallus gallus*) being the primary wild ancestor of the domestic chicken. Obviously, several continental subspecies (*G.g. gallus, G.g. murghi, G.g. spadiceus, G.g. jabouillei*) contributed to the domestic chicken's genetic makeup (Fumihito et al., 1994, Fumihito et al., 1996, Hillel et al., 2003, Liu et al., 2006 and Kanginakudru et al., 2008). Having said that, research conducted on the nuclear genome revealed that the yellow skin allele observed in domestic chickens did not originate from the Red jungle fowl (this characteristic being absent in the wild form) but rather from the Grey jungle fowl *Gallus sonneratii* (Eriksson et al., 2008), demonstrating that at least one other jungle fowl species had contributed genetic material to the domestic genome.

Investigations into mtDNA phylogeny by Liu et al. (2006) and Miao et al. (2013) revealed nine highly divergent matrilineal clades named clades A-I, whereby no breed pertained to a specific clade. Their distinct distribution patterns and expansion signatures suggest these clades may have originated in different regions. Clades A-G encompass both domestic chickens and Red jungle fowl; clade E is the most widely distributed matrilineal lineage while clades A and B are ubiquitously distributed across Eurasia but absent in Africa. Clades C, F and G are mostly identified in East Asia, Southeast Asia and South Asia. The geographical distribution of clade D is particularly interesting as it appears to mirror the dispersals of cultures practicing cockfighting. It is found in South Asia, Southeast Asia, East Asia and Africa with a frequency peaking in Southeast Asia and Oceania. Clades H and I, on the contrary, are much rarer and specific to domestic chickens; these are geographically restricted to East Asia and South Asia respectively. Finally, four additional clades named clade W-Z specific to Red jungle fowls were identified (Miao et al., 2013).

Patterning in modern and ancient mtDNA of chickens is increasingly used to tackle research questions, such as the Austronesian dispersals and the first introduction of the chicken in South America (e.g. Storey et al., 2007, Storey et al., 2010, Storey et al., 2012, Gongora et al., 2008 and Thomson et al., 2014) or the origin and genetic signatures of early fowl in Europe (Girdland Flink et al., 2014). Recently, mitogenetic analysis was also applied to test the hypothesis of early chicken husbandry in Holocene Northern China (Xiang et al., 2014).[4] Specimens analysed originate from ~11.5–10.5 ka calBP Nanzhuangtou, ~8.0–7.7 ka calBP Cishan, ~6.5–4.6 ka calBP Wangyin and from a Warring States (476-221 BCE) tomb at Jiuliandun. Amplification of a 159bp fragment of the mitochondrial cytochrome c oxidase subunit 1 (COI) gene and comparison with 196 homologous sequences from *Gallus, Phasianus, Alectoris, Lophura, Tetraophasis and Syrmaticus* allowed the conclusion that the 13 ancient sequences obtained were closer to the genus *Gallus* than to any of the other genera considered (Xiang et al., 2014). Comparison with other taxa of the subfamilies *Perdicinae* and *Phasianinae* widespread in East Asia and available on GenBank confirms the partitioning between *Gallus* and other genera.

Xiang et al. (2014) also amplified a 326bp fragment of the mitochondrial DNA control region in eight samples and obtained eight different haplotypes belonging to clades A, C, and F. Though replies questioned aspects of their methodology (Peng et al., 2015) and the likelihood of DNA survival in ancient chicken bones (Peters et al., 2015), Xiang et al., 2015a and Xiang et al., 2015b carried out a re-assessment of their data and concluded that their original findings remained valid.

Thus, taken at face value, the results of the ancient DNA analysis of early and middle Holocene galliform bone specimens (Xiang et al., 2014) directly contradict expectations based upon the climatic context and associated fauna as well as morphological re-analysis. More specifically, what we currently know about the Holocene climate and vegetation suggests that chickens should not exist in Northern China at this time, and the morphological analysis concludes that bone remains supposedly deriving from chickens are actually from pheasants. Based on a subsample of these, however, the ancient DNA data clearly demonstrates that the bones are in fact chickens and that *Gallus* must have been present in the study area since the very beginning of the Holocene. No obvious or satisfying reconciliation of these contradictory conclusions has been forthcoming.

One possible explanation might be that some specimens analysed represent late intrusives from superposing Bronze Age and/or historic layers (Table 2), but single radiocarbon dates on a galliform bone from Nanzhuangtou and Cishan each confirmed their respective archaeological ages (Xiang et al., 2014, Table S1). According to Xiang (2015b), additional finds also originate from secure early and middle Holocene contexts, but more data including the results of analyses of associated ceramics and lithics are required to substantiate this assessment. Unfortunately, no radiocarbon date is available for the galliform remains from Neolithic Wangyin, although archaeology reported an extensive late Dawenkou cemetery (~5.0–4.6 ka calBP) on top of the occupation layers attributed to the late Beixin (~6.5–6.3 ka calBP) and early Dawenkou (~6.3–5.0 ka calBP) cultures (Table 2). In sum, the fact that Northern Chinese sites often exhibit long stratigraphies explains why post-hoc human activities and bioturbation by burrowing taxa (*Rhizomys, Myospalax*, *Hystrix*, *Meles*) could have caused displacement of material culture (e.g. Horwitz et al., 2012), necessitating radiocarbon dating of key archaeological specimens. Lastly, a sequencing approach that does not rely on the amplification of PCR products from these bones but instead takes advantage of advances in massively parallel sequencing technologies will also help to verify the initial genetic results and resolve the current discordance between the genetic and all other lines of evidence.

CONCLUSIONS

Proxies generated with Holocene palaeoclimate archives agree that conditions improved under the influence of the East Asian Summer Monsoon, with higher average temperatures particularly during the middle Holocene thermal optimum ~7.8–5.3 ka calBP. Nonetheless, even during the latter warm-moist conditions alternated with cooler episodes. Despite the fact that temperatures were on average somewhat higher, the mammalian fauna in the study area remained essentially Palaearctic, complemented by few Indomalayan taxa exhibiting eurytherm properties. Arboreal pollen analyses confirm that during the thermal optimum, the northern limit of the temperate deciduous forest had shifted ca. 800 km further north (Yu et al., 2000). Ecological requirements and current distributions of the terrestrial mammalian taxa identified in the early and middle Holocene assemblages fit well into this picture. Contrary to assertions in the literature, however, habitat requirements of tropical Red jungle fowl precluded extending its native range into Northern China, the middle Yangtze River basin probably delimiting its northernmost boundary during the Holocene thermal optimum.

A zooarchaeological re-evaluation of avifaunal remains from Northern and Northwestern China (Deng et al., 2013 and Eda et al., 2016, Table 2) confirms this assessment, since it could be demonstrated that the foraging and food-producing early and middle Holocene communities inhabiting the study area exploited exclusively pheasants, not chickens. Prior to this work, the Common or Ring-necked pheasant, unquestionably the most widespread and numerous galliform taxon in the study area was almost invisible in the archaeofaunal record of Holocene China, despite the species› ecological adaptability and high population density in temperate East Asia. Neolithic galliform bone assemblages also exhibited a strong bias towards adult males, which we explain by higher predation pressure by humans due to the pheasant cocks' conspicuous habitus and behaviour as well as larger size.

Paradoxically, the results emerging from this work are diametrically opposed to those based on ancient mtDNA analyses published by Xiang et al., 2014, Xiang et al., 2015a and Xiang et al., 2015b. These authors obtained mitogenome signatures from *Gallus* exclusively and no sequences of the ubiquitous *Phasianus* at all. Is it possible that these ancient remains represent an extinct, less thermophile (sub-)species of *Gallus*once distributed throughout Northern China? Adapted to seasonal fluctuating climate, temperature and humidity, this hypothetical wild fowl population would have faced multiple challenges relative to heat, cold, plant cover, food sources, predators, photoperiodism and many other crucial factors, with corresponding effects on reproduction, growth, development, and possibly social and sexual behaviour likely resulting in a genetic make-up that differs from modern populations of wild and domestic fowl. The ancient mtDNA signatures, however, do not support this possibility, since the clades identified within the specimens (A, C and F) are widely distributed and recorded in both domestic chickens and present-day tropical Red jungle fowls. According to Xiang et al. (2014), early chicken husbandry in Northern China would fit in as well in autochthonous agricultural innovations reported from the Yellow River basin, such as millet cultivation (Barton et al., 2009, Lu et al., 2009 and Yang et al., 2012), pig husbandry (Larson et al., 2010 and Cucchi et al., 2011) and dog breeding (Yuan, 2010). Since millet was being cultivated since the early Holocene (Yang et al., 2012), potential poultry keepers would have been able to feed their chickens on staples that remain appreciated for their nutritive value today. Moreover, since this chicken-crop combination would represent a cost-efficient subsistence activity, would one not expect it spreading readily in the wake of dispersal of millet (and later rice) cultivation throughout China and beyond?

As such, millets diffused from China to Korea in the 6th millennium calBP with secure direct dates from ~5.4 ka calBP, but whether domestic fauna, such as the pig, diffused at this time or significantly later remains unclear (Fuller et al., 2008). A Neolithic transfer of chickens, however, can be excluded, since it was noted that poultry farming in Korea started more than three millennia later ~2.0 ka calBP (M. Eda, pers. comm., June 2015). This date fits remarkably well with the beginning of fowl husbandry in Japan coinciding with the middle Yayoi Period ~2.2–1.7 ka calBP (Eda and Inoué, 2011). Obviously, chicken husbandry crossed the Yellow Sea together with other agricultural innovations known from mainland China, such as wet-rice farming. West of Northern China in the Wei River and Middle and Upper Yellow River valleys, the integration of chicken husbandry into local subsistence was a late achievement as well since securely identified remains of chickens seem absent from archaeological contexts predating the Proto-Zhou culture (1200-1050 BCE) (Eda et al., 2016, Table 1). In summary, given that in the archaeological record of regions adjacent to Northern China chickens only show up in the later Bronze Age, it is difficult to conceive that beyond the study area, poultry husbandry passed unnoticed for nearly five millennia.

Finally, chicken husbandry in Neolithic Northern China would have facilitated an earlier dispersal to Europe across Central and West Asia (West and Zhou, 1988). However, the paucity of archaeofaunal analyses available for Central Asia in the 1980s precluded testing this hypothesis. Although this research gap is only being closed slowly, faunal studies published since then are consistent in that chickens represent a very late addition to the Central Asian barnyard. Extensive faunal assemblages from Iron Age South Kazakhstan dating 3.0–2.5 ka calBP, for instance, did not yield a single specimen attributable to *Gallus* (Benecke, 2003). Thus, early evidence comes from the 2.4–2.3 ka calBP fortification of Kurgansol in Bactria/Uzbekistan (Benecke, 2014). Erected during the military campaigns of Alexander the Great, it is therefore highly probable that the Kurgansol chickens were introduced from the army›s region of origin in West Asia rather than being of East Asian origin. Conceivably, intensification of irrigation practices and increasing cultivation of cereals in the oasis landscapes of Central Asia promoted the adoption of poultry farming and its subsequent development into a sustainable activity.

Overall, the hypothesis of an autochthonous domestication episode of *Gallus* in Holocene Northern China is supported by the current ancient DNA results (Xiang, 2014), but not by any other evidence. The climatic context suggests that the habitat requirements for (sub-)tropical Red jungle fowl were not present in early and middle Holocene Northern China. The zooarchaeological re-evaluation of the avifaunal record available suggests that along the middle Yangtze River and its tributaries, the presence of (wild or domestic) *Gallus* seems possible ~5.0–4.0 ka calBP, as suggested by terracotta figurines associated with the Shijiahe culture and a single bone find from a late Neolithic context (Eda et al., 2016). However, even if chickens arrived in Central China towards the end of the Middle Holocene, the early stages of fowl husbandry further north in Northern China seem to post-date ~3.5 ka calBP, when bone finds and two skeletons of chickens from the late Shang Dynasty capital of Yinxu and the contemporaneous royal burial grounds at Dasikongcun have been discovered (Table 2). The presence of *Gallus*in Shang Dynasty times is also supported by contemporaneous early Chinese writing depicting distinct characters for pheasants and chickens on oracle bones (Shen and Cao, 2001).

In conclusion, the weight of evidence still suggests that in Northern and Northwestern China, the integration of poultry husbandry into food-producing economies was a comparably late development, resulting from human-mediated dispersal of domestic flocks originating from regions further south during the Bronze Age. Having said that, additional ancient DNA, morphometric and dating programs of purported chicken remains across China and East Asia generally will no doubt resolve the current paradox, and establish the spatial and temporal origins of the world's most numerous domestic animal.

ACKNOWLEDGEMENTS

The authors are indebted to Prof. Yuan Jing, who kindly granted access to galliform remains from key archaeological sites in China housed at the Institute of Archaeology, Chinese Academy of Social Sciences, Beijing. The Coordination of Research between Europe and China (CO-REACH) BIOARCH project (J.-D. Vigne, K. Dobney) provided financial support to J.P. to conduct research at the Institute of Archaeology, Chinese Academy of Social Sciences, Beijing. H.D. was supported by the Youth foundation of Humanity and Social Science of the Ministry of Education of China (15YJC780001). G.L and O.L. were supported by the Arts and Humanities Research Council (AH/L006979/1) and the European Research Council (337574-UNDEAD). M. Eda (Hokkaido University) and S. Yu (Sun Yat-sen University, Guangzhou) kindly provided data, C. Eggers (LMU Munich) helped with Fig. 1. The manuscript benefited from the comments provided by two reviewers.

REFERENCES

1. An, Z., 2000. The history and variability of the East Asian paleomonsoon climate. Quat. Sci. Rev 19, 171e187.
2. Anhui Provincial Institute of Archaeology, 1992. Identification and study of faunal remains from the site at Shishanzi, Suixi county, Anhui. Kaogu Archaeol. 3, 253e262 (in Chinese).
3. Anyang Archaeological Team of Institute of Archaeology, Chinese Academy of Social Sciences, 1985. The excavation at Hougang in Anyang, 1979. Kaogu Xuebao (Acta Archaeol. Sin.) 1, 33e88 (in Chinese, with English summary).

4. Baoding Prefectural Administration of the Preservation of Ancient Monuments, Xushui County Administration of the Preservation of Ancient Monuments, Archaeology Department of Beijing University, History Department of Hebei University, 1992. Trial Digging at Nanzhuangtou site in Xushui County, Hebei Province. Kaogu (Archaeology) 11, 961e986 (in Chinese).
5. Baoji Archaeological Team, Baoji workstation of Shannxi Provincial Institute of Archaeology, 1993. Fulinbao at Baoji: Report on the excavation of the Neolithic Site. Cultural Relics Publishing House, Beijing (in Chinese, with English summary).
6. Barton, L., Newsome, S.D., Chen, F., Wang, H., Guilderson, T.P., Bettinger, R.L., 2009. Agricultural origins and the isotopic identity of domestication in northern China. Proc. Natl. Acad. Sci. U. S. A. 108, 5523e5528.
7. Beebe, W., 1921. In: H.F., Witherby, G. (Eds.), A Monograph of the Pheasants, vol. II
8. (London).
9. Benecke, N., 1994. Der Mensch und seine Haustiere. Theiss, Stuttgart.
10. Benecke, N., 2003. Iron age economy of the inner Asian steppe. A bioarchaeological perspective from the Talgar region in the Ili River valley, southeastern Kazakhstan. Eurasia Antiq. 9, 63e84.
11. Benecke, N., 2014. Faunal remains of Kurganzol. In: Sverchkov, L.M. (Ed.), Kurganzol e Krepost' Aleksandra Na Juge Uzbekistana. SMI-ASIA, Tashkent, pp. 170e176.
12. Bo, K., 2011. Animal use in the late phase of Neolithic in China e fauna remains of late longshan culture of Taosi site, Shanxi. In: Institute of Archaeology, Chinese Academy of Sciences (Ed.), Sandai Kaogu (Xia, Shang and Zhou Dynasty Archaeology), vol. 4. Science Press, Beijing, pp. 129e182 (in Chinese).
13. Chen, Q., 1993. In: Ji, Qingguo (Ed.), The Identification and Marks of Animal Bones from Zuojiashan Site, Nong'an County, in: the Department of Archaeology, Jilin University. Zhishi Press, Beijng, pp. 57e71 (in Chinese).
14. Chen, Z., 1994. Natural remains. In: Institute of Archaeology, Chinese Academy of Social Sciences, Archaeological Excavation and Researches in the Yin Ruins. Science Press, Beijing, pp. 415e418 (in Chinese).
15. Chen, F., Xu, Q., Chen, J., Birks, J.B., Liu, J., Zhang, S., Jin, L., An, C., Telford, R.J., Cao, X., Wang, Z., Zhang, X., Selvaraj, K., Lu, H., Li, Y., Zheng, Z., Wang, H., Zhou, A., Dong, G., Zhang, J., Huang, X., Bloemendal, J., Rao, Z., 2015. East Asian summer monsoon precipitation variability since the last deglaciation. Sci. Rep. 5, 11186. http://dx.doi.org/10.1038/srep11186.
16. Cheng, T. (Ed.), 1978. Fauna Sinica, Series Vertebrata, Aves, Vol. 4: Galliformes. Science Press, Beijing (in Chinese).
17. Cheng, T., 1987. A Synopsis of the Avifauna of China. Science Press, Beijing and Paul Parey Scientific Publishers, Hamburg, Berlin.
18. China-US Joint Archaeological Team, 2010. Study on the longshan culture of Shantaisi. Kaogu Archaeol. 10, 51e60 (in Chinese, with English summary).
19. Corbet, G.B., Hill, J.E., 1992. The Mammals of the Indomalayan Region: a Systematic Review. Natural History Museum Publications, London and Oxford University Press, Oxford.
20. CPAM, Hebei Province Relics Preservation Station and Handan Relics Preservation Station, 1981. The Cishan Site in Wu'an, Hebei Province. Kaogu Xuebao (Acta Archaeol. Sin.) 3, 303e338 (in Chinese, with English summary).
21. CPAM of Anyang Prefecture of Henan Province, 1983. Excavation of a community site of the longshan culture at Baiying, Tangyin, Henan. Pap. Chin. Archaeol. 3, 1e47 (in Chinese).
22. Crawford, R.D., 1984. Domestic fowl. In: Mason, I.L. (Ed.), The Evolution of Domesticated Animals. Longman, London, pp. 298e311.
23. Cucchi, T., Hulme-Beaman, A., Yuan, J., Dobney, K., 2011. Early Neolithic pig domestication at Jiahu, Henan Province, China: clues from molar shape analyses using geometric morphometric approaches. J. Archaeol. Sci. 38, 11e22.
24. Delacour, J., 1977. The Pheasants of the World, second ed. Spur Publications, Hindhead, UK.
25. Deng, H., Yuan, J., Song, G., Wang, C., Masaki, E., 2013. Reexamination of the domestic chicken in ancient China. Kaogu Archaeol. 6, 83e96 (in Chinese, with English summary).
26. Department of Archaeology of Northwest University, 2000. Excavation Report of An'Ban Site in Fufeng. Science Press, Beijing (in Chinese, with English summary).
27. Dobney, K., Colledge, S., Conolly, J., Manning, K., Peters, J., Shennan, S., 2013. The origins and spread of stock-keeping. In: Colledge, S., Conolly, J., Dobney, K., Manning, K., Shennan, S. (Eds.), The Origins and Spread of Domestic Animals in Southwest Asia and Europe. Left Coast Press, Walnut Creek CA, pp. 17e26.
28. Du, J., Wang, X., 2004. New archaeological work about Shang city-site at Yanshi. Kaogu Archaeol. 12, 3e12 (in Chinese, with English summary).
29. Eda, M., Inoue, T., 2011. Identi fication of Phasianidae (chicken and indigenous fowls in Japan) bones using nonmetric characteristics and its application for the reevaluation of chickens from the Yayoi period. Zoo-Archeology 28 (5), 23e33 (in Japanese, with English summary).
30. Eda, M., Lu, P., Kikuchi, H., Li, Z., Li, F., Yuan, J., 2016. Reevaluation of early Holocene chicken domestication in northern China. J. Archaeol. Sci. 67, 25e31.
31. Erbersdobler, K., 1968. Vergleichend morphologische Untersuchungen an Einzelknochen des postcranialen Skeletts in Mitteleuropa vorkommender mittelgroßer Hühnervogel. Diss. Med. Vet. (LMU Munich) € .
32. Eriksson, J., Larson, G., Gunnarsson, U., Bed'hom, B., Tixier-Boichard, M., Stromstedt, L., Wright, D.,

Jungerius, A., Vereijken, A., Randi, E., Jensen, P., € Andersson, L., 2008. Identification of the yellow skin gene reveals a hybrid origin of the domestic chicken. PLoS Genet. 4 (2), e1000010. http://dx.doi.org/ 10.1371/ journal.pgen.1000010.

33. Flad, R.K., Yuan, J., Li, S., 2007. Zooarchaeological evidence for animal domestication in Northwest China. In: Madsen, D., Chen, F.-H., Gao, X. (Eds.), Late Quaternary Climate Change and Human Adaptation in Arid China. Developments in Quaternary Science, vol. 9. Elsevier, Amsterdam, pp. 167e203.

34. Fu, Y., 2000. Study on Animal's Remains from An'Ban Site in Fufeng, Shaanxi. Department of Archaeology of Northwest University,Excavation report of An'Ban site in Fufeng. Science Press, Beijing, pp. 290e294 (in Chinese).

35. Fuller, D.Q., Qin, L., Harvey, E., 2008. A critical assessment of early agriculture in East Asia, with emphasis on Lower Yangzte rice domestication. In: Paper Presented in the International Seminar on the 'First Farmers in Global Perspective', Lucknow, India, 18-20 January, 2006. Pragdh ar a 18, pp. 17e52.

36. Fumihito, A., Miyake, T., Sumi, S., Takada, M., Ohno, S., Kondo, N., 1994. One subspecies of the red junglefowl (Gallus gallus gallus) suffices as the matriarchic ancestor of all domestic breeds. Proc. Natl. Acad. Sci. U. S. A. 91, 12505e12509.

37. Fumihito, A., Miyake, T., Takada, M., Shingu, R., Endo, T., Gojobori, T., Kondo, N., Ohno, S., 1996. Monophyletic origin and unique dispersal patterns of domestic fowls. Proc. Natl. Acad. Sci. U. S. A. 93, 6792e6795.

38. Giles, F.H., 1932. Miscellaneous notes. No. III. Migration of jungle-fowl. Nat. Hist. Bull. Siam Soc. 8, 333e334.

39. Girdland Flink, L., Allen, R., Barnett, R., Malmstrom, H., Peters, J., Eriksson, J., € Andersson, L., Dobney, K., Larson, G., 2014. Establishing the validity of domestication genes using DNA from ancient chickens. Proc. Natl. Acad. Sci. U. S. A. 111, 6184e6189.

40. Glutz von Blotzheim, U.N., Bauer, K.M., Bezzel, E., 1973. Band 5. Galliformes und Gruiformes. In: Glutz von Blotzheim, U.N. (Ed.), Handbuch der Vogel Mitte-€ leuropas. Akademische Verlagsgesellschaft, Frankfurt a. Main.

41. Gongora, J., Rawlence, N.J., Mobegi, V.A., Han, J., Alcalde, J.A., Matus, J.T., Hanotte, O., Moran, C., Austin, J.J., Ulm, S., Anderson, A.J., Larson, G., Cooper, A., 2008. IndoEuropean and Asian origins for Chilean and Pacific chickens revealed by mtDNA. Proc. Natl. Acad. Sci. U. S. A. 105, 10308e10313.

42. Groppi, L., 2006. Alligator sinensis (On-line). Animal Diversity Web. http:// animaldiversity.org/accounts/ Alligator_sinensis/ (accessed 23.06.15).

43. Hebei Provincial Cultural Relics Institute, Baoding Municipal Administration of Cultural Relics, Xushui County Office for the Preservation of Ancient Monuments, Department of History and Culture of Shanxi University, 2010. Excavation to Nanzhuangtou Site in Xushui County, Hebei in 1997. Kaogu (Archaeology), vol. 3, pp. 361e392 (in Chinese, with English summary).

44. Henan Institute of Cultural Relics and Archaeology, 1999. Wuyang Jiahu. Science Press, Beijing (in Chinese, with English summary).

45. Henan Provincial Institute of Archaeology, Henan Group of the Archaeology Team, the Office for Program of the Works in the Changjiang River Valley, 1989. Xiawanggang Site at Xichuan. Cultural Relics Publishing House, Beijing (in Chinese, with English summary).

46. Henan Working Team No1 of the Institute of Archaeology, Chinese Academy of Social Sciences, 1984. Excavation of the Neolithic Site at Peiligang. Kaogu Xuebao (Acta Archaeol. Sin.), vol. 1, pp. 23e52 (in Chinese, with English summary).

47. Higham, C.F.W., 1975. Non Nok Tha: the Faunal Remains. University of Otago, Studies in Prehistoric Anthropology, vol. 7. Dept. Anthropology, University of Otago.

48. Higham, C.F.W., Kijngam, A. (Eds.), 2010. The Origins of the Civilization of Angkor. Vol. VI. The Excavation of Ban Non Wat: the Iron Age, Summary and Conclusions. The Thai Fine Arts Department, Bangkok.

49. Higham, C.F.W., Kijngam, A. (Eds.), 2012. The Origins of the Civilization of Angkor. Vol. V. The Excavation of Ban Non Wat: the Bronze Age. The Thai Fine Arts Department, Bangkok.

50. Hillel, J., Groenen, M.A.M., Tixier-Boichard, M., Korol, A.B., David, L., Kirzhner, V.M., Burke, T., Barre-Dirie, A., Crooijmans, R.P.M.A., Elo, K., Feldman, M., Freidlin, W.P.J., M€ aki-Tanila, A., Oortwijn, M., Thomson, P., Vignal, A., Wimmers, K., Weigend, S., 2003. Biodiversity of 52 chicken populations assessed by microsatellite typing of DNA pools. Genet. Sel. Evol. 35, 533e557.

51. Ho, P.T., 1969. The loess and the origin of Chinese agriculture. Am. Hist. Rev. 75 (1), 1e36.

52. Ho, P.T., 1977. The indigenous origins of Chinese agriculture. In: Reed, C.A. (Ed.), Origins of Agriculture. Mouton, The Hague, pp. 413e484.

53. Hochmuth, M., Benecke, N., Witteyer, M., 2005. Cocks and song birds for Isis Panthea and Mater Magna: the bird remains from a sanctuary in Mogontiacum/ Mainz. In: Grupe, G., Peters, J. (Eds.), Feathers, Grit and Symbolism, pp. 319e327. Documenta Archaeobiologiae 3. Leidorf, Rahden/Westf.

54. Hoffmann, R.S., 2001. The southern boundary of the Palaearctic Realm in China and adjacent countries. Acta Zool. Sin. 47 (2), 121e131.

55. Horwitz, L., Cohen, S.L., Więçkowski, W., Mienis, H.K., Baker, J., Jastrzebska, E., 2012. Recent indian Porcupine (Hystrix indica) Burrows and their Impact on ancient faunal and Human remains: a case study from Tel Zahara (Israel). J. Taphon. 10 (2), 85e112.

56. Hou, L., 1989. Birds from early Yinxu in An'yang. Kaogu Archaeol. 10, 942e947 (in Chinese).

57. Hu, S., 2010. The animal's remains from Dongying Site in Gaoling. In: Shannxi Provincial Institute of Archaeology, School of Cultural Heritage, Northwest

University, Dongying Site in Gaoling. Science Press, Beijing, pp. 147e200 (in Chinese).

58. Hu, S., Zhang, Y., Zhang, T., 2007. The Animal's Remains from Guantaoyuan Archeological Site in Baoji City, Shaanxi, and its Paleo-environmental Significance. J. Northwest Univ. 37 (1), 115e118 (in Chinese, with English summary).
59. Hubei Provincial Museum, 2007. Jiuliandun-large Tomb of a Chu Noble in the Middle Reaches of the Yangtze. Cultural Relics Publishing House, Beijing (in Chinese, with English summary).
60. Institute of Archaeology, Chinese Academy of Sciences, The Ban P'o Museum, Xi'an, 1963. The Neolithic Village at Ban P'o, Xi'an. Cultural Relics Publishing House, Beijing (in Chinese, with English summary).
61. Institute of Archaeology, Chinese Academy of Social Sciences, 1994a. Archaeological Excavation and Researches in the Yin Ruins. Science Press, Beijing (in Chinese).
62. Institute of Archaeology, Chinese Academy of Social Sciences, 1994b. Baijiacun Site at Lintong. Bashu Press, Chengdu (in Chinese, with English summary).
63. Institute of Archaeology, Chinese Academy of Sciences, 1965. Qujialing-Jiangshan. Science Press, Beijing (in Chinese, with English summary).
64. Institute of Archaeology, Chinese Academy of Social Sciences, 1983. Beishouling at Baoji. Cultural Relics Publishing House, Beijing (in Chinese, with English summary).
65. Institute of Archaeology, Chinese Academy of Social Sciences, 1999. The Erlitou Site
66. in Yanshi. The Encyclopedia of China Publishing House, Beijing (in Chinese, with
67. English summary).
68. Institute of Archaeology, Chinese Academy of Social Sciences, 2000. Excavations of the Wangyin Site in Shandong. Science Press, Beijing (in Chinese, with English summary).
69. Jia, L., Zhang, Z., 1977. The fauna from Xiawanggang site in Xichuan, Henan. Cult. Relics 6, 41e49 (in Chinese).
70. Kanginakudru, S., Metta, M., Jakati, R.D., Nagaraju, J., 2008. Genetic evidence from Indian red jungle fowl corroborates multiple domestication of modern day chicken. BMC Evol. Biol. http://dx.doi.org/10.1186/1471-2148-8-174, 2008 Jun 10.
71. Kong, Q., 2000. The animal's remains from the Neolithic site at Qianbuxia. In: Shandong Cultural Relics Institute, Liuting Administration of the Preservation of Ancient Monuments, the Collection of Archaeology Reports on the Excavations from Highway Construction in Shandong 1997. Science Press, Beijing, pp. 103e105 (in Chinese).
72. Larson, G., Liu, R., Zhao, X., Yuan, J., Fuller, D., Barton, L., Dobney, K., Fan, Q., Gu, Z., Liu, X., Luo, Y., Lv, P., Andersson, L., Li, N., 2010. Patterns of East Asian pig domestication, migration, and turnover revealed by modern and ancient DNA. Proc. Natl. Acad. Sci. U. S. A. 107, 7686e7691.
73. Li, Y., 1974. The animal's bones from Dawenkou tombs. In: Shandong Cultural Relics Institute, Jinan Museum, Dawenkou. Cultural Relics Publishing House, Beijing, pp. 156e158 (in Chinese).
74. Li, Y., Han, D., 1963. The animal's remains from the Neolithic site at Ban P'o, Xian. In: Institute of Archaeology, Chinese Academy of Sciences, the Ban P'o Museum, Xi'an, the Neolithic Village at Ban P'o, Xian. Cultural Relics Publishing House, Beijing, pp. 255e269 (in Chinese).
75. Li, X., Zhou, W., An, Z., Dodson, J., 2003. The vegetation and monsoon variations at the desert-loess transition belt at Midiwan in northern China for the last 13 ka. Holocene 13, 779e784.
76. Li, F., Peng, L., Eda, M., Yuan, J., Zhu, Y., 2015. Study of the bird remains from the Tengjiagang site: also a retrospect and Prospects of archaeozoological studies on Chinese bird remains. Huaxia Archaeol. 15 (1), 34e40 (in Chinese).
77. Liaoning Institute of Cultural Relics and Archaeology, Benxi Museum, 1994. Machengzi e Ancient Remains from the Caves in the Upper Reaches of Taizihe River. Cultural Relics Publishing House, Beijing (in Chinese, with English summary).
78. Lignereux, Y., Peters, J., Bubien-Waluszewska, A., Sillieres, P., 1995. Sacrifices d'oiseaux en l'honneur d'Isis au 1er siecle apres Jesus-Christa Belo, en Betique (Bolonia, Cadix, Andalousie). Rev. Med. V et 146, 575 e582.
79. Liu, K.-B., 1988. Quaternary history of the temperate forests of China. Quat. Sci. Rev. 7, 1e20.
80. Liu, L., Chen, X., 2012. The Archaeology of China. From the Late Paleolithic to the Early Bronze Age. Cambridge World Archaeology. Cambridge University Press, Cambridge.
81. Liu, L., Yan, Y., Qin, X., 2001. Faunal remains of the longshan period unearthed from the 1990 excavation at the Kangija site in Lintong, Shannxi. Huaxia Archaeol. 1, 3e24 (in Chinese, with English summary).
82. Liu, Y.-P., Wu, G.-S., Yao, Y.-G., Miao, Y.-W., Luikart, G., Baig, M., Beja-Perreira, A., Ding, Z.-L., Palanichamy, M.G., Zhang, Y.-P., 2006. Multiple maternal origins of chickens: out of the Asian jungles. Mol. Phylogenet. Evol. 38, 12e19.
83. Liu, J.H., Xu, H., Qin, D.Y., Wang, H., Wang, J.H., Li, H.H., Bao, S.J., 2013. Water cycle evolution in the Haihe River Basin in the past 10,000 years. Chin. Sci. Bull. 58, 3312e3319.
84. Lu, H., Zhang, J., Liu, K., Wu, N., Li, Y., Zhou, K., Ye, M., Zhang, T., Zhang, H., Yang, X., Shen, L., Xu, D., Li, Q., 2009. Earliest domestication of common millet (Panicum miliaceum) in East Asia extended to 10,000 years ago. Proc. Natl. Acad. Sci. U. S. A. 106, 7367e7372.
85. Luoyang Hanwei City Ruins Team, Institute of Archaeology, Chinese Academy of Social Sciences, 1984. The preliminary sampling and excavation of Yanshi Shang city-site. Kaogu Archaeol. 6, 488e504 (in

Chinese).

86. Ma, D., Zhou, Y., Zhang, Y., 1955. Excavation report of Dasikong village in An'yang 1953. Kaogu Xuebao (Acta Archaeol. Sin.) 9, 25e90 (in Chinese).

87. McGowan, P.J.K., 1994. Family Phasianidae (pheasants and partridges). In: del Hoyo, J., Elliott, A., Sargatal, J. (Eds.), Handbook of the Birds of the World. Vol. 2. New World Vultures to Guineafowl. Lynx Editions, Bellaterra, pp. 434e553.

88. Miao, Y.-W., Peng, M.-S., Wu, G.-S., Ouyang, W.-N., Yang, Z.-Y., Yu, N., Liang, J.-P., Pianchou, G., Beja-Pereira, A., Mitra, B., Palanichamy, M.G., Baig, M., Chaudhuri, T.K., Shen, Y.-Y., Kong, Q.-P., Murphy, R.W., Yao, Y.-G., Zhang, Y.-P., 2013. Chicken domestication: an updated perspective based on mitochondrial genomes. Heredity 110, 277e282.

89. Neimeng Archaeological Team, Institute of Archaeology, Chinese Academy of Social Sciences, 1997. Excavation Report of Xinglongwa Site in Aohanqi, Neimenggu in 1992. Kaogu (Archaeology), vol. 1, pp. 1e26 (in Chinese).

90. O'Reilly, D.J.W., Driesch, A., von den Voen, V., 2006. Archaeology and archaeozoology of Phum Snay: a late prehistoric cemetery in northwestern Cambodia. J. Archaeol. Asia Pac. 45, 188e211.

91. Peng, M.S., Shi, N.N., Yao, Y.G., Zhang, Y.P., 2015. Caveats about interpretation of ancient chicken mtDNAs from northern China. Proc. Natl. Acad. Sci. U. S. A. 112 (16), E1970eE1971.

92. Peters, J., 1997. Hahn oder Kapaun? Zur Kastration von H€ ahnen in der Antike. Eur.

93. Poult. Sci. Arch. Geflügelk.) 61, 1e8.

94. Peters, J., 1998. Romische Tierhaltung und Tierzucht. Eine Synthese aus arch- € € aozoologischer Untersuchung und schriftlich-bildlicher Überlieferung. M. Leidorf, Rahden/Westf.

95. Peters, J., Buitenhuis, H., Grupe, G., Schmidt, K., Pollath, N., 2013. The long and € winding road: ungulate exploitation and domestication in Early Neolithic Anatolia (10000-7000 cal bc). In: Colledge, S., Conolly, J., Dobney, K., Manning, K., Shennan, S. (Eds.), The Origins and Spread of Domestic Animals in Southwest Asia and Europe. Left Coast Press, Walnut Creek CA, pp. 83e114.

96. Peters, J., Lebrasseur, O., Best, J., Miller, H., Fothergill, T., Dobney, K., Thomas, R.M., Maltby, M., Sykes, N., Hanotte, O., O'Connor, T., Collins, M.J., Larson, G., 2015. Questioning new answers regarding Holocene chicken domestication in China. Proc. Natl. Acad. Sci. U. S. A. 112 (19), E2415.

97. Qi, G., Lin, Z., An, J., 2006. The Identification Report of Faunal Remains from Dadiwan Site in Qin'an, in: the Gansu Provincial Institute of Cultural Relics and Archaeology, Dadiwan in Qin'an. Cultural Relics Publishing House, Beijing, pp. 861e910 (in Chinese).

98. Raethel, H.-S., 1988. Hühnervogel der Welt. Neumann, Neudamm € .

99. Ren, S., 1996. Several major achievements in Early Neolithic China, ca. 5000 BC. Archaeology 1996, 37e49. In Chinese, translated by Wing Kam-Cheung; ed. by B. Gordon. https://www.google.de/search?q¼Prof. þRENþShinan,þ&ie¼utf- 8&oe¼utf-8&gws_rd¼cr&ei ¼ZUwKVv75NaTgyQOdm4b4Dg. accessed 29.09.15.

100. School of Cultural Heritage, Northwest University, Shaanxi Provincial Institute of Archaeology, Chunhua County Museum, 2013. The Excavation of Zaoshugounao Site in Chunhua County, Shaanxi in 2007. Cult. Relics, vol. 2, pp. 55e66.

101. Shaanxi Provincial Institute of Archaeology, Baoji Archaeological Team, 2007. Guantaoyuan at Baoji. Cultural Relics Publishing House, Beijing (in Chinese, with English summary).

102. Shandong Archaeological Team, Institute of Archaeology, Chinese Academy of Social Sciences, Museum of Teng County, 1984. The Excavation of the Beixin Neolithic Site in Teng County, Shandong Province. Kaogu Xuebao (Acta Archaeol. Sin.), vol. 2, pp. 159e191 (in Chinese, with English summary).

103. Shandong Cultural Relics Institute, Jining Municipal Administration of Cultural Relics, Rengcheng Administration of the Preservation of Ancient Monuments, 2010. Excavation report of Yuhuangding site in Jining city, Shandong. In: Shandong Cultural Relics Institute, Haidai Archaeology 3. Science Press, Beijing, pp. 1e113 (in Chinese).

104. Shandong Cultural Relics Institute, Jinan Museum, 1974. Dawenkou. Cultural Relics Publishing House, Beijing (in Chinese).

105. Shanxi Archaeological Team of the Institute of Archaeology, Chinese Academy of Sciences, Linfen Municipal Bureau of Culture, 1980. Brief Excavation Report on Taosi Site in Xiangfen County, Shanxi. Kaogo (Archaeology), vol. 1, pp. 18e31 (in Chinese).

106. Shen, J., Cao, J., 2001. A New Compilation of Yiagu Characters. The Chinese University of Hongkong, Hongkong.

107. Smith, A.T., Xie, Y. (Eds.), 2008. A Guide to the Mammals of China. Princeton University Press, Princeton and Oxford.

108. Stebich, M., Mingram, J., Han, J., Liu, J., 2009. Late Pleistocene spread of (cool-) temperate forests in Northeast China and climate changes synchronous with the North Atlantic region. Glob. Planet. Change 65, 56e70.

109. Storey, A.A., Ramirez, J.M., Quiroz, D., Burley, D.V., Addison, D.J., Walter, R., Anderson, A.J., Hunt, T.L., Athens, J.S., Huynen, L., Matisoo-Smith, E.A., 2007.

110. Radiocarbon and DNA evidence for a pre-Columbian introduction of Polynesian chickens to Chile. Proc. Natl. Acad. Sci. U. S. A. 104, 10335e10339.

111. Storey, A.A., Spriggs, M., Bedford, S., Hawkins, S.C., Robins, J.H., Huynen, L., MatisooSmith, E., 2010. Mitochondrial DNA from 3000-year old chickens at the Teouma site. Vanuatu. J. Archaeol. Sci. 37, 2459e2468.

112. Storey, A.A., Athens, J.S., Bryant, D., Carson, M., Emery, K., deFrance, S., Higham, C., Huynen, L., Intoh,

M., Jones, S., Kirch, P.V., Ladefoged, T., McCoy, P., MoralesMuniz, A., Quiroz, D., Reitz, E., Robins, J., Walter, R., Matisoo-Smith, E., 2012. ~ Investigating the global dispersal of chickens in prehistory using ancient mitochondrial DNA signatures. PLoS One 7 (7), e39171.

113. Thomson, V.A., Lebrasseur, O., Austin, J.J., Hunt, T.L., Burney, D.A., Denham, T., Rawlence, N.J., Wood, J.R., Gongora, J., Flink, L.G., Linderholm, A., Dobney, K., Larson, G., Cooper, A., 2014. Using ancient DNA to study the origins and dispersal of ancestral Polynesian chickens across the Pacific. Proc. Natl. Acad. Sci. U. S. A. 111, 4826e4831.

114. Tomek, T., Bochenski, Z.M., 2009. A Key for the Identi fication of Domestic Bird Bones in Europe: Galliformes and Columbiformes. Institute of Systematics and Evolution of Animals, Polish Academy of Sciences, Krakow .

115. von den Driesch, A., 1976. A Guide to the Measurement of Animal Bones from Archaeological Sites. Peabody Mus. Bull. 1, Cambridge MA.

116. von den Driesch, A., 1987. Evolution und Haustiere, in: Siewing, R. (Hrsg.), Evolution. 3. Aufl. G. Fischer, Stuttgart, New York, pp. 393e414.

117. von den Driesch, A., Pollath, N., 2000. Tierknochen aus dem Mithrastempel von € Künzing, Lkr. Deggendorf. In: Schmotz, K., (Hrsg) (Eds.), Vortr€ age des 18. Niederbayerischen Arch€ aologentages. Leidorf, Rahden/Westf, pp. 145e162.

118. Walker, M.J.C., Berkelhammer, M., Bjorck, S., Cwynar, L.C., Fisher, D.A., Long, A.J., € Lowe, J.J., Newnham, J.J., Rasmussen, S.O., Weiss, H., 2012. Formal subdivision of the Holocene series/Epoch: a discussion paper by a working group of INTIMATE (integration of ice-core, marine and terrestrial records) and the subcommission on quaternary stratigraphy (International Commission on stratigraphy). J. Quat. Sci. 27, 649e659.

119. Wang, H., 2011. Animal Subsistence of the Yangshao Period in the Wei River Valley: a Case-study from the Site of Wayaogou in Shaanxi Province, China. Doctoral Thesis. University College London. http://discovery.ucl.ac.uk/1318143/ (accessed 04.03.16).

120. Wang, Y.J., Cheng, H., Edwards, R.L., An, Z.S., Wu, J.Y., Shen, C.-C., Dorale, J.A., 2001. A High-resolution absolute-dated late Pleistocene monsoon record from Hulu cave, China. Science 294, 2345e2348.

121. Wang, C., Lu, H., Zhang, J., Gu, Z., He, K., 2014. Prehistoric demographic fluctuations in China inferred from radiocarbon data and their linkage with climate change over the past 50,000 years. Quat. Sci. Rev. 98, 45e59.

122. West, B., Zhou, B.-X., 1988. Did chickens go north? New evidence for domestication. J. Archaeol. Sci. 15, 515e533.

123. Wu, J., 1990. Remains of Animal and Plant, in: Shannxi Provincial Institute of Archaeology, Longgangsi. Cultural Relics Publishing House, Beijing, pp. 40e42 (in Chinese).

124. Wu, J., 1993. Preliminary study on faunal remains from Fulinbao site. In: Fulinbao at Baoji: Report on the Excavation of the Neolithic Site. Cultural Relics Publishing House, Beijing, pp. 221e224 (in Chinese).

125. Wu, X., 2001. Zooarchaeological study on the pottery figurines of fowl from Dengjiawan site. Jianghan Archaeol. 4, 65e72 (in Chinese).

126. Xiang, H., Gao, J., Yu, B., Zhou, H., Cai, D., Zhang, Y., Chen, X., Wang, X., Hofreiter, M., Zhao, X., 2014. Early Holocene chicken domestication in northern China. Proc. Natl. Acad. Sci. U. S. A. 111, 17564e17569.

127. Xiang, H., Hofreiter, M., Zhao, X., 2015a. Reply to Peng et al.: archaeological contexts should not be ignored for early chicken domestication. Proc. Natl. Acad. Sci. U. S. A. 112, E1972eE1973.

128. Xiang, H., Gao, J., Yu, B., Hofreiter, M., Zhao, X., 2015b. Reply to Peters et al.: further discussions confirm early Holocene chicken domestication in northern China. Proc. Natl. Acad. Sci. U. S. A. 112, E2416.

129. Xu, Q., Xiao, J., Li, Y., Tian, F., Nakagawa, T., 2010. Pollen-based quantitative reconstruction of Holocene climate changes in the Daihai Lake area, inner Mongolia, China. J. Clim. 23, 2856e2868.

130. Yang, J., 2008. Fauna remains of Erlitou site, Yanshi, Henan. In: Institute of Archaeology, Chinese Academy of Social Sciences (Ed.), the Early Bronze Cultures in China: the Thematic Researches on the Erlitou Culture. Science Press, Beijing, pp. 479e539 (in Chinese).

131. Yang, D.Y., Liu, L., Chen, X., Speller, C.F., 2008. Wild or domesticated: DNA analysis of ancient water buffalo remains from north China. J. Archaeol. Sci. 35, 2778e2785.

132. Yang, X., Wan, Z., Perry, L., Lu, H., Wang, Q., Zhao, C., Li, J., Xie, F., Yu, J., Cui, T., Wang, T., Li, M., Ge, Q., 2012. Early millet use in northern China. Proc. Natl. Acad. Sci. U. S. A. 109, 3726e3730.

133. Yi, S., 2011. Holocene vegetation responses to East asian monsoonal changes in South Korea. In: Blanco, J. (Ed.), Climate Change e Geophysical Foundations and Ecological Effects. InTech, Rijeka, Croatia, pp. 157e178.

134. Yi, S., Saito, Y., Zhao, Q., Wang, P., 2003. Vegetation and climate changes in the Changjiang (Yangtze River) Delta, China, during the past 13,000 years inferred from pollen records. Quat. Sci. Rev. 22, 1501e1519.

135. Yu, G., Chen, X., Ni, J., Cheddadi, R., Guiot, J., Han, H., Harrison, S.P., Huang, C., Ke, M., Kong, Z., Li, S., Li, W., Liew, P., Liu, G., Liu, J., Liu, Q., Liu, K.-B., Prentice, I.C., Qui, W., Ren, G., Song, C., Sugita, S., Sun, X., Tang, L., Van Campo, E., Xia, Y., Xu, Q., Yan, S., Yang, X., Zhao, J., Zheng, Z., 2000. Palaeovegetation of China: a pollen data-based synthesis for the mid-Holocene and last glacial maximum. J. Biogeogr. 27, 635e664.

136. Yuan, J., 2002. The environmental Information of Ancient Animals, in: Archaeological Team of Luoyang City, Luoyang Zaojiaoshu. Science Press, Beijing, pp. 113e119 (in Chinese).

137. Yuan, J., 2003. Study on remains of aquatic and terrestrial animals. In: Institute of Archaeology, Chinese Academy of Social Sciences, Archaeological Team of Guangxi Autonomous Region, Museum of Zengpiyan

Site in Guilin, Archaeological Team of Guiling City, Zengpiyan- a Prehistoric Site in Guilin. Cultural Relics Publishing House, Beijing, China, pp. 297e341 (in Chinese).

138. Yuan, J., 2010. Zooarchaeological study on the domestic animals in ancient China. Quat. Sci. 30, 298e307 (in Chinese, with English summary).

139. Yuan, J., Li, J., 2010. Research report on faunal remains from Nanzhuangtou site in Xushui county. Hebei. Kaogu Xuebao (Acta Archaeol. Sin.) 3, 385e391 (in Chinese).

140. Yuan, J., Lv, P., Li, Z., Deng, H., Masaki, E., 2015. Restudy of the origin of domestic chicken in ancient China. Cult. Relics South. China 3, 53e57 (in Chinese, with English summary).

141. Yunnan Provincial Institute of Cultural Relics and Archaeology, Yunnan Archaeological Team of Institute of Archaeology, Chinese Academy of Social Sciences, Chengdu Institute of Cultural Relics and Archaeology, Chuxiong Museum, Yongren County Cultural Centre, 2003. Excavation of the Caiyuanzi and Mopandi Sites in Yongren, Yunnan, in 2001. Kaogu Xuebao (Acta Archaeol. Sin.) 2, 263e296 (in Chinese, with English summary).

142. Yunnan Provincial Museum, 1977. The neolithic site at Dadunzi in Yuanmou county, Yunnan Province. Kaogu Xuebao (Acta Archaeol. Sin.) 1, 43e72 (in Chinese, with English summary).

143. Zeuner, F.E., 1963. A History of Domesticated Animals. Hutchinson,, London.

144. Zhang, X., 1985. The animal's bones from Dadunzi Neolithic site in Yuanmou, Yunnan. Yunnan Relics 17, 29e34 (in Chinese).

145. Zhao, Y., Yu, Z., Chen, F.-H., An, C., 2007. Holocene vegetation and climate changes from fossil pollen records in arid and semi-arid China. In: Madsen, D., Chen, F.-H., Gao, X. (Eds.), Late Quaternary Climate Change and Human Adaptation in Arid China. Developments in Quaternary Science, vol. 9. Elsevier, Amsterdam, pp. 51e65.

146. Zhong, B., 2010. The animal's remains from Yuhuangding site in Jining City. In: Shandong Cultural Relics Institute, Haidai Archaeology, vol. 3. Science Press, Beijing, pp. 98e99 (in Chinese).

147. Zhou, B., 1981. The animal remains discovered at Cishan village, Wu'an, Hebei Province. Kaogu Xuebao (Acta Archaeol. Sin.) 3, 339e347 (in Chinese, with English summary).

148. Zhou, B., 1983a. The animal's remains from Baiying site of longshan culture in Tangyin, Henan. Pap. Chin. Archaeol. 3, 48e50 (in Chinese).

149. Zhou, B., 1983b. The Animal's Remains from Beishouling Neolithic Site in Baoji City, in: Institute of Archaeology, Chinese Academy of Social Sciences, Beishouling at Baoji. Cultural Relics Publishing House, Beijing, pp. 145e153 (in Chinese).

150. Zhou, B., 1992. Faunal remains from Nanzhuangtou site in Xushui county, Hebei Province. Kaogu Archaeol. 11, 966e967 (in Chinese).

151. Zhou, B., 1994. Identification Report on Faunal Remains from Baijiacun Site, in: Institute of Archaeology, Chinese Academy of Social Sciences, Baijiacun Site at Lintong. Bashu Press, Chengdu, pp. 123e126 (in Chinese).

152. Zhou, B., 1999. Animal Remains of Shizhaocun and Xishanping Site, in: Institute of Archaeology, Chinese Academy of Social Sciences, Shizhaocun and Xishanping. The Encyclopedia of China Publishing House, Beijing, pp. 335e339 (in Chinese).

153. Zhou, B., 2000. The animal's remains from Wangyin neolithic site in Yunzhou city, Shandong. In: Excavations of the Wangyin Site in Shandong. Science Press, Beijing, pp. 414e416 (in Chinese).

154. Zhou, W., An, Z., Dull, A.J.T., Donahue, D.J., Heads, M.J., 1998. Reappraisal of Chinese Loess Plateau stratigraphic sequences over the last 30,000 years: Precursors of an important Holocene climatic event. In: Mook, W.G., Plicht, J., van der (Eds.), Proc. 16th Int. 14C Conference. Radiocarbon, vol. 40, pp. 905e913.

CHAPTER 10

USING PALYNOLOGY TO RE-ASSESS THE DEAD SEA LAMINATED SEDIMENTS – INDEED VARVES?

Lourdes López-Merino, Suzanne A.G. Leroy, Amram Eshel, Valentina Epshtein, Reuven Belmaker, Revital Bookman

Keywords: Dead Sea, Laminated sediments, Aragonite, Air-borne pollen, Reworked pollen, Flash-flood events

INTRODUCTION

Fine-laminated lacustrine sequences have commonly proven to be annually deposited. Thus, varve-based chronologies of these sequences can be obtained (e.g. Ojala and Alenius, 2005 and Zolitschka et al., 2015). Large portions of the Dead Sea Basin (DSB) Late Quaternary sediments are laminated (Neev and Emery, 1967 and Begin et al., 1974), i.e. the Lisan Formation (70–13 ka BP; Stein and Goldstein, 2006 and Torfstein et al., 2013) and the Ze›elim Formation (<10 ka BP; Migowski et al., 2006). These laminated sections consist of mm-scale alternating detrital and authigenic aragonite laminae. Based on age-depth models and lamina counting, these laminae were assumed to be varves in most studies, i.e. rainy season-detrital *versus* summer-aragonite deposition (Neev and Emery, 1967, Begin et al., 1974, Heim et al., 1997, Migowski et al., 2004, Prasad et al., 2004, Neumann et al., 2009, Leroy et al., 2010 and Neugebauer et al., 2015). However, the exact seasonal character of the Dead Sea laminae has not been confirmed in a robust manner. This is of extreme importance for the accurate use of the DSB laminated sediments as palaeoenvironmental and palaeoclimate archives. Therefore, the aim of this paper is to re-address the nature of these laminated sediments in order to aid accurate interpretations of environmental change in the region.

Dead Sea detrital laminae are composed of a mixture of regional dust inputs and local run-off erosion products from the catchment area (Belmaker et al., 2011 and Haliva-Cohen et al., 2012). In numerous lakes, carbonate deposition is closely related to biological activity (e.g. Thompson et al., 1997 and Salmaso and Decet, 1998). In contrast, carbonate deposition in the Dead Sea is inorganic and results from the interaction between freshwater run-off and Dead Sea hypersaline brine (Katz and Kolodny, 1989 and Stein et al., 1997). The primary origin of aragonite from the diluted upper water mass is confirmed by the excellent state of preservation of the crystals and their concentration within specific layers in the laminated sequences. This supports non-overlapping times of deposition between aragonite and detrital laminae (Heim et al., 1997 and Stein et al., 1997). The commonly presumed season for aragonite precipitation is summer. The trigger is attributed to evaporation and warming of the high bicarbonate surface waters that entered via run-off during the wet season ("whitening" events) (Neev and Emery, 1967 and Stein et al., 1997). On the other hand, Barkan et al. (2001)measured carbonate system parameters in the upper water mass that formed after the heavy flooding during the extreme winter of 1992 and showed that, at least in the modern Dead Sea, aragonite precipitation occurs just after the massive input of TCO_2 during the wet season. Research so far, on geochemical and palaeolimnological parameters, has not established the exact timing of the aragonite deposition and its relation to detrital input events.

Based on the observations of Barkan et al. (2001), we hypothesise that aragonite may have not been deposited during summer as is commonly interpreted, but instead the detrital-aragonite couplets may represent flash-flood events rather than an annual cycle. Thus, laminated sequences could be formed by flash-flood events delivering sediments into the

Citation: Lourdes López-Merino, Suzanne A.G. Leroy, Amram Eshel, Valentina Epshtein, Reuven Belmaker, Revital Bookman, Using palynology to re-assess the Dead Sea laminated sediments – Indeed varves?, Quaternary Science Reviews, Volume 140, 15 May 2016, Pages 49-66, ISSN 0277-3791, http://dx.doi.org/10.1016/j.quascirev.2016.03.024.

lake followed by aragonite deposition in a climate-controlled lacustrine environment, i.e. precipitation in the drainage basin. To examine this hypothesis we performed grain-size and palynological analyses of detrital-aragonite couplets from two well-dated high-stand laminated sequences in the Ze›elim Formation: the Hellenistic-early Roman and the late 19th – early 20th centuries (Bookman (Ken-Tor) et al., 2004). Grain-size distribution provides information on sediment source (i.e. dust or watershed erosion), while palynology provides information on both seasonality (well-preserved, air-borne pollen) and sediment transport (reworked, water-borne pollen). In addition, palynological analysis was carried out on fine mud deposits collected immediately after modern flash-flood events, assuming that these deposits represent flood suspended matter.

MATERIALS AND METHODS

Study Area and Selected Laminated Sediments

The Dead Sea (Fig. 1A) is a closed, inland hypersaline lake (e.g. Niemi and Ben-Avraham, 1997) and the descendant of the larger Late Pleistocene Lake Lisan (e.g. Bookman et al., 2006 and Stein, 2014). Its main fresh water tributaries are the Jordan River and flash-floods that flow west from the Jordanian Mountains and east from the Judean Mountains. Long-term fluctuations of the Dead Sea lake level are caused by rainfall fluctuations over the watershed (Enzel et al., 2003 and Bookman et al., 2004). Rain occurs between autumn and spring and it can be either spatially localised or widespread (Dayan and Sharon, 1980). The main synoptic conditions triggering rain and dust transport to the Dead Sea area are the east Mediterranean cyclones and the Red Sea trough. The former is responsible of most winter rain and dust transport, while the latter is more related to autumn and spring dust storms (Ganor and Foner, 1996 and Dayan et al., 2007). Because the Dead Sea is located on the border between semiarid and arid climates, rainfall varies seasonally and annually and is often concentrated in intense showers that cause flash-flood events and erosion (Dayan and Morin, 2006 and Greenbaum et al., 2006).

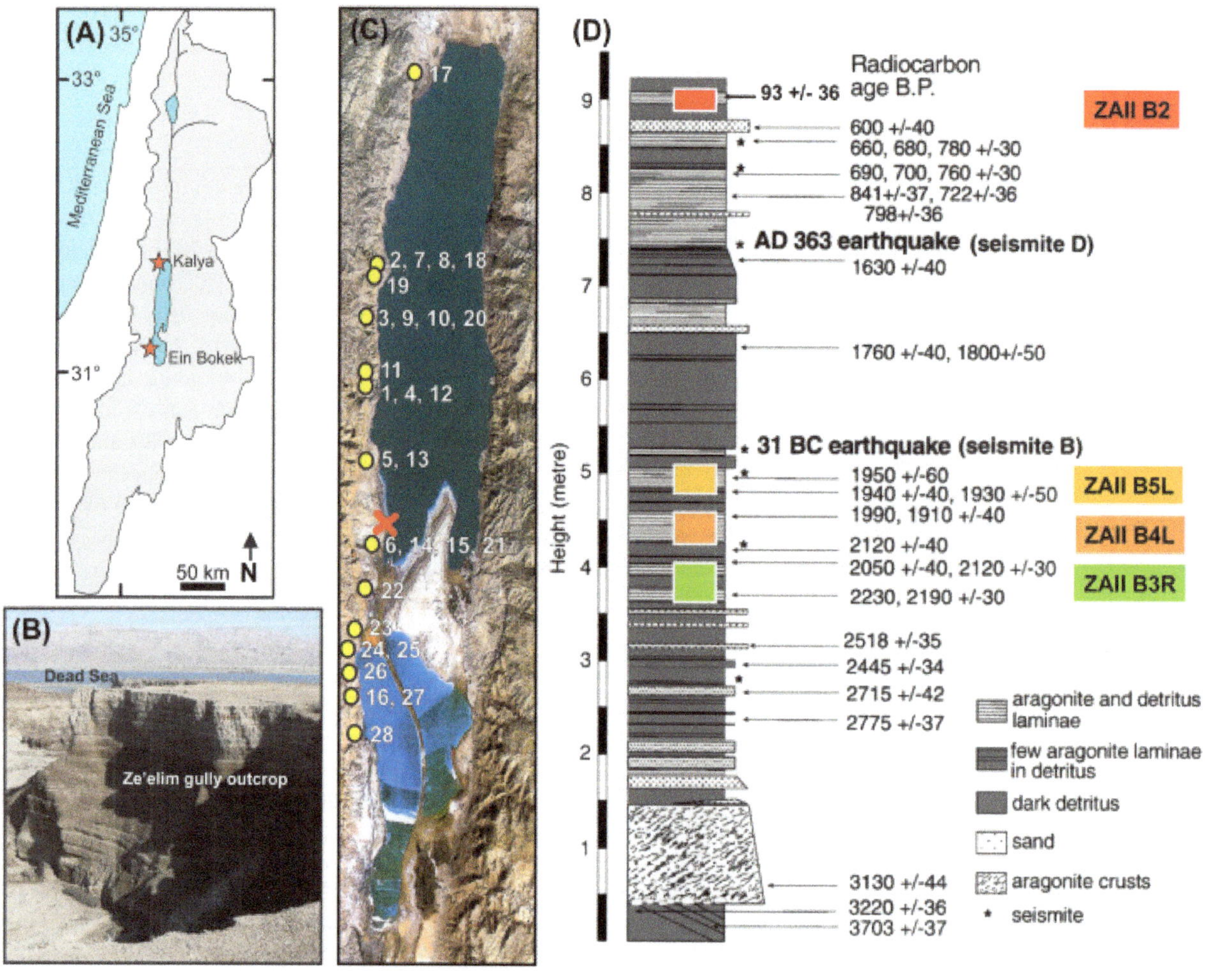

Figure 1: (A) Location map of the Dead Sea and its watershed. Red stars indicate the location of the two aeropalynological stations. (B) Photo of the Ze›elim gully outcrop sampled in this study (Photo by L. López-Merino). (C) Location map of the laminated sediment blocks at Ze›elim and flash-flood sampling sites. Red cross indicates the location of the Ze›elim outcrop. Yellow circles indicate flash-flood sampling locations (D) Sedimentary scheme of the Ze›elim outcrop profile and radiocarbon chronology (modified from Bookman (Ken-Tor) et al., 2004). ZA11B2 sediment block represent the late 19th – early 20th centuries high-stand. ZA11B3R, ZA11B4L and ZA11B5L sediment blocks represent the Hellenistic-early Roman high-stand. Aragonite crusts were deposited in a coastal to terrestrial environment, while the aragonite laminae (discussed in this study) were deposited in a lacustrine environment. (For interpretation of the references to colour in this figure legend, the reader is referred to the web version of this article.)

The Ze›elim Formation consists of the lacustrine Holocene deposits in the DSB (Fig. 1B). Extensive outcrops of the formation were described at the Ze›elim Plain, one of the largest fan deltas along the western margin of the Dead Sea (Fig. 1C). Outcrops of the Ze›elim Formation were used for palaeoclimate and palaeoseismicity reconstructions (Ken-Tor et al., 2001a, Ken-Tor et al., 2001b, Bookman et al., 2004 and Kagan et al., 2011). The reconstructed lake-level curve revealed two high-stand periods characterised by relatively continuous stable high water-levels of at least a few decades. These periods of high-stand enabled the formation of alternating detrital and aragonite laminae sequences. The two periods correspond to the Hellenistic-early Roman and the late 19th – early 20th centuries high-stands (Bookman (Ken-Tor) et al., 2004; Fig. 1D). In order to eliminate the long-term climatic influence on the pollen record and to have regularly laminated sediments, these two relatively stable high-stand periods were chosen to be the focus of this study.

Air-Borne Pollen Calendars

Although several aerobiological studies are available for the Israel-Jordan area, most of them do not provide data regarding the Dead Sea region. Thus, we focussed on pollen calendars prepared in areas under semiarid to arid conditions close to the Dead Sea. Surveys of allergenic airborne pollen data collected on the roof of Tsell Harim hotel (Ein Bokek) at the western coast of the southern Dead Sea and on the roof of the Megilot Regional Council building (Kalya) at the northern part of the Dead Sea (Fig. 1A; Waisel, unpublished report 1 and Waisel, unpublished report 2) were chosen for comparison purposes, as the Ze›elim Plain is located in between and shares semi-arid features with both. Additionally, a flowering calendar was composed using the information contained in the *Handbook of Wildflowers of Israel* (Shmida and Darom, 2000 and Shmida and Darom, 2002), paying special attention to the desert and steppe flora, along with the more frequent Mediterranean types in the pollen data.

Reported Ein Bokek and Kalya air-borne data present low total pollen concentrations during the entire year although with a well-defined spring peak. A dominance of Amaranthaceae pollen during the whole year is obvious. Poaceae pollen is also present throughout the year, peaking in spring (March–June). Another characteristic pollen type with large abundance is *Olea*, recorded in April–May at Kalya and in April–May (-June) at Ein Bokek. *Pinus* peaks in March–April in both stations. *Plantago*, another spring bloomer, peaks in March at Kalya and in April at Ein Bokek. An interesting difference between Kalya and Ein Bokek is related to *Artemisia*, which is abundant in Kalya in October–November but almost absent in Ein Bokek, likely because *Artemisia* is a steppe plant and although both Ein Bokek and Kalya are located in the desert, the latter is closer to the steppe area.

Other interesting pollen types that give information on specific flowering periods are infrequent or absent in the aerobiology studies performed in stations close to the Dead Sea. However, they are common in the palaeopalynological studies on Dead Sea Holocene sediments (Baruch, 1993, Heim et al., 1997, Neumann et al., 2007, Neumann et al., 2009, Neumann et al., 2010, Leroy, 2010, Leroy et al., 2010, Litt et al., 2012, Langgut et al., 2014 and Langgut et al., 2015a). Brassicaceae and Asteraceae have been identified in March–April in Arad, a town located in the southern Judean Desert (Kantor et al., 1966). Additionally, Brassicaceae species start their flowering period in February in Jerusalem, while Asteraceae species extend their blossom up to May and a few species also bloom in autumn (September–December) (Feinbrun et al., 1959). In fact, the number of Asteraceae and Brassicaceae species is large in Israel, and although many bloom in spring, some of them also flower in autumn. Arboreal pollen types are representative of long-distance transport to the Dead Sea region. Nonetheless, they are equally informative. *Quercus* is recorded during spring, in March–April mainly (Keynan et al., 1989). Finally, *Pinus* blooming peaks during spring (March–April) as well (Feinbrun et al., 1959).

The use of flowering calendars and palynological analysis to resolve the season in which an event occurred, i.e. a historical earthquake at central Israel (Langgut et al., 2015b), has proven to be reliable in a similar Dead Sea climate context.

Sampling

The level of the modern Dead Sea is the result of a large human-induced retreat linked to the damming of its northern tributary (Sea of Galilee) and water pumping for industrial purposes. This retreat has triggered the formation of several meter deep gullies in the Ze'elim Plain (Ben Moshe et al., 2008). Late Holocene laminated sequences are exposed in these gullies (Fig. 1B). The two aforementioned high-stand periods were identified in the radiocarbon-dated outcrop (Bookman (Ken-Tor) et al., 2004; Fig. 1D) in February 2011. One block of laminated sediments from the late 19th – early 20th centuries high-stand (ZA11B2), and three blocks of laminated sediments from the Hellenistic-early Roman high-stand (ZA11B3R, ZA11B4L, ZA11B5L) were collected (Fig. 1D). Sampling of individual, mm-scale laminae was done at the sedimentology laboratory of the University of Haifa using a scalpel. Only consecutive detrital and aragonite laminae with no suspicion of contamination or mixing with the upper and lower layers were considered. Overall, 65 detrital-aragonite couplets (130 samples) were analysed: 13 in block ZA11B2 (couplets 1 to 13), 11 in block ZA11B3R (couplets 14 to 24), 17 in block ZA11B4L (couplets 25 to 41) and 24 in block ZA11B5L (couplets 42 to 65).

Additionally, fine mud carried in suspension in flash-floods that occurred between February 2009 and June 2012, was collected from drying puddles immediately after the events. A total of 28 flash-flood samples were analysed (Fig. 1C).

Grain-Size Distribution Analysis

Grain-size analysis of the Ze'elim detrital laminae was performed at the University of Haifa using a Beckman-Coulter LS 230 laser particle size analyser over the particle size range of 0.02–2000 μm. Five replicate samples were measured for each detrital lamina. Grain-size was determined after dissolution of carbonate minerals. In block ZA11B4L and in a few samples of block ZA11B2 it was not determined due to low sample mass available. Additionally, for two thicker detrital laminae, grain-size distribution was measured in sub-samples from the top and bottom parts in order to identify settling patterns.

Palynological Analysis

Palynological extraction was carried out at Brunel University London. Around 0.5–4 ml (usually ~2 ml) of detrital material (detrital laminae and flash-flood samples) and 0.5–12 ml (usually ~5 ml) of aragonite laminae were deflocculated with a $Na_4O_7P_2$ solution (10%). Carbonate dissolution was done with concentrated HCl (35%). Elimination of silicates was obtained by HF (48%) followed by HCl. The residue was sieved through 125 and 10 μm nylon meshes and mounted on slides with glycerol. Concentration estimates (number of palynomorphs/ml of sediment) were calculated based on the addition of *Lycopodium* tablets at the beginning of the chemical treatment (Stockmarr, 1971). Palynological identification and counting was completed on light microscopes at × 400, and at × 1000 using immersion oil for more delicate identifications, and complemented by the Brunel University London pollen reference collection and atlases (Reille, 1992, Reille, 1995 and Reille, 1998). Where possible, a minimum of 200 well-preserved (air-borne), non-reworked, pollen grains were counted per sample (average = 242). Palynological diagrams were elaborated using Psimpoll 4.27 (Bennett, 2009).

As pollen found in southern Dead Sea sediments could be transported by both wind and water run-off (Horowitz et al., 1975 and Baruch, 1993), it is necessary to be sure about the air-borne component to be able to detect the timing of laminae deposition. Hence, reworked pollen grains (water-borne) were counted separately form well-preserved (air-borne) pollen grains and classified into different categories: broken, corroded, crumpled, degraded and other reworked (when two or more features were present). Broken stamina were grouped with the reworked material. On average, 618 reworked pollen grains were counted per sample. Fungal spores had a significant presence and thus were also quantified. On average, 195 fungal spores were counted per sample. Full counts of reworked pollen grains and fungal spores.

The concentration of Total well-preserved pollen grains, Total reworked pollen grains and Total fungal spores were calculated separately. The first includes the well-preserved pollen grains only (number of well-preserved pollen grains per ml of sediment), the second one includes the reworked pollen grains only (number of reworked pollen grains per ml of sediment), and the third includes the fungal spores only (number of fungal spores per mil of sediment).

Numerical Analysis

In order to unravel the seasons in which the laminae could have been deposited, Principal Components Analysis (PCA) was applied to the well-preserved palynological dataset. This approach has been confirmed as valid for detecting seasonality in palaeoenvironmental studies (i.e. Festi et al., 2015). In this case, the PCA has been performed on the transposed well-preserved pollen data matrices (samples in columns as variables, and taxa in rows as cases). This approach enables us to summarise the main palynological assemblages of co-existing taxa and their importance in each sample. Thus, the palynological composition of the samples can be compared based on co-variation patterns (López-Merino et al., 2012). Taxa showing large factor scores (i.e. larger abundances) in a given principal component explain most of the variation of the palynological dataset in samples with large factor loadings. This enables quantitative expression of the proportion of variance explained by each principal component for each sample (López-Merino et al., 2012).

Baruch (1993) presented the palynological analysis of 35 surface samples distributed along four vegetation zones of Israel: Mediterranean, transitional and steppe, desert, and the Dead Sea shore. The Mediterranean zone was dominated by arboreal taxa (*Quercus calliprinos*, *Pinus* and *Sarcopoterium spinosum*), the transitional zoned was characterised by *Sarcopoterium spinosum*, Brassicaceae and *Artemisia*, while the desert area was dominated by Brassicaceae and Asteraceae. In the pollen rain in the Dead Sea shore, Amaranthaceae pollen was overrepresented. Therefore, in order to reduce background noise and detect the main characteristic palynological associations, Amaranthaceae pollen was excluded from the analysis. Many species belonging to this pollen type bloom all yearlong and very locally. Therefore, its pollen is overrepresented and dominates the pollen assemblages regardless the season (Baruch, 1993, Waisel, unpublished report 1 and Waisel, unpublished report 2). After the exclusion of Amaranthaceae and prior to the statistical analysis data-set proportions were recalculated. Correlation matrices and varimax rotation solutions were applied to constrain the co-variation in the components. PCA was done using the IBM SPSS statistics 20 software.

RESULTS

Grain-Size

The siliciclastic fraction of the detrital laminae showed at least three different grain-size modes. Most samples had two grain-size modes, a first fine mode of 3–10 μm and a second coarser mode of 20–90 μm (Fig. 2). A ~10 μm mode was obtained on samples from collectors installed on a buoy on the Dead Sea surface during three years (1997–1999) (Singer et al., 2003). This grain-size distribution represents dust deposited over the Dead Sea similar to that of long-range transported Harmattan dust (Stahr et al., 1994) and is comparable to the grain-size distribution of air-borne dust in the Sde Boker area (Negev) (Offer et al., 1992). Thus, the fine grain-size mode in the detrital laminae siliciclastic fraction is consistent with fine dust largely wind-transported from medium to long range. Although a bimodal distribution is in accordance with Haliva-Cohen et al. (2012), the average grain-size measured for the second mode in this study is coarser, possibly due to the carbonate dissolution stage. Few samples included a minor third mode of 100–200 μm (Fig. 2) consistent with the contribution of loess (Haliva-Cohen et al., 2012), likely representing a local source.

The general grain-size distribution measured from detrital laminae suggests it is composed of recycling of local and distal sources as demonstrated in previous studies (e.g. Belmaker et al., 2014). Additionally, grain-size distribution within thicker detrital laminae shows graded bedding that suggests sediment deposition during flash-flood events. This phenomenon was shown with upward fining of the grain-size distribution in specific laminae (Fig. 2D) and was also described in petrographic thin-slides (Haliva-Cohen et al., 2012). The understanding that the detrital laminae consist of both air-borne particles and recycled local (mostly Quaternary sequences) material lead to the conclusion that the palynological analysis should separate air-borne from reworked pollen. This approach is practiced in the Dead Sea sediments at a very fine detail here for the first time.

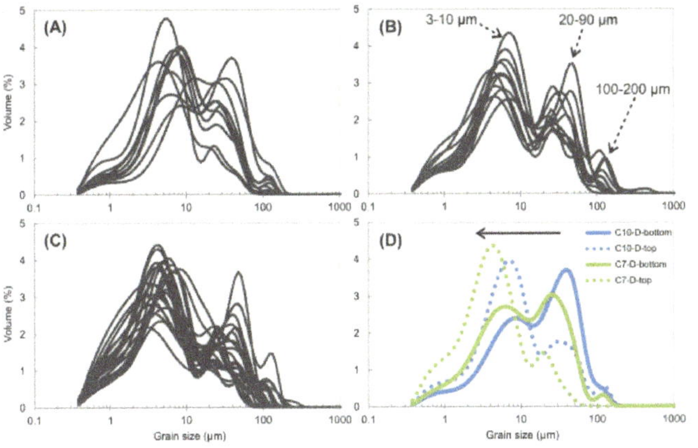

Figure 2: Grain-size distribution in detrital samples from blocks ZA11B2 (A), ZA11B3R (B) and ZA11B5L (C). Grain-size distribution of detrital laminae C7 and C10 from block ZA11B2 showing upward fining pointing to graded bedding during flash-flood events are also shown (D).

Palynology

Two prerequisites are necessary to validate the timing of laminae deposition: (i) detecting features indicative of seasons and (ii) identifying features suggestive of a link between detrital laminae and flash-flood events. These prerequisites were achieved by counting separately poorly-preserved pollen grains (reworked), mainly indicative of water transport and reworking from soils, from well-preserved pollen grains, mostly derived from air-borne transport, hence providing blooming period information.

Air-Borne: Detecting Seasonal Features

Well-preserved pollen grains were present in both flash-flood samples (Fig. 3) and detrital-aragonite couplets (Fig. 4, Fig. 5, Fig. 6 and Fig. 7). Although the number of pollen types identified is high (81 types in the flash-floods, 67 types in ZA11B2, 65 types in ZA11B3R, 70 types in ZA11B4L, and 73 types in ZA11B5L), few pollen types dominate the pollen assemblages. As expected due to the location of the Ze›elim Wadi and fan delta under semi-arid conditions, one of the main pollen types in both detrital and aragonite layers is Amaranthaceae (Baruch, 1993). Other abundant pollen types are Asteraceae liguliflorae, Asteraceae tubuliflorae, *Artemisia*, Brassicaceae, *Quercus calliprinos* t., *Pinus*, *Olea* and Poaceae (Fig. 4, Fig. 5, Fig. 6 and Fig. 7). Pollen assemblages from recent flash-flood samples show the current highly anthropised landscape with low presence of long-distance Mediterranean arboreal pollen types such as *Quercus* and *Olea*, as well as larger percentages of *Pinus* pollen due to anthropogenic afforestation (Fig. 3). Hence flash-flood samples have not been included in the statistical analysis. A preliminary PCA included samples from both high-stand periods (Hellenistic-early Roman and late 19th – early 20th centuries). PCA results separated the two periods on the basis of *Olea* (ZA11B2 *versus* ZA11B3R, ZA11B4L, ZA11B5L, data not shown). This is consistent with the intense olive tree cultivation during Hellenistic-early Roman times, a period dominated by intense arboriculture, which has no equivalent in the present (Baruch, 1993, Neumann et al., 2007, Neumann et al., 2010 and Leroy, 2010). Therefore, in order not to conceal differences among detrital and aragonite laminae due to the impact of the higher presence of *Olea* during the older high-stand period, the two high-stand periods were explored separately.

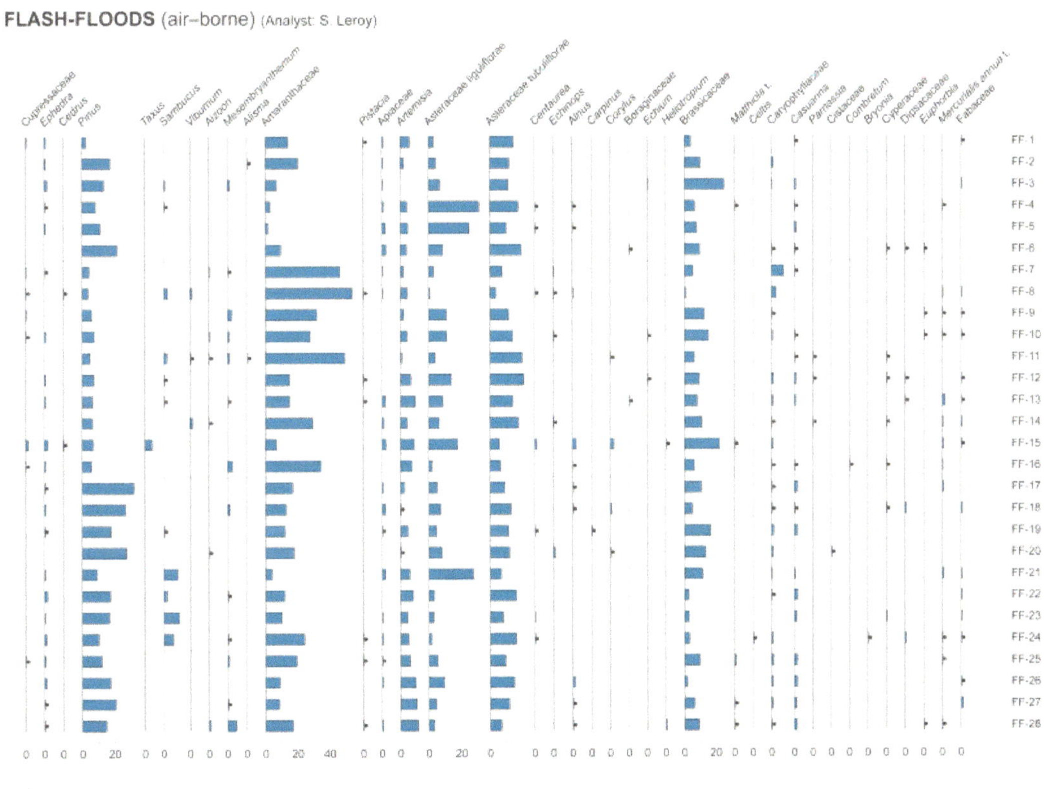

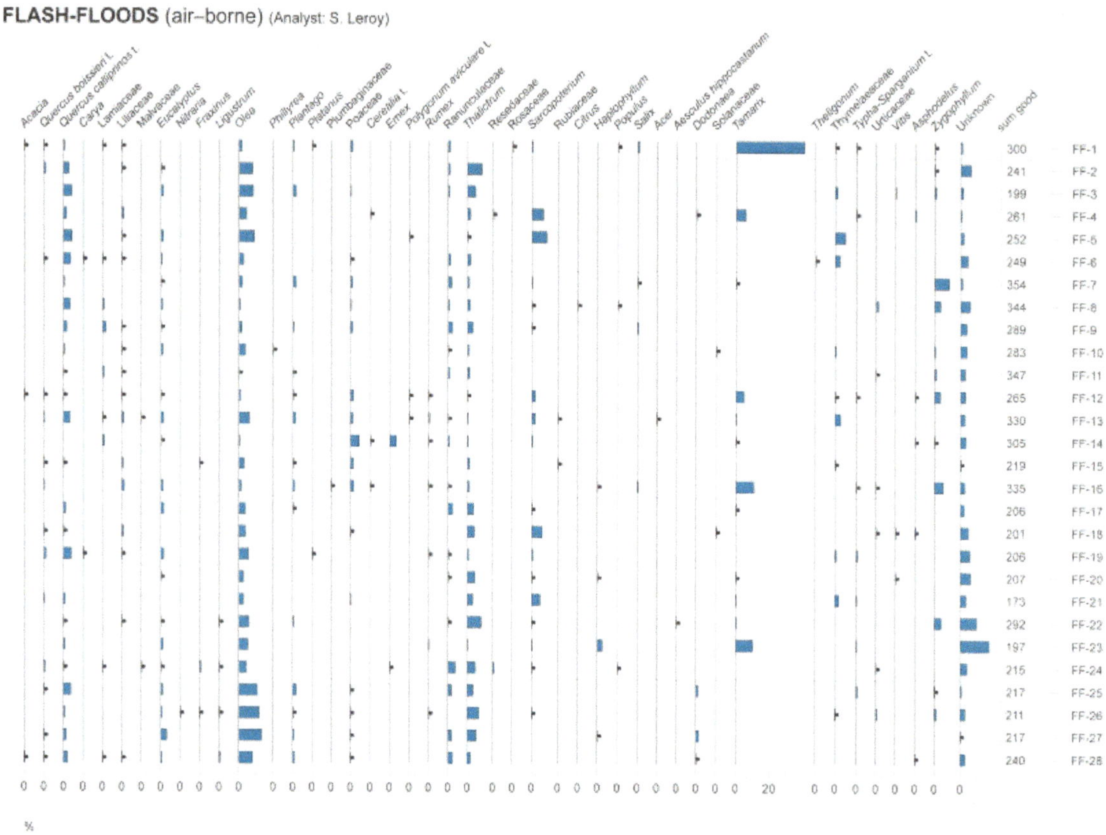

Figure 3: Percentage diagram of well-preserved pollen (air-borne) in modern flash-flood samples. Black dots represent percentages below 0.5%. Flash-flood samples are grouped by months and ordered from North to South. Details on flash-flood samples.

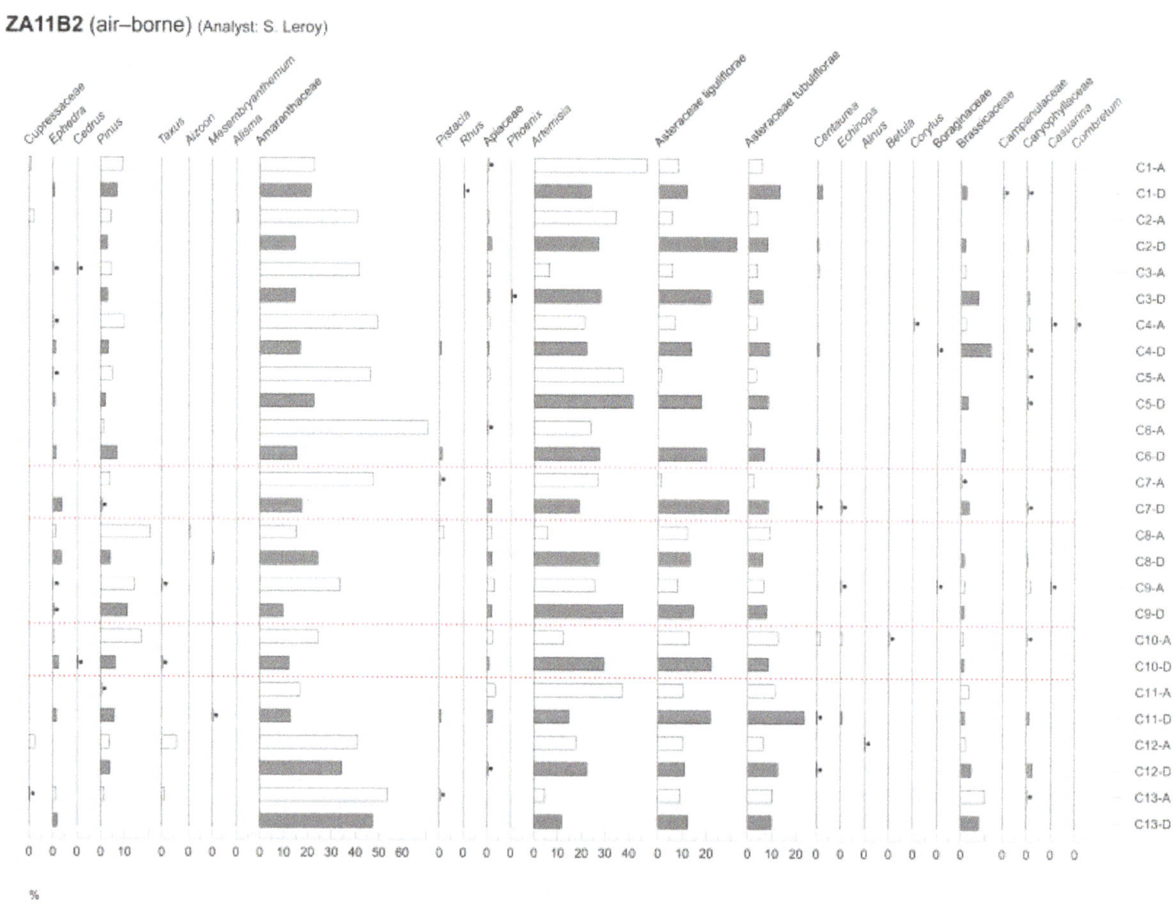

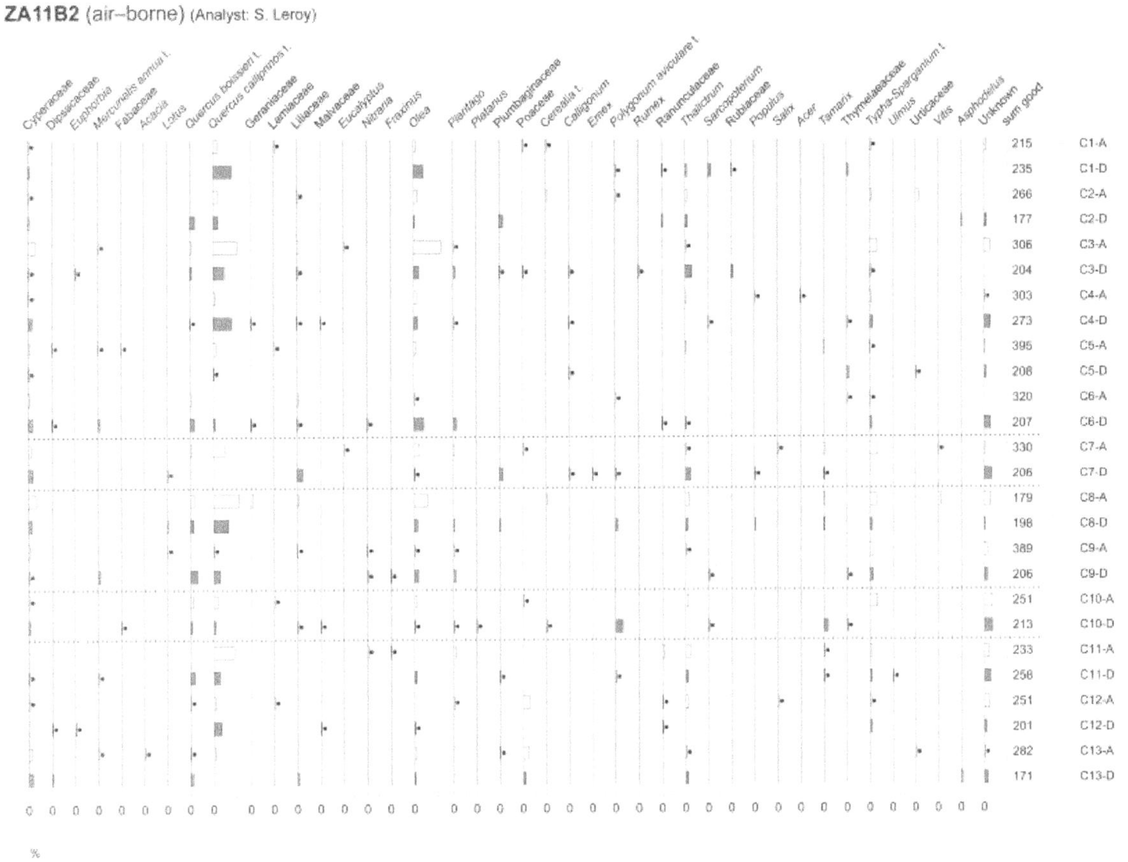

Figure 4: Percentage diagram of well-preserved pollen (air-borne) of block ZA11B2 (late 19th – early 20th century high-stand). Black dots represent percentages below 0.5%. Red lines separate non-contiguous samples. Detrital laminae are in grey and aragonite laminae in white. Numbers of the detrital-aragonite couplets. (For interpretation of the references to colour in this figure legend, the reader is referred to the web version of this article.)

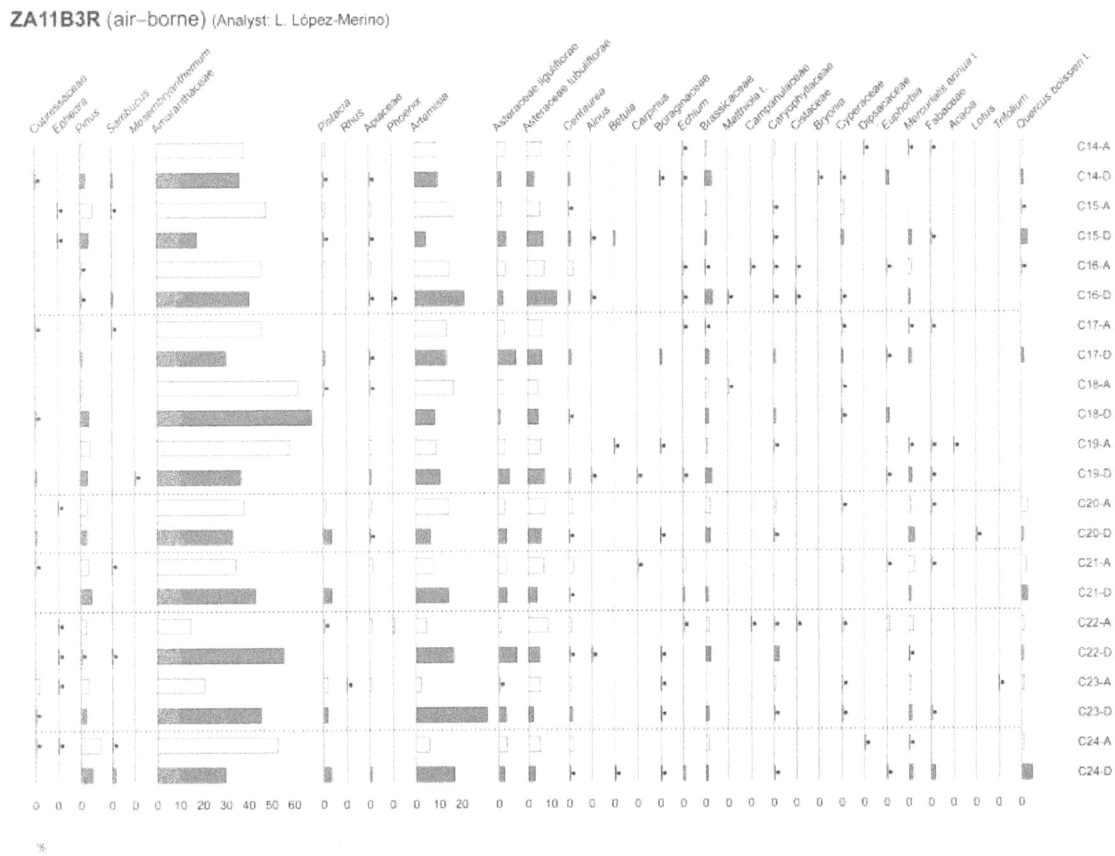

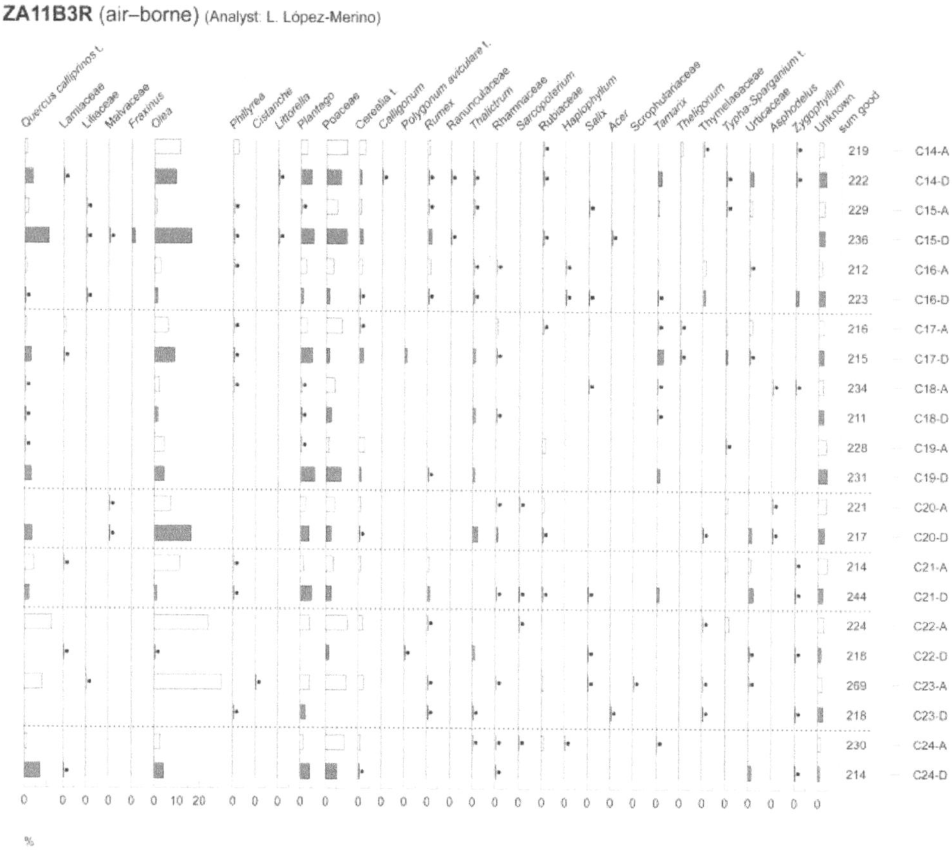

Figure 5: Percentage diagram of well-preserved pollen (air-borne) of block ZA11B3R (Hellenistic-early Roman high-stand). Black dots represent percentages below 0.5%. Red lines separate non-contiguous samples. Detrital laminae are in grey and aragonite laminae in white. Numbers of the detrital-aragonite couplets. (For interpretation of the references to colour in this figure legend, the reader is referred to the web version of this article.)

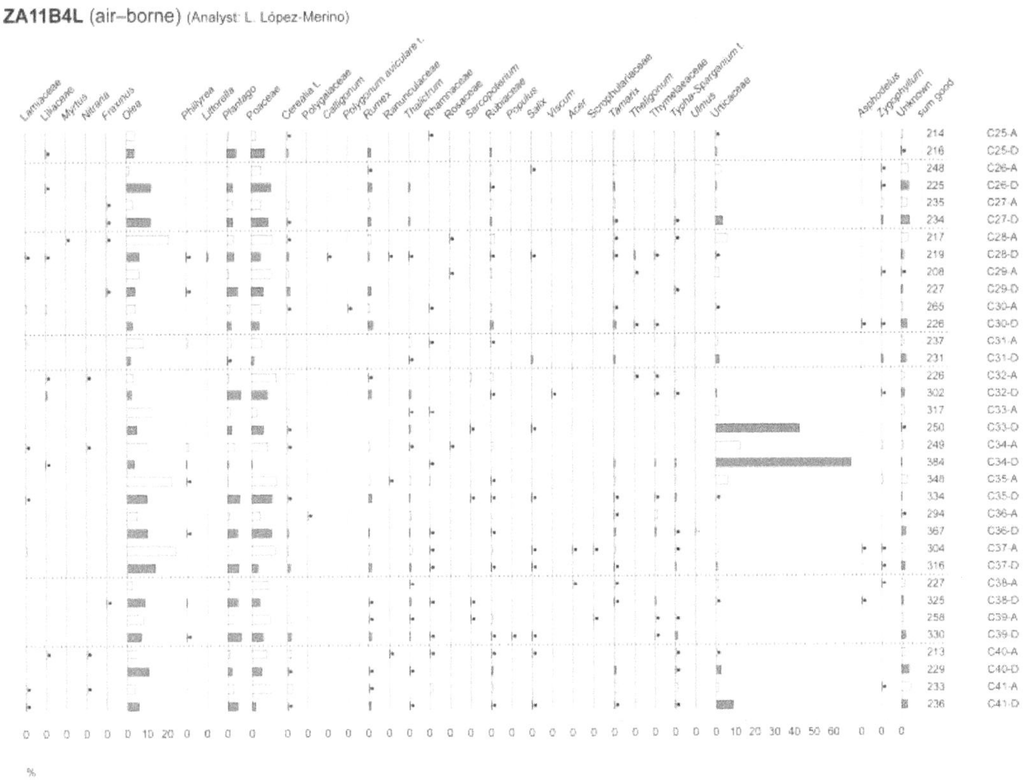

Figure 6: Percentage diagram of well-preserved pollen (air-borne) of block ZA11B4L (Hellenistic-early Roman high-stand). Black dots represent percentages below 0.5%. Red lines separate non-contiguous samples. Detrital laminae are in grey and aragonite laminae in white. Numbers of the detrital-aragonite couplets. (For interpretation of the references to colour in this figure legend, the reader is referred to the web version of this article.)

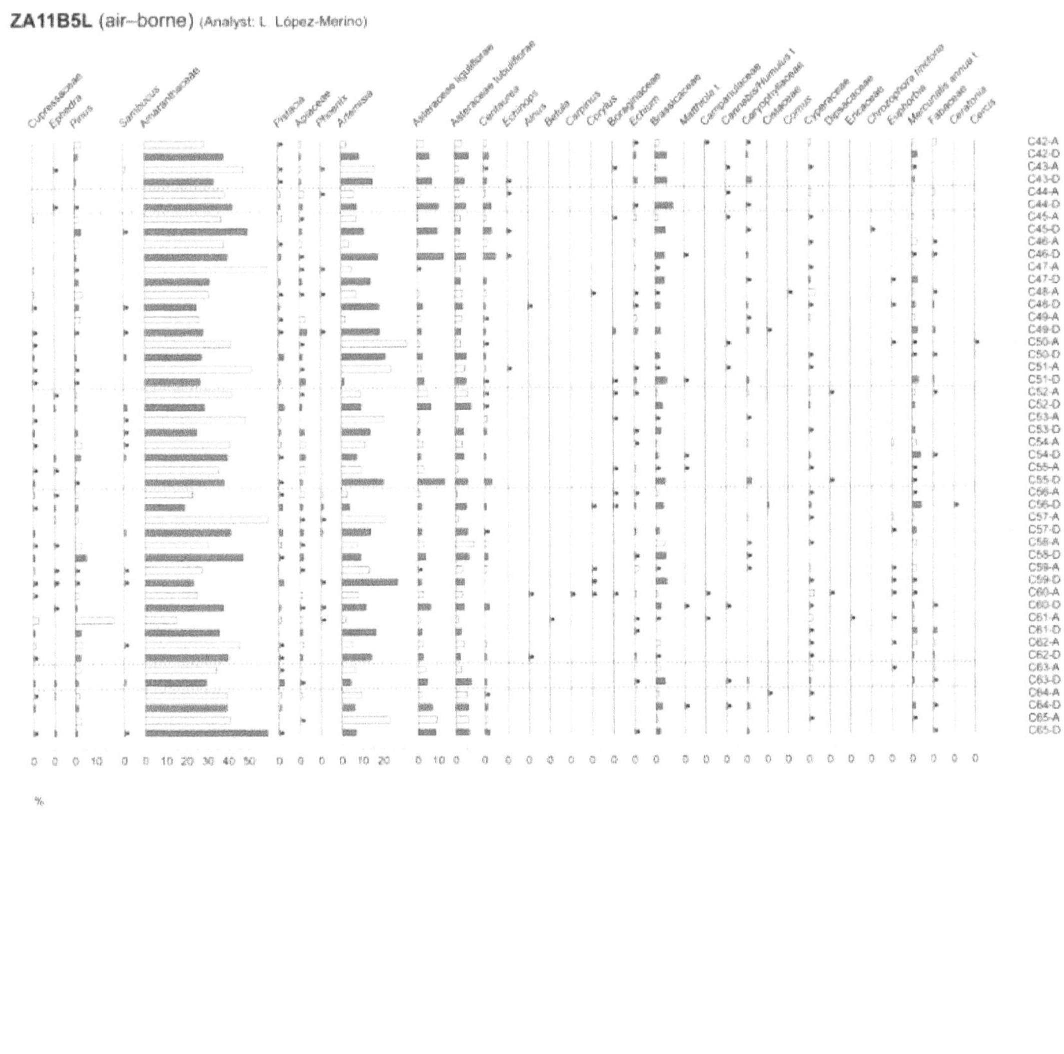

Figure 7: Percentage diagram of well-preserved pollen (air-borne) of block ZA11B5L (Hellenistic-early Roman high-stand). Black dots represent percentages below 0.5%. Red lines separate non-contiguous samples. Detrital laminae are in grey and aragonite laminae in white. Numbers of the detrital-aragonite couplets. (For interpretation of the references to colour in this figure legend, the reader is referred to the web version of this article.)

The PCA of the late 19th – early 20th centuries high-stand transposed data matrix resulted in three principal components explaining 93.6% of the total variance in the dataset (Fig. 8). $PC1_{Recent}$ explains 46.9% of the variance, with *Artemisia* commanding the largest positive factor score (Fig. 8). *Artemisia* species in the steppe area bloom in autumn, particularly in September–October (–November). Hence, $PC1_{Recent}$ seems to be a strong autumn indicator. $PC2_{Recent}$ explains 32.3% of the variance. Asteraceae liguliflorae, Asteraceae tubuliflorae and Brassicaceae present large positive factor scores. *Artemisia* also has a positive score, although moderate. On the other hand, *Pinus*, *Olea*, *Quercus calliprinos* t. and *Typha-Sparganium* t. have negative scores (Fig. 8). Asteraceae and Brassicaceae species are abundant in Israel. Although many of them bloom in spring, some bloom all year-long particularly in autumn-winter. The observation that for $PC2_{Recent}$ spring (*Pinus*, *Olea*, *Quercus calliprinos* t.) and late spring-early summer (*Typha-Sparganium* t.) bloomers present negative scores, while the autumn indicator *Artemisia* has a positive score, could be indicative of this principal component as an autumn indicator as well. This is perhaps more related to the direction of winds. Horowitz et al. (1975) pointed out that the origin and routes of winds (e.g. due to dust and rain storms) are an important factor in the pollen provenance. As mentioned above for the surveys of allergenic airborne pollen data at Ein Bokek (southern Dead Sea) and Kalya (northern Dead Sea) (Fig. 1), *Artemisia* is important in autumn at Kalya only, representing the *Artemisia*-dominated steppe ($PC1_{Recent}$). However, *Artemisia* is almost absent at Ein Bokek, representing a more desert-like vegetation like the assemblage separated by $PC2_{Recent}$. $PC3_{Recent}$ explains 14.4% of the variance. *Pinus*, *Quercus calliprinos* t. and *Olea* present large positive factor scores, while Asteraceae tubuliflorae, Asteraceae liguliflorae and *Typha-Sparganium* t. have more moderate scores (Fig. 8). The flowering period of *Pinus*, *Quercus calliprinos* t. and *Olea* extends from March to May. *Typha-Sparganium* t. flowers in June–July, while most Asteraceae species have their peak period of pollen release from March to June. Thus, $PC3_{Recent}$ is a strong indicator for spring.

The PCA of the Hellenistic-early Roman high-stand transposed data matrix resulted in five principal components explaining 92.0% of the total variance (Fig. 9). $PC1_{Hellenistc}$ explains 40.8% of the variance, with *Artemisia* accounting for the largest positive factor score (Fig. 9), likely indicating that samples dominated by $PC1_{Hellenistic}$ have an autumn-dominated pollen assemblage as indicated by $PC1_{Recent}$ for the 19th – early 20th centuries sequence. $PC2_{Roman}$ explains 33.3% of the variance, with *Olea* presenting the largest positive factor score, whereas Poaceae, *Quercus calliprinos* t. and *Plantago* show more moderate scores (Fig. 9). The species included within these pollen types are spring bloomers. Thus, $PC2_{Hellenistic}$ seems to indicate spring in assemblages with large percentages of olive tree, i.e. olive tree crops. $PC3_{Hellenistic}$ accounts for 9.6% of the variance. Asteraceae liguliflorae, Asteraceae tubuliflorae, Brassicaceae and *Centaurea* present large positive factor scores, while Apiaceae, *Artemisia* and *Tamarix* have more moderate positive scores. Poaceae and *Quercus calliprinos* t. present moderate negative scores (Fig. 9). Similar to $PC2_{Recent}$, this palynological assemblage could be related to autumn. $PC4_{Hellenistic}$ explains 5.2% of the total variance, with *Pinus*, Poaceae and *Quercus calliprinos* t. showing large positive factor scores and Asteraceae tubuliflorae, Apiaceae and *Quercus boissieri* t. more moderate positive scores. On the other hand, *Olea* has a negative factor score (Fig. 9). Similar to $PC2_{Hellenistic}$, $PC4_{Hellenistic}$ likely reflects spring, although with an assemblage without olive tree crops. Finally, $PC5_{Hellenistic}$, which accounts for the 3.1% of the total variance only, is directed by Urticaceae (Fig. 9). This principal component, indicative of spring, is important in few samples only.

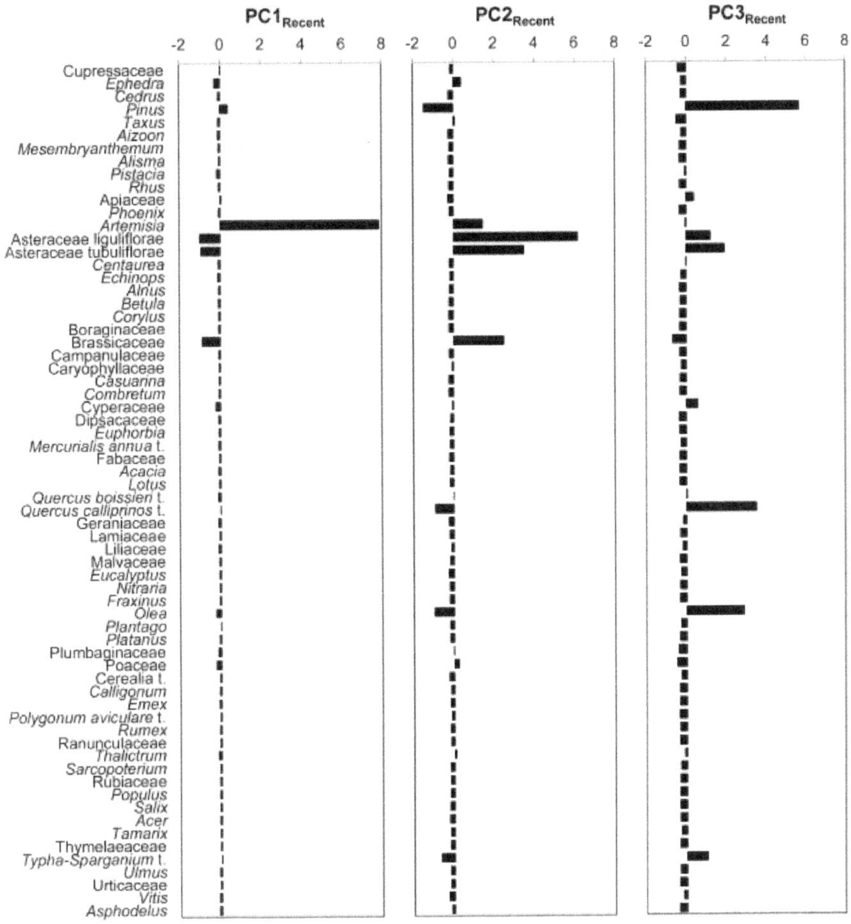

Figure 8: Factor scores of the three principal components (transposed matrix) obtained for the air-borne component of the late 19th – early 20th centuries high-stand samples (block ZA11B2).

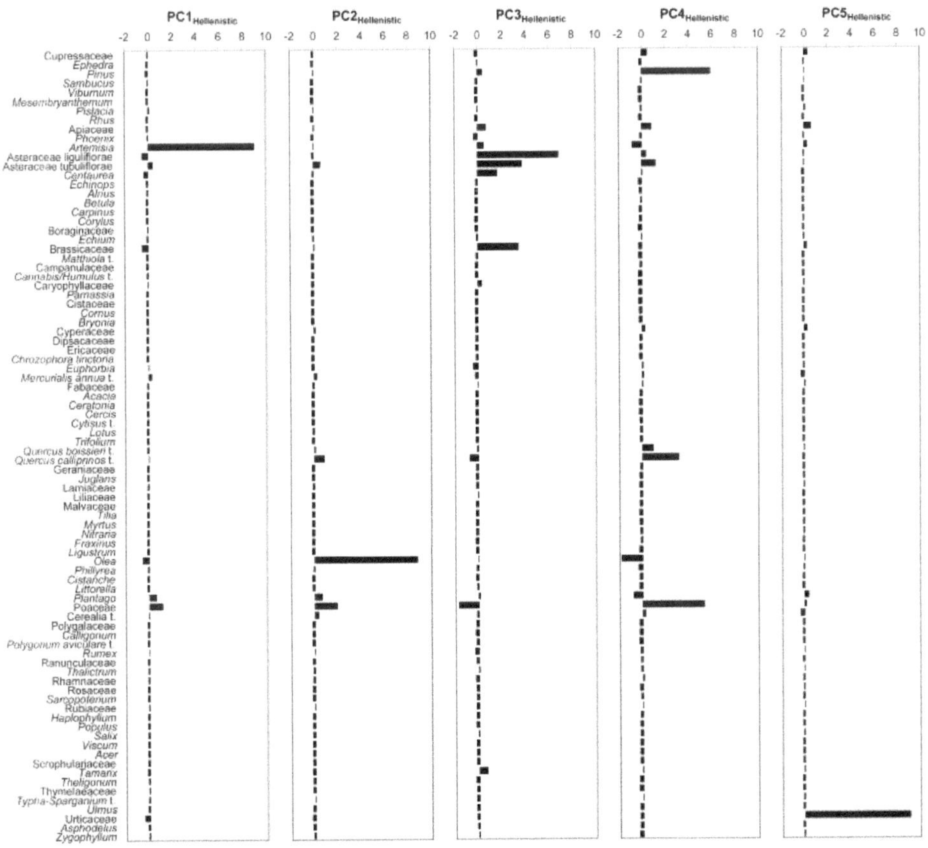

Figure 9: Factor scores of the five principal components (transposed matrix) obtained for the air-borne component of the Hellenistic-early Roman high-stand samples (blocks ZA11B3R, ZA11B4L and ZA11B5L).

Although for the late 19th – early 20th centuries high-stand, some couplets present an autumn-detrital ($PC1_{recent}$ and $PC2_{recent}$) and a spring-aragonite ($PC3_{recent}$) deposition, most of the variance of the detrital-aragonite couplets is explained by $PC1_{recent}$ and $PC2_{recent}$, pointing at an autumn deposition of the entire couplet (Fig. 10). For the Hellenistic-early Roman high-stand, many detrital-aragonite couplets present an autumn-detrital ($PC1_{Hellenistic}$, $PC3_{Hellenistic}$) and a spring-aragonite ($PC2_{Hellenistic}$, $PC4_{Hellenistic}$ and $PC5_{Hellenistic}$) deposition, while many other couplets present deposition in either autumn or spring (Fig. 10). Furthermore, eight couplets suggested a spring-detrital and a following autumn-aragonite deposition. This result appears in the 12.3% of the couplets only (8 out of 65), and it is difficult to explain, likely pointing to an anomalous result that may be related to contamination from adjacent laminae that was introduced during the difficult sampling of the mm-scale alternating laminae. Further deliberation on the meaning of seasonal deposition will be presented in the discussion.

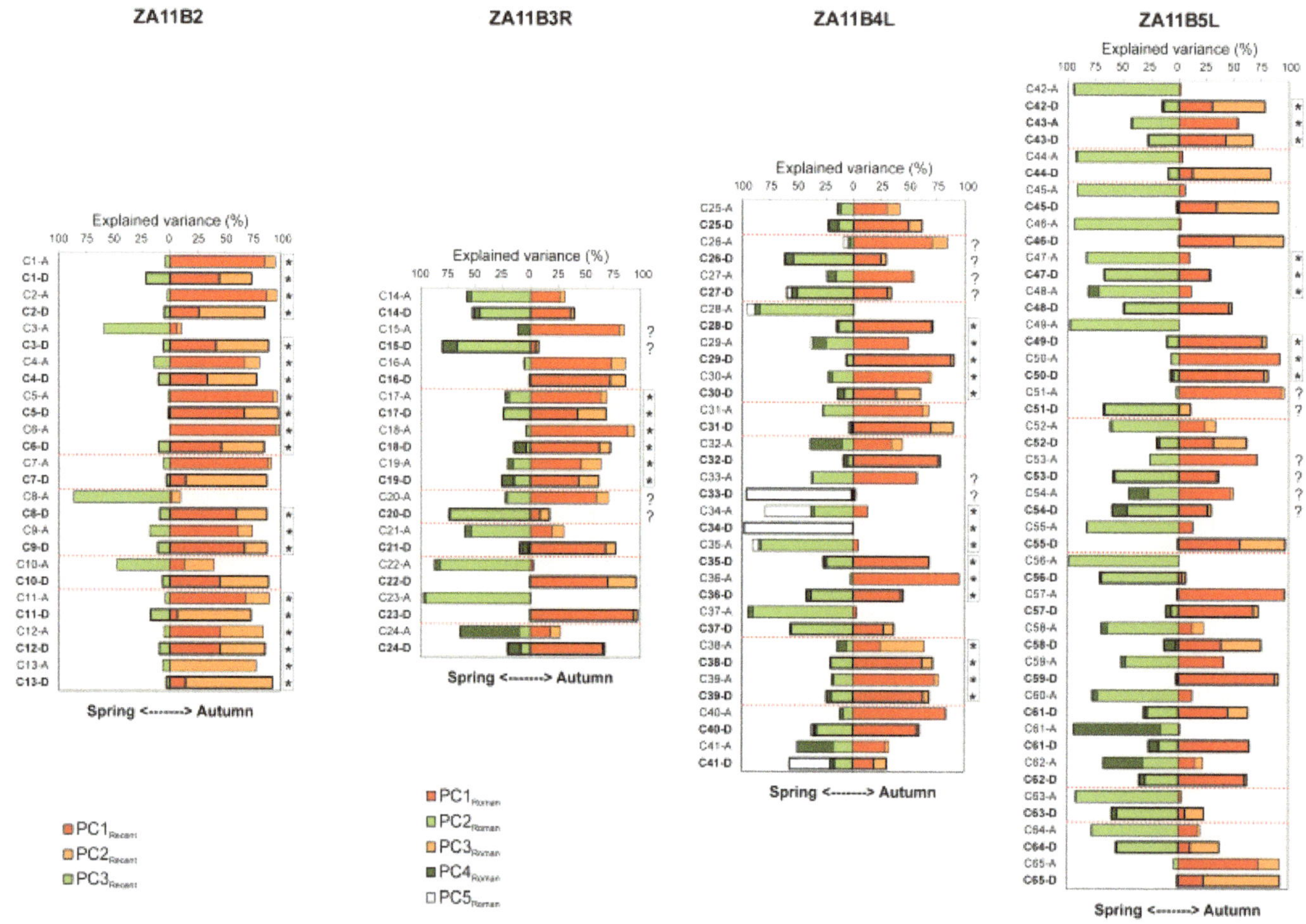

Figure 10: Percentage of explained variance (square of factor loadings x 100) of the principal components extracted for the late 19th – early 20th centuries high-stand samples ($PC1_{Recent}$, $PC2_{Recent}$ and $PC3_{Recent}$) and the Hellenistic-early Roman high-stand samples ($PC1_{Hellenistic}$, $PC2_{Hellenistic}$, $PC3_{Hellenistic}$, $PC4_{Hellenistic}$ and $PC5_{Hellenistic}$). Proportion of variance that can be explained by the extracted principal components (communalities) is shown for each lamina. The graphs can be read as a sequence of laminae deposition (detrital and following aragonite), with the percentage of the variance explained by the principal components with autumn palynological assemblages plotting to the right and by the principal components with spring palynological assemblages plotting to the left. Asterisks (*) highlight the laminae in which it is possible that more than one detrital-aragonite couplet have been deposited in one year. Question marks (?) highlight the eight couplets with an anomalous result likely due to contamination during sampling. Red lines separate non-contiguous samples. Numbers of the detrital-aragonite couplets. (For interpretation of the references to colour in this figure legend, the reader is referred to the web version of this article.)

Water-borne: linking detrital layers with flash-flood events

The concentration of well-preserved pollen grains is usually higher in detrital laminae and flash-flood samples than in aragonite laminae, although the differences are much larger when comparing the concentration of reworked grains (Fig. 11). Aragonite laminae have considerably much lower concentration of reworked pollen grains than detrital laminae and flash-floods samples (Fig. 11). This suggests that water-transported reworked pollen is more significant than wind-transported reworked pollen (which can be the main source of reworking in aragonite layers). The concentration of fungal spores is also lower in aragonite than in detrital and flash-flood samples. The combined concentrations of reworked pollen and fungal spores can be used to assess the link between the formation of detrital laminae and flash-flood events (Fig. 12). Flash-flood and detrital samples present similar log-log distributions with a stronger overlap than flash-floods and aragonite laminae, the latter having lower reworked and fungal spores concentrations and no clear distribution pattern (Fig. 12).

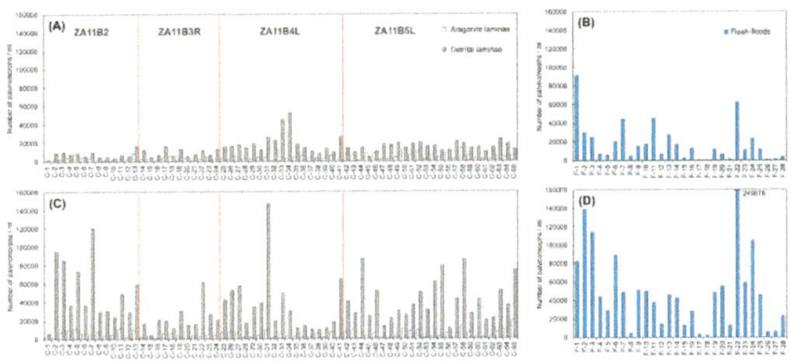

Figure 11: (A) Concentration of Total well-preserved pollen grains in the Ze›elim detrital-aragonite couplets. (B) Concentration of Total well-preserved pollen grains in modern flash-flood samples. (C) Concentration of Total reworked pollen grains in the Ze›elim detrital-aragonite couplets. (D) Concentration of Total reworked pollen grains in the modern flash-flood samples. Numbers of the detrital-aragonite couplets. Numbers of the flash-flood samples.

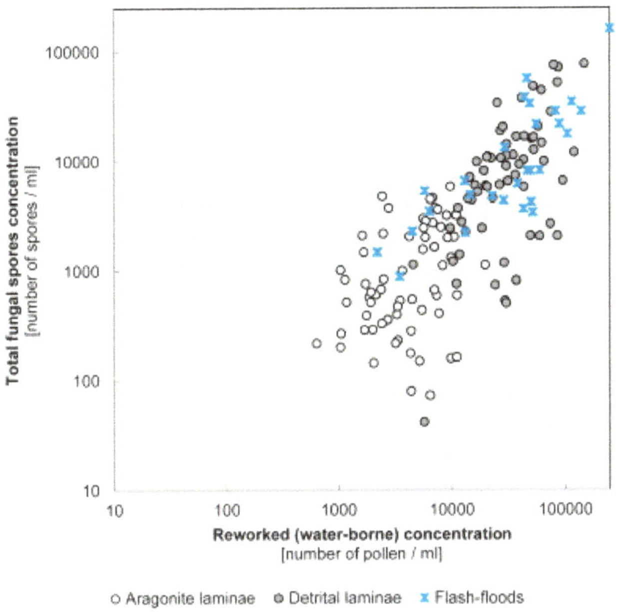

Figure 12: Concentration of Total reworked pollen grains (water-borne) versus Total fungal spores for detrital and aragonite laminae in the Ze'elim outcrop and contemporary flash-flood samples.

DISCUSSION

The Timing of Detrital-Aragonite Deposition

If Dead Sea laminae were varves in the way assumed by some authors, pollen spectra in aragonite laminae would correspond always to summer bloomers, while in detrital layers would have to correspond to autumn-winter-spring pollen assemblages. However, our palynological analysis shows that the air-borne component presents a more complex picture for the timing of aragonite deposition. Furthermore, the detailed palynological examination of the deposition season of detrital-aragonite couplets suggests three possible scenarios (Fig. 10):

- Scenario (i): both detrital and the following aragonite lamina have autumn palynological assemblages (27 out of 65 couplets, 41.5%),
- Scenario (ii): both detrital and the following aragonite lamina have spring palynological assemblages (9 out of 65 couplets, 13.8%), and
- Scenario (iii): the detrital lamina has an autumn palynological assemblage and the following aragonite lamina a spring one (21 out of 65 couplets, 32.3%).

These scenarios show no summer pollen assemblages in the aragonite laminae. One may argue that the lack of summer pollen results from the relatively low dust flux during this season (Singer et al., 2003). However, considering the rapid deposition of aragonite laminae along with the abundance of autumn pollen assemblages this argument does not stand. Thus, we conclude that the deposition of aragonite did not depend exclusively on evaporation and warming of surface waters (Neev and Emery, 1967). This indicates that aragonite precipitation occurred always during the rainy season, either at the same season of the detrital laminae deposition (scenarios i and ii) or during the following spring (scenario iii). These scenarios suggest that the deposition of aragonite laminae requires higher carbonate alkalinity in the flash-flood, hence enabling aragonite saturation without evaporation (Barkan et al., 2001).

Several of the consecutive couplets analysed showed a deposition within the same season (Fig. 10). On the one hand, consecutive autumn pollen assemblages were observed in the following couplets of the late 19th – early 20th century high-stand (block ZA11B2): couplets 13, 12 and 11; couplet 9 and the detrital layer of couplet 8; couplets 6, 5 and 4 and the detrital layer of couplet 3; and couplets 2 and 1. In the Hellenistic-early Roman high-stand consecutive autumn pollen assemblages were observed in a) couplets 19, 18 and 17 in block ZA11B3R; b) couplets 39 and 38; couplet 36 and the detrital layer of couplet 35; and couplets 30 and 29 and the detrital layer of couplet 28 in block ZA11B4L; and c) couplet 50 and the detrital layer of couplet 49; and couplet 43 and the detrital layer of couplet 42 in block ZA11B5L (Fig. 10). On the other hand, consecutive spring pollen assemblages were observed in the aragonite layer of couplet 35 and couplet 34 in block ZA11B4L; and in the aragonite layer of couplet 48 and couplet 47 in block ZA11B5L (Fig. 10).

These short deposition sequences do not allow us to determine whether they were deposited during the same year (i.e. a year with more flash-flood events) or during consecutive years in the same season. The compilation of the annual distribution of flash-floods in five Dead Sea stations (Supplementary Fig. S7) shows the occurrence of flash-flood events from October to May mainly, with few events in June and September. Thus, the modern flash-floods record shows that both situations are possible. However, more than one event a season is highly probable in the large drainages. In fact, the palynological assemblages obtained in this study for both detrital and aragonite laminae are those of the rainy season, i.e. when flash-floods are more likely to occur.

Scenario (iii), when a detrital lamina is deposited in autumn and the following aragonite one in spring, agrees with an aragonite precipitation due to mixing of flood freshwater with Dead Sea brine during the rainy season. However, in contrast to scenarios (i) and (ii), the laminae from the same couplet are deposited in different seasons within the rainy period. This scenario may also indicate a longer lag period between the freshwater arrival to the Dead Sea and the chemical deposition. The lacustrine conditions that can lead to this situation are only estimated at this stage. However, the need for sufficient accumulation of dissolved bicarbonate in the lake surface water or the effect of turbulence on chemical deposition is hypothesised as possibly critical in this interpretation.

The analyses performed in this study present for the first time direct evidence for the timing of laminae deposition in the DSB. The abovementioned scenarios suggest that both detrital and aragonite laminae were deposited in a varying seasonal patterns rather than in a strict annual cycle. Detrital layers present grain-size distributions composed of sediment recycling of rock and sediment outcrops in the Dead Sea watershed and distal sources by aeolian transport. These recycled sediments reach the Dead Sea via run-off or as direct dust deposition from the atmosphere. Run-off sediment transport as the main process is confirmed by the large presence of reworked pollen grains and fungal spores (water-borne component) in the detrital laminae compared to the aragonite laminae (Fig. 12). Erosion of soils delivered reworked pollen and fungal spores with the sediments that deposited within the detrital laminae. Aragonite precipitates from the upper surface water layer due to mixing of flood-water with Dead Sea brine (Stein et al., 1997 and Barkan et al., 2001).

This study indicates that, in contrast to previous assumptions, detrital-aragonite couplets were deposited during the rainy season as a result of flash-flood events, hence Dead Sea laminated sediments cannot be considered as varves. This result is of great importance for the accurate use of the DSB laminated sequences as palaeoenvironmental and paleoclimate archives (e.g. the ICDP Dead Sea Deep Drilling Project), and it delivers exciting new information to the Dead Sea scientists and the palaeoclimate community alike.

Palaeoenvironmental Implications and Further Research

During a flood event, water entering the Dead Sea is dispersed in a plume that floats over the dense saline waters (Nehorai et al., 2013). The suspended sediment in the plume ultimately deposits on the lake bottom. Plume sediment dispersal is a function of flood discharge, wave energy and distance from the shore. Consequently, the central deep part of the basin will record fewer flash-flood events as compared to the margin. In addition, the potential problem of reworking of deposited sediment destroying depositional laminations has to be taken into account. This may prevent, with the uncertainty of having more than one couplet deposited per year, the use of the detrital-aragonite couplets for laminated-based chronologies, i.e. if the laminated-based chronology of a studied sediment core is of long duration the number of couplets per time-unit may vary due to variations in the shore distance or/and flash-flood discharge. In addition, larger discharge carrying more sediment will deposit thicker and coarser detrital laminae. On the other hand, the higher the lake level the further the deposition location from the shore. This yields thinner and finer grain-size laminae. In brief, both parameters are climatically controlled, but result in a reverse effect on the deposition character of the detrital laminae, preventing the use of the thickness and grain-size of the laminae as a direct climatic indicator.

Barkan et al. (2001) pointed out that the aragonite precipitation rate calculated by Stein et al. (1997) for the laminated sediments of the Lisan Formation required a run-off influx six-fold higher than the contemporary run-off influx. Thus, indicating that the Lisan period was more humid than nowadays and that flash-flood events could have been more pronounced. Stein et al. (1997), however, based on previous reports (Neev and Emery, 1967 and Begin et al., 1974), interpreted the aragonite formation as precipitation due to increased evaporation in summer. Therefore, understanding the deposition processes of the laminated sediments is of considerable importance for palaeoenvironmental reconstructions. This work has provided the first palynological evidence for unravelling the timing of aragonite laminae deposition, which has been identified as occurring during the rainy season rather than summer, but further research is necessary.

CONCLUSIONS

This research attempts to identify by palynology the timing of the Dead Sea laminae deposition. Our results demonstrate that:

1. Couplets are not necessarily deposited with annual cyclicity. The air-borne component is more complex than a rainy season-detritus *versus* a summer-aragonite deposition, as all laminae presented pollen assemblages representative of the rainy season.

2. The concentration of reworked pollen grains in detrital layers and flash-flood samples is similar and much higher than in aragonite layers. Palynological concentration of reworked grains compared with fungal spores, which represent erosion and soil products, separated detritus and flash-flood samples from aragonite ones.

3. Grain-size distribution indicates that detrital laminae are composed of recycling of local and distal sources, with coarser particles deposited in the Dead Sea watershed and later transported via run-off to the lake.

4. Aragonite precipitation from the upper surface water layer does not necessarily require evaporation and warming of surface waters.

The conclusions drawn above suggest that detrital-aragonite couplets in the Dead Sea laminated sediments are most likely not varves and that the laminae deposition is related to the occurrence of flash-flood events. Whether for specific rainfall patterns (i.e. one flood a year, which is very unlikely) those laminae could provide annual information or not is something that needs to be further tested in other Dead Sea contexts, likely in long and undisturbed laminated sections with many continuous laminae such as the Lisan upper member (i.e. Prasad et al., 2004) and the Ein Feshkha sequences (Kagan et al., 2011).

Further research is necessary to understand whether the number of couplets formed due to flash-flood events depends on the distance from the shoreline, meaning that sediments sections covering the same chronology from different parts of the DSB (shoreline *versus* deep basin) could have different number of laminations. Hence, laminae counting as dating tool of Dead Sea sediments should be re-evaluated. However, these laminated sequences should be used for the reconstruction of palaeo-flash flood records that will have a significant impact on understanding the palaeo-hydrology of the DSB and its implication to high-resolution climatic interpretation.

ACKNOWLEDGEMENTS

This research was made possible through the support of the British Council's Britain-Israel Research and Academic Exchange Partnership (BIRAX), which aims to further institutional links between universities in the UK and Israel, awarded to S.A.G. Leroy and R. Bookman and entitled "Resolving the deposition pattern of the Dead Sea laminae and its implication for understanding hydro-climatic short-term variations in the Levant" (BY2/GEO/08), and partly by an Israel Science Fund grant (ISF, Grant # 1093/10) to R. Bookman. L. López-Merino was first supported by a post-doctoral fellowship (Spanish Government) at Brunel University London and later by the Brunel MINT Scheme. We thank the valuable help of Amanda Rozeik (Brunel University London) with the palynological extraction, and Uri Sha'anan and Dr Nimer Taha (University of Haifa) with the grain-size analysis and the sampling of flash-floods. We are indebted to Professor Antonio Martínez Cortizas (University of Santiago de Compostela) and Dr Stephen Kershaw (Brunel University London) for their perceptive comments on earlier drafts.

REFERENCES

1. Barkan, E., Luz, B., Lazar, B., 2001. Dynamics of the carbon dioxide system in the Dead Sea. Geochimica Cosmochimica Acta 65 (3), 355e368.
2. Baruch, U., 1993. The Palynology of Late Quaternary Sediments of the Dead Sea. Unpublished PhD Dissertation. Hebrew University of Jerusalem, Jerusalem. Hebrew with English abstract.
3. Begin, Z.B., Ehrlich, A., Nathan, Y., 1974. Lake Lisan e the Pleistocene Precursor of the Dead Sea. Geological Survey of Israel, Bulletin No. 63, pp. 1e30.
4. Belmaker, R., Lazar, B., Stein, M., Beer Jürg, B., 2011. Short residence time and fast transport of fine detritus in the Judean Desert: clues from 7 Be in settled dust. Geophys. Res. Lett. 38 (16), L16714.
5. Belmaker, R., Stein, M., Beer, J., Christl, M., Fink, D., Lazar, B., 2014. Beryllium isotopes as tracers of Lake Lisan (last glacial Dead Sea) hydrology and the laschamp geomagnetic excursion. Earth Planet. Sci. Lett. 400, 233e242.
6. Ben Moshe, L., Haviv, I., Enzel, Y., Ziberman, E., Matmon, A., 2008. Incision of alluvial channels in response to a continuous base level fall: field characterization, modeling, and validation along the Dead Sea. Geomorphology 93 (3e4), 524e536.
7. Bennett, K.D., 2009. Documentation for Psimpoll 4.27 and Pscomb 1.03. C Programs for Plotting Pollen Diagrams and Analysing Pollen Data. The 14Chrono Centre, Archaeology and Palaeoecology. Queen's University of Belfast, Belfast, UK. Available at: http://www.chrono.qub.ac.uk/psimpoll/psimpoll.html.
8. Bookman (Ken-Tor), R., Enzel, Y., Agnon, A., Stein, M., 2004. Late Holocene lake levels of the Dead Sea. Geol. Soc. Am. Bull. 127 (1e2), 555e571.
9. Bookman, R., Bartov, Y., Enzel, Y., Stein, M., 2006. Quaternary lake levels in the Dead Sea basin: two centuries of research. Geol. Soc. Am. Special Pap. 401, 155e170.
10. Dayan, U., Morin, E., 2006. Flash flood-producing rainstorms over the Dead Sea: a review. Geol. Soc. Am. Special Pap. 401, 53e62.
11. Dayan, U., Sharon, D., 1980. Meteorological parameters for discriminating between widespread and spotty storms in the Negev. Isr. J. Earth Sci. 29, 253e256.
12. Dayan, U., Ziv, B., Shoob, T., Enzel, Y., 2007. Suspended dust over South-Eastern Mediterranean and its relation to atmospheric circulations. Int. J. Climatol. 24, 1001e1011.
13. Enzel, Y., Bookman (Ken-Tor), R., Sharon, D., Gvirtzman, H., Dayan, U., Ziv, B., Stein, M., 2003. Late Holocene climates of the Near East deduced from Dead Sea level variations and modern regional winter rainfall. Quat. Res. 60 (3), 263e273.
14. Feinbrun, N., Rahat, A., Tas, J., 1959. Further studies in atmospheric pollen in Jerusalem. Bull. Res. Counc. Isr. 8D, 31e40.
15. Festi, D., Kofler, W., Bucher, E., Carturan, L., Mair, V., Grabielli, P., Oeggl, K., 2015. A novel pollen-based method to detect seasonality in ice cores: a case study from the Ortles glacier, South Tirol, Italy. J. Glaciol. 61 (229), 815e824.
16. Ganor, E., Foner, H., 1996. The mineralogical and chemical properties and the behavior of aeolian Saharan dust over Israel. In: Guerzoni, S., Chester, R. (Eds.), The Impact of Desert Dust across the Mediterranean. Kluwer Academic Publishers, Dordrecht, pp. 163e172.
17. Greenbaum, N., Ben-Zvi, A., Haviv, I., Enzel, Y., 2006. The hydrology and paleohydrology of the Dead Sea tributaries. Geol. Soc. Am. Special Pap. 401, 63e93.
18. Haliva-Cohen, A., Stein, M., Goldstein, S.L., Sandler, A., Starinsky, A., 2012. Sources and Transport Routes of Fine Detritus Material to the Late Quaternary Dead Sea Basin.
19. Heim, C., Nowaczyk, N.R., Negendank, J.F.W., Leory, S.A.G., Ben-Avraham, Z., 1997. Near east desertification: evidence from the Dead Sea. Naturwissenschaften 84, 398e401.
20. Horowitz, A., Weinstein, M., Ganor, E., 1975. Palynological determination of dust storms provenances in Israel. Pollen spores 17 (2), 223e231.
21. Kagan, E., Stein, M., Agnon, A., Neumann, F., 2011. Intrabasin paleoearthquake and quiescence correlation on the late Holocene Dead Sea. J. Geophys. Res. 116 (B4). B04311.
22. Kantor, S.Z., Frank, M., Hoch-Kantor, D., Barkai-Golan, R., Marian, D., Schachnner, E., Kessler, A., de Vries, A., 1966. Airborne allergens and clinical response of asthmatics in Arad, a new town in a desert area in Israel. J. Allergy 37 (2), 65e74.
23. Katz, A., Kolodny, N., 1989. Hypersaline brine diagenesis and evolution in the Dead Sea-Lake Lisan system (Israel). Geochem. Cosmochimica Acta 53 (1), 59e67.

24. Ken-Tor, R., Agnon, A., Enzel, Y., Stein, M., Marco, S., Negendank, J.F.W., 2001a. Highresolution geological record of historic earthquakes in the Dead Sea basin. J. Geophys. Res. 106 (B2), 2221e2234.

25. Ken-Tor, R., Stein, M., Enzel, Y., Agnon, A., Marco, S., Negendank, J.F.W., 2001b. Precision of calibrated radiocarbon ages of historic earth quakes in the Dead Sea Basin. Radiocarbon 43 (3), 1371e1382.

26. Keynan, N., Waisel, Y., Shomer-Ilan, A., Tamir, R., 1989. Forecasting pollen pollution: correlation with floral development. Ann. Allergy 63, 417e420.

27. Langgut, D., Neumann, F.H., Stein, M., Wagner, A., Joy, Kagan, E.J., Boaretto, E., Finkelstein, I., 2014. Dead Sea pollen record and history of human activity in the judean highlands (Israel) from the Intermediate bronze into the Iron ages (~2500-500 BCE). Palynology 38 (2), 280e302.

28. Langgut, D., Finkelstein, I., Litt, T., Neumann, F.H., Stein, M., 2015a. Vegetation and climate changes during the Bronze and Iron Ages (~3600-600 BCE) in the southern Levant based on palynological records. Radiocarbon 57 (2), 217e235.

29. Langgut, D., Yannai, E., Taxel, I., Agnon, A., Marco, S., 2015b. Resolving a historical earthquake date at Tel Yavneh (central Israel) using pollen seasonality. Palynology. http://dx.doi.org/10.1080/01916122.2015.1035405.

30. Leroy, S.A.G., 2010. Pollen analysis of core DS7-1SC (Dead Sea) showing intertwined effects of climatic change and human activities in the Late Holocene. J. Archaeol. Sci. 37, 306e316.

31. Leroy, S.A.G., Marco, S., Bookman, R., Miller, C.S., 2010. Impact of earthquakes on agriculture during Roman-Byzantine period from pollen records of the Dead Sea laminated sediment. Quat. Res. 73 (2), 191e200.

32. Litt, T., Ohlwein, C., Neumann, F.H., Hense, A., Stein, M., 2012. Holocene climate variability in the Levant from the Dead Sea pollen record. Quat. Sci. Rev. 49, 95e105.

33. Lopez-Merino, L., Silva Sanchez, N., Kaal, J., Lopez-Saez, J.A., Martínez Cortizas, A., 2012. Post-disturbance vegetation dynamics during the late Pleistocene and the Holocene: an example from NW Iberia. Glob. Planet Change 92e93, 58e70.

34. Migowski, C., Agnon, A., Bookman, R., Negendank, J.F.W., Stein, M., 2004. Recurrence pattern of Holocene earthquakes along the Dead Sea transform revealed by varve-counting and radiocarbon dating of lacustrine sediments. Earth Planet. Sci. Lett. 222 (1), 301e314.

35. Migowski, C., Stein, M., Prasad, S., Negendank, J.F.W., Agnon, A., 2006. Holocene climate variability and cultural evolution in the Near East from the Dead Sea sedimentary record. Quat. Res. 66 (3), 421e431.

36. Neev, D., Emery, K.O., 1967. The Dead Sea. Depositional processes and environments of evaporates. Geological Survey of Israel, Bulletin No. 41, pp. 1e147.

37. Nehorai, R., Lensky, I.M., Hochman, L., Gertman, I., Brenner, S., Muskin, A., Lensky, N.G., 2013. Satellite observations of turbidity in the Dead Sea. J. Geophys. Res. Oceans 118 (6), 3146e3160.

38. Neugebauer, I., Brauer, A., Schwab, M.J., Dulski, P., Frank, U., Hadzhiivanova, E., Kitagawa, H., Litt, T., Schiebel, V., Taha, N., Waldmann, N., 2015. Evidences for centennial dry periods at ~3300 and ~2800 cal. yr BP from micro-facies analyses of the Dead Sea sediments. Holocene 25 (8), 1358e1371.

39. Neumann, F.H., Kagan, E.J., Schwab, M., Stein, M., 2007. Palynology, sedimentology and palaeoecology of the late Holocene Dead Sea. Quat. Sci. Rev. 26, 1476e1498.

40. Neumann, F.H., Kagan, E.J., Stein, M., Agnon, A., 2009. Assessment of the effect of earthquake activity on regional vegetation e high-resolution pollen study of the Ein Feshka section, Holocene Dead Sea. Rev. Palaeobot. Palynol. 155 (1e2), 42e51.

41. Neumann, F.H., Kagan, E.J., Leroy, S.A.G., Baruch, U., 2010. Vegetation history and climate fluctuations on a transect along the Dead Sea west shore and their impact on past societies over the last 3500 years. J. Arid Environ. 74 (7), 756e764.

42. Niemi, T.M., Ben-Avraham, Z., 1997. Active Tectonics in the Dead Sea Basin. In: Niemi, T.M., Ben-Avraham, Z., Gat, J. (Eds.), The Dead Sea: The Lake and its Settings. Oxford University Press, New York, pp. 73e81.

43. Offer, Z.Y., Zangwill, A., Azmon, E., 1992. Characterization of airborne dust in the Sde Boker area. Isr. J. Earth Sci. 41, 235e245.

44. Ojala, A., Alenius, T., 2005. 10 000 years of interannual sedimentation recorded in the Lake Nautaj€arvi (Finland) clastic-organic varves. Palaeogeogr. Palaeoclimatol. Palaeoecol. 219, 285e302.

45. Prasad, S., Vos, H., Waldmann, N., Goldstein, S., Stein, M., 2004. Evidence from Lake Lisan of solar influence on decadal- to centennial-scale climate variability during marine oxygen isotope stage 2. Geology 32, 581e584.

46. Reille, M., 1992. Pollen et Spores d'Europe et d'Afrique du Nord. Laboratoire de Botanique Historique et Palynologie, CNRS, Marseille.

47. Reille, M., 1995. Pollen et spores d'Europe et d'Afrique du Nord. Supplement 1. Laboratoire de Botanique Historique et Palynologie, Marseille.

48. Reille, M., 1998. Pollen et spores d'Europe et d'Afrique du Nord. Supplement 2. Laboratoire de Botanique Historique et Palynologie, Marseille.

49. Salmaso, N., Decet, F., 1998. Interactions of physical, chemical and biological processes affecting the seasonality of mineral composition and nutrient cycling in the water column of a deep subalpine lake (Lake Garda, Northern Italy). Arch. für Hydrobiol. 142, 385e414.

50. Shmida, A., Darom, D., 2000. Handbook of Wildflowers of Israel. Desert flora. Keter Publishing House Ltd, Jerusalem.

51. Shmida, A., Darom, D., 2002. Handbook of Wildflowers of Israel. Mediterranean Flora. Keter Publishing House Ltd, Jerusalem.

52. Singer, A., Ganor, E., Dultz, S., Fischer, W., 2003. Dust

deposition over the Dead Sea. J. Arid Environ. 53, 41e59.

53. Stahr, K., Hermann, L., Jahn, R., 1994. Long distance dust transport in the SudanoSahelian Zone and the consequences for soil properties. In: Buerkert, B., Allison, B.E., Von Oppen, M. (Eds.), Wind Erosion in West Africa: the Problem and its Control. University of Hohenheim, Germany, Weikersheim, Germany, pp. 23e33. Margraf Verlag.

54. Stein, M., 2014. The evolution of Neogene-Quaternary water-bodies in the Dead Sea rift valley. In: Garfunkel, Z., Ben-Avraham, Z., Kagan, E. (Eds.), Dead Sea Transform Fault System: Reviews. Springer, Netherlands, pp. 279e316.

55. Stein, M., Goldstein, S.L., 2006. U-Th and radiocarbon chronologies of late Quaternary lacustrine records of the Dead Sea basin: methods and applications. Geol. Soc. Am. Special Pap. 401, 141e154.

56. Stein, M., Starinsky, A., Katz, A., Goldstein, S.L., Machlus, M., Schramm, A., 1997. Strontium isotopic, chemical, and sedimentological evidence for the evolution of Lake Lisan and the Dead Sea. Geochimica Cosmochimica Acta 61 (18), 3975e3992.

57. Stockmarr, J., 1971. Tablets with spores used in absolute pollen analysis. Pollen Spores 13, 614e621.

58. Thompson, J.B., Schultze-Lam, S., Beveridge, T.J., Des Marais, D.J., 1997. Whiting events: biogenic origin due to photosynthetic activity of cyanobacterial picoplankton. Limnol. Oceanogr. 42 (1), 133e141.

59. Torfstein, A., Goldstein, S.L., Kagan, E.J., Stein, M., 2013. Integrated multi-site U-Th chronology of the last glacial Lake Lisan. Geochimica Cosmochimica Acta 104, 210e231.

60. Waisel, Y., unpublished report 1. Survey of allergenic pollen and spores in the Kalya region 2002-2003. Tel Aviv University.

61. Waisel, Y., unpublished report 2. Survey of allergenic airborne pollen and spores in the Dead Sea region 1999-2002. Tel Aviv University.

62. Zolitschka, B., Francus, P., Ojala, A.E.K., Schimmelmann, A., 2015. Varves in lake sediments e a review. Quat. Sci. Rev. 117, 1e41.

CHAPTER 11

MILK TO MANDALAY: DAIRY CONSUMPTION, ANIMAL HISTORY AND THE POLITICAL GEOGRAPHY OF COLONIAL BURMA

Jonathan Saha

Keywords: Animals, Colonial Burma, Dairy, Cattle, Borders, Political geography

Writing at the end of the First World War, Colin Metcalfe Enriquez, a colonial official who served in Burma's borderworlds, outlined the difficulties of meeting visiting dignitaries' dietary expectations when in remote parts of the colony. He recalled some of the tedious culinary preparations for one particular viceregal arrival:

Take, for instance, the case of milk. Nothing is so necessary to State Visits as good, foamy milk, full of cream and so on. There was a large file relating to the particular quart I have in mind. It was to be delivered several months hence at a little river-side village where the steamer would touch at tea-time. Some say the correspondence started before the cow was born. But she was a placid old beast, and didn't mind that a bit, until the milkman put on his apron. Then she shied, and kicked the bucket over. The whole credit of the apron belongs to the Sanitary Commissioner, who had made a special study of the art of milking. But apparently he did not know this old cow, and how irreconcilable the apron made her. Her panic only increased with rehearsals. There were sterilisers, and boilers, and strainers too, including a bit of an old shirt. But these were kept beyond reach of her apprehensive glances. At the last moment, when everything was ready, a frantic order came to buy the milkman rubber gloves. Rubber gloves in a Burmese river-side village!! They were simply not to be had – not for all th[e] wealth of Thibaw [the last king of Burma], if he ever had any. [1] As it transpired, the viceroy's ship did not even land at the village. It steamed past oblivious to all the fuss. The carefully orchestrated efforts, frenetic correspondence and the cow's distress had been for nothing.

As Enriquez's passage attests, the British found it hard to get cow's milk in Burma. Despite legislation that required village headmen to provide provisions for officials and other Europeans touring the colony, fresh milk often proved difficult to acquire.[2] This was in large part because, in contrast to the rest of British India, there was not a large local market for dairy products. The experience of imperialism did little to change this situation. Although during colonial rule the consumption of milk certainly became more prevalent, it remained comparatively unpopular throughout the period. By the time of the Japanese occupation, an estimated 1.8 fluid ounces of milk per capita were drunk daily in the colony. This compared to 6.6 fluid ounces across the border in India, and to roughly forty fluid ounces in Denmark and the United Kingdom. Of the milk consumed in Burma, almost a third was imported and eighty-nine per cent of this was tinned condensed milk.[3]Deborah Valenze has characterised the global history of cow's milk as the triumphant emergence of a culturally malleable, universal commodity. She argues that it has been a story 'of [the] conquest of space, energy, and dietary preferences'.[4] It would seem that the example of colonial Burma reveals some of the limits to this campaign of conquest.

Whilst Valenze›s narrative of conquest and resistance deploys terms associated with imperialism, in what follows I consider the history of milk not in terms of the 'conquest of space', but instead in terms of the 'production of space'.[5] This is because the history of milk was imbricated in the drawing of what has been called the 'embryonic border' between Burma and the rest of British India, which eventually led to the colony's succession from the Raj in 1937.[6] Milk was caught

Citation: Jonathan Saha, Milk to Mandalay: dairy consumption, animal history and the political geography of colonial Burma, Journal of Historical Geography, Volume 54, October 2016, Pages 1-12, ISSN 0305-7488, http://dx.doi.org/10.1016/j.jhg.2016.05.016.

up in the geopolitical territorialisation of the colony, rather than simply being a conquering colonial commodity. Milk was part of this production of colonial space both figuratively and materially. In the imperial imagination the lack of cow›s milk was believed to be the result of the religious mores of Burmese Buddhists. At the same time, the introduction of dairying was associated with the encroachment of Indians into Burmese life. Colonial officials viewed it as an alien and potentially damaging intrusion. Inextricably linked to these imperial representations, were material obstacles to the introduction of large-scale dairy production in the colony. Milk offers another demonstration that, as Felix Driver and others have long argued, material encounters and imaginative representations were inseparable in the making of imperial geographies. In addition, as a definitively mammalian product for nourishing warm-blooded creatures, milk points to the importance of animals within the production of colonial space.[7]

Building on the insights of more-than-human geography, this article attempts to keep animals within its analytical framework. Over the last fifteen years animal geographers have been arguing that nonhumans are central to the production of spaces. Introducing their ground-breaking early collection on the subject, Chris Philo and Chris Wilbert called attention to how humans have attempted to confine other species in 'animal spaces', both material and imagined: the zoo, the farm, the wild. But they also went further and urged geographers to take into account the ways that animals transgressed these confinements, producing their own 'beastly places' in the process.[8] This has pushed geographers to consider how animals might be brought into their scholarship, as well as how animals might force changes within scholarship itself. Certainly, the once axiomatic centrality of human-animals in human geography has been questioned. The privileged role of agent ascribed to humans has been much critiqued through anti-humanist post-structuralism and a re-conceptualisation of agency as relational, dispersed and contingent.[9] As Henry Buller has tracked in his recent surveys of this expanding field, this has led to considerable methodological innovation. Ethnographic methods along with observational and participatory forms of multi-species research have enabled geographers to provide space in their work for nonhuman creatures as subjects for themselves, as opposed to mirrors for humanity.[10] For those of us working on the past, some of these methods are challenging to deploy without the risk of anachronism. Nevertheless, the recognition of the importance of embodied experiences and materiality, the dissolution of the nature-culture binary, and the attention being paid to how social contexts are produced, which underpin these new methods, are developments that historical geographers can build on to uncover how animals have been imbricated in imperial geographies.

Within these discipline trends, this article picks up particularly on Kersty Hobson›s call for animals to be conceived of as subjects within political geography.[11] In this case, the article argues that the politics of Burma's contentious separation from British India was informed by cattle. This does not mean attempting to represent the subjective experience of the animals involved in this history; although at appropriate moments – and when the sources allow – this is brought into the discussion. Rather, it means conceptualising cattle as 'lively commodities'. The term captures the tension in how nonhumans were recognised as subjects with particular capacities and characteristics, whilst simultaneously attending to the ways in which they were rendered as objects representing human desires and exchange values.[12] The implication of this when studying milk (or other animal-derived human food stuffs) is that historians should not neglect the living, flesh and blood means of production involved. In order to bring animals into the historical political geography of colonial Burma as subjects, the new relationships with cattle engendered by the commodification of their bodies and in the consumption of their milk need to be put at the centre of the study.

Drinking cow›s milk was an act of 'interspecies intimacy'. It was a material and imaginative encounter with another mammal that involved consuming a liquid usually expressed to feed that creature›s young. Brett Walker makes a similar point in his study of moments when humans have been attacked by carnivorous, predatory animals. He argues that these events demonstrate the unnerving human-animal intimacy of encounters in which a human is reduced to being another creature›s source of protein.[13]The consumption of milk is perhaps a more optimistic example of interspecies intimacy; although modern, intensive commercial dairy farming would suggest otherwise.[14] In Burma this intimacy was bound up with the making of boundaries between the coloniser and the colonised. As with taboos on cannibalism and eating certain animals, perceptions of milk drinking reinforced imperial mappings of cultural difference.[15] In addition, in order to facilitate milk consumption the colonial state in Burma was drawn into improving its production and distribution. To achieve this they had to marshal competing uses of cattle and maintain the welfare of the colony's ox population. In other words, the colonial state devised and enacted policies to make space for cattle based on their reading of the animals' needs. The perceived threats from Indian-owned dairy herds to the quality of milk and the health of cattle resulted in these lively commodities becoming subjects in Burma's contentious political geography as it became imagined apart from the Indian empire.

To explore the interplay between the imaginative and material processes through which cow›s milk contributed to the production of Burma›s 'embryonic border' with India, the article is divided into three sections. The first section reveals how the potentially transgressive intimacy of dairy was negotiated when British imperial actors found themselves in a culture that did not routinely consume dairy products.[16] It shows how the consumption of cow's milk fed into the production of a particular imaginative geography, described by Stephen Keck as the 'Burmascape'.[17] The second section explores the material arrangements put in place to create a network for distributing milk to allay colonial concerns over sanitation. It focuses on the colonial state's attempts to order and police space. The third section uncovers the perceived problem of herds of Indian milch cattle breeding with Burmese plough cattle. It explores how colonial policies to monitor and restrict Indian cattle were coterminous with policies to monitor and restrict Indian humans.

This is not the first study to historicise milk consumption and its absence in southern Asia. In 1970 the cultural anthropologist Frederick J. Simoons examined the history of the 'milk-line' that has been said by scholars since the nineteenth century to bisect the region. His work was highly sophisticated and deployed a complex interaction of factors to explain milk drinking and its absence: cultural, ecological, psychological and biological. He also acknowledged the presence of comparable milking practices existing on either side of the line. Nevertheless, his article reinforced the line›s analytical utility (see Fig. 1). Beginning at the Bay of Bengal, the line was said to run from 'somewhere southeast of Chittagong, roughly on the border between the Moslem Bengali and Buddhist Arakanese', it then 'runs inland, to turn northward along the western edge of the Chittagong Hill Tracts, where it separates Bengali from animistic hill tribes'.[18] In other words, the milk-line roughly maps onto the contentious geopolitical borders between contemporary Myanmar, Bangladesh and northeast India. It also reinscribed the Area Studies division between South and Southeast Asia, now considered a heuristic separation with its origins in Cold War geopolitics.[19] Recent developments in border studies have exposed the superficiality of these divisions, instead conceptualising this region as one criss-crossed with dynamic social, cultural and economic networks between various mutable ethnic identities.[20] For Simoons, though, the line had cultural meaning.

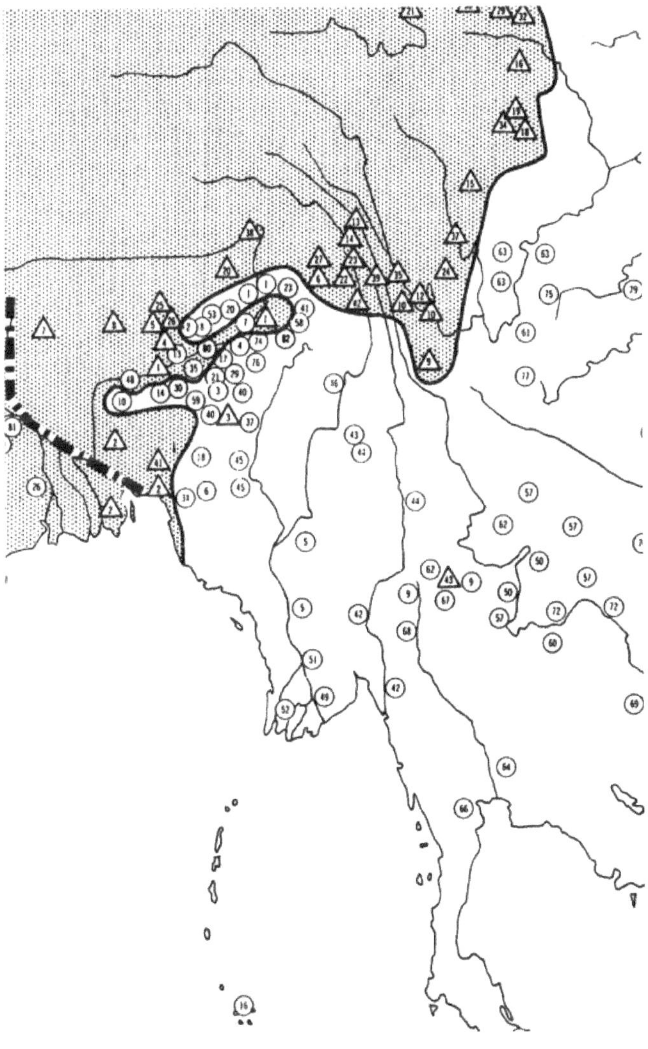

Figure 1: The milk-line in southern Asia, detail from Map 3 in F.J. Simoons, The traditional limits of milking and milk use in southern Asia, Anthropos 65 (1970) 547–593. The shaded area is 'Milking predominant', the unshaded area is 'Nonmilking predominant'.

The existence of predominantly non-milking societies on the east side of the line were 'survivals' of ancient Hindu ahimsa beliefs introduced to Southeast Asia from India. They were 'remnants' of older cultural mores.[21] As sophisticated as Simoons› work was in many respects, he relied heavily on imperial writings, often taking their observations at face value. Through this, his study reified and naturalised what remains a contentious division between South and Southeast Asia along the western borders of Myanmar. This article denaturalises this border by uncovering the colonial history of how milk became entangled in the immanent political geography of British Burma.

Drinking Milk in the 'Burmascape'

Critiquing the discursive 'othering' through which social and conceptual hierarchies are maintained is a central concern shared by both animal studies and postcolonial studies.[22] However, because of an understandable sensitivity to how imperial authorities made accusations of animality to dehumanise colonised populations in the past, until quite recently 'species' has not been much discussed in postcolonial deconstructions of race, gender, sexuality and class.[23] Nevertheless, it is evident that underlying conceptual shifts in imperial thought shaped how differences between humans and other animals, as well as differences between 'types' of humans, have been defined and explained. These differentiations were inextricably entangled and often mutually reinforcing.[24] In addition, material interactions with animals were viewed as markers of difference between the colonisers and the colonised. Imperial observers in Burma cited examples of tactile and sentimental relationships Burmese people were said to have with certain animals to represent the colony as in need of British rule.[25] Milk consumption, as a contested form of human-animal intimacy, was bound up in this politics of colonial sensibilities. As part of imperial writings on the distinctiveness of the colony's cultural landscape, milk informed the imaginative geography of Burma as a place distinct from India. Stephen Keck has argued that the turn-of-the-century writings of colonial scholar officials and travel-writers sympathetic to Burmese culture generated a particular imaginative geography that he calls the 'Burmascape'. These authors rendered Burma a 'unique geographic entity' that could be objectively observed.[26] The absence of milk drinking was simultaneously a sign of the colony›s distinctiveness (particularly from India) and a measure of its (unfavourable) difference to Europe.

Being unable to acquire milk whilst travelling Burma was a frequent gripe in imperial writings. In this it stood in contrast to the rest of British India. Where fresh cow›s milk was drunk in the colony it was associated with Indian populations, such as Sikhs employed in the military. Their consumption of milk was seen as evidence of their 'pastoral Aryan' instincts.[27] British authors' explanations for the absence of milk drinking among the Burmese in Burma often rested upon wider imperial representations of the population, and particularly of Buddhism. It was argued that as Buddhists the Burmese objected to the consumption of milk out of an exaggerated – from the British point of view – sympathy with animal life. As the travelling artist and author, Robert Talbot Kelly, wrote in 1905, in a typical passage, 'Forbidden by their religion to take life, meat seldom forms part of their diet, and to such an extreme is this principle carried that they sometimes even decline to milk their cows, who become dry in consequence'.[28] This depiction of Buddhist morality in

Burma was an inaccurate exaggeration. Vegetarianism was not widespread in the late nineteenth century or early twentieth century, instead emerging as a response to the perceived threats of colonialism to Buddhist learning and of beef eating to local ecologies.[29] As we shall see in the following sections, there were other uses to which cattle in Burma were put, particularly in agricultural labour. These uses had a higher priority than milk production. By emphasising a religiously-motivated veganism as an explanation for the alleged Burmese abstinence from dairy products, British writers were implicitly claiming a normative status for their own consumption habits.

Despite this essentialising explanation, the scarcity of milk and the novelty of drinking it among the Burmese made it an even more conspicuous commodity for the British. In some respects, milk drinking was made a strange experience. Ventriloquizing Burmese witnesses, the travel writer Geraldine Mitton claimed in 1907 that locals found the British habit of drinking cow›s milk 'disgusting'.[30] More than representing any actual Burmese perceptions, this comment demonstrates that Britons in Burma had a heightened awareness of the animal intimacy of their milk-drinking habits and were troubled by how they may have been perceived because of them. Other writers made milk drinking appear strange in this colonial context through detailing extreme examples of acquiring it from creatures whose milk was not commonly consumed. For instance, in 1900 the imperial hunters Fitz William Pollok and W. Thom recounted killing a rhinoceros cow and its calf. Because of the difficulty in getting milk they milked the dead mother, getting two bottles' worth for their trip. Apparently the taste was 'very weak and very sweet'.[31]

More unsettling for imperial sensibilities was an episode in a short anonymous piece of travel writing published in 1909, purportedly written by a Burmese dog who had been adopted by an English family holidaying in the colony.

The Headman brought us in some boiling hot tea and some 'jaggery' (blocks of coarse sugar) to sweeten it with. The Mem Sahib drank one cup and then felt so much better that she began to grumble because there was not any milk. When the Headman learnt what was wanted, he went and had a woman with a baby brought in, and told her. She was so kind that she emptied some milk out of her chest into the tea! Even then the Mem Sahib was not pleased, but the Colonel Sahib said it was a kindly act and a 'work of merit,' and she must not be ungrateful and he took the cup from her, and later I saw it was empty. I don't know who drank it, anyhow I didn't get the chance.[32]

There is a lot going on in this passage. It presents the Burmese headman and female villager as lacking in Edwardian-era, British sensibilities and a related sense of shame, locating them in the rural backwaters of the colony. At the same time, it contributed to a wider portrayal of married English women in the Indian Empire as demanding, prudish and prissy.[33] Yet, the implication that 'Colonel Sahib' had in fact drunk the human breast milk gave a transgressive undertone to the otherwise 'gentle' humour. It was transgressive because it was an adult drinking mother›s milk, but it was also transgressive because he was drinking breast milk from a Burmese woman. Other male imperial writers in Burma sexualised breastfeeding when they described witnessing it in public. It was a way of circuitously acknowledging their own feelings of sexual desire. These were descriptions of shameful, barely suppressed desires for the racialised, eroticised 'other', brought on by the common tropical threats to white masculinity of loneliness, over-exhaustion and climate.[34] The oblique way in which the colonel's consumption of the tea is hinted at perhaps speaks to this wider ambivalence of imperial curiosity and repulsion.[35] Given that drinking mother's milk was transgressive, the mistake of their Burmese host in assuming the equivalence of cow›s milk with human breast milk hints at imperial anxieties about dairy consumption. It implied a concern that in Burmese eyes drinking cow›s milk amounted to the same thing as an adult drinking mother›s milk.

Despite these accounts of the defamiliarisation of drinking cow›s milk, it was noted by some imperial writers that occasionally milk was consumed by the Burmese. This was explained – as so many aspects of colonised peoples› behaviour were when they appear to encroach on imperial habits – as a form of imperfect mimicry.[36] As yet another British travel writer, Grattan Geary, put it as early as 1886: milk was drunk 'only in imitation of European selfishness'. Geary was using milk to illustrate the creeping modernisation of Burma, a portrayal that fitted with his wider project of representing the colony as ripe for commercial exploitation.[37] In this vein, drinking milk was often read as a measure of Burma's increasing engagement with the world market. Mostly, it was the consumption of tinned condensed milk that was noted by imperial observers.[38] Writers usually recorded the northern European origin of tinned milk in their descriptions. This detail was mentioned to suggest the cosmopolitanism of milk drinking and so to deploy tinned milk as evidence of the changes to Burmese ways of life. It also reflected the dominance of Dutch, British and Swiss companies as importers of tinned milk during the first half of the twentieth century.[39] Emphasising it as a consumption habit exogenous to Burma, tinned milk was also associated with the population of Chinese merchants who had migrated to the colony to sell imported goods.[40] These portrayals fitted with the widely-held notion that the Burmascape was in decline; that the 'real' Burma was disappearing in the wake of modernity.[41]

Drinking cow's milk took on additional meanings for imperial writers in the context of colonial Burma. The perception that it was not consumed out of a religious concern for animals served to perpetuate an essentialising elision between the population of Burma and Burmese Buddhists. Buddhism had a central place in defining the colony for the British.[42] In imperial writings Indian milk drinkers, such as the Sikh soldiery, might be in Burma but they were not of Burma. British writers' reported experiences of struggling to acquire milk were not banal observations. They can be situated alongside their retelling of numerous apocryphal stories of Burmese women breastfeeding orphaned animals.[43] Through the over-determined role of Buddhism in defining the Burmascape, both were used to demonstrate that the Burmese were too close to animals. In contrast to Indian populations, who were represented as cruel to animals, the Burmese were viewed as being overly sympathetic.[44] Where milk drinking was observed, it was understood to be a form of consumerist mimicry that was part of the erosion of 'traditional' Burma. All of these strands contributed to the British imaginative geography of the colony. Imperial writings on dairy consumption – or, rather, the lack of it – in Burma reified this geography as a given entity, implicitly relying upon it to frame their writings. Burma was where you could not get milk in British India.

Disciplining Dairy Geographies

Imperial explanations for the lack of milk drinking were entangled with the material difficulties of producing cow's milk for human consumption faced in the colony. The historical experiences of the colony's cattle was one barrier to milk production. The logistics of distribution was a further problem. In addition to these there were official concerns over the hygiene of the milch cattle that were available, as well as the insanitary conditions in which milk was prepared for human consumption. The proposed solutions to these issues involved a rearrangement of space to allow for the monitoring and regulation of human-cattle interactions, particularly in urban contexts. Within the discussions over these logistical challenges, the cattle were rendered subjects through their status as 'lively commodities' – commodities valued for their capacities as living entities.[45] Milk production necessitated fostering relationships with cows that attempted to take into account their needs and preferences, however skewed interpretations of those may have been. But the difficulty of milk did not end with the cow. Once produced, the milk itself was liable to adulteration and infection necessitating state and scientific intervention.[46] Limiting the mobility of dairy cattle and removing them from urban areas through policies designed to order and police space were central to colonial schemes for improving milk production and distribution.

Colin Metcalfe Enriquez›s tongue-in-cheek account of the tribulations involved in arranging for a cow to be milked, with which I opened this paper, picks up on an important point: the relative novelty of milk drinking meant that both people and cattle were unused to the production process. Muriel Bowden, wife of a timber firm employee working in Burma in the interwar years, wrote of these problems in a letter to her mother in 1923, 'There are hundreds of cows all skin and bone but no one ever dreams of [milking] them'. She went on, 'We were trying to buy one for our own use. We did have one and the first morning the man brought half a cup full and said the cow kicked him and he wouldn›t do it anymore!' Two weeks later she reported that they had managed to purchase a cow who would 'consent' to be milked.[47] As the animal historian Erica Fudge has shown using the findings of current animal welfare science, cows unused to being milked, or that have been put under stress by unfamiliar places or people, were less likely to produce a good quantity or quality of milk.[48] Enriquez and Bowden may not have been making fanciful anthropomorphic projections when they suggested that these cows were disinclined to be milked. It was a practice they lacked experience of and it would have been performed by the untrained hands of British colonials› servants, with or without rubber gloves. It was likely that in these circumstances the cows would become anxious and upset, and that this would make milking them even more difficult.

Whilst most of the cattle herds seen by Muriel Bowden and other imperial observers would have been employed in agricultural labour, there were a number of small scale Burmese dairy operations in central Burma.[49] The details of one dairy were reported on in 1911. It was a small and highly specialised industry located near Ava, one of the pre-colonial courtly capitals. The dairy had been established before colonial rule. Here the calves were allowed to suckle from their mothers until the evenings, at which point the lactating cows would be milked by humans. Since the cattle were neither bred especially for milk production nor wholly separated from their young, the yields were low, as they were for cattle across the colony.[50] They were milked in purpose-built bamboo sheds, presumably to put the animals at ease. However, the dairy did not produce liquid milk but ghi (clarified butter). This was a luxury product that was occasionally used in cooking across the colony, as was the cottage cheese produced as a by-product of the process. [51] The Burmese name for this local ghi was tawbat (tho pat'). [52] This word was, and continues to be, used more generally to refer to European butter. Reflecting on the previous section, it seems that imperial consumption practices were understood through a preexisting food culture, rather than being seen as unfamiliar, alien, irreligious or even repulsive, as British writers assumed. These local producers may have later adapted their production to meet the new demand for condensed milk, since it was reported in 1940 that condensed milk originating from around this region was being sold in Mandalay.[53] Evidently there was an older, pre-colonial geography of dairy production centred on the political and cultural heartland of the Konbaung Dynasty.

This cottage industry notwithstanding, cattle were not widely used for dairy production. This absence was not viewed negatively by all British officials. According to some imperial writers the fact that the calves received most of their mothers› milk explained their later strength when fully-grown beasts of burden.[54] As the colony's foremost veterinary official, George H. Evans, noted in his 1905 monograph on Burma's buffaloes and cattle, '[t]his is by nature's method .... The calves therefore obtain the full benefit of the whole of their natural nourishment'.[55] The emphasis on 'natural' interactions underscored Evans› views of Burmese relations with their cattle. He thought them the best cared for cattle in the East. Although the colonial state did establish an experimental farm near Mandalay in an attempt to increase milk yields in Burmese oxen, it was of little consequence.[56] Evans went on in his report to ask whether attempts to increase milk yields were necessary at all. Using cattle for dairies was viewed as being in tension with their use as labourers in agriculture, an issue explored in the following section. Depriving the calves of their mothers› milk might add value to lactating cows, but it was believed to reduce the future labour power of the calves themselves.

By the twentieth century most of the dairy production in the colony was conducted by Indians who had migrated to Burma with their own cattle. Indeed, the presence of cattle on ships from the subcontinent was a significant contributing factor in the unpleasant and unhealthy conditions migrants experienced during this journey.[57] These herds were predominantly located in urban spaces and at railway towns. This was to meet the mostly urban demand for milk, since substantial European and Indian populations lived in the colony›s towns and cities.[58] Evans was less than complimentary about Burma's Indian population and their cattle. He viewed these animals as neglected and unhealthy, and believed their milk to have been adulterated through their owners' unclean habits. He argued that given the 'dire maladies' associated with milk produced by Indian milkmen it would be better to advise the milk-drinking foreign population in Burma to mimic the Burmese in abstaining from dairy. He, like many British imperial writers, viewed the muscular physique of Burmese men as evidence that the lack of milk appeared to have little in the way of a detrimental effect on their bodies.[59] Whilst concern about adulteration was a common one back in Britain at the time, Evans' writing suggests that in empire this concern was refracted through notions of race.[60] It was indirect contact with the bodies of Indians, through their animals, that was identified as the primary problem with the milk supply. Indian dairymen were potential vectors of disease. Drinking milk, he worried, might bring the British into proximity with 'uncleanly' Indian bodies.[61]

It was not only human health that was a concern. The regular outbreaks of rinderpest in the colony could result in large death tolls for oxen employed as plough cattle. In one outbreak that occurred in the year 1895 in Akyab district, on the border with Bengal, 1,563 cattle were registered as having died from the disease in under three months. The report into this particularly devastating epizootic was circulated among officials to learn from the lax attitude taken to monitoring the health of cattle. Foot-and-mouth was also a perennial threat, and the government issued descriptions of the disease and its treatment to the population in English and Burmese. The importance of the health of cattle was reflected in the Burmese term chosen by the British as a translation of 'veterinary medicine' in their public notices, which translated directly as 'buffalo-oxen-animal healing knowledge' (kvai nvā: tịrcchān' kụsa mhu chui□' rā paññā). [62] Despite its importance, officials struggled to identify the causes for particular outbreaks, including the Akyab episode. Nonetheless, the condition of Indian cattle and the neglect by their owners was identified as the primary disease pathway. They were said to move in large, uncontrolled herds, eating from grazing lands indiscriminately, drinking from agriculturalists› bunds until they were dry and befouling them, trespassing and trampling crops, and, when diseased, left to die in paddy fields spreading illnesses to the plough cattle.[63] In 1909 Evans, now head of the civil veterinary department, continued to raise concerns about Indian dairy herds spreading disease, particularly foot-and-mouth and rinderpest, because of their disregard for rules on trespassing. He argued that these herds were 'a great strain on the effectiveness of the operations for checking disease and on the obedience of the Burman cultivator to rules which he sees broken with impunity by natives of India'.[64] In the same year he also voiced concern about tuberculosis being spread from dairy cattle to humans through their milk. He encouraged the government to incentivise inoculations for dairy cows, but anticipated difficulties with Hindu owners, who constituted the majority of dairymen, because of their religious sensibilities.[65]

To deal with the threat colonial officials proposed schemes that coalesced around restricting the mobility of Indian cattle and regulating the milk supply in urban spaces. To restrict the mobility of dairy herds the 1871 Cattle-Trespass Act was increasingly used, although, as Evans lamented, often with little success. This legislation enabled municipal authorities to seize and impound wandering, out-of-place or unclaimed cattle.[66] Officials also used wider legislative powers, such as those in the Village Act empowering them to remove people from towns and villages, to deal with Indian cattle. One deputy commissioner recalled in correspondence from 1915 that he had once used these powers to have an entire community of Indian cattle owners ejected from a town because they had 'made the place thoroughly insanitary'.[67] To find ways of better regulating the urban milk supply, F.J. Wrath, the official who ran the government's experimental dairy farm, investigated the milk supply to Mandalay in 1917. He noted that due to the lack of adequate transport infrastructure most dairy came from cattle owned by Indians who resided in the city. On the basis of his analysis of the quality of milk in the colony, he noted that the milk of urban cows contained less fat than that of rural cows and on this basis deemed it less healthy. He also criticised what he saw as the insanitary conditions of city-dwelling cows. His proposed solution was to establish urban depots so that milk brought in from the countryside could be speedily distributed throughout the city. The scheme would offer, it was hoped, a more regulated and reliable supply of milk and thus increase consumption.[68]

No such scheme was put in place and in 1941 a report into the dairy industry across British India and Burma – now a separate colony, but still investigated within the same report – came up with almost identical proposals as a way of improving and growing the market.[69] The desire to remove cows from the city was part of wider developments in colonial governance that saw some animals pushed out of urban spaces for reasons of public health, itself a manifestation of shifts in the geography of cities that saw some nonhumans excluded.[70] In addition to this, it may have been part of a change in imperial sensibilities that desired animals to be separated from the edible products that came from their bodies by moving sites for the production of meat to less visible places.[71] The milk depot may have been serving the same function, in this regard, as the abattoir: to hide the animal materiality of food production. One attempted scheme to improve the insanitary milk supply and remove cattle from the streets was to set up dairies in already existing imperial institutions, such as prisons, hospitals and asylums. These institutions were ready-made visible, urban signs of colonial authority whose internal regimes were nonetheless hidden from view.[72] They were also enclaves of colonial medical practice.[73]For instance, the Rangoon Lunatic Asylum housed a dairy herd of at least fifty animals by 1920, despite being perpetually and acutely overcrowded with humans. It provided the inmates with milk, which was deemed to be an essential foodstuff for the maintenance of their health and made possible the asylum's regime of forced feeding. In addition, the asylum's dairy sold milk to consumers in Rangoon where it could fetch a higher price from European purchasers than that commanded by Indian dairymen in the city.[74]

Although cow's milk was widely lauded for its natural and health-giving characteristics in official circles, making and selling it was not left to occur 'naturally' in Burma.[75]Concerned with the health of human and cattle populations, officials devised numerous schemes for intervening in its production and distribution. With the opening of a medical laboratory in Rangoon in 1926, scientific examinations of the composition of milk in the colony became a routine procedure. Milk from across the colony was sent there for testing.[76] A dedicated research project into the quality of milk in the early 1930s determined that there were particular problems in Burma. The colony›s dairy herds produced milk with low levels of fat and high quantities of non-milk solids. More sophisticated adulteration techniques made identifying malfeasance difficult. The scientists undertaking the research proposed additional criteria that supplemented the newly enacted legislation for monitoring the content of foods and drugs sold in the colony.[77] The various attempts to discipline the dairy industry discussed in this section inspired policies seeking to order and police space. In the process, they reified the political geography of Burma. The specificity of the problems of milk production reinforced the notion of Burma as a place apart from the rest of British India. Furthermore, the framing of 'Indian' cattle as a problem naturalised this geography by embodying it in animals.

Breeding Racial Tensions

The rendering of cattle as lively commodities in the milk industry was seen to be in tension with their commodification in a different economic sector, the rice industry. This was overwhelmingly the most important part of Burma's colonial economy. The late nineteenth century saw a rapid expansion of the deltaic rice frontier. By the opening decades of following century the Burma delta had become the largest rice producing region in the world.[78] The importance of plough cattle was reflected in their market value, which doubled between the end of World War One and 1930. As a result of their cost and necessity, the loss of cattle was often a contributing cause of indebtedness among agriculturalists.[79] Whilst labouring cattle were essential to rice cultivation they were also vulnerable actors. Mortality rates were high for working cattle, particularly during the monsoon months. As a result, colonial officials were concerned about the supply of plough cattle. In particular, they worried that the bloodlines of the Burmese breed of oxen, apparently favoured by cultivators, were at risk. As with concerns about the spread of epizootic disease discussed in the previous section, Indian milch cattle were considered a particular threat. This imperial imperative to protect a so-called 'Burmese' breed of ox reified and naturalised Burma as a geographic entity, with Indian cattle figured as invasive.[80] These concerns were entangled with colonial policies regarding the human Indian population in the colony and with emergent Burmese nationalism.

There is not a trans-historically applicable definition of what makes a 'breed', and tracing their historical emergence is challenging. Ecological and genetic factors physically shaped animals in ways that were beyond human control, even guiding human action.[81]According to the dominant mode of thinking about breeding in the late nineteenth and early twentieth centuries, a breed was conceived of as a population that corresponded to a particular ancestry.[82] In this vein, breeding dairy livestock in Europe during this period relied upon inbreeding and selection on the basis of pedigree to achieve a purity of type. The aim was to raise consistently aesthetically-pleasing, and thus commercially recognisable, types of cow. Caution was aired about selecting purely for milk yields for fear of increasing susceptibility to disease.[83] In Burma, the predominant domesticated species of ox was Bos indicus, a species that was distinct from the European Bos taurus. Buffaloes were not widely used for dairy and became increasingly less favoured for agricultural work during the earlytwentieth century, falling as a proportion of labouring animals even further from their position in 1903, when they were already outnumbered by working oxen nearly five to one. [84] It was within the species Bos indicus that distinctions on the basis of breed were made.

Bos indicus was the same species as the oxen commonly found on the Indian subcontinent. British observers distinguished the Burmese oxen from the wider species group by defining them as a distinct breed. This was not a straightforward distinction for imperial natural historians to make. During the nineteenth century writers used different taxonomical labels for bovines and, at different times, used the terms bison, buffalo and ox in overlapping ways. [85] This confusion notwithstanding, the local ecological pressures and a long pre-colonial history of domestication would have been responsible for producing physical differences between oxen in India and Burma. The latter were described as stockier and more powerful.[86] By the twentieth century this indigenous variety was viewed by British officials as being under threat from the importation of Indian oxen, predominantly used for producing milk, that 'promiscuously' mated in the colony.[87]So, whilst differences existed between Indian and Burmese oxen, these differences were ossified in an imperial taxonomic framework organised by the idea of the breed. Based on this, imperial policy sought to protect the hereditary features that they deemed typical to the Burmese 'purebred'; defined in colonial stock breeding reports as having no more than an eighth of exotic blood in them.[88]

The colonial state had not always been concerned with protecting indigenous cattle from mixing with Indian stock. In the 1850s there were attempts made to improve the breed by importing Nellore bulls from India to sire cattle for beef, since it had proved very difficult for the military to acquire enough local slaughter cattle to meet the demands of European troops in the newly annexed portions of Lower Burma. These experiments were slow to produce results, but were not deemed to have been a total failure. Despite this, they do not appear to have been carried on into the 1860s.[89] This period represents the high point of imperial confidence in being able to improve the stock of domesticated animals in the colony. The government also attempted to establish a sheep farm at Thayetmyo, a town on the border of British Burma and still independent Upper Burma, but the scheme resulted in disastrously high mortality rates before it was abandoned in 1860.[90] After gleaning information on elephant breeding practices in the borderlands between Burma and Siam, the establishment of an elephant farm was even mooted in 1868, although the suggestion was not taken up.[91]

By the turn of the century the state was less inclined to embark on experiments in animal breeding. Instead, to improve the local breed of oxen, officials were encouraged to organise agricultural shows with prizes for the best specimens. These shows appear to have been entirely ineffectual, and in 1895 the deputy commissioner in charge of organising one show held in the delta actually forgot to turn up himself.[92] The only attempt to 'improve' Burmese oxen was at F.J. Wrath's small-scale experimental farm near Mandalay that sought to increase milk yields using indigenous animals. Based on the results of some experiments conducted in India that examined the effects of early castration, the farm attempted to encourage Burmese cattle owners to perform early castration to protect bloodlines.[93] The state's attitude towards Burmese oxen had reversed from wanting to improve the breed through crossbreeding in the 1850s, to attempting to protect its purity in the 1900s. By 1917, a colonial report resolved to protect 'purebred' draught cattle by restricting exotic and crossbred oxen to 'special areas' away from local breeding grounds.[94]

This shift in attitude to the Burmese breed was brought about by a growing recognition of the importance of cattle to the production of rice in the Burma delta. Along with good, regular monsoons, healthy cattle were vital to agricultural production in the colony.[95] The stocky, strong Burmese ox, often bred in the Shan States and other northern parts of the colony, was thought to be especially suited to labour in paddy fields, particularly in comparison to the Indian variety. Through reports into breeding and the use of cattle, Burma was imagined as being constituted of upland areas where cattle were bred and the southern deltaic region where they were worked, often heavily in trying conditions.[96]This was an animal geography that was transgressed by mobile herds of milking cattle imported from India residing along the sides of waterways and in the railway towns that ran through the spine of the colony linking the southern districts to crucial nodes in the north and on the borders. Male calves were sold by Indian dairymen to supplement their incomes, encouraging further 'crossbreeding', alongside that instigated by the oxen themselves.[97] Following the colony's transportation network, migrant Indian cattle penetrated the spaces of both local animal breeding and animal labour.

Figure 2: Young Burmese Bull – pure bred, The Supply of Plough Cattle in Burma, agricultural survey No.7 of 1928, Rangoon, 1929.

Figure 3: Evil effects of crossing – cross-bred cow, The Supply of Plough Cattle in Burma, agricultural survey No.7 of 1928, Rangoon, 1929.

To many officials, by the start of World War One the existing measures for protecting Burmese plough cattle from the 'evils' of Indian milch cattle were deemed inadequate. The push for greater controls began in 1915 with an agricultural and cooperative conference held in Mandalay. Although the extensive measures that were pushed for did not materialise, at least up until 1930, they reveal how colonial officials came to frame Indian cattle as a problem breed. The conference was attended by over nine hundred people from across Burma, including representatives of agriculturalists and state officials. It unanimously agreed that action had to be taken to protect indigenous cattle from Indian cattle. Their suggested course of action was three-pronged: taxation, prohibition and segregation. They argued that owners of milch cattle imported from India should be taxed per head of cattle, that they should be prohibited from importing and owning more than a limited number of cattle, and that they should be required to keep their animals segregated from Burmese cattle to prevent crossbreeding.[98]

The dangers highlighted at this conference, of Indian cattle crossbreeding, trespassing and spreading diseases, were further reinforced in the 1917 official report into stock breeding in the colony. This report also echoed the conference's call for immediate remedial action to be taken, along the same lines as those advocated by the delegates.[99]Although no such action was taken, the concern did not go away. Ten years later the 1928 report into the supply of plough cattle lamented that the recommendations of the 1917 report had not been acted upon. This report was particularly concerned with crossbreeding, using photographs to illustrate the 'evil effects' of this mixture (see Fig. 2 and Fig. 3). The first image shows a 'purebred' Burmese bull which has a consistent colour, prominent hump and stocky build. The 'cross-bred' cow in the second image, in contrast, has patchy colouration, a smaller hump and is a skinnier animal. Through these images the author intended to identify and reify distinctive traits that defined the Burmese breed, using the photographs to help fix bodily difference.[100] Their recommendations again drew on those made in 1915 and 1917, but added that there should be a compulsory census of Indian breeders and forced castration of 'undesirable' animals.[101]

Attitudes to Indian cattle in the colony were conterminous with attitudes to Indian people. The interventions listed above can be considered as part of a wider range of state controls placed on Indian migrants to Burma. The timing of these committees was synchronous with inquiries into the sanitary conditions that Indian workers travelled and lived in, and into their role in the spread of smallpox. At the same time that these measures were taken to monitor and regulate Indian breeds of cattle, the state introduced compulsory medical checks and vaccinations on human arrivals from the subcontinent.[102] In addition, the concerns expressed by officials contributing to these reports on cattle in Burma were indicative of British officialdom's paternalistic attitude towards the Burmese people, viewing their role as protecting the Burmese from the Indian and Chinese populations.[103] The administrative view of the colony, which by the turn of the century held it to be culturally distinct from India, was increasingly imagining it as a separate geo-political entity. Officials began planning for it to be separated from British India.

During the interwar years anti-Indian sentiments gained ground within Burmese nationalism. Burma›s initial omission from the 1919 Montagu-Chelmsford reforms, that introduced dyarchy to British India, sparked nationalist protests. Thereafter, the relationship with India was a hotly contested issue within nationalist politics. Politicians were divided between those against separation, fearing that governed separately the colony would not receive the benefits of reforms being negotiated for India, and those agitating for separation, railing against the perceived problems of uncontrolled immigration.[104] During this time, concern about mixed-race marriages between Indian men and Burmese women became prevalent. Indian migrants were figured by some as a threat to the Bama race and to Buddhism, fostering a xenophobic and misogynistic political climate. [105] There were a number of anti-Indian riots in the 1930s in which Indian casualties heavily outweighed Burmese. At the same time further immigration controls were passed. The 1935 Government of India Act was enacted in 1937 separating Burma from India and granting the colony its own constitutional arrangements for limited self-government. Whilst concerns about Indian dairymen and their cattle were not as pressing or acrimonious as those surrounding Indian creditors and low-skilled labourers, they nestled within the same political fault lines.

The colonial framing of the problems of dairy production informed nationalistic writing. A leading article in the Burmese-language daily, The Sun, addressed the issue of dairy farming in October 1915. The author attempted to encourage Burmese people to learn more about breeding dairy herds and to take it up for the sake of the country's prosperity, noting that the production of dairy was mostly in the hands of Indians with little knowledge. This meant looking after the breeding of cattle, something that was deemed at the time to be 'imperfect' (ma phrañ' jụ). However, the writer also acknowledged that for many this work was thought to be of a bad sort (dụrāciwa). [106] The lack of milk production was figured as a developmental lag, with Indians viewed as a barrier to Burma overcoming its dependence on foreign dairy products. An earlier article had made the same observations about cattle breeding in the colony, noting that Burmese people were missing out on an economic opportunity since husbandry was predominantly done by Indians.[107] An article published that same year reported on a state-led conference into the milk supply in much the same vein. Echoing colonial concerns, the paper drew attention to the problem of Indian dairymen, referring to them as Kalar, a term that could hold derogatory connotations. The article warned of strategies that these Indian dairymen could deploy to avoid being caught adulterating milk with dirty water and unclean utensils. The article also warned that the milk could be poisoned (achip') by them and no one would be able to find out. [108]

The articles published in The Sun show a close engagement with imperial policy and reports. Building on the problems raised officially, Burmese journalists translated them into economic and social concerns for the nation's development. Moving into the interwar years, it is worth noting that official complaints about crossbreeding were being voiced at the same time as anxieties over relationships between Indian men and Burmese women were becoming increasingly prominent. Although I have no evidence of the two being explicitly connected, there was a representational link made between human strength and the health of cattle through the use of the ox to represent the Bu rmese peasantry in nationalist iconography; powerful but in need of leadership. [109] The extent of the overlap between human and animal concerns in Burmese nationalist thought deserves greater study.

CONCLUSION

The construction of herds of Indian dairy cattle as an epizootic and genetic threat to the Burmese ox was entangled with the realisation of the 'embryonic border' between India and Burma. Within the intertwined discussions of dairy consumption and production, as well as the welfare of oxen, Burma was implicitly reified as a geopolitical unit. Imperial imaginings of Burmese Buddhism offered an essentialising explanation for the relative absence of milk drinking in the colony. Indian dairy herds were figured as invasive entities, naturalising a political geography embodied in the Burmese breed of oxen. Uncovering the role of cattle as lively commodities in this history draws attention to the competing ways that their value as living creatures could be realised. Dairy herds were pitted against plough cattle in a limited political ecology, particularly in the vast rice-producing delta. All this shows that in investigating the emergence of milk as a commodity in colonial Burma it is necessary be attentive to how space was produced. Burma was not a frontier in milk's otherwise inexorable 'conquest of space', as Valenze's narrative might suggest. Instead, the history of milk was entangled in the immanent geography of British Burma.

As well as shedding light on the more-than-human aspects of colonial Burma's political geography, the article is a warning against historical narratives about food that take the globe to be an unproblematic, ready-made space across which consumption habits have spread.[110] Discussions of the globalization of consumption preferences that are inattentive to theorisations of space risk reifying political geographic frameworks. The anthropologist of milk, Andrea Wiley, has emphasised that long local histories have been pivotal in how milk has become increasingly globalized, making insightful comparisons between China, India and the USA.[111] Her work is a sophisticated response to some of the more blithe commentary on the globalisation of milk, but her work still takes these national units for granted. In addition, whilst she opens space for researching at different scales, her approach does not deconstruct how the scalar levels of local and global have been produced over time. She does not attempt to excavate the history of imperialism in the making of these scales. However, as this article has shown, the material and imaginative work involved in making, selling and consuming milk was not innocent of the spatial politics of colonial rule. As such, it encourages greater recognition that the geographic entities used to track and trace milk, and other commodities, across the planet have their own histories, histories shaped by the asymmetries and territorialisations of imperial power.[112]

ACKNOWLEDGEMENTS

I am very grateful to the Arts and Humanities Research Council for providing the funding for this research (AH/L014939/1). Thanks to Stephen Legg, Rachel Johnson, William Gould, Natasha Pairaudeau and Erich DeWald for their feedback on drafts of this paper. I also received invaluable comments on earlier versions as part of the 'History of Milk in Asia' panels at the Asian Dynamics Initiative's conference, hosted by the University of Copenhagen, and at the research seminar of the Department of History at the University of Aberystwyth. The comments of the editor and the three anonymous reviewers have been exceptionally helpful in improving the article.

REFERENCES

1. C.M. Enriquez, A Burmese Loneliness:ATale of Travel in Burma, the Southern Shan States and Keng Tung, Calcutta, 1918, 245.
2. R.W. Winston, Four Years in Upper Burma, London, 1892, 182; G.W. Bird, Wanderings in Burma, London, 1897, 43; F.W.T. Pollok and W.S. Thom, Wild Sports of Burma and Assam, London, 1900, 201; G.E. Mitton, A Bachelor Girl in Burma, London, 1907, 131; Anon., A Dog's Life in Burma, Told by the Dog, London, 1909, 36, 68, 86.
3. Report on the Marketing of Milk in India and Burma (Abridged Version), Delhi and Simla, 1941, 8–11.
4. D. Valenze, Milk: A Local and Global History, New Haven, 2011, 4–7.
5. H. Lefebvre, The Production of Space, translated by D Nicholson-Smith, Oxford, 1991.
6. N. Osada, An embryonic border: racial discourses and compulsory vaccination for Indian immigrants at ports in colonial Burma, 1870–1937, Moussons 17 (2011) n.p.
7. F. Driver, Imagining the tropics: views and visions of the tropical world, Singapore Journal of Tropical Geography 25 (2004) 1–17; and for a study that argues for the material importance of animals, see J. Lorimer and S. Whatmore, After the 'King of Beasts': Samuel Baker and the embodied historical geographies of elephant hunting in mid-nineteenth-century Ceylon, Journal of Historical Geography 35 (2009) 668–689.
8. C. Philo and C. Wilbert, Animal spaces, beastly places: an introduction, in: C. Philo and C. Wilbert (Eds), Animal Spaces, Beastly Places: New Geographies of Human-Animal Relations, London, 2000, 1–36.
9. E. Fudge, A left-handed blow: writing the history of animals, in: N. Rothfels (Ed.), Representing Animals, Bloomington, 2003, 3–18; L. Nash, The agency of nature or the nature of agency? Environmental History 10 (2005) 67–69; D. Haraway, The Companion Species Manifesto: Dogs, People, and Significant Otherness, Chicago, 2003.
10. H. Buller, Animal geographies I, Progress in Human Geography 38 (2013) 308–318; H. Buller, Animal geographies II: methods, Human Geography 39 (2014) 374–384.
11. K. Hobson, Political animals? On animals as subjects in an enlarged political geography, Political Geography 26 (2007) 250–267; although, as has been recently noted, the impact of this on the sub-field of political geography has only been slight, see K. Srinivasan, Towards a political animal geography? Political Geography, in press.
12. R-C. Collard and J. Dempsey, Life for sale? The politics of lively commodities, Environment and Planning A 45 (2013) 2682–2699; N. Shukin, Animal Capital: Rendering Life in Biopolitical Times, Minnesota, 2009.
13. B.L. Walker, Animals and the intimacy of history, History and Theory 42 (2013) 45–67.
14. G. Gaard, Toward a feminist postcolonial milk studies, American Quarterly 65 (2013) 595–618.
15. J. Bourke, What It Means to Be Human: Reflections from 1791 to the Present, London, 2011, 265–330; S-W. Poon, Dogs and British colonialism: the contested ban on eating dogs in colonial Hong Kong, Journal of Imperial and Commonwealth Studies 42 (2014) 308–328.
16. Indeed, the very history of the scientific designation 'mammal' reveals the ambivalent and gendered place that natural historians gave humans within the animal kingdom. See L. Schiebinger, Why mammals are called mammals: gender politics in eighteenth-century natural history, American Historical Review 98 (1993) 382–411.
17. S.L. Keck, British Burma in the New Century, 1895–1919, Basingstoke, 2015, 28–43.
18. F.J. Simoons, The traditional limits of milking and milk use in southern Asia, Anthropos 65 (1970) 555.
19. V.L. Rafael, The cultures of area studies in the United States, Social Text 41 (1994) 91–111; W. van Schendel, R. Cribb, J.D. Sidaway and T. Than, Southeast Asia: an idea whose time is past? Journal of the Humanities and Social Sciences of Southeast Asia 168 (2012) 497–510.
20. M. Baud and W. van Schendel, Toward a comparative history of borderlands, Journal of World History 8 (1997) 211–242; W. van Schendel, Geographies of knowing, geographies of ignorance: jumping scale in Southeast Asia, Environment and Planning D: Society and Space 20 (2002) 647–668; J.C. Scott, The Art of Not Being Governed: An Anarchist History of Upland Southeast Asia, New Haven, 2009; M.W. Charney, Literary culture on the Burma–Manipur frontier in the eighteenth and nineteenth centuries, The Medieval History Journal 14 (2011) 159–181; M. Sadan, Being and Becoming Kachin: Histories beyond the State in the Borderworlds of Burma, Oxford, 2013.
21. Simoons, The traditional limits of milking and milk use in southern Asia, 547–593.
22. G. Bankoff and S. Swart, Breeds of empire and the 'invention' of the horse, in: G. Bankoff and S. Swart (Eds), Breeds of Empire: The 'Invention' of the Horse in Southeast Asia and Southern Africa, 1500–1950, Copenhagen, 2007, 6–10.
23. P. Armstrong, The postcolonial animal, Society & Animals 10 (2002) 413–419. Also see the discussion in the introduction to S. Rajamannar, Reading the Animal in the Literature of the British Raj, New York, 2012.
24. M. Radhakrishna, Of apes and ancestors: evolutionary science and colonial ethnography, Indian Historical Review 33 (2006) 1–23; J. Vetter, Wallace's other line: human biogeography and field practice in the eastern colonial tropics, Journal of the History of Biology 39 (2006) 89–123; S. Sivasundaram, Imperial transgressions: the animal and human in the idea of race, Comparative Studies of South Asia, Africa and the Middle East 35 (2015) 156–172.

25. For a fuller discussion, see J. Saha, Among the beasts of Burma: animals and the politics of colonial sensibilities, c.1840–1950, Journal of Social History 48 (2015) 933–955.
26. Keck, British Burma in the New Century, 30–33.
27. V.C.S. O'Connor, The Silken East: A Record of Life and Travel in Burma, Volume 1, London, 1904, 288.
28. R.T. Kelly, Burma, Painted and Described, London, 1905, 248. For more on Kelly, and on imperial travel writing on Burma more widely, see S.L. Keck, Picturesque Burma: British travel writing 1890–1914, Journal of Southeast Asian Studies 35 (2004) 387–414.
29. Turner, Saving Buddhism: The Impermanence of Religion in Colonial Burma, Honolulu, 2014, 91–93; M.W. Charney, Demographic growth, agricultural expansion, and livestock in the lower Chindwin in the eighteenth and nineteenth centuries, in: G. Bankoff and P. Boomgaard (Eds), A History of Natural Resources in Asia: The Wealth of Nature, Basingstoke, 2007, 227–244.
30. Mitton, A Bachelor Girl in Burma, 131.
31. Pollok and Thom, Wild Sports of Burma and Assam, 470–473.
32. Anon., A Dog's Life in Burma, 68.
33. Sen, Woman and Empire: Representations in the Writings of British India, 1858–1900, London, 2002.
34. Saha, Among the beasts of Burma; W. Anderson, The trespass speaks: white masculinity and colonial breakdown, American Historical Review 102 (1997) 1343–1370. The politics of cross-racial breastfeeding might also be considered as part of the context for understanding this transgression. It was a common practice to employ Indian wet-nurses for British children in the Raj. However, using Rudyard Kipling's Jungle Book stories, it has been argued that literary portrayals of receiving milk represented Oedipal desires for the absent and forbidden wet-nurse mother figure, see J. Hotchkiss, The jungle of Eden: Kipling, wolf boys, and the colonial imagination, Victorian Literature and Culture 29 (2001) 435–449.
35. I am not suggesting that these were innate desires, but rather endorse Ann Laura Stoler's argument that colonialism engenders these taboos that require policing, see A.L. Stoler, Race and the Education of Desire: Foucault's History of Sexuality and the Colonial Order of Things, Durham, N.C, 1995.
36. H.K. Bhabha, Of mimicry and man: the ambivalence of colonial discourse, October 28 (1984) 125–133. For a discussion of this discursive strategy in colonial Burma, see P. Edwards, Half-cast: staging race in British Burma, Postcolonial Studies 5 (2002) 279–295.
37. G. Geary, Burma, After the Conquest, Viewed in Its Political, Social, and Commercial Aspects, from Mandalay, London, 1886, 18–19.
38. O'Connor, The Silken East, 183, 666.
39. British Library, India Office Records [hereafter IOR], V/25/500/268: F.D. Odell, Market Surveys of Burma Crops: An Introductory Note, Bulletin No. 32 of 1936, Rangoon, 1936, 10.
40. A.M.R. Hart, Picturesque Burma: Past & Present, London, 1897, 253.
41. Keck, British Burma in the New Century, 71–93.
42. See, for instance, J.G. Scott, The Burman: His Life and Notions, third edition, London, 1910; H. Fielding-Hall, The Soul of a People, third edition, London, 1899.
43. For further details, see Saha, Among the beasts of Burma, 918–919.
44. P. Chakrabarti, Beasts of burden: animals and laboratory research in colonial India, History of Science 48 (2010) 125–152.
45. Collard and Dempsey, Life for sale; D. Haraway, When Species Meet, Minneapolis, 2008, 45–67.
46. P. Atkins, Liquid Materialities: A History of Milk, Science and the Law, Farnham, 2010.
47. IOR MSS EUR/F575: A. Bowden and I. Adams (Eds), Ever Your Loving Mully: Letters from Burma of Muriel Bowden to Her Mother Alice Britten in England, 1922-24, 2008, 1 July 1923, 15 July 1923.
48. E. Fudge, Milking other men's beasts, History and Theory 52 (2013) 13–28.
49. IOR V/26/540/1: Report of the Stock Breeding Committee Burma 1917, Rangoon, 1918, 51.
50. IOR V/27/540/20: G.H. Evans, Monograph on the Cattle and Buffaloes of Burma Calcutta,, 1905, 18.
51. IOR V/25/500/264: A. McKerral, Sagaing District, Department of Agriculture, Burma, Agricultural Surveys No. 2 Rangoon, 1911, 27–28.
52. Transliterated according to J. Okell, with U Saw Tun and Daw Khin Mya Swe, Burmese (Myanmar): An Introduction to the Script, DeKalb, 1994, Appendix 3, Transliteration: the alphabet with transliteration equivalents.
53. Report on the Marketing of Milk, 26.
54. H. Yule, A Narrative of the Mission Sent by the Governor-General of India to the Court of Ava in 1855, with Notices of the Country, Government, and People, London, 1858, 2; Mitton, A Bachelor Girl in Burma, 131.
55. IOR V/27/540/20: Evans, Monograph on the Cattle, 15.
56. IOR V/26/540/1: Report of the Stock Breeding Committee, 57.
57. Report of the Committee Appointed to Enquire into the Allegation of Inconvenience and Hardship Suffered by Deck Passengers Travelling between Burma and India, Rangoon, 1918.
58. IOR V/26/540/1: Report of the Stock Breeding Committee, 37.
59. IOR V/27/540/20: Evans, Monograph on the Cattle, 32.
60. P.J. Atkins, Sophistication detected: or, the adulteration of the milk supply, 1850–1914, Social History 16 (1991) 317–339.
61. IOR V/27/540/20: Evans, Monograph on the Cattle, 32. This may have been part of the 'affective wall' that Anglo-Indians erected between themselves and Indian environments, see E.M. Collingham, Imperial Bodies:

The Physical Experience of the Raj, c.1800–1947, Cambridge, 2001.

62. National Archives of Myanmar, Yangon [hereafter NAM], 1/15(E) 13842, 1894 File No. 1V-1: 27 February 1894. This was in contrast to the rest of British India where militarily useful animals such as horses and camels were the priority, see S. Mishra, Beasts, murrains, and the British Raj: reassessing colonial medicine in India from the veterinary perspective, 1860–1900, Bulletin of the History of Medicine 85 (2011) 587–619.

63. IOR V/26/540/1: Report of the Stock Breeding Committee, 38.

64. NAM 1/15(E) 3200, 1909 File No. 2C-5: 27 October 1909.

65. NAM 1/15(E) 3198, 1909 File No. 2C-3: 25 August 1909.

66. IOR V/27/540/18, 1910: The Cattle Trespass Act Manual [with correction slips], Rangoon, 1910.

67. NAM 1/15(E) 7594, 1915 File No. 2C-3: 18 November 1915.

68. IOR V/25/500/267: F.J. Warth, The Mandalay Milk Supply, Department of Agriculture, Burma, Bulletin No. 15 Rangoon, 1916, 1–3, 6, 8–11, 14–16.

69. Report on the Marketing of Milk.

70. D. Arnold, Pollution, toxicity and public health in metropolitan India, 1850–1939, Journal of Historical Geography 42 (2013) 124–133; C. Philo, Animals, geography, and the city: notes on inclusions and exclusions, Environment and Planning D: Society and Space 13 (1995) 655–681.

71. R.J.H. Woods, From colonial animal to imperial edible: building an empire of sheep in New Zealand, ca. 1880–1900, Comparative Studies of South Asia, Africa and the Middle East 35 (2015) 117–136.

72. D. Arnold and C. Anderson, Envisioning the colonial prison, in: F. Dikotter and I. Brown (Eds), Cultures of Confinement: A History of the Prison in Africa, Asia and Latin America, Ithaca, 2007, 304–331.

73. D. Arnold, Colonizing the Body: State Medicine and Epidemic Disease in Nineteenth-Century India, Berkeley, 1993.

74. See Report on the Rangoon Lunatic Asylum, for the year 1884, Rangoon, 1885, and subsequent years. Although it is not stated in the reports the numbers by the 1920s must have been higher than fifty, since sixty-four animals were sold between 1919 and 1923 without being replaced and milk was still being produced, see Report on the Lunatic Asylums of Burma, for the triennium 1921–1923, Rangoon, 1924. For more on the management of the asylum, see J. Saha, Madness and the making of a colonial order in Burma, Modern Asian Studies 47 (2013) 406–435.

75. For instance, it is referred to as the 'most perfect single food known to mankind' and its status as a 'natural' product that cannot be artificially reproduced is praised in Report on the Marketing of Milk, ii, 9, 36.

76. See Second Annual Report of the Harcourt Butler Institute of Public Health, Rangoon, For the Year 1927, Rangoon, 1928, and following reports.

77. E.H. Bunce, Investigations on milk standards under the Burma Food and Drugs Act, 1928, Analyst 57 (1932) 449–453; Anon., Report on the Burma Food and Drugs Acts, 1928 with recommendations for food standards, Analyst 58 (1933) 759–761.

78. M. Adas, The Burma Delta: Economic Development and Social Change on an Asian Rice Frontier, 1852–1941, Madison, 1974; I. Brown, A Colonial Economy in Crisis: Burma's Rice Delta and the World Depression of the 1930s, London, 2005. However, it must be recognised that this expansion had pre-colonial roots, see: M. Adas, Imperialist rhetoric and modern historiography: the case of Lower Burma before and after conquest, Journal of Southeast Asian Studies 3 (1972) 175–192; P.A. Coclanis, Southeast Asia's incorporation into the world rice market: a revisionist view, Journal of Southeast Asian Studies 24 (1993) 251–267.

79. C. Siok-Hwa, The Rice Industry of Burma, 1852–1940, Kuala Lumpur, 1968, 45–48.

80. For histories of how breeding animals reified colonial cultures, see G. Bankoff, A question of breeding: zootechny and colonial attitudes toward the tropical environment in the late nineteenth-century Philippines, The Journal of Asian Studies 60 (2001) 413–437; S. Cheang, Women, pets, and imperialism: the British Pekingese dog and nostalgia for old China, Journal of British Studies 45 (2006) 359–387; A. Skabelund, Breeding racism: the imperial battlefields of the 'German' shepherd dog, Society & Animals 16 (2008) 354–371.

81. S. White, From globalized pig breeds to capitalist pigs: a study in animal cultures and evolutionary history, Environmental History 16 (2011) 94–120; Woods, From colonial animal to imperial edible.

82. Bankoff and Swart, Breeds of empire and the 'invention' of the horse, 11.

83. B. Theunissen, Breeding without Mendelism: theory and practice of dairy cattle breeding in the Netherlands 1900–1950, Journal of the History of Biology 41 (2008) 637–676.

84. IOR V/26/540/1: Report of the Stock Breeding Committee, 45.

85. J. Crawfurd, Quadrupeds and other animals of Burma (1827); M.W. Charney (Eds), SOAS Bulletin of Burma Research 3 (2005) 530–533; Rev. F. Mason, The Natural Productions of Burmah, or Notes on the Fauna, Flora, and Minerals of the Tenasserim Province and Burman Empire, Maulmain, 1850, 247–250; E. Blyth, Catalogue of Mammals and Birds of Burma, Hertford, 1875, 47–49.

86. Existing domestication was noted by British imperial writers, see Mason, The Natural Productions of Burmah, 248; and it was deemed a mark of 'civilization' by some, see J. Crawfurd, On the relation of the domesticated animals to civilization, Transactions of the Ethnological Society of London 2 (1863) 387–468. For more on imperial views on domestication, see D. Brantz, The domestication of Empire: human-animal relations at the intersection of civilization, evolution, and acclimatization in the nineteenth century, in: K. Kete (Ed.), A Cultural

History of Animals in the Age of Empire, Oxford, 2007, 74–94.

87. IOR V/26/540/1: Report of the Stock Breeding Committee, 37–38; The Supply of Plough Cattle in Burma, Agricultural Survey No.7 of 1928 Rangoon, 1929, 5, 31, 38, 53.

88. IOR V/26/540/1: Report of the Stock Breeding Committee, 37.

89. NAM 1/1(A) 244, 1856 File No. 66; NAM, 1/1(A) 246, 1856 File No. 68; 1/1(A) 309, 1857 File No. 38.

90. NAM 1/1(A) 116, 1854 File No. 60; NAM, 1/1(A) 342, 1858 File No. 15; NAM, 1/1(A) 479, 1859 File No. 81; NAM, 1/1(A) 510, 1860 File No. 10.

91. NAM 1/1(A) 1400, 1868 File No. 508: 18 May 1868; 4 April 1868.

92. NAM 1/1(E) 13895, 1895 File No. 5A-1: 5 June 1895.

93. IOR V/26/540/1: Report of the Stock Breeding Committee, 57.

94. IOR V/26/540/1: Report of the Stock Breeding Committee, 37.

95. This was true for much of British India, see S. Mishra, Cattle, dearth, and the colonial state: famines and livestock in colonial India, 1896–1900, Journal of Social History 46 (2013) 989–1012.

96. For the pre-colonial roots of one of these breeding sites, see Charney, Livestock in the lower Chindwin.

97. IOR V/26/540/1: Report of the Stock Breeding Committee, 38, 50, 56.

98. NAM 1/15(E) 7594, 1915 File No. 2C-3: 3 September 1915.

99. IOR V/26/540/1: Report of the Stock Breeding Committee, 35–57.

100. Anderson, Legible Bodies.

101. Supply of Plough Cattle in Burma, 5. Kaur, Indian labour, labour standards, and workers' health in Burma and Malaya, 1900–1940, Modern Asian Studies 40 (2006) 425–475; Osada, An embryonic border.

102. This attitude is apparent in imperial writings throughout the early twentieth century. See, for instance, H.T. White, A Civil Servant in Burma, London, 1913; M. Collis, Trials in Burma, London, 2008.

103. For a comprehensive overview, see R. Taylor, The State in Myanmar, London, 2008, 175–218.

104. C. Ikeya, Refiguring Women, Colonialism, and Modernity in Burma, Honolulu, 2011.

105. The Sun (Suriyā), 16 October 1915, 11.

106. The Sun (Suriyā), 1 April 1915, 12.

107. The Sun (Suriyā), 1 June 1915, 13.

108. See, for instance, the cover of the nationalist newspaper, New Burma, 23 February 1938.

109. See, for a clear example of this, E. Frazão, B. Meade and R. Anita, Converging patterns in global food consumption and food delivery systems, Amber Waves 6 (2008) 22–29; for historical approaches that raise the issue of Eurocentrism and consider global interconnectivity, but do not historicise global space, see the introductory essays to R. Grew, Food in Global History, London, 2008, and J-L. Flandrin and M. Monatanari, Food: A Culinary History, New York, 2013.

110. A.S. Wiley, Transforming milk in a global economy, American Anthropologist 109 (2007) 666–677; A.S. Wiley, Milk for 'growth': global and local meanings of milk consumption in China, India, and the United States, Food and Foodways 19 (2011) 11–33.

111. Burton, Not even remotely global? Method and scale in world history, History Workshop Journal 64 (2007) 323–328; S. Hodges, The global menace, Social History of Medicine 25 (2012) 719–728; J.D. Sidaway, C.Y. Woon and J.M. Jacobs, Planetary postcolonialism, Singapore Journal of Tropical Geography 35 (2014) 4–21.

CHAPTER 12 TESTING THE AMINOSTRATIGRAPHY OF FLUVIAL ARCHIVES: THE EVIDENCE FROM INTRA-CRYSTALLINE PROTEINS WITHIN FRESHWATER SHELLS

K.E.H. Penkman, R.C. Preece, D.H. Keen, D. Maddy, D.C. Schreve, M.J. Collins

INTRODUCTION

Across western Europe many large-scale river terrace sequences preserve sedimentary archives that appear to reflect fluvial system responses driven by climate change at orbital (Milankovitch) frequencies (Bridgland et al., 2004; Bridgland 2006; Bridgland and Westaway, 2007), but the precise timing of events requires accurate dating of the deposits. Such geochronological age estimates are provided by various means: radiocarbon, U-series, electron spin resonance (ESR), optically stimulated luminescence (OSL) and amino acid geochronology, allowing correlation with stages of the oxygen isotope framework of the global ocean, which in turn is correlated with variability in the orbital parameters of obliquity and precession (Imbrie et al., 1984).

Amino acid racemization (AAR) depends on the slow interconversion (racemization) of l-amino acids, the basic building blocks of protein, into an equilibrium mixture of l- and d-amino acids in fossils over time. Even in an optimal set of samples, with closed system behaviour and identical (internal) chemical environments for degradation, aminostratigraphic correlation between sites relies on the samples sharing an equivalent temperature history (e.g. Wehmiller et al., 2000). Bowen et al. (1989) provided the first comprehensive aminostratigraphic framework of non-marine deposits in southern Britain, based upon the extent of a single AAR reaction (A/I epimerization) in mollusc shells, proposing correlations of the deposits with the marine oxygen isotope record. This framework has been refined with the addition of further analyses and the inclusion of further sites (e.g. Bowen, 1999 and Bowen, 2000) and modified most notably in the case of Hoxne (Bowen, 2003) as the result of new ESR age estimates (Grün and Schwarcz, 2000).

This paper presents the results of improvements in sample pre-treatment and new analytical methods (Penkman, 2005; Penkman et al., 2007). Instead of using only isoleucine, as in previous work, multiple amino acids are analysed from the intra-crystalline fraction of *Valvata*, *Bithynia* and *Corbicula* shells. For the period under scrutiny, the last ~450 ka, it confirms that there is a relationship between the age of terraces in the Lower Thames and the extent of racemization.

MATERIALS AND METHODS

Materials

Rates of racemization may vary significantly between species (Miller and Hare, 1975; Lajoie et al., 1980). For marine taxa, amino acid ratios of fast, moderate and slow racemizers (Miller and Mangerud, 1985) are often standardized to a common reference species (e.g. Bowen et al., 1985; Bowen and Sykes, 1988). Racemization rates also vary between non-marine taxa (Hughes, 1987) and although they were standardized for the study of the Somme and Seine terraces (Bates, 1993), it was thought that insufficient data were available for a study across different environments in southern England and assumed that rates of A/I epimerization were approximately the same (Bowen et al., 1989). In this study we test this assumption on the intra-crystalline fraction by investigating four taxa abundant

Citation: K.E.H. Penkman, R.C. Preece, D.H. Keen, D. Maddy, D.C. Schreve, M.J. Collins, Testing the aminostratigraphy of fluvial archives: the evidence from intra-crystalline proteins within freshwater shells, Quaternary Science Reviews, Volume 26, Issues 22–24, November 2007, Pages 2958-2969, ISSN 0277-3791, http://dx.doi.org/10.1016/j.quascirev.2007.06.034.

in fluvial sediments throughout Europe and commonly used for aminostratigraphy. Three of these are prosobranch gastropods: *Valvata piscinalis* (Müller), which in NW Europe occurs commonly in both temperate and cold stage sediments, and *Bithynia tentaculata* (L.) and *B. troschelii*, which are known only from interglacial or interstadial deposits. The fourth species is the bivalve *Corbicula fluminalis* (Müller), which in Britain is known only from interglacial sediments but not from those of Last Interglacial age (Keen, 1990; Meijer and Preece, 2000). The systematics of *Corbicula* are complex and attribution of the Pleistocene shells to *C. fluminalis* is provisional (Meijer and Preece, 2000). An earlier Tertiary species, *Corbicula obovata* (J. Sowerby) was also analysed to test whether AAR values were increased in samples significantly older than the Pleistocene material in question. Previous studies have shown that ratios from *Corbicula* vary systematically, depending on whether the samples were taken from the umbonal region or the outer margin (West et al., 1999); all samples in this study have been taken from the outer margin. *V. piscinalis*, *B. tentaculata*, *C. fluminalis* and *C. obovata* shells have therefore been analysed from a range of sites, detailed in the Online Supplement.

Methods

All samples were prepared using the procedures of Penkman (2005) and Penkman et al. (2007). In brief, each sample was powdered and bleached for 48 h with 12% NaOCl to isolate the intra-crystalline fraction. Bleaching removes unwanted (contaminating or partly leached) proteins from gastropod samples (Penkman et al., 2007), reducing within-site variability. Some samples were also analysed unbleached, to enable comparison with previous methods. Two subsamples were then taken: one fraction was directly demineralized and the free amino acids analysed (referred to as the 'free amino acid' fraction; FAA; 'F'), and the second was treated with 7 M HCl under N_2 at 110 °C for 6 h or 24 h to release the peptide-bound amino acids, thus yielding the 'total hydrolysable amino acid' concentration (THAA; 'H'). The shorter hydrolysis time (6 h) did not appear to break all the peptide bonds; therefore, in subsequent experiments, 24 h hydrolysis was used. Samples were then dried and rehydrated for Reverse Phase High Pressure Liquid Chromatography (RP-HPLC) analysis with l-homo-arginine as an internal standard.

The amino acid compositions of the samples were analysed in duplicate by RP-HPLC using fluorescence detection, following a modified version of the method of Kaufman and Manley (1998). A sample is injected and mixed online with the derivitising reagent (*n*-iso-l-butyryl l-cysteine (IBLC) and *o*-phthaldialdehyde (OPA)). The amino acids are separated on a C_{18} HyperSil BDS column at 25 °C using a gradient elution of three solvents: sodium acetate buffer, methanol and acetonitrile. During preparative hydrolysis both asparagine and glutamine undergo rapid irreversible deamination to aspartic acid and glutamic acid, respectively (Hill, 1965). It is therefore not possible to distinguish between the acidic amino acids and their derivatives and they are reported together as Asx and Glx. Error bars for the duplicate analyses are not included in the figures, for the sake of clarity, but an alternative set of figures, with error bars, is included in the Online Supplement. Although some samples analysed would not pass the criteria for closed system protein (Preece and Penkman, 2005), all data obtained is shown. Statistical analysis was performed using Minitab v.14.

Pre-Treatment Developed to Isolate a Closed System

In order to overcome problems arising from burial diagenesis, it is argued that the protein system must remain closed from synthesis to analysis (Brooks et al., 1990; Collins and Riley, 2000; Miller et al., 2000). In a closed system, diagenetic reactions of indigenous biomolecules should be predictable. Consequently the distribution between original molecules and their degradation products can theoretically be used to interpret the diagenetic history and to identify samples contaminated by exogenous amino acids and/or those that have lost part of their original protein. A closed system has the further advantage of a consistent chemical environment, including narrow pH range, as well as consistent ion and water concentrations (Towe, 1980; Sykes et al., 1995). The success of avian eggshell as a material for amino acid geochronology has been attributed to its approximation to a closed system (Brooks et al., 1990; Miller et al., 2000). Mollusc shell, more widely available as dating material, does not form a completely closed system (Miller and Hare, 1980; Roof, 1997), retaining only 60% of its original protein in experimental studies. However, a component of the protein within mollusc shells does appear to be effectively 'closed' and immune from environmental effects. This 'intra-crystalline' fraction can be isolated by bleaching, removing the inter-crystalline matrix and any exogenous amino acids (Towe, 1980). Preliminary results of bleaching in terrestrial gastropods have provided improved datasets (Sykes et al., 1995), as have results from a larger investigation on both modern Holocene aquatic shells and their Pleistocene counterparts (Penkman et al., 2007). Indeed the latter study demonstrated for the first time that the intra-crystalline fraction behaved as a closed system, both experimentally and in fossil shells. Bleach pre-treatment removes any partly degraded/leached and contaminating proteins, providing a closed-system protein fraction with more tractable kinetics. As a consequence of the ability to isolate a closed system in gastropod shells, it has been possible to measure routinely not only the THAA (all the amino acids present in the system, the conventional fraction for analysis), but also to obtain reliable data from the FAA fraction (those amino acids that are not peptide-bound). Only in a well-behaved shell will there be a consistent relationship between the data from these two fractions, as the FAA fraction is much more readily lost from the shell than the peptide-bound amino acids. This approach is useful as it is possible to sample different fractions of protein diagenesis from within the same sample, therefore obviating the need for extensive screening of the data based on THAA values only (Bowen, 2000).

Advantages of Analysing Multiple Amino Acids

Previous work only utilised the extent of racemization (epimerization) of one amino acid, l-isoleucine to d-alloisoleucine, although some 12 other amino acids were resolved and used as a check on HPLC chromatograms (Bowen, 2000). However, a further refinement used here involves a new chromatographic technique, which uses reverse-phase separation with fluorescence detection. This now permits both the discrimination of multiple chiral amino acids and their detection at low concentrations from smaller sample sizes (Kaufman and Manley, 1998). The advantages offered by this technique are twofold. Firstly, the number of degradation reactions that can be used to interpret the diagenetic history is increased, so that fast-racemizing amino acids can be used on shorter timescales and slower ones on longer timescales. Secondly, although the reaction rates are different, they should be correlated, providing a check on the integrity of the samples (Preece and Penkman, 2005). The l and d isomers of 10 amino acids were routinely detected, but the amino acids studied in detail were those whose both dand l enantiomers were well resolved: Asx, Glx, serine (Ser), alanine (Ala) and valine (Val).

Testing the Techniques: the Terrace Sequence of the Thames Valley

The amino acid results obtained in this study need to be tested against a secure stratigraphical template. The terrace sequence of the Thames valley is arguably one of the best studied in Europe and provides the longest and potentially most-complete sequence of Pleistocene sediments in Britain. Opinions differ about the attribution of certain sites in the Thames valley to specific stages, and especially how they correlate with the marine oxygen isotope record (Gibbard, 1985 and Gibbard, 1994; Bridgland, 1994), but there is no doubt from their location in the terrace staircase about their relative stratigraphical positions. All the Thames sites analysed are younger than the Anglian Stage, normally attributed to Marine Oxygen Isotope Stage (MIS) 12 (Bowen, 1999) during which the Thames was glacially diverted into its present valley. In the Middle and Lower Thames four terraces contain interglacial deposits that are thought to have accumulated during MIS 11, 9, 7 and 5e (Bridgland, 1994). This model is supported by several lines of evidence, including the original amino acid A/I data, molluscan and mammalian biostratigraphy (e.g. Bowen et al., 1989 and Bowen et al., 1995; Preece, 1995; Schreve, 2001). This sequence provides a valuable stratigraphical template against which to gauge the new amino acid data. As an additional check on the veracity of the data from the Thames, new amino acid analyses have also been undertaken on sites outside of the Thames valley that have good independent evidence of age, some of which also contain important Palaeolithic archaeology. Although the majority of sites are fluvial, lacustrine sites are also included, as this technique has the potential to correlate a range of freshwater sediments. Critical evidence of their age is given in the Online Supplement.

The locations of the Lower Thames sites under test are given in Fig. 1, with details as follows:

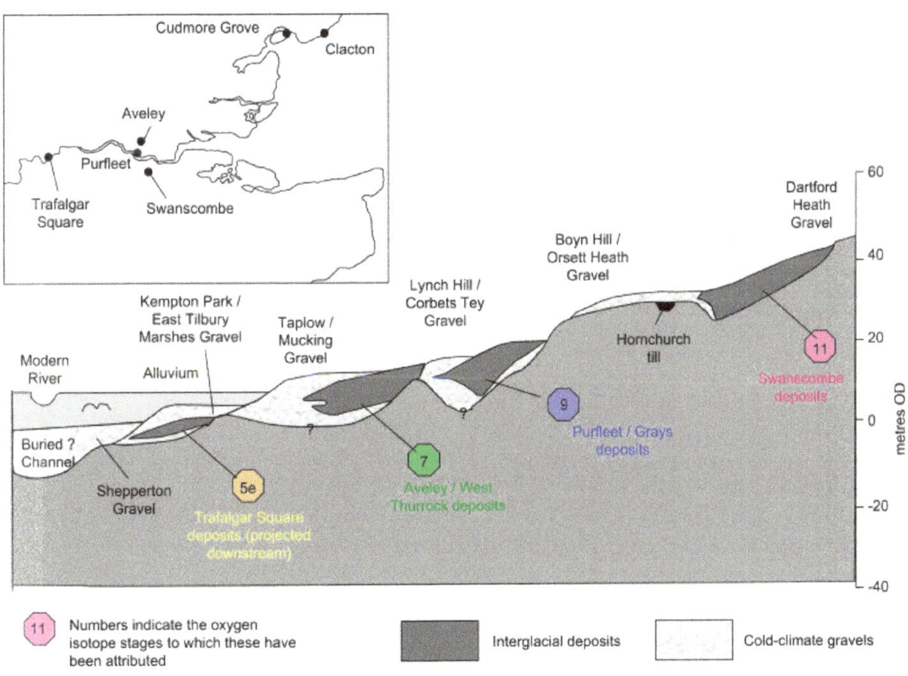

Figure 1: Lower Thames terrace staircase (after Bridgland, 1994), with inset showing location of Thames sites.

Swanscombe: Dierden's Pit (Ingress Vale; TQ 595748) is part of the complex of sites near Swanscombe. The shelly gravels have yielded the so-called 'Rhenish fauna' including *Theodoxus danubialis*, *Viviparus diluvianus* and *Valvata naticina* (Kerney, 1971; see also Schreve et al., 2007), which allows correlation with the Middle Gravels of the nearby Barnfield Pit sequence (Grid ref: TQ 595745), the horizon containing the celebrated hominin skull (Ovey, 1964; Kerney, 1971). The geomorphological and stratigraphical positions of these gravels within the Orsett Heath Formation (Bridgland, 1994) show that they post-date the Anglian glaciation (Gibbard, 1985; Bridgland, 1994). A/I amino acid geochronology, molluscan and mammalian biostratigraphy support a Hoxnian (MIS 11) age (Kerney, 1971; Turner and Kerney, 1971; Bowen et al., 1989 and Bowen et al., 1995; Bridgland, 1994; Preece, 1995; Schreve, 2001).

Purfleet: Two quarries on the northern bank of the Thames at Purfleet (Bluelands Pit, TQ 560784, and Greenlands Pit, TQ 568785) contain highly fossiliferous gravel, sand and intertidal silt deposits of up to 5 m in thickness (Hollin, 1977). The Purfleet Member is recognised as an independent unit, within the Corbets Tey Gravel Formation, by Bridgland (1994), with tentative correlation to MIS 9 indicated by the A/I amino acid geochronology, lithostratigraphy, terrace stratigraphy and biostratigraphy (Bowen et al., 1995; Bowen, 2000; Schreve et al., 2002).

Aveley/Lion Pit Tramway Cutting: The Aveley Member stratotype is at Sandy Lane Quarry (TQ 552808), now infilled, where the fossiliferous sediments consist of gravel, sand, silt and detritus mud, overlying bedrock or earlier Pleistocene sediments (Hollin, 1977; Gibbard, 1994; Bridgland, 1994). The deposits at Aveley were originally correlated with the Ipswichian (West, 1969), but their elevation above that of the Trafalgar Square sediments, together with different mammalian assemblages, led Sutcliffe, 1964 and Sutcliffe, 1975 and Sutcliffe and Bowen (1973) to suggest that they were older than the Ipswichian. The terrace succession model of Bridgland (1994; but see Gibbard, 1994 for an alternative view) places the interglacial sediments within the Mucking Formation, the third of four terraces of the Lower Thames thought to post-date the Anglian. The Lion Pit tramway cutting at West Thurrock, Essex (TQ 598783) exposes fluvial gravel and sand up to 9 m in thickness (Schreve et al., 2006) that overlie bedrock at 4 m O.D. (Gibbard, 1994). This site also lies in the Mucking Formation of the Lower Thames (Bridgland, 1994). Support for a correlation with MIS 7 for both sites has come from A/I amino acid geochronology (Bowen et al., 1989) and biostratigraphy based on molluscs (Keen, 1990) and mammals (Schreve, 2001).

Trafalgar Square: The stratotype of the Trafalgar Square Member is at Canada House, Trafalgar Square, London (TQ 292805) (Bowen, 1999). The thick interglacial deposits, up to 12 m of gravel, sand, silt and detritus mud, are part of the Kempton Park Formation (Bridgland, 1994 and Bridgland, 2006), formerly the 'Upper Floodplain Terrace' of the Thames. The highly fossiliferous deposits vary in nature from shelly sands to organic muds and are correlated by means of biostratigraphy (Franks, 1960; Sutcliffe and Bowen, 1973; Gibbard, 1985; Preece, 1999) and aminostratigraphy (Bowen et al., 1989; Bridgland, 1994) with the Ipswichian and therefore with MIS 5e (Bowen et al., 1989; Bowen, 1999).

RESULTS

The two separate analyses made on each individual sample, the FAA and THAA, should show a strong positive correlation for each amino acid d/l. This is because the extent of racemization in the THAA is, in large part, driven by peptide-bond hydrolysis, which in turn liberates FAA (Miller et al., 2000). A lack of the expected correlation between FAA and THAA will reveal samples in which the closed system may have been compromised (e.g. Preece and Penkman, 2005). Plotting the extent of racemization of THAA against FAA presents the samples in relative aminostratigraphical order based on their d/l's, with more degraded samples having higher d/l values for both the FAA and THAA fractions (e.g. Fig. 2, Fig. 3 and Fig. 4). Occasionally concentrations were too low for detection in one fraction; in those cases the data from one fraction plot along the zero-axes.

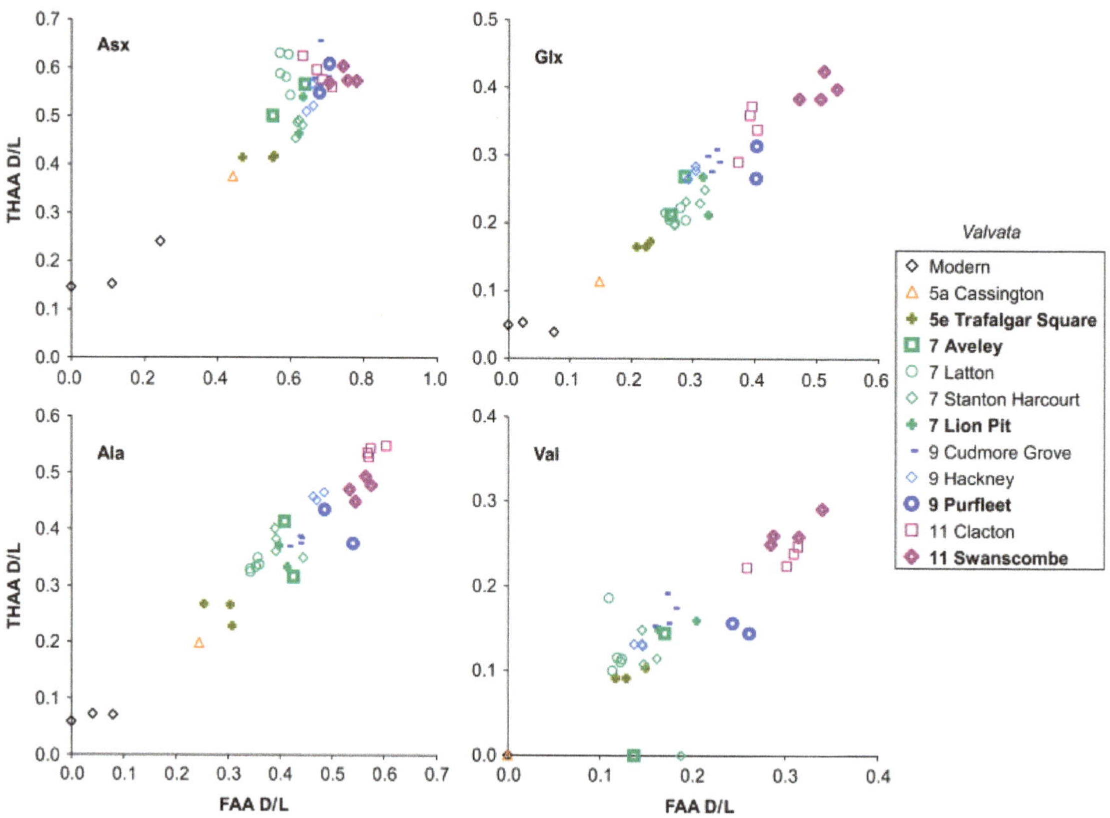

Figure 2: THAA vs. FAA for Asx, Glx, Ala and Val d/l in the bleached (intra-crystalline) fraction of *Valvata piscinalis*from the Lower Thames terrace sequence and other sites in the Thames/Medway system.

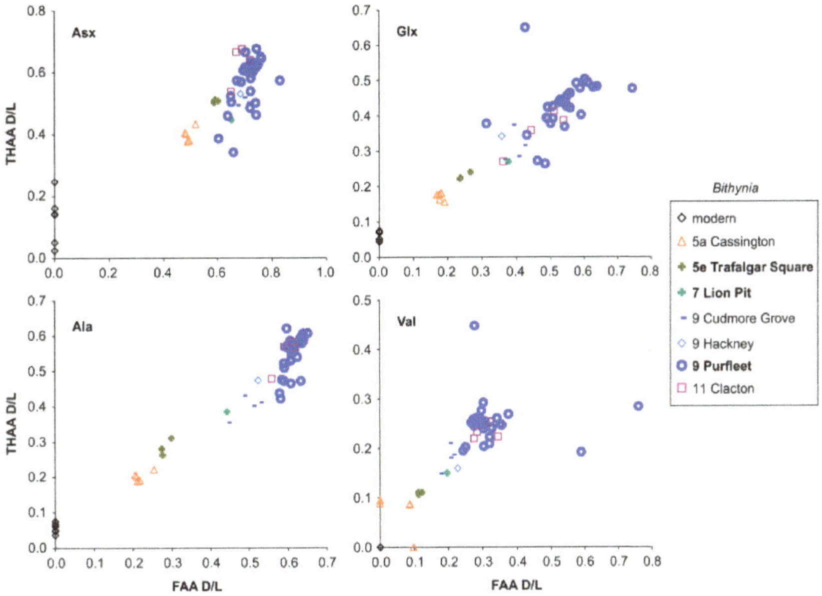

Figure 3: THAA vs. FAA for Asx, Glx, Ala and Val d/l in the bleached (intra-crystalline) fraction of *Bithynia tentaculata* shells from the Lower Thames terrace sequence and other sites in the Thames/Medway system. Although some samples fail the closed system test of high correlation between FAA and THAA, all data are shown; note the large spread in data from Purfleet.

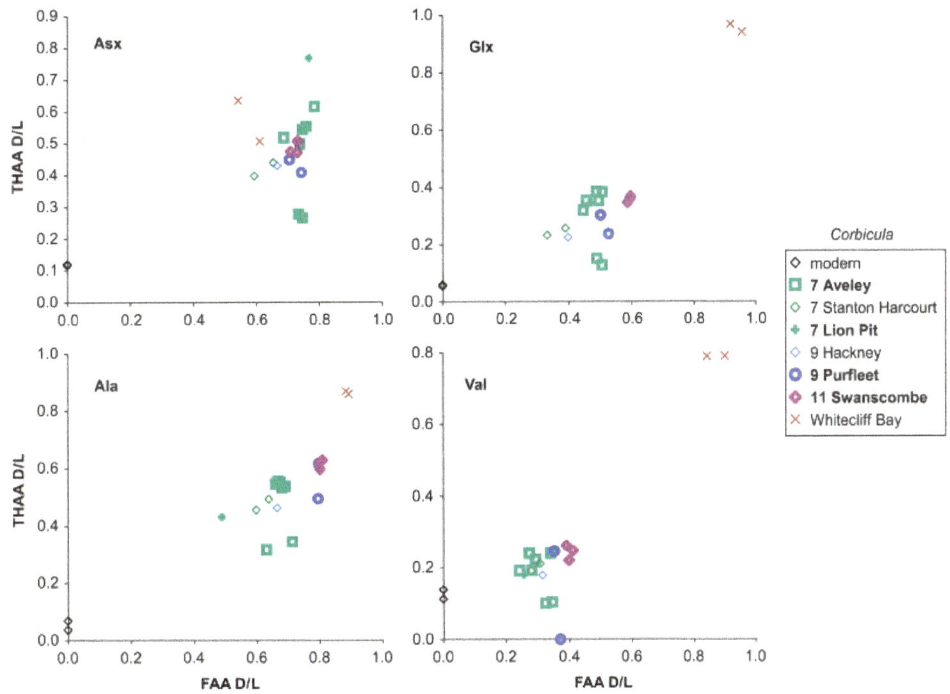

Figure 4: THAA vs. FAA for Asx, Glx, Ala and Val d/l in the bleached (intra-crystalline) fraction of *Corbicula* sp. from the Pleistocene Thames terrace sequence, other sites in the Thames/Medway system and the Oligocene/Eocene Bembridge Marls at Whitecliff Bay, Isle of Wight.

Species Effect

The results from *Corbicula* are discussed in more detail below but, within the gastropod shells, differences were found in the absolute d/l values within the intra-crystalline fraction of *Bithynia* and *Valvata* samples from the same site (Online Supplement Fig. 1). Along with the problem of small sample sizes, amino acid data has upper and lower limits (0 and 1) and therefore statistical tests must be applied with caution. However, tests for normality were undertaken on each dataset and two types of statistical tests applied in order to provide a useful insight alongside the interpretation based on the graphical data alone. For each amino acid within each fraction, 2-tailed *t*-tests (for normally distributed data) and Mann–Whitney tests (for non-normal data) were undertaken on the data from four Thames sites: 14/32 2-tailed *t*-tests and 10/32 Mann–Whitney tests showed a significant difference (Online Supplement Worksheet 4; the MIS 9 site at Cudmore Grove was substituted for Purfleet due to the diagenetic problems at Purfleet discussed below). However, the differences do not remain consistent as the samples increase in age and although the data from *Bithynia* is usually slightly higher, this is not true in all cases. The intra-crystalline proteins extracted from the two different gastropod species are sufficiently different to alter the rates of protein breakdown and hence observed d/l. It is therefore necessary to develop aminostratigraphic frameworks for each individual species.

AAR from Thames Sites

Fig. 2, Fig. 3 and Fig. 4 show the THAA vs. FAA for the intra-crystalline (bleached) data for Asx, Glx, Ala and Val for the Lower Thames sites of Trafalgar Square, Aveley/Lion Pit, Purfleet and Swanscombe. As a further check, data is presented from additional sites in the Thames system (Cassington, Latton, Stanton Harcourt, Cudmore Grove, Hackney and Clacton). For each of the amino acids, the modern samples lie at the lowest FAA and THAA values. The strong positive correlation observed in the bleached gastropod THAA vs. FAA graphs (Fig. 2 and Fig. 3) indicate that the extent of protein degradation in their shells is an indication of the relative age of the sites.

Within a closed system AAR is driven by the extent of protein degradation, itself a function of the time/temperature history of a sample. Less degradation occurs during cold stages and as a consequence we have used statistical tests on the *V. piscinalis* shells to establish whether it is possible to use the extent of AAR to discriminate between interglacial shells found within different terrace levels from the Lower Thames terrace staircase. For each amino acid within each fraction (FAA and THAA), both 1-tailed and 2-tailed *t*-tests and Mann–Whitney tests were performed on the Trafalgar Square, Aveley/Lion Pit, Purfleet and Swanscombe data. If the result of the *t*-test produces a $p<0.05$, this enables discrimination between the two sites at a 95% confidence level. The 1-tailed *t*-test assumes prior knowledge of the stratigraphy of the sites whereas the 2-tailed *t*-test assumes no prior knowledge and therefore provides a more useful guide of the resolving power of this technique when applied to sites of unknown age. The full results are presented in the Online Supplement but are also summarized in Table 1: in general, discrimination is better for the younger sites. The data tend to support the model of Bridgland (1994), with deposits present in the Lower Thames sequence intermediate in age between the Ipswichian (MIS 5e) and the Hoxnian (MIS 11), although the level of resolution in the older material (>MIS 7) allows this to be stated with less certainty. The results reveal that there is sufficient coherence within a single dataset to enable statistical discrimination between sites, something that McCarroll (2002) had urged racemization researchers to address.

Table 1: Summary of statistical test results (Minitab), using both 1- and 2-tailed *t*-tests (assuming normal distribution) and Mann–Whitney tests (assuming non-normal distribution)

	Is Trafalgar Square younger than Aveley/Lion Pit?		Is Aveley/Lion Pit younger than Purfleet?		Is Purfleet younger than Swanscombe?	
	Yes	No	Yes	No	Yes	No
Valvata (1-tailed *t*-test)	8	0	6	2	7	1
Valvata (2-tailed *t*-test)	8	0	5	3	6	2
Valvata (1-tailed MW-test)	8	0	5	2	7	1
Valvata (2-tailed MW-test)	8	0	4	4	7	1

The results of each duplicate analysis are included in order to provide a statistically significant sample size.

Number in "Yes" column represents the number of amino acid (AA) fractions that enable discrimination between the two Thames sites in question, and also supports their stratigraphical position at a 95% confidence level.

Number in "No" column represents the number of AA fractions that do not enable discrimination between the two sites.

For example: In the 2-tailed *t*-tests, analysis of *Valvata* shell results in eight out of eight AA fractions supporting the hypothesis that Trafalgar Square is distinguishable and younger than Aveley/Lion Pit; only six out of eight support the hypothesis that Swanscombe is older than Purfleet.

Full details of the statistical tests are given in the Online Supplement Worksheet 3.

Corbicula has previously been identified as yielding problematic amino acid data at some sites, notably at Purfleet (Miller et al., 1979; Bowen et al., 1995), although unbleached shells have provided A/I values consistent with other species at other sites (Miller et al., 1979; Hughes, 1987; Bowen et al., 1989). However, bleaching did not appear to improve the data obtained from this bivalve in fossil samples (Penkman et al., 2007). As can be seen (Fig. 4), the d/l values of none of the amino acids selected enable discrimination between the MIS 7 and MIS 9 sites. The samples from Aveley show a large scatter, but the d/l values from this site generally exhibit some of the highest values, despite the likelihood that it is younger than both Hackney and Purfleet. However, high d/l values consistent with a Tertiary age were obtained from *C. obovata* from the Bembridge Marls at Whitecliff Bay, Isle of Wight (see Gale et al., 2006 for discussion of age).

Purfleet

In contrast to other sites studied in the Thames Valley, which have yielded A/I results on whole shell consistent with their relative stratigraphic context, the sites at Purfleet have previously given amino acid ratios seriously overestimating their age (Miller et al., 1979; Bowen et al., 1989) although subsequent analyses from *Valvata* were consistent with the stratigraphical position of the site (Bowen et al., 1995). The data from the intra-crystalline fraction of *Bithynia* and *Valvata* show a large scatter (Fig. 2 and Fig. 3). Carbonate concretions ('race') occur at the site, formed not only around stones, but also around freshwater mussels, leaving the internal casts of the shells (Schreve et al., 2002). This clearly demonstrates that bicarbonate has been leached and re-deposited. Such post-depositional diagenetic processes are likely to be implicated in the anomalous amino acid results obtained.

AAR from All Sites

Further data from other sites in southern England are included in Fig. 5 and provide a wider context. The points from each site have been coloured according to independent geochronological age estimates that correlate them with oxygen isotope stratigraphy (see Online Supplement for details; for all four of the amino acids and the unbleached dataset see Online Supplement Figs. 2–6). As previously reported, bleaching reduces the variation within a single site compared to unbleached samples (Online Supplement Figs. 2–6; for example note the decrease in scatter in the Hoxne dataset upon bleaching, Online Supplement Fig. 4), simplifying interpretation of the data. It is therefore this bleached intra-crystalline dataset that is interpreted in detail here.

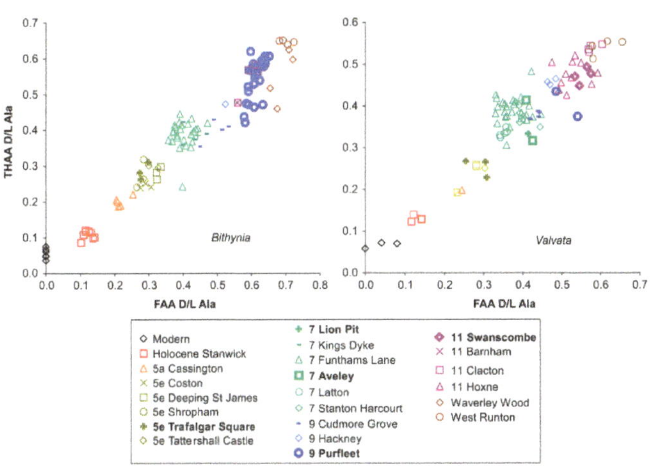

Figure 5: THAA vs. FAA for Ala d/l in the bleached (intra-crystalline) fraction from shells of *Bithynia tentaculata* and *B. troschelii* (left) and *Valvata piscinalis* (right). The colouring represents the independent age evidence for each site: black=modern, red=Holocene, orange=MIS 4/5a, yellow=Ipswichian/MIS 5e, green=MIS 7, blue=MIS 9, purple=Hoxnian/MIS 11, brown=pre-MIS 12 'Cromerian Complex'.

There is an increase in the extent of racemization with time in the bleached gastropod shells (Fig. 2, Fig. 3 and Fig. 5). As expected, there has been little racemization within cold stages, as the reaction rates of protein breakdown would have slowed at these lower temperatures. In samples correlated with part of MIS 7 age or younger, the extent of racemization is sufficient to differentiate the sites already separated by lithostratigraphy and geomorphology (Bridgland, 1994). MIS 7 is fairly well differentiated from later stages in some amino acids. In the case of Asx, which undergoes rapid racemization, there is a clustering of data beyond MIS 5, but in slower racemizing amino acids such as Glx, Ala and Val, discrimination is better. Sites older than MIS 7 show higher levels of racemization, but separation is increasingly less clear with age (Fig. 5; Online Supplement Figs. 2–3).

DISCUSSION

Debate over the existence of two or more interglacials in the UK record since the Anglian (MIS 12) is of longstanding (Bowen, 1978; Shotton et al., 1983) and research has focused on the Thames fluvial deposits in an attempt to solve this question. Gibbard, 1985 and Gibbard, 1994 proposed that only two interglacials were represented in the Thames terraces after the Anglian diversion: the Hoxnian (now correlated with MIS 11), such as at the site at Swanscombe, and the Ipswichian (correlated with MIS 5e) represented by the Trafalgar Square deposits. Sediments at intermediate elevations, such as those at Purfleet and Aveley, were also attributed to the Ipswichian by Gibbard (1994), their additional height explained as due either to deposition by Thames tributaries or to sea-level rise at this time. Bridgland (1994), on the other hand, identified four interglacials in the same post-diversion sequence, as did the A/I aminostratigraphical model of Bowen et al. (1989).

The applicability of the new AAR technique reported here also shows that the amino acid data are consistent with the stratigraphical position of the terraces and support the model of four post-Anglian interglacials. Good separation is observed between the younger sites and it may be possible to separate events at the Oxygen Isotope sub-stage level up to and including MIS 7. Older events are not separated with such clarity but the method does not fail completely, with increasingly higher racemization values in the slower racemizing amino acids.

Interspecies variation, as reported in other studies of whole shell (e.g. Lajoie et al., 1980), was also observed in the intra-crystalline fraction. Somewhat counter-intuitively, small, relatively thin-shelled gastropods yield more consistent data than the much more robust shells of the bivalve *Corbicula*, although we suspect that this is due to the different protein compositions of the different ultra-structural layers in this species (Penkman et al., 2007).

The different amino acids that contribute to the entrapped protein within the biomineral undergo racemization at different rates, reflecting differences in the mechanisms of isomerization and the varying stability of their respective peptide bonds. Consequently, by analysing multiple amino acid dl ratios, the overall range of discrimination between terraces is extended. By combining these different d/l values, it is possible to assess the overall state of protein degradation with a greater degree of confidence. This technique has the advantage in that it draws information from multiple amino acids and also from the extent of racemization in both FAA and THAA, therefore providing multiple isochrons. Unusual within-amino acid patterns are identifiable, because although racemization rates differ between individual amino acids, they should be highly correlated in a closed system and therefore examination of d/l patterns within a single sample provides an element of quality control.

Despite the advantages of the simple bleaching preparation step to isolate intra-crystalline amino acids, it is in itself insufficient to yield data from gastropod shells able to clearly resolve sites older than MIS 7, due to increased levels of within-site variability. Recently the ability to discriminate events within the 'Cromerian Complex' using a different biomineral, the calcitic operculum from one of the gastropods studied here, *Bithynia* sp., has been presented (Penkman et al., accepted). The striking difference in resolution between gastropod shells, which are made of aragonite, and the opercula, suggests that the diagenetic transformation of aragonite may be partially responsible for the poor resolution observed in older gastropod samples; higher concentrations of intra-crystalline amino acids within the opercula may also contribute to improved discrimination. A previous study has shown that isoleucine epimerization measurements on unbleached *Bithynia* opercula from five sites, including Trafalgar Square and Aveley, also yielded lower and more consistent ratios (Hughes, 1987). The level of natural variability within a horizon has significant impacts on the level of resolution that can be obtained from the amino acid data and the low levels of variability observed within the bleached opercula in both Cromerian (Penkman et al., accepted) and younger material, including that from the problematic site at Purfleet (Fig. 6) indicate that analysis of this biomineral has the potential to increase the level of resolution of this technique.

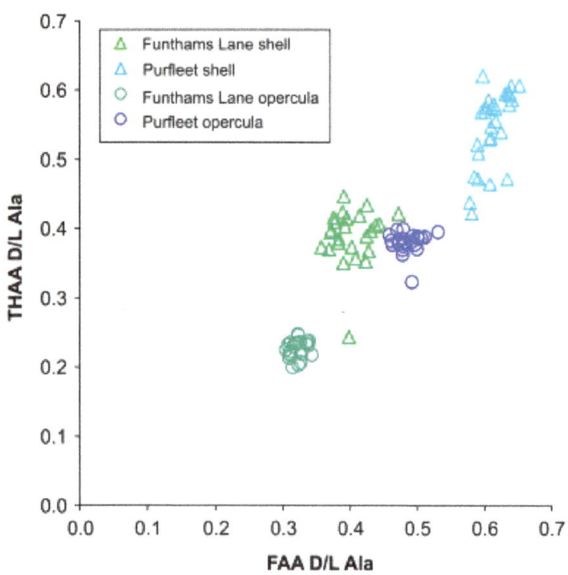

Figure 6: Left: THAA vs. FAA d/l Ala for bleached *Bithynia* shells and opercula from Funthams Lane and Purfleet. Each of the measurements is from a single shell or operculum. Note the variation in relative rates between the two biominerals and the much greater consistency in the measurements obtained from opercula. The outlying shell sample has an amino acid concentration two times that of the other shells, indicating the importance of adequate bleaching (Penkman et al., 2007).

CONCLUSIONS

Using the new technique of intra-crystalline amino acids analysed by RP-HPLC, it is shown that the extent of degradation in skeletal protein from freshwater gastropods makes an important contribution to the interpretation of fluvial deposits. The organic matter entrapped within the biomineral (intra-crystalline fraction) is believed to be a more reliable fraction for analysis than the whole shell (Sykes et al., 1995; Penkman et al., 2007), with the preparatory bleaching step removing both secondary contamination and any residual (inter-crystalline) organic matrix, which can degrade and leach at an unpredictable rate over time. Analysis of multiple amino acids in both the FAA and THAA fractions shows an increase in the extent of protein degradation with time in commonly occurring gastropod shells, enabling correlations to be made between both fluvial and other freshwater sites using AAR. The racemization data reported is different from earlier work that examined d/l values from the THAA of whole mollusc shells containing both intra- and inter-crystalline material. However, the same separation of the Lower Thames terraces has emerged.

The levels of AAR in gastropod shells allow distinction between sites correlated with MI stages and substages up to and including MIS 7. In gastropod shells from deposits higher within the terrace sequence (i.e. older), the extent of racemization is greater still, but the levels of natural variability are higher. The increase in variability may be due to aragonitic shells undergoing diagenetic alteration to calcite and it is possible that with careful selection of unaltered carbonates, the aminostratigraphic framework for these gastropods could be extended further. Discrimination between sites and the level of natural variability within sites promise to be much better if diagenetically stable calcitic opercula are used. In this study lithostratigraphy and geomorphology of river terraces have provided a stratigraphical framework for the testing of the utility of intra-crystalline AAR for geochronology, but it is apparent that there is value in including AAR as a routine analysis for the study of fluvial records.

ACKNOWLEDGEMENTS

Thanks to Rebecca Briant, David Bain, David Bridgland, Harry Langford and Simon Lewis for providing some of the samples used in this study, and Berni Bowler, Rob Hunter and Ian Harrison for help with the analyses. The two referees are gratefully acknowledged for their helpful comments and improvements to the manuscript. Professor David Keen sadly passed away on 16 April 2006 and it is a great pity that he did not see this work published. This research was supported by NERC and the Wellcome Trust. This paper is a contribution to IGCP-449 'Global Correlation of Late Cenozoic fluvial deposits', to IGCP 518 'Fluvial deposits as evidence for climate change and landscape evolution in the Late Cenozoic' and to FLAG Focus 1.

REFERENCES

1. Bates, M.R., 1993. Quaternary aminostratigraphy in Northwestern France. Quaternary Science Reviews 12, 793–809.
2. Bowen, D.Q., 1978. Quaternary Geology: a stratigraphic framework for multidisciplinary work. Pergamon, Oxford, 221pp., Russian translation, Moscow, 272pp., 2nd edition 1985.
3. Bowen, D.Q., 1999. A revised correlation of Quaternary deposits in the British Isles, Geological Society Special Report, No. 23.
4. Bowen, D.Q., 2000. Revised aminostratigraphy for land–sea correlations from the northeastern North Atlantic margin. In: Goodfriend, G.A., Collins, M.J., Fogel, M.L., Macko, S.A., Wehmiller, J.F. (Eds.), Perspectives in Amino Acid and Protein Geochemistry.

Oxford University Press, Oxford, pp. 253–262.

5. Bowen, D.Q., 2003. Uncertainty in Oxygen Isotope Stage 11 sea level: an estimate 1372 m above low water from Great Britain, In: Droxler, A., Poore, R.Z., Burkle, L.H. (Eds.), Earth's Climate and Orbital Eccentricity: The Marine Isotope Stage 11, Geophysical Monograph 137, American Geophysical Union, pp. 131–144.
6. Bowen, D.Q., Sykes, G.A., 1988. Correlation of marine events and glaciations on the northeast Atlantic margin. Philosophical Transactions of the Royal Society B 318, 619–635.
7. Bowen, D.Q., Sykes, G.A., Reeves, A., Miller, G.H., Andrews, J.A., Brew, J.S., Hare, P.E., 1985. Amino Acid Geochronology of raised beaches in southwest Britain. Quaternary Science Reviews 4, 279–318.
8. Bowen, D.Q., Hughes, S., Sykes, G.A., Miller, G.H., 1989. Land–sea correlations in the Pleistocene based on isoleucine epimerization in non-marine mollusks. Nature 340, 49–51.
9. Bowen, D.Q., Sykes, G.A., Maddy, D., Bridgland, D.R., Lewis, S.G., 1995. Aminostratigraphy and amino acid geochronology of English lowland valleys: the Lower Thames in context. In: Bridgland, D.R., Allen, P., Haggart, B.A. (Eds.), The Quaternary of the Lower Reaches of the Thames, Field Guide. Quaternary Research Association, Durham, pp. 61–63.
10. Bridgland, D.R., 1994. Quaternary of the Thames, Geological Conservation Review Series No. 7.
11. Bridgland, D.R., 2006. The Middle and Upper Pleistocene sequence in the Lower Thames: a record of Milankovitch climatic fluctuation and early human occupation of southern Britain. Proceedings of the Geologists' Association 117, 281–305.
12. Bridgland, D., Westaway, R., 2007. Climatically controlled river terrace staircases: a worldwide Quaternary phenomenon. Geomorphology, doi:10.1016/j.geomorph.2006.12.032.
13. Bridgland, D.R., Maddy, D., Bates, M., 2004. River terrace sequences: templates for Quaternary geochronology and the marine–terrestrial correlation. Journal of Quaternary Science 19, 203–218.
14. Brooks, A.S., Hare, P.E., Kokis, J.E., Miller, G.H., Ernst, R.D., Wendorf, F., 1990. Dating Pleistocene archaeological sites by protein diagenesis in ostrich eggshell. Science 248, 60–64.
15. Collins, M.J., Riley, M.S., 2000. Amino acid racemization in biominerals, the impact of protein degradation and loss. In: Goodfriend, G.A., Collins, M.J., Fogel, M.L., Macko, S.A., Wehmiller, J.F. (Eds.), Perspectives in Amino Acid and Protein Geochemistry. Oxford University Press, Oxford, pp. 120–142.
16. Franks, J.W., 1960. Interglacial deposits at Trafalgar Square, London. New Phytologist 59, 145–152.
17. Gale, A.S., Higgett, J.M., Pa¨like, H., Laurie, E., Hailwood, E.A., Hardenbol, J., 2006. Correlation of Eocene–Oligocene marine and continental records: orbital cyclicity, magnetostratigraphy and sequence stratigraphy of the Solent Group, Isle of Wight, UK. Journal of the Geological Society of London 163, 401–415.
18. Gibbard, P.L., 1985. The Pleistocene history of the Middle Thames Valley. Cambridge University Press, Cambridge.
19. Gibbard, P.L., 1994. Pleistocene History of the Lower Thames Valley. Cambridge University Press, Cambridge.
20. Gru¨n, R.G., Schwarcz, H.P., 2000. Revised open system U-series/ESR age calculations for teeth from Stratum C at the Hoxnian Interglacial type locality, England. Quaternary Science Reviews 19, 1151–1154.
21. Hill, R.L., 1965. Hydrolysis of proteins. Advances in Protein Chemistry 20, 37–107.
22. Hollin, J.T., 1977. Thames interglacial sites, Ipswichian sea levels and Antarctic ice surges. Boreas 6, 33–52.
23. Hughes S.A., 1987. The aminostratigraphy of British Quaternary nonmarine deposits. Unpublished Ph.D. Thesis, University of Wales, Aberystwyth.
24. Imbrie, J., Shackleton, N.J., Pisias, N.G., Morley, J.J., Prell, W.L., Martinson, D.G., Hays, J.D., Macintyre, A., Mix, A.C., 1984. The orbital theory of Pleistocene climate: Support from a revised chronology of the marine d18O record. In: Berger, A. (Ed.), Milankovitch and Climate, Part 1. Reidel, Hingham MA, pp. 269–305.
25. Kaufman, D.S., Manley, W.F., 1998. A new procedure for determining DL amino acid ratios in fossils using reverse phase liquid chromatography. Quaternary Science Reviews 17, 987–1000.
26. Keen, D.H., 1990. Significance of the record provided by Pleistocene fluvial deposits and their included molluscan faunas for palaeoenvironmental reconstruction and stratigraphy: case studies from the English Midlands. Palaeogeography, Palaeoclimatology, Palaeoecology 80, 25–34.
27. Kerney, M.P., 1971. Interglacial deposits at Barnfield Pit, Swanscombe, and their molluscan fauna. Journal of the Geological Society, London 127, 69–93.
28. Lajoie, K.R., Wehmiller, J.F., Kennedy, G.L., 1980. Inter- and intrageneric trends in apparent racemization kinetics of amino acids in Quaternary molluscs. In: Hare, P.E., Hoering, T.C., King, Jr., K. (Eds.), Biogeochemistry of Amino Acids. Wiley, New York, pp. 305–340.
29. Miller, G.H., Mangerud, J., 1985. Aminostratigraphy of European marine interglacial deposits. Quaternary Science Reviews 4, 215–278.
30. McCarroll, D., 2002. Amino-acid geochronology and the British Pleistocene: secure stratigraphical framework or a case of circular reasoning? Journal of Quaternary Science 17, 647–651.
31. Meijer, T., Preece, R.C., 2000. A review of the occurrence of Corbicula in the Pleistocene of North-West Europe. Geologie en Mijnbouw— Netherlands Journal of Geoscience 79, 241–255.
32. Miller, G.H., Hare, P.E., 1975. Use of amino acid reactions in some Arctic marine fossils as stratigraphic and chronologic indicators. Carnegie Institute of Washington Yearbook 74, 612–617.
33. Miller, G.H., Hare, P.E., 1980. Amino acid

geochronology: integrity of the carbonate matrix and potential of molluscan fossils. In: Hare, P.E., Hoering, T.C., King, Jr., K. (Eds.), Biogeochemistry of Amino Acids. Wiley, New York, pp. 415–443.

34. Miller, G.H., Hollin, J.T., Andrews, J.T., 1979. Aminostratigraphy of UK Pleistocene deposits. Nature 281, 539–543.

35. Miller, G.H., Hart, C.P., Roark, E.B., Johnson, B.J., 2000. Isoleucine epimerization in eggshells of the flightless Australian birds Genyornis and Dromaius. In: Goodfriend, G.A., Collins, M.J., Fogel, M.L., Macko, S.A., Wehmiller, J.F. (Eds.), Perspectives in Amino Acid and Protein Geochemistry. Oxford University Press, Oxford, pp. 161–181.

36. Ovey, C.D., 1964. The Swanscombe Skull: a survey of research on a Pleistocene site. Occasional Paper No. 20, Royal Anthropological Institute.

37. Penkman, K.E.H., 2005. Amino acid geochronology: a closed system approach to test and refine the UK model. Unpublished Ph.D. Thesis, University of Newcastle.

38. Penkman, K.E.H., Kaufman, D.S., Maddy, D. and Collins, M.J., 2007. Closed-system behaviour of the intra-crystalline fraction of amino acids in mollusc shells. Quaternary Geochronology, doi:10.1016/j.quageo.2007.07.001.

39. Penkman K.E.H., Preece, R.C., Keen, D.H., Collins, M.J., accepted. Amino acid geochronology of the type Cromerian of West Runton, Norfolk, UK. Quaternary International.

40. Preece, R.C., 1995. Mollusca from interglacial sediments at three critical sites in the Lower Thames. In: Bridgland, D.R., Allen, P., Haggart, B.A. (Eds.), The Quaternary of the Lower Reaches of the Thames, Field guide. Quaternary Research Association, Durham, pp. 53–62.

41. Preece, R.C., 1999. Mollusca from Last Interglacial fluvial deposits of the River Thames at Trafalgar Square, London. Journal of Quaternary Science 14 (1), 77–89.

42. Preece, R.C., Penkman, K.E.H., 2005. New faunal analyses and amino acid dating of the Lower Palaeolithic site at East Farm, Barnham, Suffolk. Proceedings of the Geologists' Association 116, 363–377.

43. Roof, S., 1997. Comparison of isoleucine epimerization and leaching potential in the molluskan genera Astarte, Macoma, and Mya. Geochimica et Cosmochimica Acta 61, 5325–5333.

44. Schreve, D.C., 2001. Differentiation of the British late Middle Pleistocene interglacials: the evidence from mammalian biostratigraphy. Quaternary Science Reviews 20, 1693–1705.

45. Schreve, D.C., Bridgland, D.R., Allen, P., Blackford, J.J., Gleed-Owen, C.P., Griffiths, H.I., Keen, D.H., White, M.J., 2002. Sedimentology, palaeontology and archaeology of late Middle Pleistocene River Thames terrace deposits at Purfleet, Essex, UK. Quaternary Science Reviews 21, 1423–1464.

46. Schreve, D.C., Harding, P., White, M.J., Bridgland, D.R., Allen, P., Clayton, F., Keen, D.H., Penkman, K.E.H., 2006. A Levallois knapping site at West Thurrock, Lower Thames, UK: its Quaternary context, environment and age. Proceedings of the Prehistoric Society 72, 21–52.

47. Schreve, D.C., Keen, D.H., Limondin-Lozouet, N., Auguste, P., Santisteban, J.I., Ubilla, M., Matoshko, A., Bridgland, D.R., 2007.

48. Progress in faunal biostratigraphy of Late Cenozoic fluvial sequences during IGCP 449. Quaternary Science Reviews 26, in press.

49. Shotton, F., Sutcliffe, A.J., Bowen, D.Q., Currant, A.P., Coope, G.R., Harmon, R.S., Shackleton, N.J., Stringer, C., Turner, West, R.G., Wymer, J., 1983. Interglacials after the Hoxnian in Britain. Quaternary Newsletter 39, 19–25.

50. Sutcliffe, A.J., 1964. The mammalian fauna. In: Ovey, C.D. (Ed.), The Swanscombe Skull: a survey of research on a Pleistocene site. Occasional Paper No. 20, Royal Anthropological Institute, pp. 85–111.

51. Sutcliffe, A.J., 1975. A hazard in the interpretation of glacial–interglacial sequences. Quaternary Newsletter 17, 1–3.

52. Sutcliffe, A.J., Bowen, D.Q., 1973. Preliminary report on excavations at Minchin Hole. Pengelly Cave Studies Trust 21, 12–25.

53. Sykes, G.A., Collins, M.J., Walton, D.I., 1995. The significance of a geochemically isolated intracrystalline fraction within biominerals. Organic Geochemistry 23, 1059–1065.

54. Towe, K.M., 1980. Preserved organic ultrastructure: an unreliable indicator for Paleozoic amino acid biogeochemistry. In: Hare, P.E., Hoering, T.C., King, Jr., K. (Eds.), Biogeochemistry of Amino Acids. Wiley, New York, pp. 65–74.

55. Turner, C., Kerney, M.P., 1971. A note on the age of the freshwater beds of the Clacton Channel. Journal of the Geological Society of London 127, 87–93.

56. Wehmiller, J.F., Stecher, H.A., York, L.L., Friedman, I., 2000. The thermal environment of fossils: effective ground temperatures at aminostratigraphic sites on the US Atlantic Coastal Plain. In: Goodfriend, G.A., Collins, M.J., Fogel, M.L., Macko, S.A., Wehmiller, J.F. (Eds.), Perspectives in Amino Acid and Protein Geochemistry. Oxford University Press, Oxford, pp. 219–250.

57. West, R.G., 1969. Pollen analysis from interglacial deposits at Aveley and Grays, Essex. Proceedings of the Geologists' Association 80, 271–282.

58. West, R.G., Andrew, R., Catt, J.A., Hart, C.P., Hollin, J.T., Knudsen, K.- L., Miller, G.F., Penney, D.N., Pettit, M.E., Preece, R.C., Switsur, V.R., Whiteman, C.A., Zhou, L.P., 1999. Late and Middle Pleistocene deposits at Somersham, Cambridgeshire, UK: a model for reconstructing fluvial/estuarine depositional environments. Quaternary Science Reviews 18, 1247–1314.

CHAPTER 13 POTENTIAL OF LOMAS FOG OASIS ENVIRONMENTS FOR PRECERAMIC HUNTER–GATHERERS UNDER PAST ENSO MODES

David Beresford-Jones, Alexander G. Pullen, Oliver Q. Whaley, Justin Moat, George Chauca, Lauren Cadwallader, Susana Arce, Alfonso Orellana, Carmela Alarcón, Manuel Gorriti, Patricia K. Maita, Fraser Sturt, Agathe Dupeyron, Oliver Huaman, Kevin J. Lane, Charles French

Keywords: Lomas palaeoenvironments, Holocene, ENSO, Hunter–gatherers, Origins of agriculture, South coast Peru

INTRODUCTION

"*Climatically ... the Peruvian coast is in a state of delicate balance, and even a slight change of oceanic circulation pattern affecting the humidity of the air would be sufficient to cause the retreat of the lomas.*" [EdwardLanning, 1963: 369.]

The coast of Peru is one of the world's driest deserts and yet for much of the year its air is pregnant with water vapour, a paradox due to the rigid stratification of its tropical air masses above the cold Humboldt Current offshore. Along the seaward Andean foot-slopes this creates the unique ecological phenomenon of *lomas* – 'oases born of mists' (Dollfus, 1968, cited by Rostworowski, 1981: 34, *our translation*) – in which vegetation flourishes during the austral winter, before fading back to barren desert in the summer. The role that these lomas fog oases played in the long history of human ecology that was to lay the foundations here for one of humanity's few independent hearths of agriculture and cradles of "pristine" civilisation, has been much debated.

Over the long trajectory of the Middle Preceramic Period (c. 8000–4500 BP) hunter–gatherers here exploited a variety of ecological niches, juxtaposed along this otherwise extremely arid littoral, including: the riparian oases along the floodplains of rivers arising in the Andean hinterlands; estuarine wetlands; and not least, one of the world's richest marine ecosystems sustained by deep upwelling offshore. Explaining how, and *why*, human exploitation shifted between these different ecologies through time is key to understanding increasing sedentism and the emergence of agriculture here. And there is a dichotomy in interpretation of precisely how important lomas fog oases were to these changes: illuminated by only a few detailed, well-reported investigations along almost 2500 km of Pacific coastline, most notably in Chilca (e.g. Engel, 1987 and Quilter, 1991), and Quebrada de los Burros in the far south (e.g. Lavallée and Julien, 2012). This remains a fundamental issue, not least because lomas environments – highly sensitive both to climate shifts and to human impact – were seemingly abandoned at the end of the Middle Preceramic Period: a critical time period during which intensified exploitation of marine resources, alongside the cultivation of cotton and food plants on river floodplains (Moseley, 1975), laid the foundations for larger scale populations and the first emergence of Andean social complexity.

Early investigators, particularly struck by the density of lithic artefacts encountered in the lomas, asserted that these environments were the prime resource for Early Holocene hunter–gatherers on the Peruvian coast, and moreover, that they had formerly been far more extensive. Lanning, 1963 and Lanning, 1967 and Engel (1973) believed that the lomas had provided rich resources for transhumant hunter–gatherers living there during winter months: what Engel (1981: 24) called the 'fog oasis situation'. They further interpreted the occurrence of large expanses of desiccated vegetation and snail shells outside the limits of today›s lomas formations as

Citation: David Beresford-Jones, Alexander G. Pullen, Oliver Q. Whaley, Justin Moat, George Chauca, Lauren Cadwallader, Susana Arce, Alfonso Orellana, Carmela Alarcón, Manuel Gorriti, Patricia K. Maita, Fraser Sturt, Agathe Dupeyron, Oliver Huaman, Kevin J. Lane, Charles French, Potential of lomas fog oasis environments for Preceramic hunter–gatherers under past ENSO modes, Quaternary Science Reviews, Volume 129, 1 December 2015, Pages 196-215, ISSN 0277-3791, http://dx.doi.org/10.1016/j.quascirev.2015.10.025.

evidence of their former extent. Indeed, Lanning (1967: 51) believed that ever since their first occupation in the Terminal Pleistocene the lomas had been in constant retreat, as the fog belt lifted due to changes in the Humboldt Current, contracting to a tenth of their original extent by the time of their abandonment around 4450 BP.

Soon, however, this model of lomas ecology in the early prehistory of Peru came to be challenged as a 'transparent adventure in environmental determinism' (Craig and Psuty, 1968: 126). Its critics (*inter alia* Parsons, 1970, Craig, 1985 and Craig, 1992) argued, as Lynch (1971: 140) summarises, that 'lomas vegetation and animal life could never have been important to man, that the greater previous extent of lomas vegetation has not been established and that … there is no reason to believe that the climate of coastal Peru has been substantially different from the present during postglacial times'. Relict extensions of desiccated vegetation and snails were, it was suggested, the product of occasional El Niño Southern Oscillation ('ENSO') climatic perturbations, rather than of permanently more extensive lomas. Indeed, some of these extraordinarily preserved formations of desiccated *Tillandsia* date far back into the Pleistocene (Hesse, 2014). Many see little evidence for any significant long-term climatic change throughout the Holocene on the Peruvian coast (e.g. Craig and Psuty, 1968, Parsons, 1970, Craig, 1985, Craig, 1992 and Wells and Noller, 1999).

Instead, other factors were now invoked to explain the changes in archaeological patterning observed to occur at the end of the Middle Preceramic, including over-exploitation of fragile lomas ecologies driven by population growth (e.g. Patterson and Moseley, 1968: 124; Cohen, 1978b and Weir and Derring, 1986), and indeed technological change. For Moseley (1975), argued that rather than diminution of lomas resources driving people to exploit marine resources, it was an enhanced ability to do so, thanks to the introduction of cotton agriculture. Moreover, Lanning›s model had failed to take into account the effects of rising sea levels on the archaeological pattering observed on the central Peruvian coast (Richardson, 1998), and new findings suggested far greater time depths for maritime adaptations (Sandweiss, 2009). Such refinements, however, still leave the ultimate cause of the abandonment of the central Peruvian lomas to be fully explained.

Meanwhile, there have been significant advances in palaeoclimatic research. Broad scale climatic changes are thought to have occurred in Peru throughout the retreat of the glaciers that mark the Pleistocene–Holocene transition and the stabilisation of sea levels that followed. These may have significantly affected monsoonal circulation and tropical rainfall of Amazonian origin over the Andes (e.g. Betancourt et al., 2000 and Mächtle and Eitel, 2012). Most palaeoclimatic reconstruction, however, using proxy record from both on and off shore locations across the Andes, has focused on the history of ENSO variation, and its effect on the Humboldt current, along the Pacific coast of Peru during the Holocene (e.g. Rollins et al., 1986, Sandweiss et al., 1996, Sandweiss et al., 2004, Rodbell et al., 1999, Moy et al., 2002, Sandweiss and Kelley, 2012, Etayo-Cadavid et al., 2013, Loubere et al., 2013, Carré et al., 2012, Carré et al., 2013 and Carré et al., 2014). On the Peruvian coast itself where 'standard palaeoclimatic records are absent or underdeveloped' (Sandweiss, 2003: 23), these often entail proxy data recovered from archaeological deposits.

In this paper we revisit the 'lomas dichotomy' in the light of new investigations of the lomas formations of the Peruvian south coast: defined herein as the 250 km of Pacific coastline encompassed by the Pisco, Ica, Río Grande de Nazca and Acarí river valleys. These valleys have a distinctive geomorphology, climate and hydrology from those of Peru›s north and central coasts (Beresford-Jones, 2011), and from those further south to the Chilean border (the 'far south'). Consequently, for much of Peru›s prehistory, they also have a shared and distinctive cultural trajectory, commonly labelled 'south coast' by archaeologists (e.g. Willey, 1971: 78).

Our findings arise out of collaborations between Cambridge University›s One River Project and the Royal Botanical Gardens Kew, as follows:

- New investigations of the botany and ecology of the highly endemic and lomas formations of this region. To date these are largely unstudied, and indeed the few references to them (Parsons, 1970 and Oka and Ogawa, 1984: 115; Craig, 1985) betray inaccuracies that underplay their resource potential.
- New archaeological investigations of these south coast lomas and their associated littoral to reveal an entire Preceramic landscape. Although the south coast archaeological record comprises some of Peru's most famous cultural manifestations, not least Paracas and Nasca, their Preceramic predecessors have not been much investigated since the pioneering work of Engel in the 1950's.

The paucity of previous scientific investigation here is, in part, because the lomas of the south coast are relatively remote from modern settlement as the course of the Pan-American highway is diverted much further inland than elsewhere along Peru's littoral. Yet this has also better preserved their fragile biodiversity and their Preceramic archaeological record, and, as we will see, conveys certain clarities to the interpretations of its ancient human ecology.

We will use these findings to reassess the importance of lomas resources to Preceramic human ecology on this previously understudied part of the Peruvian coast, revisiting the question of why changes took place through time in the context of the latest, detailed Holocene palaeoclimatic reconstructions specific to the south coast.

THE ECOLOGY OF LOMAS FOG OASES

There are two sources of water along the arid Peruvian coast, each arising in distinct seasons. The first arises from the small portion of intense precipitation over the Amazon Basin during the austral summer that overcomes the formidable rain shadow of the Andean cordillera, to drain their western flanks in seasonal streams and rivers (Prohaska, 1973). The second is more ephemeral – at least in non-ENSO years – arising in the austral winter, between around August and November, as a deepening inversion layer saturated with moisture over cold seas is blown inland by trade winds. Where this marine stratus inversion layer encounters the coastal Andean flanks between 300 and 1000 m asl, orographic cooling causes fogs to coalesce, resulting in fine, horizontally moving drizzles (*garuás* or *camanchacas*). This moisture is sustained by reduced evaporation due to stratus cloud cover producing 'occult precipitation' and fostering formations of fog-drip vegetation known as lomas. Since vegetation creates the interception surface area on which fog condenses and is trapped, this growth acts greatly to increase the amount of water that would be deposited on barren slopes (Engel, 1973, Walter, 1973, Oka, 1986 and Muenchow et al., 2013). As this occurs, lomas formations blossom across the desert with intense green meadows and pasture.

The term 'lomas' is used colloquially in Peru to describe the Andean foot slope and coastal cordillera all along the Pacific, so that ecologists, geographers and archaeologists have come to use the term idiosyncratically, as Craig (1985: 28) notes, for various distinct ecological niches, including desert scrub found at higher altitudes and vegetation along erratic quebrada watercourses. Here, we use the term *sensu stricto* to mean those plant communities which rely exclusively on fog: depauperate for much of the year, dominated by *Tillandsia*, but flourishing with lush, ephemeral herbaceous growth during the winter months (Dillon et al., 2003).

Such fog oases are to be found along parts of the western coast of South America between ~6°S and 30°S (Manrique et al., 2010). In Peru there are at least 70 discrete localities supporting lomas, occupying a total of some 8000 km^2 (Péfaur, 1982, Dillon et al., 2003 and Dillon, 2011). Their vegetation communities are clearly delimited ecosystems, unique within the context of South American ecology and floristic composition (Stafford, 1939 and Dillon, 2011). They consist of varying mixtures of annuals, short-lived perennial and woody vegetation, represented by a total of some 850 species in 385 genera and 83 families (Dillon, 2011). They are home also to a variety of animal life (Péfaur, 1981 and Zeballos et al., 2000).

Around 40% of the plant species in the lomas communities of the south coast can be classed as perennials and maintain themselves through the dry season by a variety of vegetative, starch-rich, underground storage organs ('geophytes', Fig. 6B and C) such as roots, corms, tubers and bulbs (in the case of the 15% that are monocots). Indeed, we find that many plants previously classed here as annuals turn out, on excavation, to be perennials, arising from deep corms on contractile roots, or simply re-sprouting from apparently dead leafless stems, supplied by enlarged roots or stems.

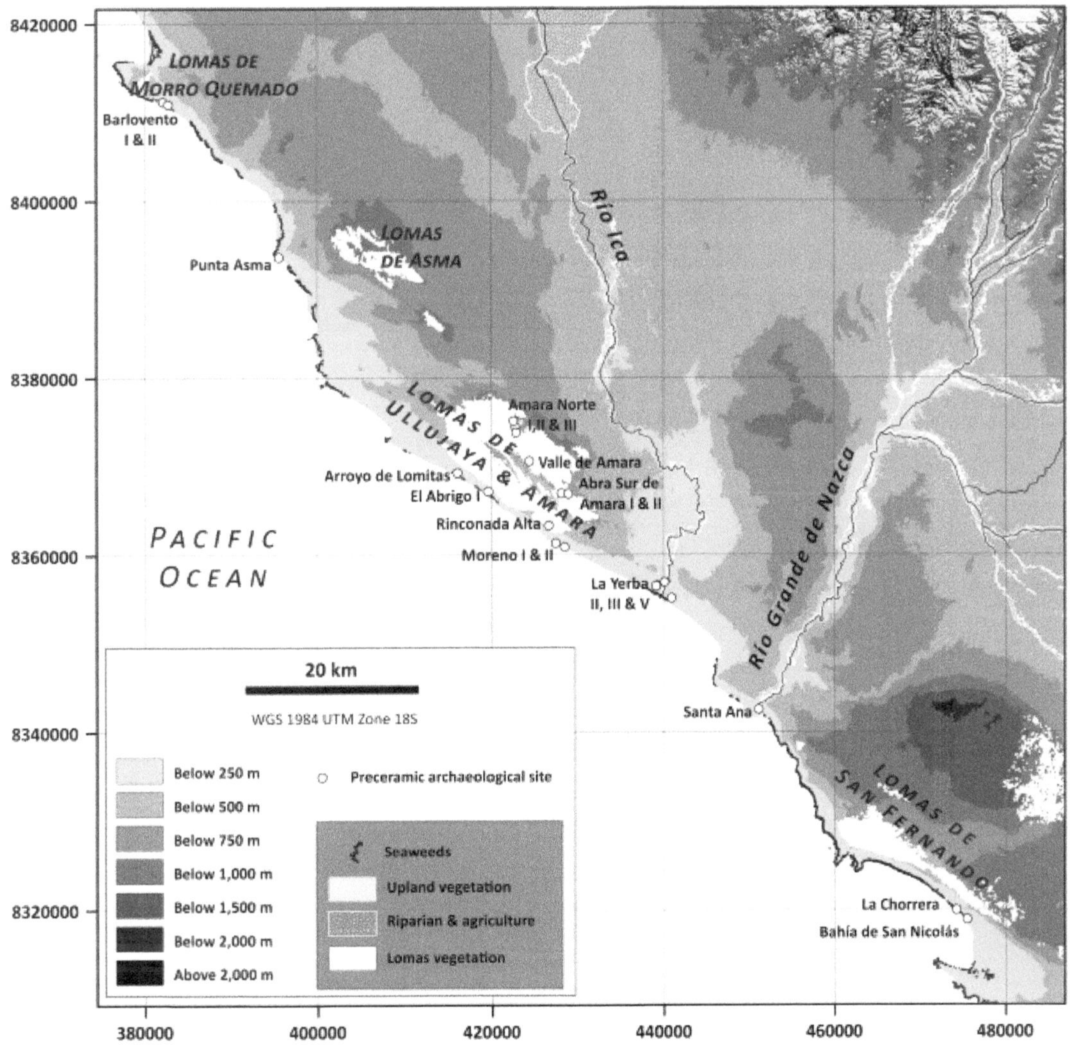

Figure 1: Map of the south coast showing Preceramic archaeological sites together with lomas and other vegetation associations derived as described in Section 3.

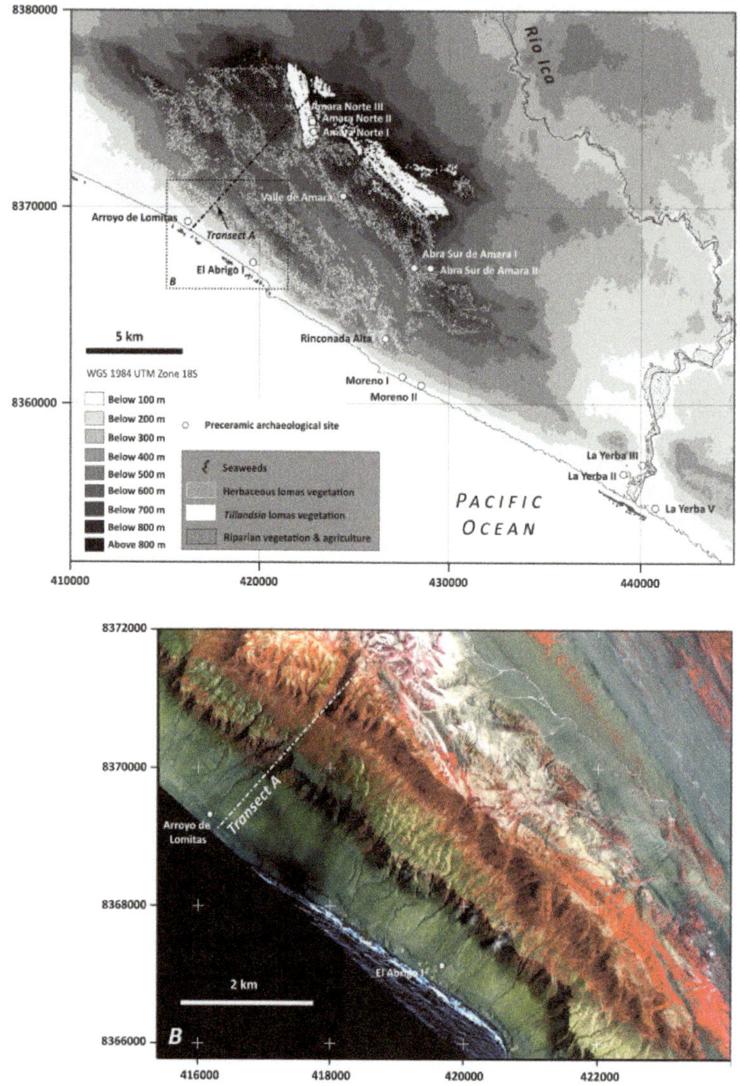

Figure 2: Map of the lomas of Amara and Ullujaya showing Preceramic archaeological sites together with lomas vegetation associations derived as described in the text, and Transect A across which RBG Kew carried of systematic collections of herbarium vouchers in plots. Inset B shows a detail of the Amara lomas using NIR to show the high density of lomas vegetation as dark grey (red in online version) on 10 February 2010 during a moderate El Niño, as described in Section 3 (For interpretation of the references to colour in this figure legend, the reader is referred to the web version of this article.).

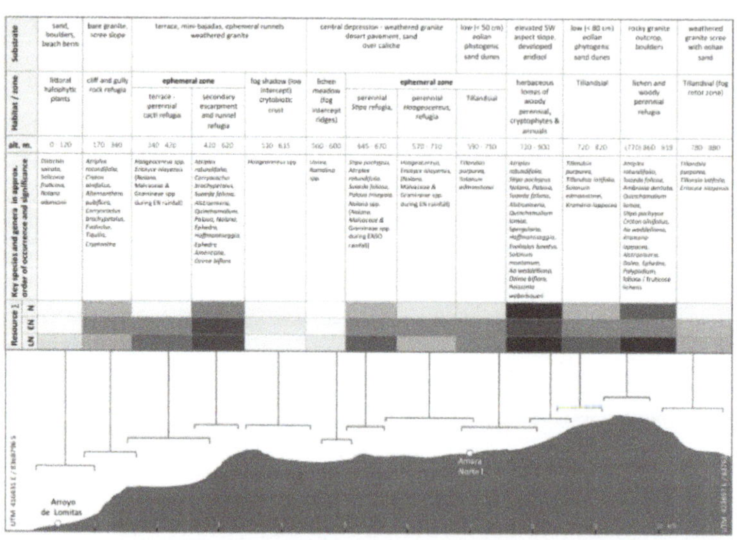

Figure 3: Transect A across the Amara lomas (see Fig. 2) showing key vegetation associations and their altitudinal ranges, together with total lomas resource values (1–7 tonal values: light = low, dark = high), relative to both the climatic mode and other resources during a given lomas year (where N is a normal year, EN an El Niño year and LN a La Niña year), as described in Section 3.

Figure 4: Details of selected Preceramic sites along the Amara lomas foreshore.

Figure 5: Details of selected Preceramic sites in the Amara lomas.

Top to bottom:
A Lomas de Amara, August 2013.
B&C Lomas de Amara tubers, *Oxalis* sp. and *Solanum montanum*, respectively.
D Guanao (*Lama guanicoe*) footprints, lomas de Amara, August 2013.
E&F Edible *Haageocereus* sp. cactus fruit, lomas de San Fernando.
G Live *Bostryx conspersus* snail on Atriplex sp., lomas de Amara, August 2013.

Figure 6: Selected south coast lomas resources.

When sustained winter fogs arrive, and depending upon the intensity and duration of the fog season, these perennials re-sprout, thereby increasing moisture capture and acting as 'nurse plants' for annuals to germinate anew from the previous year›s seed production. Once winter gives way to spring and increasing amounts of sunshine, all the lomas comes into flower simultaneously to maximise rapid pollination and set seed, many of which are adapted to resist high summer temperatures and lie strewn across the desert surface, awaiting the return of winter fogs.

Lomas vegetation is highly sensitive to ENSO climatic perturbations and indeed, certain lomas plants depend on the periodic recurrence of El Niño to propagate themselves (Caviedes, 1984, Manrique et al., 2010 and Dillon, 2011). Importantly, however, our research shows that the relationship between lomas ecosystem productivity and ENSO is rather complex: while some parts of the lomas ecosystem wax under particular ENSO variations, others wane. We return to consider the implications of this later.

Within lomas formations vegetation occurs in distinct belts whose composition is determined by variations in moisture availability, which are predominantly a function of altitude, but also of slope, aspect and type of ground surface in relation to the intensity, direction and velocity of coastal fog (Ono, 1986, Rundel et al., 1991 and Muenchow et al., 2013). The most diverse herbaceous fog meadow flourishes with winter fogs in a belt immediately beneath the altitude at which the inversion layer intercepts the land (Péfaur, 1982 and Muenchow et al., 2013). Along its margins, particularly further inland where conditions are more arid, the lomas of the south coast are characterised by transitional belts occur composed of low-density areas of cacti, and high-density formations of *Tillandsia* species, which can extend over vast tracts of desert (Péfaur, 1982, Rundel and Dillon, 1998, Whaley et al., 2010 and Hesse, 2012).

Unlike most lomas which occur along the foothills of the Andes themselves, those of the south coast are separated from the main Andean cordillera by an expanse of uplifted desert sedimentary tableland around 60 km wide (Beresford-Jones, 2011). Here, lomas formations – from north to south: Morro Quemado, Asma, Amara-Ullujaya and San Fernando – occur along the seaward flanks of the Coastal Cordillera (Fig. 1). These fragments of granite batholith run along the Pacific coast for almost 150 km, and rise abruptly out of the ocean to attain heights of around 800 m, 1000 m and 1791 m asl, in the Lomas of Morro Quemado, Amara-Ullujaya, and San Fernando, respectively. These south coast lomas formations thus lie almost on the littoral, and quite far from the parallel inland courses of the north-south flowing Río Ica, and the most westerly of the Río Grande de Nazca›s several tributaries.

NEW INVESTIGATIONS IN THE LOMAS OF THE SOUTH COAST

From 2012 to 2014 Cambridge University's One River Project and the Royal Botanical Gardens Kew, collaborated on new investigations of the lomas of the south coast through joint and several fieldwork entailing:

1. RBG Kew undertook a spatial analysis of vegetation and floristic composition in the lomas of San Fernando and Amara by systematic collection of herbarium vouchers in plots across transects to identify species and develop understanding of the ecology, biodiversity across the south coast lomas. An important purpose of this work is to produce baseline species and vegetation maps to support conservation delimitation by Peru's Servicio Nacional de Áreas Naturales Protegidas por el Estado ('SERNANP').

Fig. 1 and Fig. 2 show vegetation maps derived from two sources: Landsat 8 imagery from March 2014 to March 2015 supplemented with GeoEye imagery (October/November 2011) for the San Fernando region. Landsat imagery was processed to give Normalised Different Vegetation Index (NDVI, Tucker, 1979) for each month, these were stacked and the maximum NDVI over this yearly period calculated. This maximum NDVI imagery was thresholded to pull though the herbaceous lomas and the *Tillandsia* vegetation. Vegetation in the San Fernando region did not show up in the Landsat imagery, probably because the seasonal lomas growth 2014–2015 was very poor. Thus lomas vegetation for the San Fernando region was derived from GeoEye imagery in the same way described for Landsat. These two vegetation maps were overlaid to give Fig. 1 and Fig. 2.

Fig. 2B shows a detail of the Amara lomas from a Near infra-red false colour composite (NIR, red and green) image, showing high density of seasonal herbaceous lomas vegetation as dark grey, using GeoEye imagery from 10 February 2010, during an ENSO event entailing moderately increased sea surface temperature anomalies off the coast of Peru. (For interpretation of the references to colour in this figure legend, the reader is referred to the web version of this article.)

Fig. 3 shows the vegetation associations recorded across transect A (see Fig. 2) of the lomas de Amara. The relative extent of the vegetation units, bands or overlapping zones, is indicated with bars corresponding with the vegetation characterisation derived from herbarium voucher collections, drone flights, satellite imagery and other fieldwork including walked transects, seed analysis and root test pits. The figure also shows aggregated lomas resource values under a normal fog season and under El Niño and La Niña climatic modes.

2. Cambridge University›s One River Project carried out new investigations of the Preceramic archaeology associated with the lomas of the south coast (Arce et al., 2013 and Arce et al., 2015). We follow in the footsteps of Frédéric Engel, 1981 and Engel, 1991, whose seminal work here in the 1950›s identified 14 Preceramic sites along the littoral between Morro Quemado and the Bahía de San Nícolas, eight of which he radiocarbon dated. As well as investigating many of these in much more detail and subjecting them to modern sampling, we turned our attention also to the lomas hinterlands, in which we identify another seven previously unknown Preceramic sites. At many of these sites we carried out investigations of exposed profiles, new excavations and surface collections, during which we sampled for flotation to extract organic remains and heavy fraction components, geochemistry, monoliths for micromorphology, radiocarbon dating, etc.

Fig. 1 and Fig. 3 and Table 1, presents the results of this updated and augmented archaeological survey of the Preceramic landscape along the south coast between the Morro Quemado and the Bahía de San Nícolas, in which all radiocarbon dates are now presented calibrated using the ShCal13 curve (Hogg et al., 2013) in OxCal version 4.2 (Bronk Ramsey, 2009).

Table 1: Preceramic sites associated with south coast lomas (between 14° 19′ and 15° 15′ S)

Site	Location (WGS84 18L)		Alt. (m)	Site code	Associated lomas formation	Site Ecotone ^{14}C yr BP	Radiocarbon dates					Reference
	Easting	Northing					Error (±)	Cal BP[a]		Lab key		
Barlovento I	382165	8411166	24	14b X – 205 A	Morro Quemado	Sandy littoral						
Barlovento II	382278	8411100	18	14b X – 203	Morro Quemado	Sandy littoral						
Punta Asma	395589	8393590	10	15a II-504	Asma	Rocky littoral						
Amara Norte III	422765	8374451	698	ORP-605	Amara-Ullujalla	Tillandsia lomas						
Amara Norte II	422808	8374327	693	ORP-610	Amara-Ullujalla	Tillandsia lomas						
Amara Norte I	422855	8373815	600	ORP-585	Amara-Ullujalla	Tillandsia lomas	4325	30	4963	4728	OxA-32350	
Valle de Amara	424436	8370553	720	ORP-611	Amara-Ullujalla	Atriplex lomas						
Arroyo de Lomitas	416205	8369229	37	15a VI-525	Amara-Ullujalla	Rocky littoral	4360	95	5286	4587	I-7774	Engel, 1991: 56
El Abrigo	419684	8367142	48	15a VI-550	Amara-Ullujalla	Rocky littoral	8835	145	10,200	9539	I-7772	Engel, 1991: 56

Abra Sur de Amara I	428183	8366898	547	ORP-609	Amara-Ullujalla	Atriplex lomas			–	–		
Abra Sur de Amara II	429011	8366874	649	ORP-602	Amara-Ullujalla	Herbaceous lomas			–	–		
Rinconada Alta	426689	8363298	340	ORP-619	Amara-Ullujalla	Atriplex lomas			–	–		
Moreno I	427500	8361271	25	15a IX-300	Amara-Ullujalla	Sandy littoral			–	–		
Moreno II	428550	8360900	25	15a IX-301	Amara-Ullujalla	Sandy littoral			–	–		
La Yerba I	439213	8356533	20	15b VII-50	Amara-Ullujalla	River estuary	7520	300	9015	7696	Birm-511	Engel, 1991: 56
La Yerba I (SU 1024)	439213	8356533	20	ORP-209	Amara-Ullujalla	River estuary	6375	39	7415	7165	OxA-30913	
La Yerba II	439123	8356430	21	15b VII-45 & 100	Amara-Ullujalla	River estuary	6470	110	7571	7030	I-7773	Engel, 1991: 56
La Yerba II (SU 1015)	439123	8356430	21	ORP-207	Amara-Ullujalla	River estuary	5988	33	6881	6674	OxA-29975	
La Yerba II (SU 1079)	439123	8356430	21	ORP-208	Amara-Ullujalla	River estuary	6029	32	6936	6735	OxA-29944	
La Yerba II (level 5)	439123	8356430	21	ICA-IN	Amara-Ullujalla	River estuary	5940	45	6944	6739	OS-60556	Carré et al., 2014: Table S2
La Yerba III	440115	8356906	25	15b VII-55	Amara-Ullujalla	River estuary	5430	140	6485	5893	Birm-513	Engel, 1991: 56
La Yerba III (SU 9020)	440115	8356906	25	15b VII-55	Amara-Ullujalla	River estuary	5381	33	6270	5996	OxA-32290	
La Yerba III (SU 9009)	440115	8356906	25	15b VII-55	Amara-Ullujalla	River estuary	5290	33	6179	5920	OxA-32291	
La Yerba V b	440924	8355047	10	15b VII-25	Amara-Ullujalla	Sandy littoral	6150	120	7264	6678	I-3560	Engel, 1991: 56
Santa Ana	451155	8342556	25	15b VII-8 & 19	Mainsa	River estuary	4720	120	5650	4983	–	Engel, 1981: 61
La Chorrera	474338	8319857	21	15b XI-50	San Fernando	Rocky littoral			–	–		
Bahía de San Nicolás	475573	8318822	25	16a II-40	San Fernando	Sandy littoral	5560	80	6694	6029	–	Engel, 1981: 59

Note

[a]2σ OxCal v4.2.4 Bronk Ramsey, (2009); r:5 SHCal13 atmospheric curve (Hogg et al., 2013).

[b]Also called 'El Campo'. Its position, 'one kilometre south of the southern bank of the river, on a terrace found 8.5 meters above present sea level' (Engel, 1981: 20), is recorded differently in Engel, 1991: 155 (Fig. 127) and Engel, 1981: 122 (Fig. 43). We cannot find this site, despite much searching. Either its position is described erroneously, or it has been destroyed.

THE POTENTIAL OF SOUTH COAST LOMAS TO PRECERAMIC HUMAN ECOLOGY

The Preceramic archaeological patterning recorded on the south coast naturally reflects the differential distribution of many resources of food, fuel and raw material resources. Yet the first essential that dictates where people settle and move in such an extremely arid landscape is, of course, fresh water (e.g. Erlandson, 2001). What is most striking about the archaeological site distribution shown on Fig. 1 is that, while the largest and most visible Preceramic archaeological remains are, unsurprisingly, to be found at the mouths of the Ica and Nazca Rivers, many Preceramic sites lie scattered far beyond the river courses: and yet always intimately associated with the coastal lomas formations. Indeed, as we will see, *all* of these Preceramic sites, regardless of their locations, contain significant lomas-derived resources.

Sites on River Estuaries

The largest and most visible Middle Preceramic archaeological remains between Morro Quemado and Bahía de San Nícolas lie at the mouths of the Ica and Nazca Rivers. A number of sites on the Río Ica estuary date to early in the Middle Preceramic Period (Table 1), the largest and most visible of which are La Yerba II and La Yerba III. Their midden deposits are dominated by evidence of hunted and gathered marine resources: sea mammals, birds, fish and molluscs, particularly *Mesodesma donacium* surf clams that, until the 1997/98 El Niño, proliferated in vast beds just at the surf lines of the adjacent beaches. They also include many other organic components enjoying excellent preservation through desiccation, not least remains of plants: including, *inter alia*, seaweeds and Cyperaceae rhizomes; woven reeds and wooden posts in the vestiges of windshelters and houses; plant fibres spun and plied into lines and nets; and bottle gourd (*Lagenaria* sp., radiocarbon dated to 6936–6735 cal BP, see Table 1, OxA-29944) and both wild (*Vigna* sp.), and domesticated beans (*Phaseolus lunatus*, radiocarbon dated to 6270–5996 cal BP, see Table 1, OxA-32290): all presumably cultivated on the river floodplain. We interpret these sites on the estuary to be the logistical basecamps (*sensu*Binford, 1980) of seasonally mobile hunter–gatherers, although there is significant evidence of increasing sedentism at La Yerba III. They are located at the confluence of different habitats: riparian woodland, river estuaries, sandy beach, rocky headlands and an ecologically diverse lomas mosaic. Each could provide seasonally abundant and predictable resources, accessed at different times of the year by different members of society (men, women and children), according to varying levels of energy expenditure and hazard.

Upstream of the river estuaries, few Middle Preceramic sites are reported on the south coast (Isla, 1990 and Vogt, 2011; Gorbahn, 2013), though doubtless other remains have been obscured by subsequent agricultural and geomorphological changes. Far beyond the river courses, however, our investigations reveal other component parts of the Middle Preceramic landscape: all of which are directly dependent on lomas to provide key resources, not least water and fuel. These are of two general types.

Sites along the Lomas Littoral

The first group of Preceramic sites such as Arroyo de Lomitas, Abrigo I and Bahía de San Nícolas, all identified originally by Engel, 1981 and Engel, 1991, are located along the lomas foreshore, thereby simultaneously granting immediate access to marine and lomas resources (Fig. 1 and Fig. 4).

Our investigations of the dense, stratified midden deposits at Arroyo de Lomitas, including structured hearths (Fig. 4E and F) suggest that ground water sources here were sufficient to sustain small groups of hunter–gatherers for stretches of time. These middens are dominated by marine resources of the adjacent rocky littoral, particularly gastropods, mussels and seaweeds. Today itinerant fishermen call the site '*Agua Dulce*'(or '*Pozo de los Chilenos*' on the *Instituto Geográfico Nacional Peruano* map, Sheet 1741), and affirm that water here can always be obtained by excavating above the tidal margin where the arroyo emerges onto the beach. Indeed, *all* these Preceramic sites along the coast are associated with watercourses arising from in the lomas (Strong, 1957: 8, Engel 1981, 1991: 58).

The abrupt topography and impermeable granite massif of the Coastal Cordillera, rising out of the sea to over 1000 m, forms a vast and ideal surface for fog condensation and consequent run-off. Emerging from the seaward bluffs of the Amara and San Fernando lomas are multiple arroyos, deeply incised into the uplifted Holocene marine terrace, that stand testament to periodically significant surface flows. The range of clastic materials within the discrete water-lain deposits of these arroyos are evidence of erratic, high-energy flows perhaps during ENSO events (Fig. 4G). Yet layers of finer alluviums attest to water flows even during normal years (cf. Fontugne et al., 1999) and their courses are lined with perennial vegetation in the form of *Atriplex rotundifolia*, (Chenopodiaceae), *Croton alnifolius* (Euphorphiaceae) and *Tiquilia* spp. (Boraginaceae), a variety of small herbs such as *Alternanthera* sp. (Amaranthaceae) and *Parietaria debilis* (Urticaceae), and large relict stands of the cactus *Corryocactus* aff. *brachypetalus*, whose size is evidence of significant, if sporadic, water availability.

Unlike most coastal lomas contexts, wherein the two sources of water in this coastal desert, described in Section 2 above, are inevitably conflated, here, the *only* possible sources of fresh water beyond the river courses derive from condensing winter sea fogs. These fog-fed arroyo courses enabled Preceramic human habitation all along this arid littoral, far from its river courses, but in immediate proximity to both lomas and particular marine resources.

Logistical Field Camps within the Lomas

While Engel successfully located those Preceramic sites along the coast, he failed to find a 'single human settlement' within the south coast lomas (Engel, 1981: 27, hardly surprising when the remote lomas de Amara alone cover some 400 km^2). Yet such sites do exist, conspicuous largely as large accumulations of land snail shells (Fig. 1 and Fig. 5).

Preservation of other organic remains here is poor because of winter humidity. In lomas elsewhere Craig (1992) interprets similar accumulations of land snails to be natural death assemblages. Here, however, the Middle Preceramic anthropogenic origin of Amara Norte I is unequivocally revealed by stratigraphic contexts (Fig. 5E) also including remains of marine shell and sea urchin, grinding stones (Fig. 5F) and hearth charcoals radiocarbon dated to 4963–4728 cal BP (OxA-32350, Table 1). Moreover, these middens are surrounded by scatters of obsidian projectile points (Fig. 5B) and flaking debitage from the later stages of lithic reduction such as the retouching of preforms, left lying exposed on the surface by wind deflation of these upland landscapes.

We identify three clusters of such sites deep within the lomas of Amara — at Amara Norte, Valle de Amara and Abra Sur de Amara (Fig. 2) — which we interpret to be logistical field camps (*sensu* Binford, 1980): the vestiges of multiple, short-term forays in winter to hunt game and gather lomas specific resources, particularly plant geophytes and land snails. Elsewhere in the archaeological record such sites are rare but not unknown (cf. Engel, 1981 and Larrain et al., 2001).

Occult precipitation is sometimes sufficient to form seasonal standing ponds of water in the San Fernando and Amara lomas. In historical times atmospheric moisture has been 'harvested' within lomas by means of fabrics or nets and collecting vessels, yielding up to 10 L per day per m^2 of fabric (Klemm et al., 2012). There is no evidence that Middle Preceramic fabric technology allowed such water harvesting, but, during the foggy winter months – also the time when the lomas ecology is at its most productive – even fog drip collected from overhanging rocks eventually yields useful quantities of water (Larrain et al., 2001).

Doubtless drinking water was transported during the Middle Preceramic using bottle gourd and/or animal skins, both in evidence at the logistical basecamps on the river estuaries. Nonetheless, our observations of today›s lomas hydrology, together with a pattern of substantial Middle Preceramic archaeological deposits (see Fig. 1 and Fig. 2), at locations very widely dispersed from the river courses (Fig. 4 and Fig. 5A), strongly suggest that, fresh water derived entirely from coastal fog ecosystems was sufficient, at least seasonally, to sustain small groups of hunter–gatherers. We next assess the potential of other lomas resources to Preceramic hunter–gatherers, in the light of our ecological and archaeological investigations of the south coast lomas.

Lomas Faunal Resources

Today, despite the depauperate or ephemeral nature of much of its vegetation, the south coast lomas formations sustain a remarkably rich animal life including larger mammals such as wild guanaco camelids (*Lama guanicoe*, Palomino, 2006), two species of desert fox (*Lycalopex* spp.), many rodents, lizards and birds, and invertebrates including insects and snails. Indeed, snails are the single most conspicuous and ubiquitous element of lomas resources found in the Middle Preceramic archaeological contexts of the south coast: evidence of their importance to human diet.

There appear to be two species of snail today found on Ica›s lomas formations, the most common of which, we tentatively identify as *Bostryx reentsi* (Philippi, 1851; Bram Breure, personal communication). These snails are only active during the lomas growing season (Fig. 6G), aestivating in groups during the dry months (Ramírez, 1988 and Ramírez et al., 2003). We observe live snails in numbers grazing on the lush herby lomas vegetation during the fog season, but also in transition belts, eating algae that grows on *Tillandsia* and aestivating on their stems or beneath their roots (Craig, 1985).

Archaeological sites such as Amara Norte I and Abra Sur de Amara I, amidst these *Tillandsia* transition belts in the lomas de Amara, between 550 and 700 m asl, have middens dominated by snail shells (Fig. 5C and D). Meanwhile, sites along the littoral such as Arroyo de Lomitas, Abra I and La Yerba II, otherwise full of the evidence of marine resources, also all contain lomas snails, in some contexts in considerable quantities (Fig. 8F).

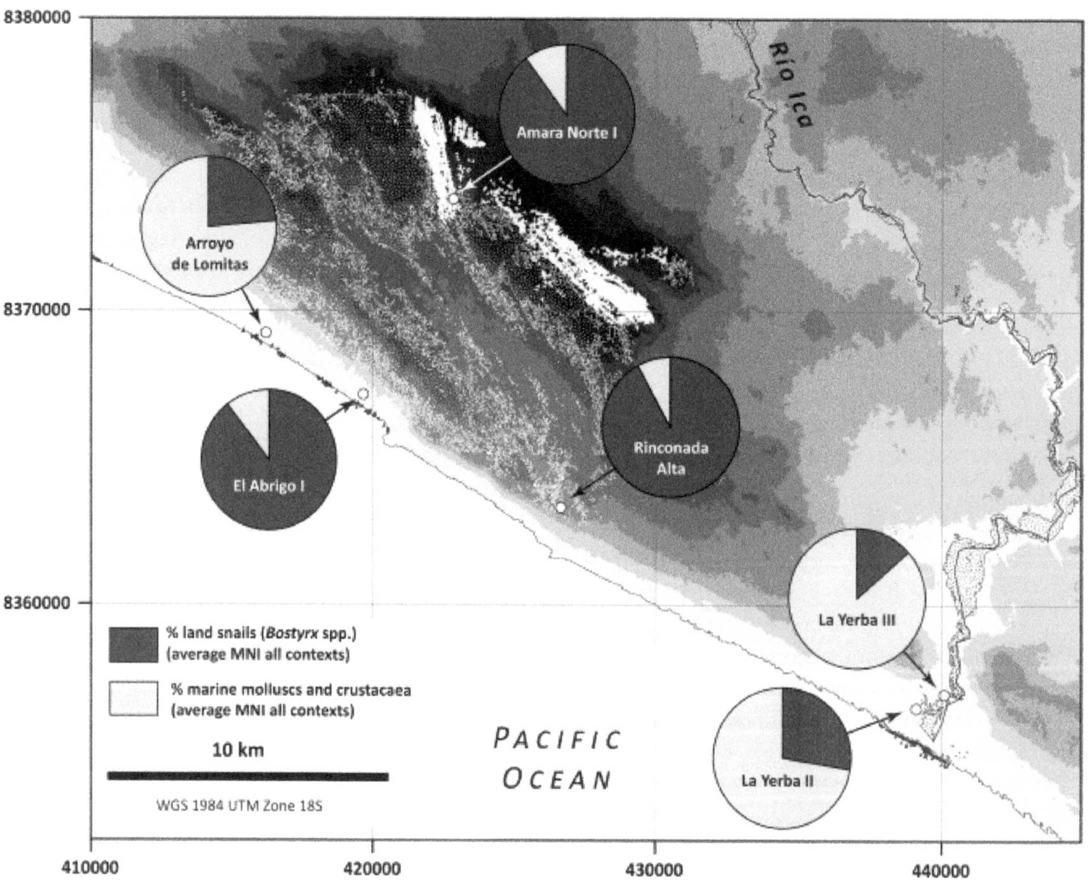

Figure 7: Land snails (*Bostyrx* spp.) as a proportion of total mollusk and crustacean assemblage (by average Minimum Number of Individuals 'MNI') across various south coast Middle Preceramic archaeological sites.

Top to bottom:
A Middle Precamic contexts, Trench 3, La Yerba II. Occupation layer SU 1079 dated 6,936 – 6,735 Cal YBP.
B La Yerba II under excavation, 2013.
C Stone grinding mortar (*'batan'*), La Yerba II, SU 1001, presumably for processing starchy plant foods.
D Examples of butchered grey deer (*Odocoileus virginianus*) bone, La Yerba II, SU 1001.
E Examples of butchered, charred camelid vertebra, presumably guanaco, (*Lama guanicoe*), La Yerba II, SU 1001.
F Examples of *Bostryx* sp. land snails, La Yerba II, SU 1005. Snails comprise 28% (by NMI) of the total mollusc and crustacean assemblage at La Yerba II..
G Desiccated *Haagocereus* sp. cactus fruit, La Yerba II, SU 1072.
H Desiccated *Oxalis* sp. tuber fragment, La Yerba II, SU SU 1079.
I Desiccated *Krameria* sp. seed, La Yerba II, SU 1072.

Figure 8: Selected lomas resources in Preceramic archaeological contexts at the site of La Yerba II, on the Río Ica estuary, south coast Peru.

In Mesolithic contexts worldwide snails have made a significant contribution to daily calorific and protein requirements (e.g. Rizner et al., 2005). Clearly this was the case during the Middle Preceramic in the lomas of the south coast of Peru. Relative to marine molluscs individual *Bostryx* yielded little, difficult to extract, meat; yet they were easily gathered in great quantities during winter months and could then – unlike marine molluscs – be stored fresh for months because of their ability to aestivate through sealing their shells with a mucus epiphragm (Ramírez, 1988). The significance of snails to Preceramic diets may well have varied therefore according to the seasonal availability of other resources, particularly during periodic collapses in marine resources induced by El Niño.

Most of the land snail assemblages in these contexts are whole and show little evidence of heat alteration. But many shells in the Amara Norte I middens show fine puncture holes suggesting that their meat was extracted using cactus thorn inserted through the epiphragm, or through the thin shell itself (see Fig. 5D). Since the taphonomies of terrestrial and marine molluscs in such ancient archaeological contexts should be similar – relative to those of other classes of organic remains – we use the proportion one to the other, measured by minimum number of individuals ('MNI'), as a proxy measure of the relative importance of lomas resources in particular contexts. Thus averages of such measures offers a useful, albeit crude, way to compare the use of lomas resources between individual site locations during the Middle Preceramic (see Fig. 7).

Meanwhile, the importance of other lomas faunal resources to south coast Middle Preceramic hunter–gatherers can be inferred from direct evidence such as animal bone – though with considerable taphonomic bias between different sites – but also from indirect evidence in the form, for instance, of the stone tools used to hunt and process animals.

Today tiny relict populations of guanaco still live in the south coast lomas of San Fernando and Amara (Fig. 6D). They are shy and more or less constrained to particularly remote parts. In the past these animals, catholic feeders on either browse or graze, may have moved more freely between the winter lomas and summer river valley oases, which may, in turn, have determined rounds of hunter–gatherer transhumance (Custred,

1979). Alongside predators such as puma (*Puma concolor*), guanaco and grey deer (*Odocoileus virginianus*) were once far more numerous in the lomas and the formerly wooded valleys of the Peruvian coast (Baied and Wheeler, 1993, Wheeler, 1995, Wheeler et al., 2011 and Beresford-Jones, 2011). Today there are no deer on the south coast of Peru, outside of captivity. Yet the Middle Preceramic archaeological record here includes evidence of both deer and camelids (presumably guanaco), together with many other vestiges of the attendant hunting activities in the lomas (Lavallée and Julien, 2012).

Before the development of woven textile technologies in the Late Preceramic, these lomas mammals would have provided skins and tendons vital for clothing and bindings. Guanaco hides were much esteemed for clothing by Yaghan hunter–gatherers of Tierra de Fuego because, unlike the hides of marine mammals – the more common prey of both Yaghan and Middle Preceramic hunters – guanaco skin is thin and flexible, so that it conforms readily to the contours of the human body, and has heavy hair, making it is far better for shedding water and keeping out the cold (Lothrop, 1928). The La Yerba II contexts contain good evidence for hide preparation in the form of numerous scraper tools made of *Choromytilus* shell and plants potentially used for curing leather (see below).

Preservation conditions at sites within the lomas itself are poor, though a few fragments of ungulate bone were recovered from Amara Norte I. At La Yerba II, 17% (by MNI) of an animal bone assemblage dominated by marine mammals and birds, are camelid and deer. Whereas all parts of deer were recorded, camelid bones were mostly (85%) of meatier hind leg parts, suggesting that these large animals were butchered off-site, perhaps near the kill sites within the lomas.

Obsidian and other stone tools, dominated by projectile points prepared for hafting, are found throughout the lomas of the south coast. Small stone structures high on bluffs in the lomas and near guanaco trails today, such as Abra Sur de Amara II, may represent hunting stands (cf. Larrain et al., 2001). Indeed, sites deep within the lomas associated with the gathering of snails such as Amara Norte I, are surrounded by scatters of obsidian lithics including many projectile points, presumably for hunting game (Fig. 5B).

The variety of obsidian points found in the lomas, are likely a palimpsest of hunting activities over time. Different lithic morphologies may also represent different target species, or hunting strategies. We suggest, however, that the majority of this lithic activity dates to late in the Middle Preceramic because those sites on the coast that date to that period, such as La Yerba III and Amara Norte I, have abundant, comparable obsidian lithic remains within stratified deposits, whereas earlier sites such as La Yerba II, contain only a few small obsidian flakes.

Lomas Flora as Food Resources

The importance of plant foods to hunter–gatherer diets outside extreme latitudes has long been recognised (e.g. Cordain et al., 2002). As noted above, many lomas perennials have evolved to cope with cycles of long aridity followed by intensive, growing seasons through underground storage organs (geophytes), several of which would have provided a valuable source of edible starch for the hunter–gatherers of the Middle Preceramic (Engel, 1987). Significant examples we have recorded in the lomas of the south coast include: the tuberous roots of *Aa weddelliana* (Orchidaceae), only recently recorded elsewhere on the Peruvian coast (Trujillo and Delgado Rodríguez, 2011); the roots of *Alstroemeria* sp. (Alstroemeriaceae), several closely related species of which are recorded as providing food for humans in the lomas of Chile where they are known as '*chuño*' (Puga, 1921 and Muñoz-Schick and Moreira-Muñoz, 2003); and the corms of various *Oxalis* (Oxalidaceae) species including *O. lomana* (Stafford, 1939, reports them tasting like cob-nuts, Fig. 6B). Several varieties of true and wild potatoes are found in lomas vegetation elsewhere in Peru (Sandra Knapp, personal communication) although only *Solanum montanum*, a highly variable species distantly related to potatoes (Bennett, 2008), is recorded today in the lomas of the south coast (Fig. 6C). Like all Solanaceae tubers, these are bitter and must be processed, but are recorded as a minor food source in historical times (Hooker, 1827). Edible lomas geophytes are at their most nutritious both before and after periods of active growth but remain available all year round, and are easily collected because plants such as *S. montanum* occur in concentrated patches and are not deeply rooted (Cohen, 1978a and Ochoa, 1998).

Edible seeds, greens and fruits of lomas plants likely provided seasonal sources of important vitamins and micronutrients. In particular almost all lomas cacti species produce palatable fruit and likely contributed significantly to diet for parts of the year (Cadwallader et al., 2012). *Corryocactus brachypetalus* which, uniquely, grows on the rocky seaward bluffs of the lomas has the largest fruit, with a pale pulp tasting not unlike apple or gooseberry. Two species of *Haageocereus* cactus (*decumbens* and aff. *tenuis*), generally found in the transitional vegetation belts of the inter lomas plain in Amara (Fig. 6E and F), both produce crops of sweet juicy fruit in January once the winter fogs have receded, whilst the fruit of *Cumulopuntia sphaerica* and *Eriosyce islayensis* are edible but somewhat less palatable.

In the desiccated conditions of the La Yerba II Middle Preceramic occupation contexts at the Río Ica estuary evidence for the use and consumption of lomas plants include *Oxalis*sp. tubers, and cactus fruit and seeds (cf. *Haageocereus* spp., Fig. 8G). Indirect evidence for the processing of starchy plant foods include stone grinding mortars ('*batanes*') both at La Yerba II and III (Fig. 8C), and also at Amara Norte I in the heart of the lomas (Fig. 5F), where plant remains themselves are only preserved through charring. Undifferentiated charred parenchyma tissue suggestive of starchy plant foods is evident in the contexts of all these Middle Preceramic sites. *Haageocereus* spines found at La Yerba III in carefully tied bundles, were presumably for used for making fishhooks (cf.Engel, 1984).

A range of lomas plants are also known today for their antibacterial or other medicinal properties including *Ephedra* spp. and *Plantago* spp. (Villagrán and Castro, 2003) and the astringent *Krameria lappacea*, used for treating inflammation, gastrointestinal illnesses and other medical purposes (Simpson, 1989 and Brack Egg, 1999). *Krameria*also has a very high tannin content so that it is sometimes used today for curing animal hides. Its remains are identified in the Middle Preceramic contexts of La Yerba II (Fig. 8I). Most of these archaeological contexts also contain many pale, finely-spun fishing yarns and net fragments. Engel

(1973) suggests, before cotton, such yarns may have been spun from that the downy seed head fibres of *Tillandsia* sp.

Many of these lomas plants are also important to the diet of the animals hunted here during the Preceramic. Guanaco graze on leguminous herbaceous lomas plants such as *Astragalus*, as well as the roots and tubers of perennial plants that they dig up, grasses, cacti and cacti fruit. In both winter and summer they can also persist on lichens and the *Tillandsia* that dominate such large parts of the south coast lomas (Raedeke and Simonetti, 1988 and Reus et al., 2009).

Lomas Flora as Fuel Resources

Fuel would have been required at these Preceramic sites for cooking, opening clam shells in bulk (Waselkov, 1987: 100), curing animal skins, roasting to make certain lomas tubers edible and indeed for warmth because, though in the tropics, the south coast of Peru is remarkably chilly in winter, when temperatures fall into single figures and strong winds blow cold, damp fogs inshore off the Pacific. Our data loggers record a minimum temperature during winter in the south coast lomas of 7 °C.

The evidence from Middle Preceramic sites such as La Yerba II and III, adjacent to the river, suggest the use of estuary and woodland species as fuel. But lomas vegetation must have been the primary source of fuel at those sites that are distant from the vegetated river courses, because ethnographic studies suggest that heavy expendable resources like fuel wood are almost always gathered within a radius of not more than two hours walking (Willis and Van Andel, 2004: 2371) Substantial fragments of wood charcoal are evident in the structured hearths at Arroyo de Lomitas (Fig. 4E). Although driftwood might have contributed to the overall fuel economy, today it is scare along this rocky shore. More exposed sites within the lomas such as Amara Norte I have poorly preserved microcharcoal remains, but blackened and fire cracked hearthstones here are evidence of repeated fires.

Today there are no trees and few woody perennial shrubs on the south coast lomas. Those few that do persist, such as *Ephedra americana, A. rotundifolia* and *C. alnifolius*, are like all lomas woody species, slow growing, yielding high calorific value fuel, but making them prone to overexploitation. Elsewhere, Preceramic charcoal assemblages have been taken as suggesting such overexploitation (Weir and Derring, 1986). Certainly, the charcoals in the hearths of Preceramic sites such as Arroyo de Lomitas, suggest that the south coast lomas once supported more woody vegetation. Large areas of these lomas are today dominated by *Tillandsia* spp., which, elsewhere, has been noted as a fuel in the Preceramic (Moseley and Willey, 1973, Weir and Derring, 1986 and Quilter, 1991). Its downy seed head fibres would also have made effective tinder (Pullen, 2005). Yet most *Tillandsia* too are slow-growing (Craig, 1985) and their removal may have promoted destabilisation of dune systems (Hesse, 2012: 35), evident in landscape deflation about sites such as Amara Norte I.

Although the story of Middle Preceramic human ecology in the lomas of the south coast is one of continuity over millennia, change through time is also evident in this archaeological record. We turn next to what that evidence might contribute to the vexed question of what caused such changes.

CHANGE THROUGH TIME

Lomas environments, like other arid area ecologies, are fragile and peculiarly sensitive to even mild climatic perturbations (Masuda, 1985), and to human impact (Dillon et al., 2003).

As discussed, the climatic factor of proximate relevance to lomas is the periodic perturbations along the Pacific coast of Peru of the El Niño Southern Oscillation ('ENSO') phenomenon. ENSO is defined by sea surface temperature ('SST') anomalies, described recently according to two spatial modes of variability: maximum SST anomalies localized in the central Pacific ('CP' mode), entailing negatively skewed SST anomalies off the coast of Peru and strong 'La Niña' events; or, maximum SST anomalies localized in the eastern Pacific ('EP' mode), entailing strong 'El Niño' events (e.g. Carré et al., 2014).

A long-standing model of ENSO history based on proxy data from archaeological sites on the north coast of Peru (Sandweiss et al., 1996, Sandweiss and Kelley, 2012 and Etayo-Cadavid et al., 2013) suggests, for north of 12^0 S, some four millennia of ENSO suppression during the Early Holocene, between c. 9000 and 5800 BP; after which there was a low-frequency of *El Niño* until 3000 BP, when the modern, higher frequency *El Niño*regime became established (cf. Rodbell et al., 1999, Moy et al., 2002 and Loubere et al., 2013).

Recently, however, refinements specific to the south coast have been proposed to this model based on $\delta^{18}O$ as a proxy for sea surface temperature recorded in surf clam shells in archaeological deposits, including from La Yerba II (Carré et al., 2012 and Carré et al., 2014), central Pacific corals (Cobb et al., 2013) and Galapagos foraminifera (Koutavas and Joanides, 2012). Together these suggest that for over five millennia, between 9600 and 4500 BP, mean annual SSTs were significantly lower than today, especially in southern Peru; that before around 8000 BP ENSO variance was skewed towards El Niño (EP mode), whereas between 7500 and 6700 BP variation was skewed towards CP mode with more frequent and intense La Niña events; and that there was a period of substantial reduction of ENSO variance was between 5000 and 4000 BP. How then might this revised model of ENSO history have impacted upon the south coast lomas and is it reflected in its archaeological record?

Variances in ENSO spatial mode and amplitude affect relative land and sea temperatures that drive convection wind regimes and govern other sea–atmosphere interactions, such as the production of aerosols and condensation nuclei, and thereby the production, intensity (water droplet size) and duration of occult precipitation in lomas. In northern and central Peru El Niño years prompt 'enormous blooming events' (Muenchow et al., 2013: 564) and population explosions of snails (Ramírez et al., 2003) in lomas (see also Dillon et al., 2003). Further south too they bring increased lomas vegetation, due to 'advection fog with a higher moisture content' over warmer waters (Eichler and Londoño, 2013: 1, see also Muñoz-Schick et al., 2001, Manrique et al., 2010 and Manrique, 2011). Dillon (2011) and Manrique (2011) record increases in the primary productivity of southern Peru's lomas, measured by plant density and cover, by thirteen times, during the very strong 1997–98 El Niño event, and three times during the moderate 2010 event, respectively. Moreover, El Niño effects usually fall in the austral summer, so producing continuity between the phases of normal lomas winter florescence. The effects of La Niña events on Peruvian lomas formations are less straightforward. Inland, colder oceanic conditions during La Niña produce dryer conditions. Yet, immediately along the coast, La Niña creates more persistent fogs (Garreaud et al., 2008, Manrique et al., 2010, Muenchow et al., 2013 and Eichler and Londoño, 2013), because the cooler humid air masses that it brings are more prone to condensation (McPhaden et al., 2006). Muenchow et al. (2013: 564) record 'abundant blooming events, high species diversity and high species coverage', albeit during a single La Niña year, in Casma. Nonetheless, the relationship between increased fog and lomas vegetation is far from simple, not least because it is governed also by changes in the altitude and temperature of the fog layer (Manrique et al., 2010, Eichler and Londoño, 2013 and Muenchow et al., 2013). Consequently lomas plant species have evolved into these moisture regimes and topographically delimited niches to form a complex mosaic of vegetation. Persistent La Niña conditions may, for instance, promote xerophytic species and widen transitional vegetation belts at the expense of herbaceous fog meadow (Latorre et al., 2011).

Fig. 3 and Fig. 9 shows that for the lomas of the south coast the impact of ENSO on lomas resources is bimodal: El Niño tending to produce a higher above ground biomass in ephemeral zones, while La Niña tend to produce a higher below ground biomass in fog zones in the form of geophytes. In sum we infer that past multi-centennial periods of increased ENSO variance and intensity – of either La Niña or El Niño – would have bolstered the extent, volume and fog trapping height of lomas vegetation, thereby positively influencing lomas hydrology and producing a self-sustaining microclimate. This, in turn, would have greatly augmented the biomass of both the floral and faunal resources exploited therein by Preceramic hunter–gatherers. Different ENSO modes would promote different parts of the lomas ecosystem and so entail different subsistence strategies, for instance, by focus on hunting guanaco and gathering snails as their populations soared due to expanded herbaceous lomas vegetation during periods of increased El Niño variation, or, by focus on gathering the starchy geophytes of lomas perennials when they thrived thanks to the more persistent fogs in epochs of increased La Niña variation.

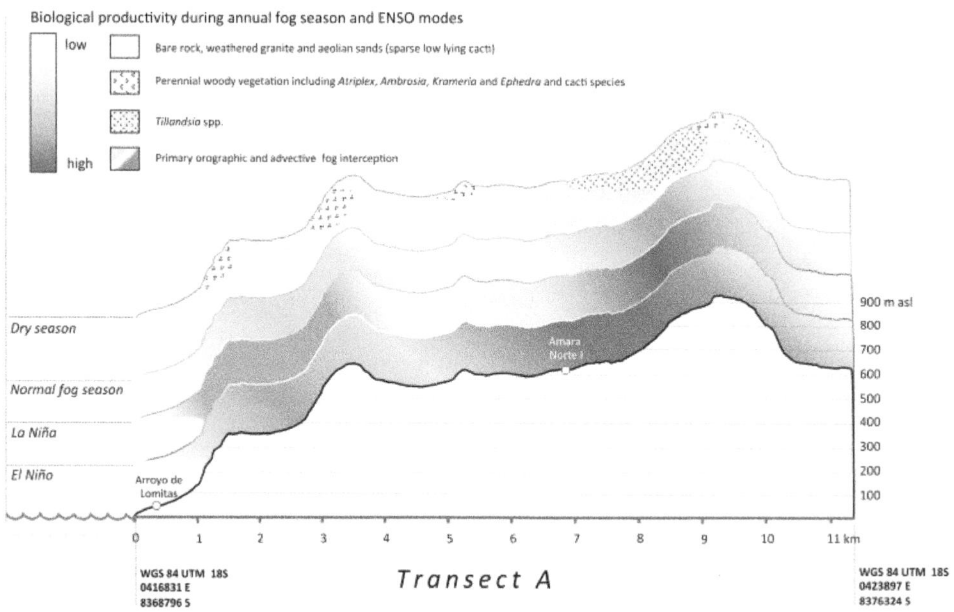

Figure 9: Conceptual model of lomas biological productivity across Transect A in Amara lomas, showing: (i) Dry season (January–June) – when seed, stem and root storage provide resources; extant woody and herbaceous perennial lomas ('WHPL') and *Tillandsia* dominated vegetation (ii) Normal fog season(occurring annually or biennially, July–December) – occult precipitation stimulating plant growth and reproduction in high diversity annual and WHPL species. (iii) El Niño – rare rainfall events triggering non-seasonal ephemeral herbaceous species of low diversity to expand across open pampas into *Tillandsia*/spp. areas and merging refugia (iv) La Niña – sustained fog and lower temperature (reducing evaporation) increasing growth and expanding margins of WHPL.

By contrast, long durations of suppressed ENSO variance would be inimical to lomas ecosystems. Indeed, without regeneration through periodic ENSO events, certain lomas species may disappear completely over time (Dillon, 2011). It would also increase their vulnerability to human impact. Vegetation in lomas acts overall, not as a water consuming, but as a water-*producing* factor because of the greatly enhanced surface area it provides for the condensation of fog water. This is particularly true of taller woody shrubs and trees which increase fog condensation in lomas by up to six-times (Ellenberg, 1958) so that their removal, say for fuel, acts in self-enhancing feedback to reduce lomas hydrology and soil humidity, and thus the growth and germination of other herbaceous plants and the entire lomas biomass (Engel, 1973, Walter, 1973, Oka, 1986 and Muenchow et al., 2013).

Fig. 10 brings together the archaeological record for the south coast lomas (Table 1) and the latest ENSO climate history data for the same region, as synthesised by Carré et al., 2012 and Carré et al., 2014. We make five observations from this combination of data, in light of the ecological information already presented.

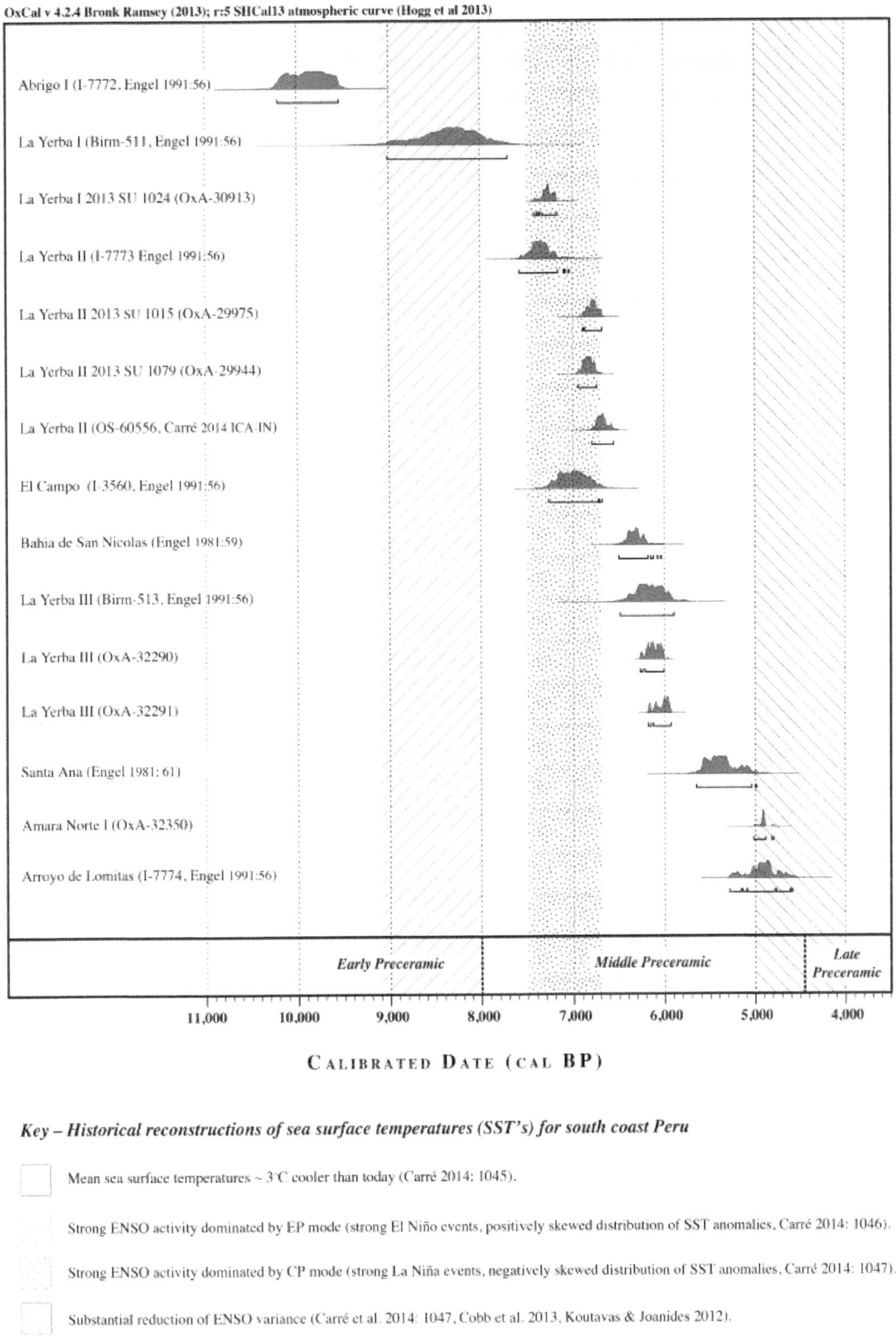

Figure 10: Comparison of ENSO history (after Carré et al., 2014) with dates of Preceramic archaeological sites on the south coast of Peru, all calibrated using the ShCal13 curve (Hogg et al., 2013) in OxCal version 4.2 (Bronk Ramsey, 2009).

Firstly, Fig. 10 shows that, over the broadest scale, all the Preceramic archaeological sites[1] intimately associated with the south coast lomas were occupied during five millennia in which, as Carré et al. (2014: 1045) summarise, 'mean annual SST was significantly lower ... than today, especially in southern Peru ~3 °C cooler [implying] an increase in the intensity of coastal upwelling'. It was this that underpinned conditions of higher oceanic productivity, and augmented lomas hydrology and biomass due to more persistent fogs (e.g. Carré et al., 2009), which sustained hunter–gatherers in Engel's (1984) 'fog oasis situation' for this enormous stretch of time on the south coast. Secondly, there was, seemingly, a millennia of increased El Niño (EP mode) before 8000 BP and the start of the Middle Preceramic Period, which would have expanded lomas vegetation extent and biomass, but simultaneously inflicted persistent shocks upon certain marine resources critical to Preceramic subsistence. Only one site at the estuary, La Yerba I, is dated by Engel (1991: 154) to this epoch, and following calibration, entails a margin of uncertainty of almost a millennia (9015–7696 Cal yrs BP, see Table 1, Fig. 10). Engel provides few details and our own investigations of the same site – so far as that can be ascertained – suggest that both its date and archaeological assemblage are, in fact, similar to the adjacent site of La Yerba II (see Table 1, Fig. 10, Arce et al., 2013). If so, there is little visible archaeological evidence of human occupation of the south coast during this period, though that also entails factors of eustatic sea-level change and stabilisation, germane to our next observation.

Thirdly, within the broad sweep of five millennia of colder seas, the start of the Middle Preceramic period is marked by the founding of several highly visible archaeological sites, most notably La Yerba II (7571–7030 Cal yrs BP) on the Río Ica estuary. This coincides with a multi-centennial period of enhanced La Niña activity (see Fig. 10), entailing even colder seas and foggier lomas: conditions reflected in the particularly cold-water ecology of parts of the La Yerba II mollusc assemblage, such as *Tegula atra* and *Choromytilus chorus*. Also, at this time eustatic sea levels stabilised, after which shoreline progradation began forming the sandy beach habitat of the easily gathered *Mesodesma* surf clams that so characterise the La Yerba II middens. It is surely no coincidence then that the establishment of the Middle Preceramic way of life here apparently takes place during an epoch of abundant and predictable oceanic and lomas resources.

Fourthly, over the five hundred years to around 6000 BP, the archaeological sites on the river estuaries show evidence of increased sedentism and a broadening of the resource base along a spectrum of mixed hunting-farming subsistence (cf. Gorbahn, 2013). Whereas sites like La Yerba II are characterised by temporary areesh wind-shelters made of reeds (cf. Quilter, 1989), later-dated sites at the river estuaries, such as La Yerba III and Santa Ana (6281–6006 Cal yrs BP and 5650–4983 Cal yrs BP), have evidence of more permanent and substantial villages and structured mortuary deposition (see Engel, 1981: 20–21, 1991: 157, cf. the 'Encanto phase' elsewhere, see Lanning, 1963 and Patterson and Moseley, 1968). These later sites have much greater quantities of obsidian, indicating far wider spheres of interaction since the nearest sources are 250 km away in the highlands, notably at Quispisisa (Tripcevich and Contreras, 2011). They also have the first evidence here for food agriculture, grown in the seasonally humid silts of the adjacent river floodplains. We recover fully domesticated lima beans (*Phaseolus lunatus*), while Engel (1981: 20) reports both *Phaseolus* beans and jícama (*Pachyrhizus tuberosus*), in the contexts of La Yerba III.

Last, but not least, Fig. 10 suggests that as the long epoch of cooler sea temperatures ended around 4500 BP, synchronous with a Holocene minimum of ENSO variance lasting a thousand years to 4000 BP: each inimical to lomas ecosystems and to water sources they fed along the littoral, so too did a Middle Preceramic way of life that had turned for millennia here about the lomas seasons and their rich ocean littoral – the 'fog oasis situation' – draw to a close.

Indeed, the significance of lomas resources to human settlement here is made stark by the observation that there are *no* Preceramic archaeological sites along the coast between Morro Quemado and Bahía de San Nícolas dated to after 4450 BP, the date more widely construed to mark the end of the Middle Preceramic Period elsewhere in Peru (Quilter, 1991). For there seems little reason to suppose that a shift to an average sea temperatures akin to modern conditions, together with suppressed ENSO activity, would have been greatly detrimental to many marine resources, not least *Mesodesma*clams, long a key dietary component here. And yet, even at the river estuaries, the archaeological record appears silent for the subsequent Late Preceramic Period [2]. Following the demise of the 'fog oasis situation', therefore, we speculate that increased reliance upon agriculture in the Late Preceramic necessitated relocation of settlement inland, into the riparian basins of the south coast rivers.

Elsewhere, on the north and central coasts of Peru this critical change heralded the florescence of greater population densities and monumental civilization after 4500 BP (e.g. Dillehay et al., 2004). This did not happen on the south coast, probably because of is distinctive geomorphological configuration. For unlike the river valleys to the north, with their broad alluvial deltas and wide ocean frontages granting easy access simultaneously to rich marine and agricultural resources, the river systems of the south coast comprise scattered riparian basins down long river courses, diverted and separated from the sea by the same coastal lomas formations that had been the prime theatre of human ecology during the Middle Preceramic.

Those lomas environments continued to be exploited seasonally for snails, plants and game and for grazing domesticated animals throughout later archaeological time periods and indeed, well into historical times (Rostworowski, 1981, Masuda, 1985, Larrain et al., 2001 and Beresford-Jones et al., 2011). Paths through the south coast lomas were still traversed for access to marine resources along the littoral (see coastal surveys by Engel, 1981, Engel, 1991 and Carmichael, 1991). Yet it also seems clear that for those later times, lomas and marine resources were strictly supplementary (Cadwallader et al., 2012 and Carmichael et al., 2014), never again dictating the rounds of human existence as they had during the Middle Preceramic Period.

CONCLUSIONS

Plainly it was the cold, upwelling ocean, with its bounty of almost inexhaustible protein sources, that chiefly sustained the Middle Preceramic hunter–gatherers of the south coast of Peru. Yet because those marine resources were distributed in hotspots all along this littoral, it was in fact the lomas formations on the granite coastal massif, and their fresh water sources, that defined the setting of human ecology at this time and thereby the patterning of human occupation and corridors of movement along the south coast between Morro Quemado and Bahía de San Nícolas: Engel's (1981: 24) 'fog oasis situation' (see Fig. 1). The south coast lomas offer unique insight into the capacity of past lomas ecosystems to support seasonal hunter–gatherer occupation because of their separation from the resources and water sources of the Andean foothills.

Lomas fauna and flora provided significant (Fig. 7), and seasonally critical, components of Middle Preceramic diet, most notably plant tubers, ungulates and storable land snails. Moreover, lomas provided critical fuel, medicinal and raw material resources. Our findings refute any view that lomas environments were not important 'could never have been important to man', or that they have not altered much through the course of the Holocene. Indeed, given that lomas ecosystems include wild relatives of the Andean potato and tomato (*Solanum* spp.) and papaya (*Carica* sp.) together with guanaco, the wild relative of llama and alpaca camelids, we suggest that the role of lomas in key Andean domestication processes merit more consideration than has hitherto been the case, not least because those processes for camelids and tubers have often been seen as going hand in hand (e.g. Pearsall, 2008: 113).

Setting the latest model for ENSO variance based upon $\delta^{18}O$ isotope records (Carré et al., 2014), alongside the archaeological patterning here, now defined with greater chronological precision, shows striking correlation between the Middle Preceramic occupation throughout the south coast lomas and a long epoch of significantly colder seas, with implications for increased intensity of coastal upwelling and more persistent lomas fogs. These include logistical field camps spread far along the lomas littoral alongside fog-fed arroyo watercourses, and previously undiscovered sites targeting gathered and hunted resources deep within the lomas itself.

Within that five millennia of colder seas, the first Middle Preceramic occupations were founded at the river estuaries during a multi-centennial period of enhanced La Niña activity, entailing even more abundant ocean and lomas resources, and coinciding too with the time at which eustatic sea-levels stabilised and shoreline progradation began to form the beach habitat necessary for abundant, easily collected *Mesodesma* clam resources that dominate the middens of these sites. These logistical basecamps at the river estuary offered complex hunter–gatherers access to a mosaic of diverse, highly productive environments [3].

Eventually, the millennia of Middle Preceramic existence defined by the 'fog oasis situation' and the epoch of colder seas drew to a close, during a Holocene minimum of ENSO variance inimical to lomas ecosystems and the water sources they sustained. Thus this latest data from the south coast upholds Lanning's (1963: 369) perspicacious assertion, made half a century ago now, that climate change – specifically linked to alterations in oceanic circulation (i.e. ENSO) – accounts for significant change in the resource potential of lomas ecosystems through the Holocene, and indeed, for why human ecology shifted out of the 'fog oasis situation'.

Yet compelling though this vision of climate-induced change in human ecology during the Middle Preceramic is, it does not preclude other factors from explanations of why the 'fog oasis situation' came to be abandoned, not least human impact, to which such climate changes would have exposed lomas ecologies. Indeed, since vegetation in lomas environments, and in particular its slow-growing, easily over-exploited woody vegetation, acts to catalyse fog condensation, such perturbations and human impact each precipitate the effects of the other. Today there are no trees and few woody species in the south coast lomas. Yet charcoal in Middle Preceramic hearths throughout this fog oasis situation attest to this not having been the case in the distant past. More importantly still, this model of climatically-induced lomas resource depression at the close of the Middle Preceramic does *not* seem to explain the emergence of agriculture in the Andes, as Lanning and others had also proposed.

Over the five hundred years between La Yerba II and III, the archaeological record of the south coast shows evidence for many of those changes widely recognised to precede the emergence of agriculture in many parts of the world, for which Flannery (1969) coined the term 'Broad Spectrum Revolution'. These include more permanent architecture, structured mortuary deposition perhaps denoting territoriality, much more extensive trade or exchange networks implied by obsidian quantities, and a widening use of resources to include floodplain farming of high-protein *Phaseolus* beans. Yet throughout this time, the 'fog oasis situation' still prevailed (Fig. 10).

We conclude therefore that on the south coast of Peru a Broad Spectrum Revolution unfolded, *not* through population pressure in deteriorating environments (e.g. Patterson and Moseley, 1968 and Cohen, 1978b), but rather as an outcome of resource *abundance*prevailing in the 'fog oasis situation' throughout the Middle Preceramic Period. Just as is now envisaged for many parts of the world (Arnold, 1996: 98; Zeder, 2012: 258), it was a combination of abundance and seasonal predictability that enabled increasingly complex Middle Preceramic hunter–gatherers here to reduce mobility by settling in logistically optimal locations at the confluence of multiple eco-zones at the river estuaries.

ACKNOWLEDGEMENTS

We thank the *Ministerio de Cultural del Perú* for granting permission for archaeological fieldwork (Resolución Directoral N° 933-2012-DGPC-VMPCIC/MC, 19 December 2012 and N° 386-2014-DGPA-VMPCIC/MC, 22 August 2014) and the export of samples for dating; Don Alberto Benavides Ganoza and the people of Samaca for facilitating fieldwork; the Leverhulme Trust (grant number RPG-117) and the late Don Alberto Benavides de la Quintana (grant number RG69428) and the McDonald Institute for Archaeological Research for funding Cambridge University›s *One River Archaeological Project*, and the NERC Radiocarbon facility (grant number NF/2013/2/2) for funding radiocarbon dating. We also thank the *Servicio Nacional* Forestal y de Fauna Silvestre (SERFOR) and the *Servicio Nacional de Áreas Naturales Protegidas por el Estado*(SERNANP), Peru for permits for the *Proyecto Kew Perú* to carry out botanical and ecological survey, and Delsy Trujillo, Eric Ramírez, Consuelo Borda and other participants of the *Proyecto Kew Perú: Conservación, Restauración de Hábitats y Medios de Vida Útiles, Ica, Peru.*

REFERENCES

1. Arce, S., Pullen, A.G., Huaman, O., Chauca, G.E., Beresford-Jones, D.G., 2013. Proyecto de investigacion arqueologica Samaca. Informe de los trabajos realizadosdurante la temporada 2013. Presentado al Ministerio de Cultura Lima Diciembre 2013. Ica, Peru.
2. Arce, S., Pullen, A.G., Chauca, G.E., Beresford-Jones, D.G., 2015. Proyecto de investigacion arqueologica Samaca. Informe de los trabajos realizados durante la temporada 2013. Presentado al Ministerio de Cultura Lima Diciembre 2014. Ica, Peru.
3. Arnold, J.E., 1996. The archaeology of complex hunteregatherers. J. Archaeol. Method Theory 3, 77e126.

4. Baied, C.A., Wheeler, J.C., 1993. Evolution of high Andean Puna ecosystems: environment, climate, and culture change over the last 12,000 years in the Central Andes. Mt. Res. Dev. 13, 145.

5. Beresford-Jones, D.G., 2011. Lost Woodlands of the Ancient Nasca. Oxford University Press, Oxford.

6. Beresford-Jones, D.G., Whaley, O.Q., Alarcon, C., Cadwallader, L., 2011. Two millennia of changes in human ecology: archaeobotanical and invertebrate records from the lower Ica Valley, south coast Peru. Veg. Hist. Archaeobotany 20, 273e292.

7. Betancourt, J.L., Latorre, C., Rech, J.A., Quade, J., Rylander, K.A., 2000. A 22,000-Year record of monsoonal precipitation from Northern Chile's Atacama Desert. Science 289, 1542e1546.

8. Binford, L.R., 1980. "Willow Smoke and Dogs Tails": hunteregatherer settlement systems and archaeological site formation. Am. Antiq. 45, 4e20.

9. Brack Egg, A., 1999. Diccionario Enciclopedico De Plantas Útiles Del Perú. Centro de Estudios Regionales Andinos, Cusco.

10. Bronk Ramsey, C., 2009. Bayesian analysis of radiocarbon dates. Radiocarbon 51, 337e360.

11. Cadwallader, L., Beresford-Jones, D.G., Whaley, O.Q., O'Connell, T.C., 2012. The signs of maize? A reconsideration of what d13C values say about palaeodiet in the Andean Region. Hum. Ecol. 40, 487e509.

12. Carmichael, P., 1991. Prehistoric Settlement of the Ica-grande Littoral, Southern Peru. Research Report to the Social Sciences and Humanities Research Council of Canada (Unpublished report).

13. Carmichael, P., Kennedy, B.V., Cadwallader, L., 2014. Coastal but not littoral: marine resources in the Nasca diet. Nawpa Pacha J. Andean Archaeol. 34, 1 ~ e24.

14. Carre, M., Klaric, L., Lavall ee, D., Julien, M., Bentaleb, I., Fontugne, M., Kawka, O., 2009. Insights into early Holocene hunteregatherer mobility on the Peruvian Southern Coast from mollusk gathering seasonality. J. Archaeol. Sci. 36, 1173e1178.

15. Carre, M., Azzoug, A., Bentaleb, I., Chase, B.M., Fontugne, M., Jackson, D., Ledru, M.-P., Maldonado, A., Sach, J.P., Schauer, A.J., 2012. Mid-Holocene mean climate in the south eastern Pacific and its influence on South America. Quat. Int. 253, 55e66.

16. Carre, M., Sachs, J.P., Schauer, A.J., Rodrguez, W.E., Ramos, F.C., 2013. Reconstructing El Nino-Southern Oscillation activity and ocean temperature seasonality from ~ short-lived marine mollusk shells from Peru. Palaeogeogr. Palaeoclimatol. Palaeoecol. 371, 45e53.

17. Carre, M., Sachs, J.P., Purca, S., Schauer, A.J., Braconnot, P., Falcon, R.A., Julien, M., Lavallee, D., 2014. Holocene history of ENSO variance and asymmetry in the eastern tropical Pacific. Science 345, 1045e1048.

18. Caviedes, C.N., 1984. El Nino 1982 ~ e83. Geogr. Rev. 267e290.

19. Cobb, K.M., Westphal, N., Sayani, H.R., Watson, J.T., Di Lorenzo, E., Cheng, H., Edwards, R.L., Charles, C.D., 2013. Highly variable El Nino-Southern Oscillation ~ throughout the Holocene. Science 339, 67e70.

20. Cohen, M.N., 1978a. Archaeological plant remains from the central coast. Nawpa ~ Pacha J. Andean Archaeol. 16, 23e50.

21. Cohen, M.N., 1978b. Population pressure and the origins of agriculture. In: Browman, D.L. (Ed.), Advances in Andean Archaeology. Mouton Publishers, The Hague, pp. 91e132.

22. Cordain, L., Eaton, S.B., Brand Miller, J., Mann, N., Hill, K., 2002. The paradoxical nature of hunteregatherer diets: meat-based, yet non-atherogenic. Eur. J. Clin. Nutr. 56, S42.

23. Craig, A.C., Psuty, N.P., 1968. The Paracas Papers, 1(2). Occasional Publication No. 1, Dept. of Geography, Florida Atlantic University, Boca Raton.

24. Craig, A.K., 1992. Archaeological occurences of Andean Landsnails. Andean Past. 3, 127e135.

25. Craig, A.K., 1985. Cis-Andean environmental transects: Late Quaternary ecology of Northern and Southern Peru. In: Masuda, S., Shimada, I., Morris, C. (Eds.), Andean Ecology and Civilization. An Interdisciplinary Perspective on Andean Ecological Complementarity. University of Tokyo Press, Tokyo, pp. 23e44.

26. Custred, G., 1979. Hunting technologies in Andean culture. J. Soc. Americanistes 66, 7e19.

27. Dillehay, T.D., Bonavia, D., Kaulicke, P., 2004. The first settlers. In: Silverman, H. (Ed.), Andean Archaeology. Blackwell, Oxford, pp. 16e34.

28. Dillon, M.O., 2011. Floristic checklist of the Peruvian lomas formations. Arnaldoa 18, 07e32.

29. Dillon, M.O., Nakazawa, M., Gonzales, S.L., 2003. The lomas formations of coastal Peru: composition and biogeographic origin, in: El Nino in Peru: biology and ~ culture over 10,000 years, Fieldiana (Botany). Field Mus. Nat. Hist. 1e9.

30. Dollfus, O., 1968. Le Perou. Travaux et Memoires de l 'Institut des Hautes Etudes de l'Amerique Latine. Universite de Paris, Paris.

31. Eichler, T.P., Londono, A.C., 2013. ENSO impacts on lomas formation in South Coastal ~ Peru: Implications for the Pliocene? Adv. Meteorol. 2013, 1e7.

32. Ellenberg, H., 1958. Nebeloasen in der Kustenwuste Perus. Universitatsven, Bonn.

33. Engel, F., 1973. New facts about pre-Columbian life in the Andean Lomas. Curr. Anthropol. 14, 271e280.

34. Engel, F.A., 1981. Prehistoric Andean Ecology: Man, Settlement and Environment in the Andes, vol. 2. The Deep South. Humanities Press, Hunter College, New York.

35. Engel, F.A., 1984. Prehistoric Andean Ecology. Man, Settlement and Environment in the Andes. In: Stone Typology, vol. 3. Humanities Press, Hunter College, New York.

36. Engel, F., 1987. De las Begonias al Maíz. La Universidad Nacional Agraria Del Perú, Lima, Perú.

37. Engel, F.A., 1991. Un Desierto en Tiempos Prehispanicos. La Univerisdad Nacional Agraria Del Peru, Lima, Perú.

38. Erlandson, J.M., 2001. The archaeology of aquatic adaptations: paradigms for a new millennium. J. Archaeol. Res. 9, 287e349.

39. Etayo-Cadavid, M.F., Andrus, C.F.T., Jones, K.B., Hodgins, G.W.L., Sandweiss, D.H., 2013. Marine radiocarbon reservoir age variation in Donax obesulus shells from northern Peru: Late Holocene evidence for extended El Nino. Geol. (Boulder) 41, ~ 599e602.

40. Flannery, K.V., 1969. Origins and ecological effects of early domestication in Iran and the Near East. In: Ucko, P.J., Dimbleby, G.W. (Eds.), The Domestication and Exploitation of Plants and Animals. Duckworth, London, pp. 73e100.

41. Fontugne, M., Usselmann, P., Lavallee, D., Julien, M., Hatte, C., 1999. El Nino vari- ~ ability in the Coastal Desert of Southern Peru during the Mid-Holocene. Quat. Res. 52, 171e179.

42. Garreaud, R., Barichivich, J., Christie, D.A., Maldonado, A., 2008. Interannual variability of the coastal fog at Fray Jorge relict forests in semiarid Chile. J. Geophys. Res. 113.

43. Gorbahn, H., 2013. The Middle Archaic site of Pernil Alto, southern Peru: The beginnings of horticulture and sedentariness in mid-Holocene conditions. Dialogo Andino 41, 61e83.

44. Hesse, R., 2012. Spatial distribution of and topographic controls on Tillandsia fog vegetation in coastal southern Peru: remote sensing and modelling. J. Arid Environ. 78, 33e40.

45. Hesse, R., 2014. Three-dimensional vegetation structure of Tillandsia latifolia on a coppice dune. J. Arid Environ. 109, 23e30.

46. Hogg, A.G., Hua, Q., Blackwell, P.G., Niu, M., Buck, C.E., Guilderson, T.P., Heaton, T.J., Palmer, J.G., Reimer, P.J., Reimer, R.W., Turney, C.S.M., Zimmerman, S.R.H., 2013.

47. SHCal13 Southern hemisphere calibration, 0e50,000 Years cal BP. Radiocarbon 55 (4), 1889e1903.

48. Hooker, W.J., 1827. Witheringia montana. Mountain Witheringia or St. Lorenzo potato. Bot. Mag. 54 (1), 2768.

49. Isla, J.C., 1990. La Esmeralda: Una ocupacion del Periodo Arcaico en Cahuachi, Nasca. Garceta Arqueol. Andina 6, 67e80.

50. Klemm, O., Schemenauer, R.S., Lummerich, A., Cereceda, P., Marzol, V., Corell, D., Heerden, J., Reinhard, D., Gherezghiher, T., Olivier, J., Osses, P., Sarsour, J., Frost, E., Estrela, M.J., Valiente, J.A., Fessehaye, G.M., 2012. Fog as a fresh-water resource: overview and perspectives. AMBIO 41, 221e234.

51. Koutavas, A., Joanides, S., 2012. El Nino-Southern Oscillation extrema in the Holo- ~ cene and Last Glacial Maximum. Paleoceanography 27, PA4208.

52. Lanning, E.P., 1963. A pre-agricultural occupation on the Central Coast of Peru. Am. Antiq. 360e371.

53. Lanning, E., 1967. Peru Before the Incas. Prentice Hall, Englewood Cliffs, N.J.

54. Larrain, H., Sagredob, E., Perez, L., Gonzalezc, B.A., Cerecedad, P., Osses, P., Velasquez, F., Lazaro, P., Navarro, M., 2001. New evidence of human habitation and hunting activity at the fog oasis of Alto Patache, south of Iquique, Chile. In: Proceedings of the 2nd International Conference on Fog and Fog Collection. Saint John's Canada.

55. Latorre, C., Gonzalez, A.L., Quade, J., Farina, J.M., Pinto, R., Marquet, P.A., 2011. ~ Establishment and formation of fog-dependent Tillandsia landbeckii dunes in the Atacama Desert: evidence from radiocarbon and stable isotopes. J. Geophys. Res. 116.

56. Lavallee, D., Julien, M. (Eds.), 2012. Prehistoria de la costa extremo-sur del Perú: los pescadores arcaicos de la Quebrada de los Burros (1000e7000 a. P.). Instituto Frances de Estudios Andinos, Lima.

57. Lothrop, S.K., 1928. The Indians of Tierra Del Fuego, Contributions from the Museum of the American Indian Heye Foundation. Lancaster Press, INC, Lancaster, PA.

58. Loubere, P., Creamer, W., Haas, J., 2013. Evolution of the El Nino-Southern Oscilla- ~ tion in the late Holocene and insolation driven change in the tropical annual SST cycle. Glob. Planet. Change 100, 129e144.

59. Lynch, T.F., 1971. Preceramic Transhumance in the Callejon de Huaylas, Peru. Am. Antiq. 36, 139e148.

60. Machtle, B., Eitel, B., 2012. Fragile landscapes, fragile civilizations: how climate € determined societies in the pre-Columbian south Peruvian Andes. Catena 103, 62e73.

61. Manrique, R., Ferrari, C., Pezzi, G., 2010. The influence of El Nino Southern Oscil- ~ lation (ENSO) on fog oases along the Peruvian and Chilean coastal deserts. In: Fifth International Conference on Fog and Fog Collection and Dew. Climatology Working Group, p. 30.

62. Manrique, R., 2011. El Nino Southern Oscillation and its Effect on Fog Oases along ~ the Peruvian and Chilean Coastal Deserts (Unpublished Doctoral thesis). Alma Mater Studiorum Universita di Bologna. http://dx.doi.org/10.6092/unibo/amsdottorato/3436.

63. Masuda, S., 1985. Algae collectors and lomas. In: Masuda, S., Shimada, I., Morris, C. (Eds.), Andean Ecology and Civilization. An Interdisciplinary Perspective on Andean Ecological Complementarity. University of Tokyo Press, Tokyo, pp. 233e250.

64. McPhaden, M.J., Zebiak, S.E., Glantz, M.H., 2006. ENSO as an integrating concept in earth science. Science 314, 1740e1745.

65. Moseley, M.E., 1975. The Maritime Foundations of Andean Civilization. Cummings Publishing Company, Menlo Park, California.

66. Moseley, M.E., Willey, G.R., 1973. Aspero, Peru: a re-examination of the site and its implications. Am. Antiq. 38, 452.

67. Moy, M., Seltzer, G.O., Rodbell, D.T., Anderson, D.,M., 2002. Variability of El Nino/ ~ Southern Oscillation activity at millennial timescales during the Holocene epoch. Nature 420, 159e162.

68. Muenchow, J., Brauning, A., Rodríguez, E.F., von Wehrden, H., 2013. Predictive € mapping of species

richness and plant species' distributions of a Peruvian fog oasis along an altitudinal gradient. Biotropica 45, 557e566.

69. Munoz-Schick, M., Pinto, Z.R., Mesa, A., Moreira, A., 2001. Fog oases during the El ~ Nino Southern Oscillation (ENSO) 1997-1998, in the coastal hills south of ~ Iquique, Tarapac a region, Chile. Rev. Chil. Hist. Nat. 74 (2), 389e405.

70. Munoz-Schick, M., Moreira-Mu ~ noz, A., 2003. Alstroemerias de Chile. Taller La Era ~ (Santiago, Chile).

71. Ochoa, C.M., 1998. Ecogeography and breeding potential of the wild Peruvian tuberbearing species of Solanum. Econ. Bot. 52, 3e6.

72. Oka, S., 1986. On trial measurements of the moisture in fog on Loma Ancon e in relation to an investigation into the conditions required for development of Lomas communities. In: Ono, M. (Ed.), Taxonomic and Ecological Studies on the Lomas Vegetation in the Pacific Coast of Peru. Tokyo Metropolitan University Makino Herbarium, Tokyo, pp. 41e51.

73. Oka, S., Ogawa, M., 1984. The Distribution of Lomas Vegetation and its Climatic Environments along the Pacific Coast of Peru. Geographical Reports of Tokyo Metropolitan University. Tokyo Metropolitan University Department of Geography, Tokyo.

74. Ono, M., 1986. Definition, classification and taxonomic significance of the lomas vegetation. In: Taxonomic and Ecological Studies on the Lomas Vegetation in the Pacific Coast of Peru. Tokyo Metropolitan University Makino Herbarium, Tokyo, pp. 5e14.

75. Palomino, V., 2006. Report only relict population of the Peruvian guanaco (Lama guanicoe) left in the coastal desert of Las Lomas de San Fernando, Ica 2003.

76. Presented at the Fortalecer el futuro de los Camelidos: Una oportunidad para crecer. IV Congreso Mundial sobre Camelidos 11-15 Octubre de 2006 Santa María, Catamarca, Argentina, pp. 1e4.

77. Parsons, M.H., 1970. Preceramic subsistence on the Peruvian coast. Am. Antiq.

78. 292e304.

79. Patterson, T.C., Moseley, M.E., 1968. Late Preceramic and Early ceramic cultures of the Central Coast of Peru. Nawpa Pacha: J. Andean Archaeol. 6, 115 ~ e133.

80. Pearsall, D.M., 2008. Plant Domestication and the shift to agriculture in the Andes. In: Silverman, H., Isbell, W.H. (Eds.), Handbook of South American Archaeology. Springer, New York, pp. 105e120.

81. Pefaur, J.E., 1981. Composition and Phenology of Epigeic animal communities in the Lomas of Southern Peru. J. Arid Environ. 4, 31e42.

82. Pefaur, J.E., 1982. Dynamics of plant communities in the lomas of southern Peru. Vegetatio 49, 163e171.

83. Philippi, R.A., 1851. Centuria quarta Testaceorum novorum. Z. Malakozoologie 8 (1), 30.

84. Prohaska, F.J., 1973. New evidence on the climatic controls along the Peruvian coast. In: Amiran, D.H.K., Wilson, A.W. (Eds.), Coastal Deserts: Their Natural and Human Environments. University of Arizona Press, Tucson, Arizona, pp. 91e107.

85. Puga, F., 1921. El chuno del ligtu. Rev. Chil. Hist. Nat. 25, 313 ~ e320.

86. Pullen, A.G., 2005. Fire and Cognition in the Palaeolithic (Unpublished PhD thesis). University of Cambridge.

87. Quilter, J., 1991. Late Preceramic Peru. J. World Prehist. 5, 387e438.

88. Quilter, J., 1989. Life and Death at Paloma, Society and Mortuary Practices in a Preceramic Peruvian Village. University of Iowa Press, Iowa City.

89. Raedeke, K.J., Simonetti, J.A., 1988. Food habits of Lama guanicoe in the Atacama Desert of northern Chile. J. Mammol. 69, 198e201.

90. Ramírez, R., 1988. Morfologia y biologia de Bostryx conspersus (Sowerby) (Mollusca, Bulimulidae) en las lomas costeras del Peru Central. Rev. Bras. Zool. 5, 609e617.

91. Ramírez, R., Cordova, S., Caro, K., Duarez, J., 2003. Response of a Land Snail Species (Bostyrx conspersus) in the Peruvian Central Coast lomas ecosystem to the 1982e1983 and 1997e1998 El Nino Events. In: Haas, J., Dillon, M.O. (Eds.), El ~ Nino in Peru: Biology and Culture over 10,000 Years, Fieldiana (Botany). Field ~ Museum of Natural History, Chicago, pp. 1e9.

92. Reus, M.L., Campos, C.M., Giannoni, S.M., Ortiz, S.G., Laspina, C.A., Navas, A.C., Peco, B., 2009. Guanaco vs. Donkey: what Are They Eating at Ischigualasto Park (San Juan)? Presented at the X International Mammalogical Congress SAREM, Mendoza.

93. Richardson III, J.B., 1998. Looking in the right places: pre-5,000 BP Maritime adaptations in Peru and the changing environment. Rev. Arqueol. Am. 33e56.

94. Rizner, M., Vukosavljevic, N., Miracle, P., 2005. The paleoecological and paleodietary significance of edible land snails (Helix sp.) across the Pleistocene-Holocene transition on the eastern Adriatic Coast. In: McCartan, S., Schulting, R., Warren, G., Woodman, P. (Eds.), Mesolithic Horizons Papers Presented at the Seventh International Conference on the Mesolithic in Europe, Belfast 2005. Oxbow, Oxford, pp. 527e532.

95. Rodbell, D.T., Seltzer, G.O., Anderson, D.,M., Abbott, M.B., Enfield, D.,B., Newman, J.,H., 1999. An ~15,000-Year record of El Nino-driven alluviation in ~ Southwestern Ecuador. Science 283, 516e520.

96. Rollins, H.B., Richardson, J.B., Sandweiss, D.H., 1986. The birth of El Nino: geo- ~ archaeological evidence and implications. Geoarchaeology 1, 3e15.

97. Rostworowski, M., 1981. Recursos Naturales Renovables y Pesca, Siglos, XVI y XVII. Instituto de Estudio Peruano, Lima, Perú.

98. Rundel, P.W., Dillon, M.O., 1998. Ecological patterns in the Bromeliaceae of the lomas formations of Coastal Chile and Peru. Plant Syst. Evol. 212, 261e278.

99. Rundel, P.W., Dillon, M.O., Palma, B., Mooney, H.A., Gulmon, S.L., Ehleringer, J.R., 1991. The Phytogeography and ecology of the Coastal Atacama and Peruvian Deserts. ALISO 13, 1e49.

100. Sandweiss, D.H., 2003. Terminal Pleistocene through

Mid-Holocene archaeological sites as paleoclimatic archives for the Peruvian coast. Palaeogeogr. Palaeoclimatol. Palaeoecol. 194, 23e40.

101. Sandweiss, D.H., 2009. Early fishing and inland monuments: challenging the maritime foundations of Andean civilization? In: Marcus, J., Williams, P.W. (Eds.), Andean Civilization. A Tribute to Michael Moseley, Monograph, vol. 63. Cotsen Institute of Archaeology, University of California, Los Angeles, pp. 39e54.

102. Sandweiss, D.H., Kelley, A.R., 2012. Archaeological contributions to climate change research: the archaeological record as a paleoclimatic and paleoenvironmental archive. Annu. Rev. Anthropol. 41, 371e391.

103. Sandweiss, D.H., Richardson III, J.B., Reitz, E.J., Rollins, H.B., Maasch, K.A., 1996. Geoarchaeological evidence from Peru for a 5000 years BP onset of El Nino. ~ Science 273, 1531e1533.

104. Sandweiss, D.H., Maasch, K.A., Chai, F., Andrus, C.F.T., Reitz, E.J., 2004. Geoarchaeological evidence for multidecadal natural climatic variability and ancient Peruvian fisheries. Quat. Res. 61, 330e334.

105. Simpson, B.B., 1989. Krameriaceae. Flora Neotropica 49, 1e108.

106. Stafford, D., 1939. On the Flora of Southern Peru. Proc. Linn. Soc. Lond. 151, 172e181.

107. Strong, W.D., 1957. Paracas, Nazca, and Tiahuanacoid cultural relationships in South Coastal Peru. Memoirs Soc. Am. Archaeol. 13, 1e43.

108. Tripcevich, N., Contreras, D.A., 2011. Quarrying evidence at the Quispisisa Obsidian Source, Ayacucho, Peru. Lat. Am. Antiq. 22, 121e136.

109. Trujillo, D., Delgado Rodríguez, A., 2011. Aa from lomas formations. A new Orchidaceae record from the desert coast of Peru. Lankesteriana 11 (1), 33e38.

110. Tucker, C.J., 1979. Red and photographic infrared linear combinations for monitoring vegetation. Remote Sens. Environ. 8 (2), 127e150.

111. Villagran, C., Castro, V., 2003. Ciencia indígena de los Andes del norte de Chile. Editorial Universitaria, Santiago, Chile.

112. Vogt, B., 2011. Las Brujas, Bajo Río Grande (Perú). In: Jahresbericht 2010, vol. 1. Deutsches Archaologisches Institut, pp. 307 € e308.

113. Walter, H., 1973. Die Vegetation der Erde in oko-physiologischer Betrachtung: Die € tropischen und subtropischen Zonen. Fischer-Verlag, Stuttgart (Stuttgart).

114. Waselkov, G.A., 1987. Shellfish gathering and shell midden archaeology. Adv. Archaeol. Method Theory 10, 93e210.

115. Weir, G.W., Derring, J.P., 1986. The Lomas of Paloma: human-environment relations in a Central Peruvian fog oasis: archaeobotany and palynology. In: Matos, M.,R., Turpin, S.,A., Eling, H.,H. (Eds.), Andean Archaeology: Papers in Memory of Clifford Evans. Institute of Archaeoogy, University of California, Los Angeles, pp. 18e44.

116. Wells, L.E., Noller, J.S., 1999. Holocene coevolution of the physical landscape and human settlement in Northern Coastal Peru. Geoarchaeology 14, 755e789.

117. Whaley, O.Q., Orellana, A., Perez, E., Tenorio, M., Quinteros, F., Mendoza, M., Pecho, O., 2010. Plantas y Vegetacion de Ica, Perúe Un recurso para su restauracion y conservacion. Royal Botanic Gardens, Kew.

118. Wheeler, J.C., 1995. Evolution and present situation of the South American Camelidae. Biol. J. Linn. Soc. 54, 271e295.

119. Wheeler, J.C., Roque, D.H., Castillo, D., Porter, A.M., Umire, Y.I., Anchea, A.V., Hawkins, A.R., Huayhua, R.R., Hernandez, L.M., Le on, J.M.A., 2011. Estudio de línea base, uso de habitat y plan de manejo del guanaco informe final (enero e noviembre 2011). Instituto de Investigacion y Desarrollo de Camelidos Sudamericanos CONOPA, Lima, Perú.

120. Willey, G.R., 1971. An Introduction to American Archaeology, vol. 2. Prentice-Hall, South America: New Jersey.

121. Willis, K.J., Van Andel, T.H., 2004. Trees or no trees? The environments of central and eastern Europe during the Last Glaciation. Quat. Sci. Rev. 2369e2387.

122. Zeballos, H., Villegas, L., Gutierrez, R., Jimenez, P., 2000. Vertebrados de las lomas de atiquipa y mejía, sur del Peru. Rev. Ecol. Latinoam. 7, 11e18.

123. Zeder, M.A., 2012. The broad spectrum revolution at 40: resource diversity, intensification, and an alternative to optimal foraging explanations. J. Anthropol. Archaeol. 31, 241e264.

CHAPTER 14 NITROGEN DYNAMICS IN TURBIC CRYOSOLS

Birgit Wild, Jörg Schnecker, Jiří Bárta, Petr Čapek, Georg Guggenberger, Florian Hofhansl, Christina Kaiser, Nikolaj Lashchinsky, Robert Mikutta, Maria Mooshammer, Hana Šantrůčková, Olga Shibistova, Tim Urich, Sergey A. Zimov, Andreas Richter

Keywords: Arctic, Tundra, Cryoturbation, Soil organic matter, Ecological stoichiometry, Nitrogen transformation, Protein depolymerization, Nitrogen mineralization, Nitrification, Nitrogen availability

INTRODUCTION

Arctic soils are commonly affected by cryoturbation, i.e., by a mixing of soil layers due to freeze–thaw processes, and are thus often characterized by horizons of poorly decomposed material subducted into and surrounded by mineral subsoil (Bockheim, 2007 and Tarnocai et al., 2009). Cryoturbated soils are estimated to store 581 Gt of organic carbon that is currently protected from fast decomposition (Tarnocai et al., 2009). Although cryoturbated horizons (Ojj or Ajj) are chemically similar to organic (O) or mineral topsoil (A) horizons, they are usually several hundred to thousand years older (Xu et al., 2009 and Hugelius et al., 2010), suggesting that C mineralization is slowed down in cryoturbated horizons. Likewise, N mineralization has been found to be lower in cryoturbated than in all other horizons, i.e., in organic (O), mineral topsoil (A), and mineral subsoil (B) horizons (Kaiser et al., 2007). Cryoturbation thus lead to retardation of soil organic matter (SOM) decomposition in general, and of N mineralization in particular, and this retardation cannot be explained by lower subsoil temperatures alone (Kaiser et al., 2007).

The rate limiting step for soil N cycling is the breakdown of N-rich, high molecular weight organic compounds by extracellular enzymes, especially the depolymerization of proteins (Schimel and Bennett, 2004, Geisseler et al., 2010 and Jones and Kielland, 2012). The resulting oligopeptides and amino acids are taken up by microorganisms and are further mineralized to NH_4^+ which, in turn, is the substrate for nitrification (Jones and Kielland, 2012). Nitrogen mineralization is considered an overflow mechanism, i.e., only an excess of N that cannot be used to build up biomass (because another element is limiting) will be mineralized (Schimel and Bennett, 2004). Thus, gross N mineralization rates are expected to increase with increasing N availability, and, consequently, microbial N use efficiency (NUE; the proportion of N taken up that is incorporated into microbial biomass) will decrease.

Nitrogen availability in high latitude ecosystems is generally low (Schimel and Bennett, 2004), but is predicted to increase with global warming. Higher soil temperatures were shown to increase soil microbial activity, resulting in increased net N mineralization, and thus N availability (Nadelhoffer et al., 1991, Hobbie, 1996 and Rustad et al., 2001). Additionally, the functional composition of tundra vegetation is expected to shift (Elmendorf et al., 2012), which is likely to alter the availability of N for soil microorganisms by changing patterns in plant N uptake, litter decomposability and competition for N in the rhizosphere (Wookey et al., 2009). An increase in N availability can stimulate plant and microbial growth (e.g., Hobbie et al., 2002 and Sistla et al., 2012), and also SOM decomposition. In a long-term fertilization study in Alaskan tundra, Mack et al. (2004)observed that fertilization not only increased plant production, but also SOM decomposition, overall result-

Citation: Birgit Wild, Jörg Schnecker, Jiří Bárta, Petr Čapek, Georg Guggenberger, Florian Hofhansl, Christina Kaiser, Nikolaj Lashchinsky, Robert Mikutta, Maria Mooshammer, Hana Šantrůčková, Olga Shibistova, Tim Urich, Sergey A. Zimov, Andreas Richter, Nitrogen dynamics in Turbic Cryosols, Soil Biology and Biochemistry, Volume 67, December 2013, Pages 85-93, ISSN 0038-0717, http://dx.doi.org/10.1016/j.soilbio.2013.08.004.

ing in a net loss of C from the system. This suggests that the emission of CO_2 from Arctic soils could be amplified by increased N availability. The effect of N additions on SOM decomposition is, however, highly variable. Across sites, and across soil horizons, both stimulation and inhibition of SOM decomposition by N addition have been observed (Lavoie et al., 2011). To predict the overall effects of N availability on SOM decomposition in Arctic soils in a future climate, an in-depth understanding of all steps of N cycling in the soil profile is necessary. So far most studies have exclusively focused on N mineralization and nitrification, but the depolymerization step of highmolecular weight organic N, which has been suggested to be rate-limiting, has not received much attention.

While many studies have investigated N dynamics in organic and mineral topsoil horizons, not much is known about mineral and cryoturbated horizons in the subsoil. In spite of the large amounts of poorly decomposed SOM in cryoturbated horizons, and in spite of the importance of N for SOM decomposition, we are only aware of one study investigating N dynamics in cryoturbated horizons. This study showed a reduction in N mineralization rates in cryoturbated compared to organic horizons (Kaiser et al., 2007). These low N mineralization rates could either have been the consequence of reduced protein depolymerization rates, or of a higher N demand of the microbial community. In the latter case, protein depolymerization would be similar in all horizons, but N mineralization would be lower in cryoturbated horizons, indicating that the microbial community needed a higher proportion of the available N for growth, and thus had a higher NUE.

We here report on microbial N transformations and NUE across different horizons in the active layer of tundra soil. We hypothesized that the observed reduction of gross N mineralization in cryoturbated horizons compared to organic horizons was due to a reduction in protein depolymerization, not due to a higher NUE of the microbial community. Since the capacity to depolymerize proteins constitutes a property of the microbial community, we further investigated the microbial community structure in different soil horizons, with the goal to identify groups that may be responsible for individual N transformations.

To achieve these goals, we sampled Turbic Cryosols (Turbels) at three sites in Siberia and Greenland and measured gross rates of protein depolymerization, microbial amino acid uptake, N mineralization, NH_4^+ uptake and nitrification using an array of ^{15}N pool dilution approaches. Additionally, we estimated the microbial community composition using phospholipid fatty acids (PLFAs) as biomarkers.

MATERIAL & METHODS

Sampling Sites

We compared soils from three different sites: (1) The heath tundra site was located in eastern Greenland close to the Zackenberg Research Station (typical tundra subzone; 74°29′ N, 20°32′ W) on sedimentary bedrock (sandstone). It was dominated by *Cassiope tetragona*, *Vaccinium uliginosum*, *Dryas octopetala*, *Salix arctica*, and *Carex* sp., with lichens between the dwarf shrubs. (2) The shrub tundra site was in north-eastern Siberia close to the town of Chersky (southern tundra subzone; 68°45′ N, 161°36′ E) on aeolian late Pleistocene sediment. It was located in a shrubby moss lichen tundra dominated by *Betula exilis*, *Vaccinium uliginosum*, *Flavocetraria nivalis*, *Flavocetraria cucullata* and *Aulacomnium turgidum*, with sparse *Larix gmelinii* trees. (3) The tussock tundra site was approximately 80 km north of Chersky (southern tundra subzone; 69°26′ N, 161°44′ E) on aeolian late Pleistocene sediment. It was dominated by *Eriophorum vaginatum*, *Carex lugens*, *B. exilis*, *Vaccinium vitis-idaea*, *Aulacomnium turgidum*, and *Dicranum* sp.

All soils were classified as Turbic Cryosols according to the World Reference Base for Soil Resources (IUSS Working Group WRB, 2007) or Turbels according to the US Soil Taxonomy (Soil Survey Staff, 1999). The active layer at the time of sampling was 47 cm (heath tundra, Greenland), 73 cm (shrub tundra, Siberia; under frost boils) and 72 cm (tussock tundra, Siberia; under frost boils).

At each site, we took samples from the active layer of three replicate soil pits. We sampled organic layers (O, including OA horizons) from the soil surface (further termed "organic horizon"). We then took samples from pockets of organic material (Ojj) or mineral topsoil (Ajj) that was buried within the subsoil. Such buried horizons, caused by cryoturbations, i.e., by subduction of organic material or mineral topsoil into the subsoil due to freeze–thaw processes (Bockheim, 2007), are common in permafrost soils. Sampled pockets of cryoturbated material were between 20 and 50 cm from the soil surface (further termed "cryoturbated horizon"). We further sampled the mineral soil surrounding the cryoturbated pockets (A, AB, B or Cg horizons, at depths from 10 to 60 cm from the surface). These horizons are collectively termed "mineral horizon". Living roots were carefully removed and samples were kept cool until analysis.

All samplings took place in August 2010. We are aware of the fact that N cycling and N availability show seasonal variation in Arctic soils (e.g., Weintraub and Schimel, 2005a, Weintraub and Schimel, 2005b and Edwards et al., 2006). Data presented here do not reflect these variations, but represent N cycling in the late growing season.

Basic Characterization and Nutrient Concentrations

Concentrations of ammonium and nitrate were determined photometrically in 1 M KCl extracts following Kandeler and Gerber (1988) and Miranda et al. (2001), respectively. Total dissolved nitrogen (TDN) was measured with a DOC/TN analyzer (Elementar LiquiToc II or Shimadzu TOC-$V_{CPH/CPN}$/TNM-1) in 0.5 M K_2SO_4 or 1 M KCl extracts (samples from Siberia and Greenland, respectively), and dissolved organic N (DON) was calculated by subtracting NH_4^+ and NO_3^- from TDN. Extraction with K_2SO_4 or KCl was previously shown to yield similar DON recovery (Jones and Willett, 2006). For determination of inorganic phosphate, soil samples were extracted with 0.5 M $NaHCO_3$(Olsen et al., 1954) and measured photometrically with the molybdate-ascorbic acid method (Murphy and Riley, 1962). Contents of organic C and total N were determined in dried and ground samples using elemental analysis-isotope ratio mass spectrometry (EA-IRMS), either with a CE Instrument EA 1110 elemental analyzer coupled to a Finnigan MAT DeltaPlus IRMS with a Finnigan MAT ConFlo II Interface, or with an Isoprime EA-IRMS system. Samples from Siberia contained traces of carbonate and were acidified in HCl atmosphere and neutralized over NaOH before EA-IRMS analysis. For determination of total P content, samples were amended with a mixture of concentrated $HClO_4$ and HNO_3 (1:4), heated stepwise to 160 °C and 220 °C for digestion, cooled to room temperature, filtered (Whatman 40 ashfree cellulose filter) and measured with inductively coupled plasma-optical emission spectrometry (ICP-OES, Perkin Elmer Optima 3000 XL) against external standards. Ratios of C, N and P were calculated as mass ratios. pH values were determined in suspensions of dried soil in de-ionized water (1:2.5; weight:volume).

Gross Rates of N Transformations

Gross rates of protein depolymerization and amino acid uptake were determined using a ^{15}N pool dilution assay as described by Wanek et al. (2010), with slight modifications to account for the low amino acid concentrations in soils. Briefly, a mixture of 20 ^{15}N labeled amino acids (>98 at%, Spectra and Cambridge Isotope Laboratories) was dissolved in 10 mM $CaSO_4$ and added to 2 g of field-moist soil in duplicates. Per sample, 500 µl solution containing 2.5 µg total amino acids were applied. After incubation for 10 or 30 min at 7 °C, activities were stopped with 19.5 ml of 10 mM $CaSO_4$ containing 3.7% of formaldehyde. Samples were extracted for 5 min and either allowed to settle for 10 min (both Siberian sites) or centrifuged for 5 min at 10 845 g (Greenland heath tundra site). Samples were filtered through synthetic wool and GF/C filters (Whatman), and loaded on cation exchange cartridges (OnGuard II H 1cc cartridges, Dionex, cleaned with 3 M ammonia and re-generated with 1 M HCl before use). After application of the samples, cartridges were washed with 10 ml distilled water, stabilized with 5 ml 5% methanol and stored cool until elution. With each batch of samples, blanks and amino acid standards were processed to correct for losses due to ion exchange. After elution of the amino acids from the cartridges with 30 ml 3 M ammonia, an internal standard was added to the samples (1 µg of nor-valine, nor-leucine and para-chloro-phenylalanine each, Sigma–Aldrich). Samples were dried with rotary evaporation, re-dissolved in 1.5 ml 20% ethanol and dried in a SpeedVac system. Finally, samples were derivatized with chloroformate before analysis with a gas chromatography–mass spectrometry system (GC–MS), consisting of a CTC autosampler (CTC Analytics) and a Trace GC Ultra coupled to a quadrupole mass spectrometer (DSQ II; Thermo Scientific). Two µl of sample were injected in splitless mode (injector temperature 270 °C), separated on an Equity-1701 column (30 m × 0.25 mm × 1 µm; Sigma–Aldrich) with 1 ml min^{-1} He as carrier gas (GC method: 105 °C for 1 min, 6 °C min^{-1} to 135 °C, 3 °C min^{-1} to 180 °C, 20 °C min^{-1} to 260 °C, 260 °C for 35 min) and detected in Selected Ion Monitoring mode. Concentrations of alanine, valine, leucine, isoleucine, proline, tryptophane, phenylalanine and tyrosine were calculated using calibrations against external standards, and the abundance of ^{15}N in each of these amino acids was calculated based on peak areas of fragments characteristic for ^{14}N and ^{15}N (for fragments see Wanek et al., 2010), using a calibration against standards of different ^{15}N abundance.

Gross rates of N mineralization (ammonification), NH_4^+ uptake and nitrification were determined as described by Kaiser et al. (2011) by adding 500 µl ^{15}N labeled $(NH_4)_2SO_4$(N mineralization and NH_4^+ uptake) or KNO_3 (nitrification) to duplicates of 2 g of field-moist soil (0.125 mM, 10 at%, Sigma–Aldrich). Samples were incubated for 4 h and 24 h at 7 °C, extracted with 13 ml of 2 M KCl for 30 min, and filtered through ashfree filter paper (Whatman 40 ashfree cellulose filter). To stabilize the extracts, 20 µl 5 mM phenylmercuric acetate were added, and samples were frozen until further processing. For N mineralization and NH_4^+ uptake, NH_4^+ was diffused into acid traps and measured with an EA-IRMS system consisting of a CE Instrument EA 1110 elemental analyzer coupled to a Finnigan MAT DeltaPlus IRMS with a Finnigan MAT ConFlo II Interface. For nitrification, NH_4^+ was removed from the samples and NO_3^- converted to NH_4^+ before diffusion into acid traps and EA-IRMS analysis (Mooshammer et al., 2012). Gross rates were based on differences in concentration and isotopic composition of NH_4^+, NO_3^- or amino acids between two time points (e.g., 4 h and 24 h) and were calculated according to the equations described in Wanek et al. (2010).

As an indicator for microbial N limitation, we calculated the efficiency of microorganisms to use amino acid N for biomass growth (N use efficiency, NUE) by comparing gross rates of amino acid uptake and N mineralization:

NUE = (gross amino acid uptake − gross N mineralization)/ gross amino acid uptake.

Phospholipid Fatty Acid (PLFA) Analysis

For analysis of PLFAs, samples were stored frozen (Greenland heath tundra) or in RNAlater (both Siberian sites; Schnecker et al., 2012). Phospholipid fatty acids were extracted from 1 g of soil with chloroform/methanol/citric acid buffer and purified on silica columns (LC-Si SPE, Supelco) using chloroform, acetone and methanol (Frostegård et al., 1991; with the modifications described by Kaiser et al., 2010). After addition of methyl-nonadecanoate as internal standard, PLFAs were converted to fatty acid methyl esters (FAMEs) by alkaline methanolysis. Samples were analyzed on a Thermo Trace GC with FID detection: 1 µl per sample was injected in splitless mode (injector temperature 230 °C) and separated on a DB-23 column (Agilent; GC method:70 °C for 1.5 min, 30 °C min^{-1} to 150 °C, 150 °C for 1 min, 4 °C min^{-1} to 230 °C, 230 °C for 15 min) with 1.5 ml min^{-1} He as carrier. Individual FAMEs were identified using qualitative standard mixes (37 Comp. FAME Mix and Bacterial Acid Methyl Esters CP Mix, Supelco) and quantified by comparison with the internal standard. We used 18:1ω9, 18:2ω6,9 and 18:3ω3,6,9 fatty acids as biomarkers for fungi, i15:0, a15:0, i16:0, i17:0 and a17:0 for gram positive bacteria, cy17:0 (9/10), cy19:0 (9/10), 16:1ω5, 16:1ω7, 16:1ω9 and 18:1ω7 for gram negative bacteria, and 14:0, 15:0, 16:0, 17:0, 18:0, 20:0, i14:0, 16:1ω10, 16:1ω11, 17:1ω6 and 10Me16:0 as unspecific markers (Kaiser et al., 2010).

Statistics

To test for significant differences between sites and horizons, we applied two-way ANOVAs and Tukey HSD tests (after transformation, if necessary) or Kruskal–Wallis tests with unpaired Mann Whitney *U* tests as post-hoc tests if normal distribution and homoscedasticity could not be achieved. We additionally performed a Principal Component Analysis including gross N transformation rates (per g total N), pH values and the relative abundance of microbial groups. Samples with missing values were omitted from the Principal Component Analysis. Correlations were tested using Spearman's rank correlations. All statistics were performed in R 2.15 (R Development Core Team, 2012).

RESULTS

Organic C, total N, and total P, as well as the C/N ratio, decreased significantly from organic to cryoturbated and mineral horizons (Tables 1 and 2). pH-Values were in the range of 4.3–5.5 for organic horizons and significantly higher in cryoturbated and mineral horizons (5.1–6.1 and 5.3–6.4, respectively). Microbial biomass (estimated as the total amount of PLFAs per g DW) decreased significantly from organic to cryoturbated and mineral horizons (Supplementary Fig. 1, Table 2), mainly due to differences in SOM content.

Table 1: C, N and P content, C/N and N/P ratios (mass ratios), and pH values of organic, cryoturbated (cryot.) and mineral horizons of heath tundra (Greenland), tussock tundra (Siberia) and shrub tundra (Siberia). Values represent means (±standard error)

Site	Horizon	C (%)	N (%)	P (%)	C/N	N/P	pH
Heath	Organic	15.76 (0.86)	0.93 (0.06)	0.102 (0.006)	17.14 (1.95)	9.18 (0.39)	5.29 (0.10)
	Cryot.	9.35 (0.95)	0.65 (0.00)	0.112 (0.007)	14.46 (1.46)	5.82 (0.37)	6.02 (0.02)
	Mineral	3.94 (1.01)	0.29 (0.08)	0.074 (0.025)	13.81 (0.61)	4.86 (1.57)	6.16 (0.11)
Tussock	Organic	22.15 (1.57)	1.00 (0.05)	0.155 (0.019)	22.38 (2.18)	6.68 (1.09)	5.13 (0.05)
	Cryot.	15.30 (4.84)	0.81 (0.18)	0.112 (0.011)	18.23 (1.76)	7.36 (1.81)	5.57 (0.28)
	Mineral	2.08 (0.40)	0.14 (0.02)	0.046 (0.039)	14.33 (0.40)	9.40 (7.42)	5.50 (0.02)
Shrub	Organic	26.63 (5.98)	0.96 (0.18)	0.115 (0.005)	27.46 (1.61)	8.36 (1.57)	4.64 (0.23)
	Cryot.	9.69 (2.57)	0.64 (0.20)	0.121 (0.084)	15.46 (0.88)	8.01 (3.90)	5.80 (0.29)
	Mineral	2.50 (0.85)	0.16 (0.03)	0.061 (0.022)	14.66 (2.96)	4.55 (2.66)	5.54 (0.19)

Table 2: Significance of differences between sites (heath tundra, Greenland; tussock tundra, Siberia; shrub tundra, Siberia) or between horizons (organic; cryoturbated; mineral), derived from two-way ANOVA with Tukey HSD test, or Kruskal–Wallis tests with Mann–Whitney *U* tests as post-hoc. Different letters indicate $p < 0.05$, with "a" denoting the highest values

Parameter	Unit	Between sites				Between horizons			
		Sign.	Heath	Tussock	Shrub	Sign.	Organic	Cryot.	Mineral
C	%	n.s.				***	a	b	c
N	%	n.s.				***	a	a	b
P	%	n.s.				*	a	ab	b
C/N	$g\ C\ g^{-1}\ N$	n.s.				***	a	b	b
N/P	$g\ N\ g^{-1}\ P$	n.s.				n.s.			
pH		**	a	b	b	***	b	a	a
Total PLFAs	$\mu mol\ g^{-1} DW$	n.s.				***	a	b	c
TDN	$\mu g\ N\ g^{-1} DW$	n.s.				***	a	b	b
DON	$\mu g\ N\ g^{-1} DW$	n.s.				**	a	b	b
NH_4^+	$\mu g\ N\ g^{-1} DW$	*	b	ab	a	*	a	ab	b
NO_3^-	$\mu g\ N\ g^{-1} DW$	n.s.				n.s.			
DON	% of TDN	**	a	a	b	n.s.			
Inorganic phosphate	$\mu g\ P\ g^{-1} DW$	**	b	a	a	n.s.			
Gross protein depolymerization	$\mu g\ N\ g^{-1} DW\ d^{-1}$	n.s.				***	a	b	b
Gross amino acid uptake	$\mu g\ N\ g^{-1} DW\ d^{-1}$	n.s.				***	a	b	b

Gross N mineralization	μg N g^{-1}DW d^{-1}	***	a	b	a	***	a	b	b
Gross NH_4^+uptake	μg N g^{-1}DW d^{-1}	*	ab	b	a	***	a	a	b
Gross nitrification	μg N g^{-1}DW d^{-1}	n.s.				**	a	b	b
Gross protein depolymerization	mg N g^{-1}N d^{-1}	*	b	ab	a	*	a	b	ab
Gross amino acid uptake	mg N g^{-1}N d^{-1}	n.s.				*	a	b	ab
Gross N mineralization	mg N g^{-1}N d^{-1}	***	a	b	a	***	a	b	a
Gross NH_4^+uptake	mg N g^{-1}N d^{-1}	**	b	b	a	n.s.			
Gross nitrification	mg N g^{-1}N d^{-1}	**	b	ab	a	*	a	b	a
NUE		**	b	a	b	n.s.			
Principal Component 1		***	b	a	a	***	a	b	a
Principal Component 2		*	b	a	b	n.s.			

Levels of significance: ***, $p < 0.001$; **, $p < 0.01$; *, $p < 0.05$; n.s., not significant.

In all horizons, the pool of dissolved N was dominated by organic N (Fig. 1). The relative contribution of DON to the TDN pool was similar in all horizons at each site, although absolute concentrations were significantly lower in mineral than in organic horizons (Table 2). We did, however, observe significant variations in the composition of the TDN pool between sites. Dissolved organic N accounted for 85 ± 3% (mean ± standard error of all horizons) and 86 ± 5% of the TDN pool at the heath tundra and tussock tundra sites, but only for 58 ± 7% at the shrub tundra site (Fig. 1). Also concentrations of inorganic phosphate varied significantly between sites. Inorganic phosphate was lower by a factor of ten at the heath tundra site in Greenland than at both Siberian sites, but did not show significant differences between horizons (Fig. 1).

Gross rates of protein depolymerization, amino acid uptake, N mineralization, NH_4^+uptake and nitrification, expressed per g DW, were generally highest in organic horizons, and decreased significantly from organic to cryoturbated and mineral horizons (Supplementary Fig. 2, Table 2). All N transformation rates were significantly and positively correlated with C and N content (Table 3). In order to assess horizon-specific differences in transformation rates that were not related to differences in SOM content, we calculated all gross N transformation rates per g total soil N. Cryoturbated horizons still exhibited significantly lower gross rates of protein depolymerization, amino acid uptake, N mineralization and nitrification per g total N than organic horizons, on average accounting for 16% (protein depolymerization), 32% (amino acid uptake), 27% (N mineralization) and 31% (nitrification) of the respective rates in organic horizons. In the case of N mineralization and nitrification, rates in cryoturbated horizons were even significantly lower than in mineral horizons (Fig. 2, Table 2).

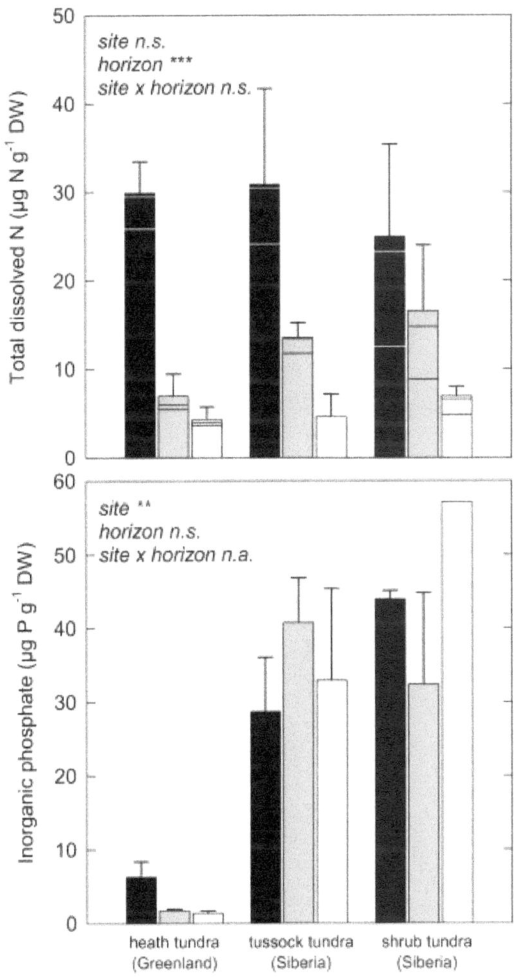

Figure 1: Upper panel: Total dissolved N in soil extracts of organic (black bars), cryoturbated (gray bars) and mineral (white bars) horizons of three tundra sites. Bars are separated into DON (lowest part), NH_4^+ (middle part) and NO_3^- (upper part). Concentrations of NH_4^+ and DON could not be determined in the mineral horizons of the tussock tundra site. Lower Panel: Concentrations of inorganic phosphate in soil extracts of organic

(black bars), cryoturbated (gray bars) and mineral (white bars) horizons of three tundra sites. All bars represent means ± standard error. Levels of significance: ***, $p < 0.001$; **, $p < 0.01$; *, $p < 0.05$; n.s., not significant; n.a., not analyzed (two-way ANOVA or Kruskal–Wallis test)

Table 3: Correlation analysis of soil C and N content (in % of DW), microbial abundances (in % of total PLFAs) and gross rates of protein depolymerization (Protein depol.), microbial amino acid uptake (AA uptake), N mineralization, microbial NH_4^+ uptake and nitrification, measured with a set of ^{15}N pool dilution approaches. For correlation with soil C and N content, gross N transformation rates were expressed in μg N g^{-1} DW d^{-1}. For correlation with microbial groups, and with each other, gross N transformation rates were corrected for the differences in SOM content between soil horizons and expressed in mg N g^{-1} N d^{-1}. Significance of correlations and correlation coefficients were determined using Spearman's rank correlations

	Protein depol.	AA uptake	N mineralization	NH_4^+ uptake	Nitrification
C	+0.577**	+0.621**	+0.445*	+0.565**	+0.639**
N	+0.519*	+0.541**	+0.418*	+0.539**	+0.650**
Fungi	+0.499*	+0.540**	+0.367	+0.176	+0.428*
Gram negative	+0.368	+0.483*	+0.304	+0.126	+0.528*
Gram positive	−0.305	−0.396	−0.166	−0.092	−0.276
AA uptake	+0.792***				
N mineralization	+0.237	+0.264			
NH_4^+ uptake	+0.356	+0.244	+0.532**		
Nitrification	+0.597**	+0.399	+0.424*	+0.394	

Levels of significance: ***, $p < 0.001$; **, $p < 0.01$; *, $p < 0.05$.

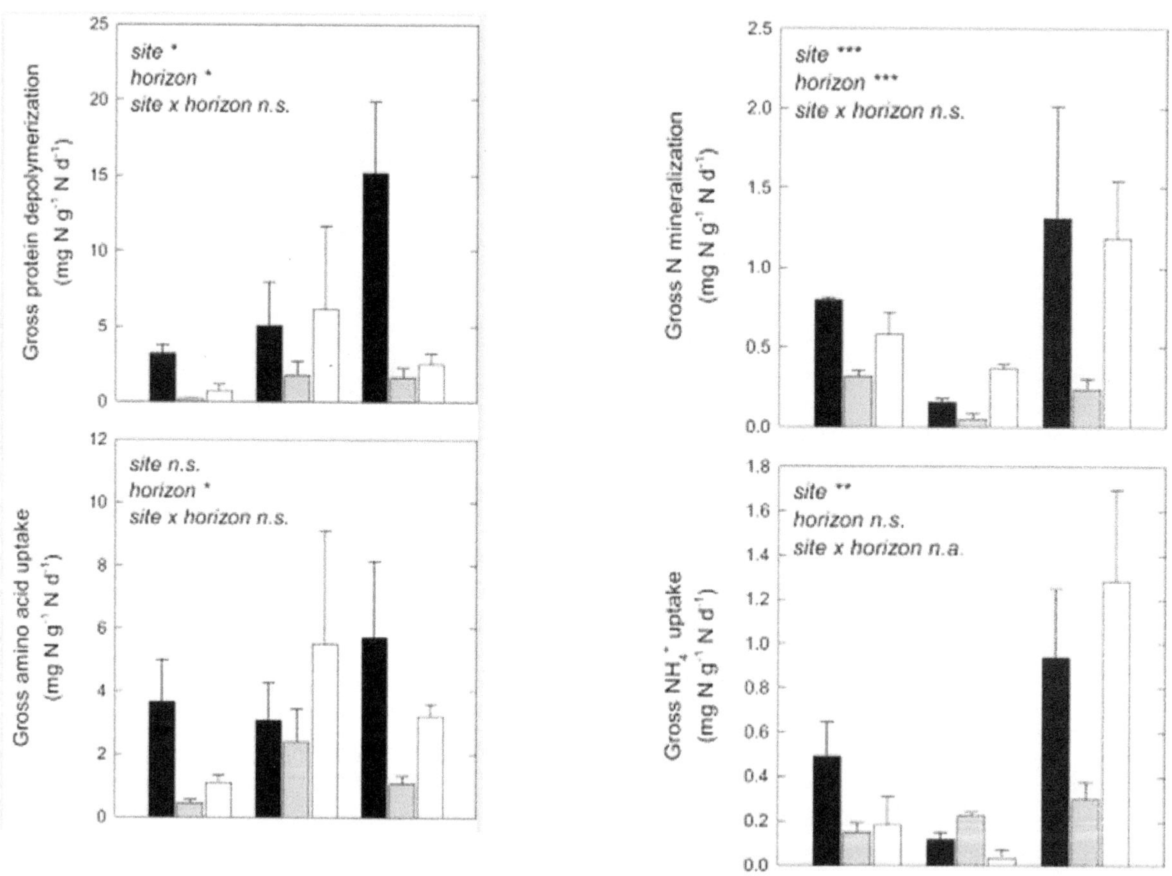

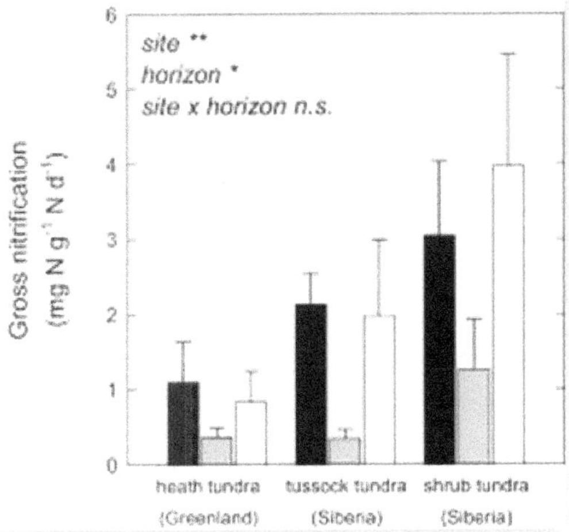

Figure 2: Gross rates of protein depolymerization, microbial amino acid uptake, N mineralization, microbial NH_4^+uptake and nitrification (per g total N), of organic (black bars), cryoturbated (gray bars) and mineral (white bars) horizons of three tundra sites. Rates were measured using a set of ^{15}N pool dilution approaches. Bars represent means ± standard error. Levels of significance: ***, $p < 0.001$; **, $p < 0.01$; *, $p < 0.05$; n.s., not significant; n.a., not analyzed (two-way ANOVA or Kruskal–Wallis test).

In spite of the differences in absolute N transformation rates, NUE did not differ significantly between horizons (Table 2), and was not correlated with soil C/N ratio ($p = 0.130$, $R^2 = 0.10$). We did, however, observe significant differences in NUE between sites. NUE was highest at the tussock tundra site in Siberia, where 90% of amino acid N taken up was incorporated into microbial biomass, and only 10% were mineralized to NH_4^+. NUE was lower at both other sites, with 66% incorporation at the shrub tundra site in Siberia, and 51% at the heath tundra site in Greenland (Fig. 4).

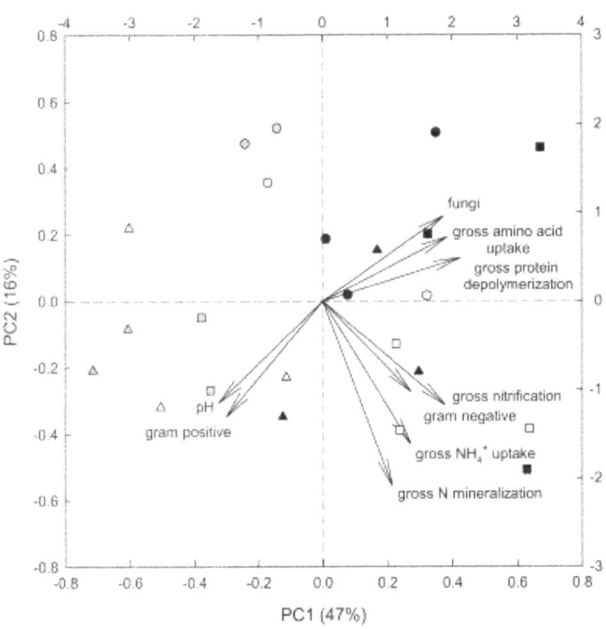

Figure 3: Ordination of organic (black), cryoturbated (gray) and mineral (white) horizons of heath tundra (Greenland; triangles), tussock tundra (Siberia; circles) and shrub tundra (Siberia; squares) using Principal Component Analysis. Data include gross rates of protein depolymerization, microbial amino acid uptake, N mineralization, microbial NH_4^+ uptake and nitrification (per g total N to correct for differences in SOM content between horizons), as well as pH values and the relative abundances of fungi, gram positive and gram negative bacteria (in % of total PLFAs).

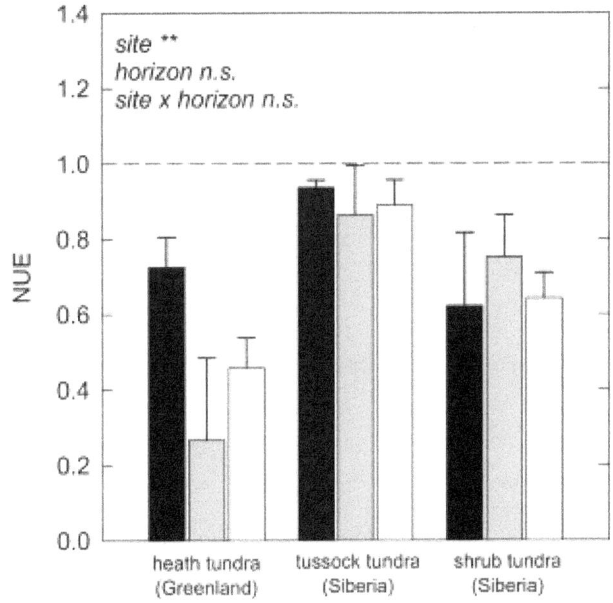

Figure 4: Nitrogen use efficiency (NUE) of organic (black bars), cryoturbated (gray bars) and mineral (white bars) horizons of three tundra sites. NUE was calculated as the proportion of amino acid N taken up by microorganisms that was not mineralized to NH_4^+. Bars represent means ± standard error. Levels of significance: ***, $p < 0.001$; **, $p < 0.01$; *, $p < 0.05$; n.s., not significant (two-way ANOVA).

To investigate possible relationships between N transformations and microbial community composition across horizons and sites, we performed a Principle Component Analysis, including gross N transformation rates (per g total N), the relative abundances of fungi, gram negative and gram positive bacteria (in % of total PLFAs), and pH values (Fig. 3). Principal Component 1 accounted for 47% of the variation in the data set, and was positively connected to all gross N transformation rates (factor loadings 0.42 for protein depolymerization, 0.38 for amino acid uptake, 0.21 for N mineralization, 0.27 for NH_4^+ uptake and 0.38 for nitrification), and thus represents an aggregated parameter of N transformation activity. Principal Component 1 separated all cryoturbated horizons and the mineral horizons of the Greenland heath tundra site from organic and the Siberian mineral horizons. These findings were supported by a two-way ANOVA that showed significantly lower values of Principal Component 1 for cryoturbated horizons compared to organic and mineral horizons, and for the Greenland heath tundra site compared to both Siberian sites (Table 2). Principal Component 2 (16% of variation), in contrast, was positively connected to gross rates of protein depolymerization and microbial uptake of amino acids (factor loadings 0.13 and 0.19), but negatively connected to gross rates of N mineralization, NH_4^+ uptake and nitrification (−0.55, −0.43 and −0.31), thus separating organic from inorganic N transformation processes. Principal Component 2 was also significantly correlated with NUE which reflects the allocation of amino acid N to mineralization

($p = 0.004$, $R^2 = 0.34$). The microbial groups of fungi, gram negative and gram positive bacteria contributed differently to Principal Components 1 and 2. Principal Component 1 was positively connected to the relative abundances of fungi and gram negative bacteria (factor loadings 0.37 and 0.27), and negatively to gram positive bacteria (−0.30). Principal Component 2 was positively connected to fungi (0.26) and negatively to both gram negative and gram positive bacteria (−0.27 and −0.35).

We further investigated possible relationships between individual N transformation processes and microbial groups using Spearman›s rank correlations, and found that gross rates of protein depolymerization, amino acid uptake and nitrification (per g total N) were significantly correlated with the relative abundance of fungi, and that gross rates of amino acid uptake and nitrification were significantly correlated with gram negative bacteria (Table 3).

DISCUSSION

High-latitude systems are usually characterized by low N availability that limits both plant and microbial growth (Hobbie et al., 2002 and Sistla et al., 2012). Plants and microorganisms rapidly immobilize all reactive N forms small enough for uptake, in particular amino acids (Jones and Kielland, 2002 and Näsholm et al., 2009) and oligo-peptides (Hill et al., 2011 and Farrell et al., 2013), and incorporate the N into their biomass, with a minimum of microbial overflow mineralization to NH_4^+, and further transformation to NO_3^-. Although organic N forms represent a major source of N for both plants and microorganisms in the Arctic, we know little about the steps controlling their availability, i.e., protein depolymerization.

Nitrogen Cycling in Cryoturbated Horizons

In the Arctic soils studied, gross rates of protein depolymerization, but also microbial amino acid uptake, N mineralization and nitrification were significantly lower in cryoturbated than in organic horizons, on average accounting for only 26% of the rates in organic horizons (Fig. 2). This corroborates our hypothesis that the whole sequence of N transformations, starting with the rate-limiting step of protein depolymerization, is decelerated in cryoturbated compared to organic horizons. Protein depolymerization limits the amount of amino acids available for microbial uptake, and with it the potential for microbial growth and N mineralization. Severe N deficiency can even limit the production of extracellular enzymes that depolymerize complex organic compounds (Weintraub and Schimel, 2005a, Wallenstein et al., 2009 and Sistla et al., 2012), including N-containing macromolecules such as proteins. While we did not directly estimate N availability in cryoturbated horizons, lower protein depolymerization rates compared to organic horizons point to a reduced N availability in cryoturbated horizons. We therefore suggest that the slow decomposition of cryoturbated SOM (Kaiser et al., 2007, Xu et al., 2009 and Hugelius et al., 2010) might be connected to N limitation of enzyme production. In this case, an increase in protein depolymerization with climate change (Weedon et al., 2011 and Brzostek et al., 2012) could facilitate the decomposition of cryoturbated SOM, and lead to higher CO_2 emissions from Arctic soils.

Microbial Communities and Nitrogen Transformations

The differences in N cycling across soil horizons and sites were likely caused, at least in part, by differences in composition and N demand of the microbial communities. The rate-limiting step in N cycling, protein depolymerization, requires specific enzymes, i.e., proteases, that are produced by a range of bacteria and fungi. We here found that gross protein depolymerization rates were significantly correlated with the relative abundance of fungi (Table 3). Fungi are able to produce a wide range of extracellular enzymes (Baldrian et al., 2011 and Schneider et al., 2012) and are involved in the degradation of many complex organic molecules including cellulose and lignin (de Boer et al., 2005 and Strickland and Rousk, 2010). Protein breakdown in particular has been assigned to ecto- and ericoid mycorrhizal fungi in high latitude systems (Read and Perez-Moreno, 2003). A low abundance of fungi in cryoturbated horizons might thus not only contribute to the low protein depolymerization rates, but also generally to the retarded decomposition of cryoturbated SOM. Protein depolymerization was closely correlated with microbial amino acid uptake (Table 3). We found a similar close correlation for N mineralization and NH_4^+ uptake. This suggests that the microbial uptake of amino acids and NH_4^+ was limited by the respective production rates (protein depolymerization and N mineralization), indicating a high demand of the microbial biomass for N. A tight coupling of production and consumption rates of both amino acids and NH_4^+ has already been demonstrated for decomposing beech leaf litter (Mooshammer et al., 2012).

While N mineralization obviously limited the amount of NH_4^+ available for microbial uptake, N mineralization itself was not correlated with any upstream process such as protein depolymerization or amino acid uptake, or with any microbial group (Table 3). Nitrogen mineralization is the microbial de-amination of organic N and excretion of NH_4^+. The potential for N mineralization thus depends on the uptake of organic N by microorganisms. Microorganisms, however, can directly control the amount of N mineralized, and will only mineralize an excess of N that is not needed for growth or other cellular processes (Schimel and Bennett, 2004). Actual N mineralization rates therefore reflect both N availability and N demand for growth by the microbial community. Nitrification, in contrast, was significantly correlated with the relative abundances of gram negative bacteria and fungi (Table 3). Nitrification requires a specific set of enzymes that oxidize ammonium over nitrite to nitrate, and is restricted to specific microbial groups. Autotrophic nitrification was found in few groups of archaea and gram negative bacteria (Hayatsu et al., 2008, Schleper, 2010 and Alves et al., 2013). The abundance of gram negative bacteria might therefore influence nitrification rates, as already demonstrated for savanna and forest soils (Balser and Firestone, 2005). Heterotrophic nitrification, in contrast, is more widespread among microorganisms, but was mainly connected to fungi, particularly in acidic soils (Hayatsu et al., 2008).

Nitrogen use Efficiency of the Microbial Community

In addition to low protein depolymerization rates in cryoturbated horizons, the observed reduction in N mineralization may also have been caused by a higher allocation of the available N to growth, and thus by a higher NUE. Our results demonstrate, however, that microbial NUE did not differ significantly between horizons (Fig. 4), indicating that the microbial communities in all horizons, including the cryoturbated ones, had a similar demand for N. Furthermore, NUE was not correlated with soil C/N ratios, suggesting that the decrease in C/N ratio from organic to mineral horizons (Table 1) was likely offset by changes in C and N availability with soil depth. Montané et al. (2007) showed for a mountain grassland soil, that while persistence of soil C was constant across the soil profile, persistence of soil N increased with soil depth, probably due to differences in chemical composition or in binding to soil minerals (for a recent review on SOM persistence see Schmidt et al., 2011).

NUE was generally rather high (as expected for N-limited systems), but differed significantly between sites (Fig. 4). NUE was significantly higher and gross N mineralization significantly lower at the tussock tundra site in Siberia than at the heath tundra site in Greenland, indicating that the microbial community at the tussock tundra site needed a higher proportion of the available N for growth. Size and composition of the TDN pool, as well as gross protein depolymerization and amino acid uptake rates were similar at both sites (Table 2), so it can be assumed that N availability for microorganisms was similar – pointing to the fact that N was not the main limiting element at the Greenland heath tundra site.

Since concentrations of inorganic phosphate were significantly lower at the Greenland heath tundra than at both Siberian sites (Fig. 1), we suggest that microbial growth at the Greenland heath tundra site could have been rather limited in P. Microorganisms have to maintain a rather constrained C:N:P ratio and have only limited capacities to store excess C, N or P in the biomass. Therefore, P limitation should lead to increased N mineralization as an overflow mechanism (Tezuka, 1990 and Sterner and Elser, 2002). Although P limitation is usually considered characteristic for old, highly weathered soils, for example in the tropics (Cleveland et al., 2002), there is increasing evidence that P is an important co-limiting and in some cases even the main limiting element in Arctic soils (Shaver and Chapin, 1995, Shaver et al., 1998, Giesler et al., 2002, Hartley et al., 2010 and Giesler et al., 2012). In the case of the heath tundra site, we found that although concentrations of inorganic phosphate were significantly lower than at the Siberian sites, total P content was similar, indicating that differences in the sorption of P to mineral soil particles may have been responsible for the low P availability at this site.

CONCLUSIONS

We found significant differences in N transformation rates and microbial N use efficiency between sites and horizons, demonstrating that the respective microbial communities differed in nutrient limitation. Across all sites, N cycling was slower in cryoturbated compared to organic horizons, starting with protein depolymerization, which is rate-limiting in N cycling. Our results thus indicate that microbial communities have a lower capacity to break down proteins in cryoturbated compared to organic horizons, likely due to differences in community composition (e.g., a lower abundance of fungi). Overall, our study suggests that burial of organic material by cryoturbation into the subsoil leads to changes in soil N transformations, which in turn may contribute to the observed retarded decomposition of cryoturbated SOM by altering N availability for microbial decomposers.

ACKNOWLEDGMENTS

This work was funded by the Austrian Science Fund (FWF) as a part of the International Program CryoCARB (Long-term Carbon Storage in Cryoturbated Arctic Soils; FWF – I370-B17). Jörg Schnecker received a Julius-Payer-Stipendium of the Austrian Society for Polar Research.

REFERENCES

1. Alves, R.J.E., Wanek, W., Zappe, A., Richter, A., Svenning, M.M., Schleper, C., Urich, T., 2013. Nitrification rates in Arctic soils are associated with functionally distinct populations of ammonia-oxidizing archaea. The ISME Journal 7, 1620e1631.
2. Baldrian, P., Vorísková, J., Dobiásová, P., Merhautová, V., Lisá, L., Valásková, V., 2011. Production of extracellular enzymes and degradation of biopolymers by saprotrophic microfungi from the upper layers of forest soil. Plant and Soil 338, 111e125.
3. Balser, T.C., Firestone, M.K., 2005. Linking microbial community composition and soil processes in a California annual grassland and mixed-conifer forest. Biogeochemistry 73, 395e415.
4. Bockheim, J.G., 2007. Importance of cryoturbation in redistributing organic carbon in permafrost-affected soils. Soil Science Society of America Journal 71, 1335e 1342.
5. Brzostek, E.R., Blair, J.M., Dukes, J.S., Frey, S.D., Hobbie, S.E., Melillo, J.M., Mitchel, R.J., Pendall, E., Reich, P.B., Shaver, G.R., Stefanski, A., Tjoelker, M.G., Finzi, A.C., 2012. The effect of experimental warming and precipitation change on proteolytic enzyme activity: positive feedbacks to nitrogen availability are not universal. Global Change Biology 18, 2617e2625.
6. Cleveland, C.C., Townsend, A.R., Schmidt, S.K., 2002. Phosphorus limitation of microbial processes in moist tropical forests: evidence from short-term laboratory incubations and field studies. Ecosystems 5, 680e691.
7. de Boer, W., Folman, L.B., Summerbell, R.C., Boddy, L., 2005. Living in a fungal world: impact of fungi on soil bacterial niche development. Fems Microbiology Reviews 29, 795e811.
8. Edwards, K.A., McCulloch, J., Kershaw, G.P., Jefferies, R.L., 2006. Soil microbial and nutrient dynamics in a wet Arctic sedge meadow in late winter and early spring. Soil Biology & Biochemistry 38, 2843e2851.
9. Elmendorf, S.C., Henry, G.H.R., Hollister, R.D., Björk, R.G., Bjorkman, A.D., Callaghan, T.V., Collier, L.S., Cooper, E.J., Cornelissen, J.H.C., Day, T.A., Fosaa, A.M., Gould, W.A., Grétarsdóttir, J., Harte, J., Hermanutz, L., Hik, D.S., Hofgaard, A., Jarrad, F., Jónsdóttir, I.S., Keuper, F., Klanderud, K., Klein, J.A., Koh, S., Kudo, G., Lang, S.I., Loewen, V., May, J.L., Mercado, J., Michelsen, A., Molau, U., MyersSmith, I.H., Oberbauer,

S.F., Pieper, S., Post, E., Rixen, C., Robinson, C.H., Schmidt, N.M., Shaver, G.R., Stenström, A., Tolvanen, A., Totland, Ø., Troxler, T., Wahren, C.-H., Webber, P.J., Welker, J.M., Wookey, P.A., 2012. Global assessment of experimental climate warming on tundra vegetation: heterogeneity over space and time. Ecology Letters 15, 164e175.

10. Farrell, M., Hill, P.W., Farrar, J., DeLuca, T.H., Roberts, P., Kielland, K., Dahlgren, R., Murphy, D.V., Hobbs, P.J., Bardgett, R.D., Jones, D.J., 2013. Oligopeptides represent a preferred source of organic N uptake: a global phenomenon? Ecosystems 16, 133e145.

11. Frostegård, Å., Tunlid, A., Bååth, E., 1991. Microbial biomass measured as total lipid phosphate in soils of different organic content. Journal of Microbiological Methods 14, 151e163.

12. Geisseler, D., Horwath, W.R., Joergensen, R.G., Ludwig, B., 2010. Pathways of nitrogen utilization by soil microorganisms e a review. Soil Biology & Biochemistry 42, 2058e2067.

13. Giesler, R., Esberg, C., Lagerström, A., Graae, B., 2012. Phosphorus availability and microbial respiration across different tundra vegetation types. Biogeochemistry 108, 429e445.

14. Giesler, R., Petersson, T., Högberg, P., 2002. Phosphorus limitation in boreal forests: effects of aluminum and iron accumulation in the humus layer. Ecosystems 5, 300e314.

15. Hartley, I.P., Hopkins, D.W., Sommerkorn, M., Wookey, P.A., 2010. The response of organic matter mineralisation to nutrient and substrate additions in sub-arctic soils. Soil Biology & Biochemistry 42, 92e100.

16. Hayatsu, M., Tago, K., Saito, M., 2008. Various players in the nitrogen cycle: diversity and functions of the microorganisms involved in nitrification and denitrification. Soil Science and Plant Nutrition 54, 33e45.

17. Hill, P.W., Farrar, J., Roberts, P., Farrell, M., Grant, H., Newsham, K.K., Hopkins, D.W., Bardgett, R.D., Jones, D.L., 2011. Vascular plant success in a warming Antarctic may be due to efficient nitrogen acquisition. Nature Climate Change 1, 50e53.

18. Hobbie, S.E., 1996. Temperature and plant species control over litter decomposition in Alaskan tundra. Ecological Monographs 66, 503e522.

19. Hobbie, S.E., Nadelhoffer, K.J., Högberg, P., 2002. A synthesis: the role of nutrients as constraints on carbon balances in boreal and arctic regions. Plant And Soil 242, 163e170.

20. Hugelius, G., Kuhry, P., Tarnocai, C., Virtanen, T., 2010. Soil organic carbon pools in a periglacial landscape: a case study from the central Canadian Arctic. Permafrost and Periglacial Processes 21, 16e29.

21. IUSS Working Group WRB, 2007. World Reference Base for Soil Resources 2006. First Update 2007. FAO, Rome.

22. Jones, D.L., Kielland, K., 2002. Soil amino acid turnover dominates the nitrogen flux in permafrost-dominated taiga forest soils. Soil Biology & Biochemistry 34, 209e219.

23. Jones, D.L., Kielland, K., 2012. Amino acid, peptide and protein mineralization dynamics in a taiga forest soil. Soil Biology & Biochemistry 55, 60e69.

24. Jones, D.L., Willett, V.B., 2006. Experimental evaluation of methods to quantify dissolved organic nitrogen (DON) and dissolved organic carbon (DOC) in soil. Soil Biology & Biochemistry 38, 991e999.

25. Kaiser, C., Fuchslueger, L., Koranda, M., Gorfer, M., Stange, C.F., Kitzler, B., Rasche, F., Strauss, J., Sessitsch, A., Zechmeister-Boltenstern, S., Richter, A., 2011. Plants control the seasonal dynamics of microbial N cycling in a beech forest soil by belowground C allocation. Ecology 92, 1036e1051.

26. Kaiser, C., Koranda, M., Kitzler, B., Fuchslueger, L., Schnecker, J., Schweiger, P., Rasche, F., Zechmeister-Boltenstern, S., Sessitsch, A., Richter, A., 2010. Belowground carbon allocation by trees drives seasonal patterns of extracellular enzyme activities by altering microbial community composition in a beech forest soil. New Phytologist 187, 843e858.

27. Kaiser, C., Meyer, H., Biasi, C., Rusalimova, O., Barsukov, P., Richter, A., 2007. Conservation of soil organic matter through cryoturbation in Arctic soils in Siberia. Journal of Geophysical Research, G02017.

28. Kandeler, E., Gerber, H., 1988. Short-term assay of soil urease activity using colorimetric determination of ammonium. Biology and Fertility of Soils 6, 68e72.

29. Lavoie, M., Mack, M.C., Schuur, E.A.G., 2011. Effects of elevated nitrogen and temperature on carbon and nitrogen dynamics in Alaskan arctic and boreal soils. Journal of Geophysical Research-Biogeosciences 116.

30. Mack, M.C., Schuur, E.A.G., Bret-Harte, M.S., Shaver, G.R., Chapin, F.S., 2004. Ecosystem carbon storage in arctic tundra reduced by long-term nutrient fertilization. Nature 431, 440e443.

31. Miranda, K.M., Espey, M.G., Wink, D.A., 2001. A rapid, simple spectrophotometric method for simultaneous detection of nitrate and nitrite. Nitric Oxide-Biology and Chemistry 5, 62e71.

32. Montané, F., Rovira, P., Casals, P., 2007. Shrub encroachment into mesic mountain grasslands in the Iberian peninsula: effects of plant quality and temperature on soil C and N stocks. Global Biogeochemical Cycles 21.

33. Mooshammer, M., Wanek, W., Schnecker, J., Wild, B., Leitner, S., Hofhansl, F., Blöchl, A., Hämmerle, I., Frank, A.H., Fuchslueger, L., Keiblinger, K.M., Zechmeister-Boltenstern, S., Richter, A., 2012. Stoichiometric controls of nitrogen and phosphorus cycling in decomposing beech leaf litter. Ecology 93, 770e782.

34. Murphy, J., Riley, J.P., 1962. A modified single solution method for determination of phosphate in natural waters. Analytica Chimica Acta 27, 31e36.

35. Nadelhoffer, K.J., Giblin, A.E., Shaver, G.R., Laundre, J.A., 1991. Effects of temperature and substrate quality

on element mineralization in six arctic soils. Ecology 72, 242e253.

36. Näsholm, T., Kielland, K., Ganeteg, U., 2009. Uptake of organic nitrogen by plants. New Phytologist 182, 31e48.

37. Olsen, S.R., Cole, C.V., Watanabe, F.S., Dean, L.A., 1954. Estimation of Available Phosphorus in Soils by Extraction with Sodium Bicarbonate. In: U.S. Department of Agriculture Circular, vol. 939, pp. 1e19.

38. R Development Core Team, 2012. R: a Language and Environment for Statistical Computing. R Foundation for Statistical Computing, Vienna, Austria.

39. Read, D.J., Perez-Moreno, J., 2003. Mycorrhizas and nutrient cycling in ecosystems e a journey towards relevance? New Phytologist 157, 475e492.

40. Rustad, L.E., Campbell, J.L., Marion, G.M., Norby, R.J., Mitchell, M.J., Hartley, A.E., Cornelissen, J.H.C., Gurevitch, J., GCTE-NEWS, 2001. A meta-analysis of the response of soil respiration, net nitrogen mineralization, and aboveground plant growth to experimental ecosystem warming. Oecologia 126, 543e562.

41. Schimel, J.P., Bennett, J., 2004. Nitrogen mineralization: challenges of a changing paradigm. Ecology 85, 591e602.

42. Schleper, C., 2010. Ammonia oxidation: different niches for bacteria and archaea? The ISME Journal 4, 1092e1094.

43. Schmidt, M.W.I., Torn, M.S., Abiven, S., Dittmar, T., Guggenberger, G., Janssens, I.A., Kleber, M., Kögel-Knabner, I., Lehmann, J., Manning, D.A.C., Nannipieri, P., Rasse, D.P., Weiner, S., Trumbore, S.E., 2011. Persistence of soil organic matter as an ecosystem property. Nature 478, 49e56.

44. Schnecker, J., Wild, B., Fuchslueger, L., Richter, A., 2012. A field method to store samples from temperate mountain grassland soils for analysis of phospholipid fatty acids. Soil Biology and Biochemistry 51, 81e83.

45. Schneider, T., Keiblinger, K.M., Schmid, E., Sterflinger-Gleixner, K., Ellersdorfer, G., Roschitzki, B., Richter, A., Eberl, L., Zechmeister-Boltenstern, S., Riedel, K., 2012. Who is who in litter decomposition? Metaproteomics reveals major microbial players and their biogeochemical functions. The ISME Journal 6, 1749e1762.

46. Shaver, G.R., Chapin, F.S., 1995. Long-term responses to factorial, NPK fertilizer treatment by Alaskan wet and moist tundra sedge species. Ecography 18, 259e 275.

47. Shaver, G.R., Johnson, L.C., Cades, D.H., Murray, G., Laundre, J.A., Rastetter, E.B., Nadelhoffer, K.J., Giblin, A.E., 1998. Biomass and CO2 flux in wet sedge tundras: responses to nutrients, temperature, and light. Ecological Monographs 68, 75e 97.

48. Sistla, S.A., Asao, S., Schimel, J.P., 2012. Detecting microbial N-limitation in tussock tundra soil: implications for Arctic soil organic carbon cycling. Soil Biology and Biochemistry 55, 78e84.

49. Soil Survey Staff, 1999. Soil Taxonomy: a Basic System of Soil Classification for Making and Interpreting Soil Surveys, second ed. Natural Resources Conservation Service. U.S. Department of Agriculture Handbook 436.

50. Sterner, R.W., Elser, J.J., 2002. Ecological Stoichiometry. Princeton University Press, Princeton, New Jersey.

51. Strickland, M.S., Rousk, J., 2010. Considering fungal:bacterial dominance in soils e methods, controls, and ecosystem implications. Soil Biology & Biochemistry 42, 1385e1395.

52. Tarnocai, C., Canadell, J.G., Schuur, E.A.G., Kuhry, P., Mazhitova, G., Zimov, S., 2009. Soil organic carbon pools in the northern circumpolar permafrost region. Global Biogeochemical Cycles 23, GB2023.

53. Tezuka, Y., 1990. Bacterial regeneration of ammonium and phosphate as affected by the carbonenitrogenephosphorus ratio of organic substrates. Microbial Ecology 19, 227e238.

54. Wallenstein, M.D., McMahon, S.K., Schimel, J.P., 2009. Seasonal variation in enzyme activities and temperature sensitivities in Arctic tundra soils. Global Change Biology 15, 1631e1639.

55. Wanek, W., Mooshammer, M., Blöchl, A., Hanreich, A., Richter, A., 2010. Determination of gross rates of amino acid production and immobilization in decomposing leaf litter by a novel 15N isotope pool dilution technique. Soil Biology & Biochemistry 42, 1293e1302.

56. Weedon, J.T., Aerts, R., Kowalchuk, G.A., van Bodegom, P.M., 2011. Enzymology under global change: organic nitrogen turnover in alpine and sub-Arctic soils. Biochemical Society Transactions 39, 309e314.

57. Weintraub, M.N., Schimel, J.P., 2005a. Seasonal protein dynamics in Alaskan arctic tundra soils. Soil Biology & Biochemistry 37, 1469e1475.

58. Weintraub, M.N., Schimel, J.P., 2005b. The seasonal dynamics of amino acids and other nutrients in Alaskan Arctic tundra soils. Biogeochemistry 73, 359e380.

59. Wookey, P.A., Aerts, R., Bardgett, R.D., Baptist, F., Bråthen, K.A., Cornelissen, J.H.C., Gough, L., Hartley, I.P., Hopkins, D.W., Lavorel, S., Shaver, G.R., 2009. Ecosystem feedbacks and cascade processes: understanding their role in the responses of Arctic and alpine ecosystems to environmental change. Global Change Biology 15, 1153e1172.

60. Xu, C.H., Guo, L.D., Ping, C.L., White, D.M., 2009. Chemical and isotopic characterization of size-fractionated organic matter from cryoturbated tundra soils, northern Alaska. Journal of Geophysical Research-Biogeosciences 114.

CHAPTER 15

LATEGLACIAL AND HOLOCENE CLIMATE AND ENVIRONMENTAL CHANGE

Xiaosen Zhang, Jane Reed, Bernd Wagner, Alexander Francke, Zlatko Levkov

Keywords: Palaeoclimate, Diatom, Lake Dojran, Mediterranean, Balkans, Younger Dryas, Holocene

INTRODUCTION

The Mediterranean region is a primary global climate response hotspot (Giorgi, 2006) and a hotspot of biodiversity (Myers et al., 2000 and Mittermeier et al., 2004). It is a transitional zone climatically influenced both by the mid-latitude westerlies and the Subtropical High pressure (anticyclone) belt, with the North Atlantic Oscillation (NAO) modulating winter precipitation and the migration of the Intertropical Convergence Zone (ITCZ) affecting summer drought (Lionello et al., 2006). It stretches longitudinally from the North Atlantic Ocean to continental Eurasia. It has a diversity of landscapes linked to spatial and altitudinal variation in climatic factors.

Palaeoenvironmental analysis offers potential to improve understanding of Mediterranean climate change, but the clear definition of regional contrasts is still elusive. An early review of lake-level reconstruction proposed an east–west contrast during the Holocene (Harrison and Digerfeldt, 1993). More recently, Roberts et al., 2008 and Roberts et al., 2011a confirmed this, defining a marked contrast during the Holocene to the east and west of a line running through the Balkans, southern Italy and Tunisia, based on stable isotope data and model output; on a centennial-decadal timescale, the complexity of regional patterns was also demonstrated in an east–west contrast between the northern Iberian Peninsula and central Turkey (Roberts et al., 2012). In contrast, Magny et al. (2013) proposed a north–south divide around ca. 40°N during the Holocene in the central Mediterranean from carbonate-based lake-level reconstruction. Peyron et al. (2013) supported this from pollen-based quantitative reconstruction of summer precipitation, and also proposed a similar pattern in the Aegean Sea. This is coherent with a north–south contrast in fire activity in the western Mediterranean (Vannière et al., 2011).

From the foregoing, the southern Balkans is a key location for understanding Mediterranean climate change, being located at the juncture of the proposed boundaries between west–east and north–south contrasting climate and hydrological domains. The southern Balkans is particularly complex, and patterns and mechanisms of climate and environmental change are still poorly understood. The complexity of palaeoenvironments is indicated, for example, by discrepancies in vegetation reconstruction between adjacent sites such as Lake Ioannina (northwestern Greece) and Nisi Fen (northern Greece) (Lawson et al., 2004 and Lawson et al., 2005), and between Lake Gramousti at low altitude and Rezina Marsh at high altitude in northwestern Greece (Willis, 1992a).

Here, we build on previous multi-proxy palaeoclimate research in investigating the Lateglacial and Holocene record of Lake Dojran (Macedonia/Greece), by using diatom analysis as a strong proxy for lake levels and trophic status to strengthen interpretation based on sedimentological and geochemical data from the same core (Francke et al., 2013). In interpretation of Holocene limnological change in terms of palaeoclimate shifts *versus* the influence of local catchment dynamics, we exploit extant regional palynological data for vegetation change, comprising late-Holocene pollen data from a separate littoral Dojran sequence (Athanasiadis et al., 2000) and chronologically-robust Holocene pollen data from the highlands and lowlands in the southern Balkans (Kotthoff et al., 2008a, Tonkov et al., 2008 and Tonkov et al., 2013). Adopting a novel approach, the importance of seasonality in

Citation: Xiaosen Zhang, Jane Reed, Bernd Wagner, Alexander Francke, Zlatko Levkov, Lateglacial and Holocene climate and environmental change, Quaternary Science Reviews, Volume 103, 1 November 2014, Pages 51-66, ISSN 0277-3791, http://dx.doi.org/10.1016/j.quascirev.2014.09.004.

driving Holocene climate change is assessed by reference to the summer and winter latitudinal temperature gradient (LTG) model (Davis and Brewer, 2009), which incorporates variation in the Subtropical High pressure and Arctic Oscillation (AO). For clarity, this is expanded upon in Section 2. We also compare with proxy data from the northeastern Mediterranean (the Balkans, Italy and Anatolia) (Fig. 1).

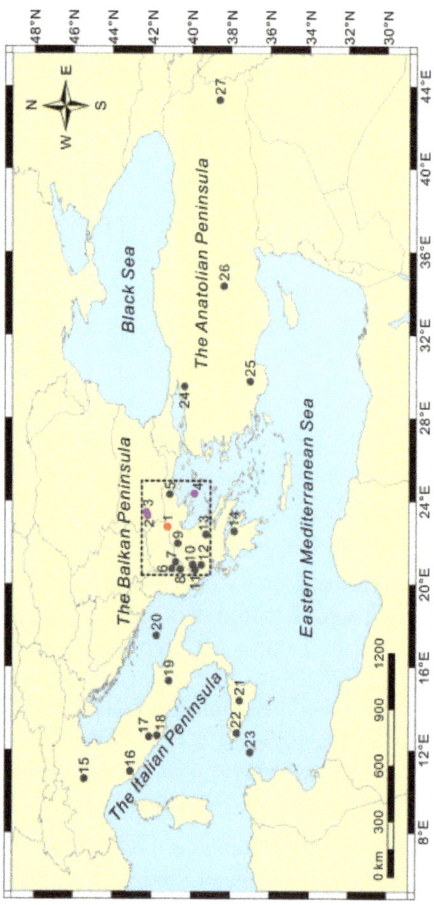

Figure 1: Map of the northeastern Mediterranean showing the locations of the study site (*red*), relevant palynological records with robust chronologies (*purple*) and other palaeoenvironmental records (*black*) referred to in this paper. 1. Lake Dojran (this paper; Athanasiadis et al., 2000 and Francke et al., 2013), 2. Lake Trilistnika (Tonkov et al., 2008), 3. Lake Ribno (Tonkov et al., 2013), 4. SL152 (Kotthoff et al., 2008a, Kotthoff et al., 2008b, Kotthoff et al., 2011 and Dormoy et al., 2009), 5. Tenaghi Philippon (Müller et al., 2011), 6. Lake Ohrid (Wagner et al., 2009 and Leng et al., 2010), 7. Lake Prespa (Aufgebauer et al., 2012, Panagiotopoulos et al., 2013, Leng et al., 2013 and Cvetkoska et al., 2014), 8. Lake Maliq (Bordon et al., 2009), 9. Nisi Fen (Lawson et al., 2005), 10. Rezina Marsh (Willis, 1992a and Willis, 1992b), 11. Lake Gramousti (Willis, 1992a), 12. Lake Ioannina (Frogley et al., 2001, Lawson et al., 2004, Wilson et al., 2008 and Jones et al., 2013), 13. Lake Xinias (Digerfeldt et al., 2007), 14. Lake Stymphalia (Heymann et al., 2013), 15. Lake Frassino (Baroni et al., 2006), 16. Lake Accesa (Drescher-Schneider et al., 2007 and Peyron et al., 2011), 17. Valle di Castiglione (Di Rita et al., 2013), 18. Lake Albano (Guilizzoni et al., 2002), 19. Lago Grande di Monticchio (Allen et al., 2002), 20. MD90-917 (Combourieu-Nebout et al., 2013), 21. Lake Pergusa (Sadori and Narcisi, 2001, Sadori et al., 2008 and Magny et al., 2012), 22. Lake Preola (Magny et al., 2011), 23. MD04-2797 (Desprat et al., 2013), 24. Lake Iznik (Roeser et al., 2012), 25. Lake Gölhisar (Eastwood et al., 2007), 26. Eski Acıgöl (Roberts et al., 2001 and Turner et al., 2008), 27. Lake Van (Wick et al., 2003 and Litt et al., 2009). The dashed-line rectangle shows the range of Fig. 2. (For interpretation of the references to colour in this figure legend, the reader is referred to the web version of this article.)

REVIEW OF HOLOCENE CLIMATIC FORCING

Summer Climate Mode and the Subtropical High Pressure

The character and influence of the Holocene summer insolation maximum across the Mediterranean is a topic for vigorous ongoing debate (Tzedakis, 2007). In a major review, Tzedakis (2007) argued that the enhanced African monsoon did not extend to the Mediterranean, and the monsoonal effect has been mainly indirect in terms of Nile discharge and runoff along the North African coast. On the other hand, the northward-migrated, strengthened North Atlantic Subtropical High pressure in response to high summer insolation blocked westerly moisture penetration into the Mediterranean (Desprat et al., 2013 and Magny et al., 2013). Tzedakis (2007) also suggested that the enhanced Indian monsoon may have accentuated summer aridity in the eastern Mediterranean. Modelling experiments show little sign of summer precipitation in spite of high summer insolation (Brayshaw et al., 2011). Magny et al. (2013) discussed the influence of ice sheets in the Northern Hemisphere on humidity in the Mediterranean, and suggested that the rapid melting of ice sheets and associated fresh water forcing in the North Atlantic Ocean contribute to summer aridity during the earliest Holocene in the south-central Mediterranean.

Davis and Brewer (2009) reconstructed Holocene changes in the latitudinal temperature gradient (LTG) based on differences in pollen-based area-average temperatures between northern (55–70°N) and southern Europe (35–45°N). The LTG reflects combined effects of the differential heating of insolation (the latitudinal insolation gradient, LIG) and the high-latitude cooling influence of remnant ice sheets (Davis and Brewer, 2009). The Subtropical High pressure is located at ca. 36°N when the summer LTG is zero (Davis and Brewer, 2009). A weaker (more positive) summer LTG drives a northern position of the Subtropical High pressure (Davis and Brewer, 2009), and the tendency would be towards much drier summers than today. The quantitative reconstruction of the Subtropical High pressure may be relevant to moisture change in the northeastern Mediterranean. The additional, area-average temperature reconstruction for southeastern Europe (Davis et al., 2003), supported by other regional syntheses (Finné et al., 2011 and Abrantes et al., 2012), may be relevant in terms of evaporation and its effect on moisture balance.

Winter Climate Mode and the Arctic Oscillation (AO)

Modelling experiments show high winter precipitation during the early Holocene (Brayshaw et al., 2011), which is coherent with a southward-shifted storm track and responds to weak winter insolation (Desprat et al., 2013). Desprat et al. (2013) also suggested that the rapid melting of ice sheets in the Northern Hemisphere and associated reorganisation of atmospheric circulation contribute to high winter precipitation during the early Holocene in the south-central Mediterranean. Roberts et al., 2008 and Roberts et al., 2011a and Magny et al. (2013) suggested that this is probably attributed to winter cyclogenesis and local precipitation. However, it is inconsistent with the positive AO/NAO and resultant aridity in the north-central Mediterranean (Magny et al., 2013). Data-model comparison shows that models underestimate the role of AO/NAO (Gladstone et al., 2005, Brewer et al., 2007 and Mauri et al., 2013), and Davis and Brewer (2009) proposed that models overestimate low-latitude warming in summer and high-latitude warming in winter.

The Mediterranean exhibits a dry climate during the positive phase of AO/NAO and more precipitation during their negative phase (Martinson et al., 2000 and Wanner et al., 2001). However, AO has a larger horizontal scale (Thompson and Wallace, 1998), and NAO alone cannot account for changes in winter precipitation in Turkey (Jones et al., 2006). In terms of precipitation, the northeastern Mediterranean is distinguished from the Levant and southeastern Mediterranean (Felis and Rimbu, 2010), and from the NAO-highly related Iberia and western Mediterranean (Roberts et al., 2012). The prominent role of AO/NAO in influencing Holocene hydroclimatic change across the Mediterranean is revealed not only at the centennial scale (e.g. Lamy et al., 2006) but also at the millennial scale (e.g. Davis and Stevenson, 2007 and Fletcher et al., 2013). The AO index is zero when the winter LTG is zero (Davis and Brewer, 2009). A weaker (more positive) winter LTG drives a more positive AO, and the tendency would be towards dry winters. The AO quantitative reconstruction (Davis and Brewer, 2009) may be more relevant than the NAO in interpretation of northeastern Mediterranean climatic records.

SITE DESCRIPTION

Lake Dojran (41°12'N, 22°44E, 144 m a.s.l.), a transboundary lake between Macedonia and Greece, sits within a karstic basin formed initially by a combination of Tertiary volcanic and tectonic activities, and the catchment sediments are largely composed of Quaternary alluvial and limnetic materials (Sotiria and Petkovski, 2004). In the catchment, the highland to the north is close to the Pirin and Rila Mountains in southwestern Bulgaria, and the lowland to the south is open to the Thessaloniki Plain and northern Aegean Sea (Fig. 2). The lake basin is surrounded by the Belasica (or Belles, Kerkini) Mountain (1847 m a.s.l.) to the north, the Kroussia (or Krusa, Dysoron) Mountain (766 m a.s.l.) to the east, the Boskija (or Boska) Mountain (714 m a.s.l.) to the northwest, and the Dab Mountain (689 m a.s.l.) to the southwest (Fig. 3; Sotiria and Petkovski, 2004). The lake is fed by small rivers, creeks and springs, with most of the runoff originating from the Belasica Mountain. Water loss is currently through evaporation and probably groundwater outflow, but during previous phases of high lake level, surface outflow was possible at the southern end of the lake through the Doiranity (or Ayiak) River, which drained into the Vardar (or Axios) River and then into the Aegean Sea (Sotiria and Petkovski, 2004). The lake water is essentially fresh (see below), suggesting groundwater throughflow in karstic aquifers. This is supported by hydrogeological investigations (Sotiria and Petkovski, 2004).

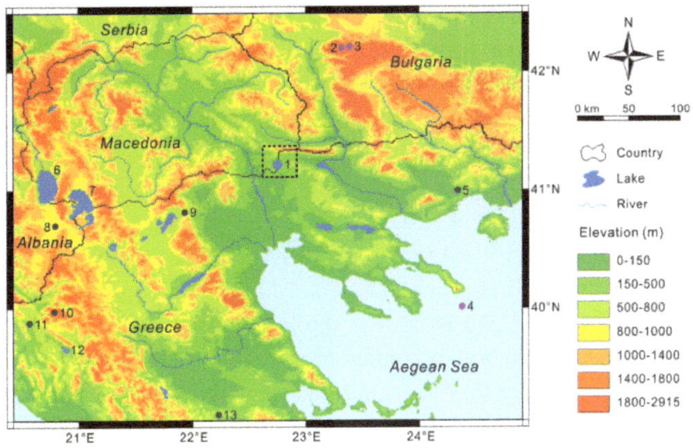

Figure 2: Map of the adjacent region of Lake Dojran (Macedonia, northern Greece and southwestern Bulgaria) with the sites 1–13 in Fig. 1. The dashed-line rectangle shows the range of Fig. 3.

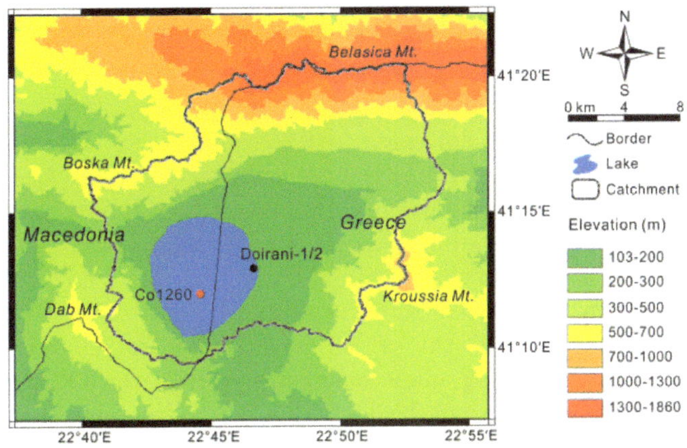

Figure 3: Map of the catchment of Lake Dojran with the coring sites Co1260 (this paper; Francke et al., 2013) and Doirani-1/2 (Athanasiadis et al., 2000).

The local climate regime is a hot and dry summer (June–September) and mild and humid winter (November–February) (Sotiria and Petkovski, 2004). In the lowlands of the catchment evergreen (*Quercus coccifera* L.) and deciduous oaks (*Quercus pubescens*Willd., *Quercus frainetto* Ten. and *Quercus dalechampii* Ten.) are the dominant trees, and at higher altitudes (>1000 m a.s.l.) beech forest (*Fagus moesiaca* Cz. and *Fagus orientalis* Lipsky) is dominant with a few scattered fir stands (*Abies borisii-regis*Mattf.) (Athanasiadis et al., 2000). Reed beds occupy the fringe of the lake, and submerged plants are common in the littoral zone. Recorded maximum lake level ranged from 7.9 to 10.0 m in 1951–1987, declined to 3.7 m in 2002 due to water abstraction practices and more intensive agriculture (Griffiths et al., 2002), and recovered to 6.7 m in 2010 due to an increase in rainfall, decrease in water use and additional water transfer from the Gjavato wells into the lake (Popovska and Bonacci, 2008 and Stojov, 2012). Total phosphorus ranged from 15 to 130 µg l^{-1} in 1953–1960 (Sotiria and Petkovski, 2004), with consistently higher minimum values of >50 µg l^{-1} reported since 1996 and an occasional hypereutrophic state (Temponeras et al., 2000, Lokoska et al., 2006 and Tasevska et al., 2010). Conductivity ranged from 0.4 to 0.6 mS cm^{-1} in 1974–1988, increased to 1.5 mS cm^{-1} in 2002 (Sotiria and Petkovski, 2004), and declined to 0.8 mS cm^{-1} in 2010 (Lešoski et al., 2010). This is within the freshwater range of eutrophic lake water. Ionic concentration may be influenced slightly by mineral-rich spring input; some springs are of high conductivity (Levkov, unpublished data), but others are extremely fresh (33 μS cm^{-1}; Griffiths et al., 2002). Dominance by fresh spring inflow is consistent with low $\delta^{13}C$ data of total dissolved inorganic carbon (−7.9 to −13.0‰ VPDB) in springs (Griffiths et al., 2002 and Francke et al., 2013), suggesting that the effect of the karstic catchment is minor (Leng and Marshall, 2004) and the groundwater influence on lake-water conductivity and salinity is insignificant. Lake Dojran is shallow and monomictic, and has a very simple, flat-bottomed morphometry (Francke et al., 2013), suggesting that moisture balance and water chemistry are sensitive to and possibly respond linearly to climate and environmental change (Gasse et al., 1997 and Fritz, 2008).

Two previous studies have been made of Holocene climate and environmental change in Lake Dojran. One is based on sedimentological and geochemical data (Francke et al., 2013) from the same core as our current study, and the other is based on pollen data from a separate littoral core Doirani-1/2 which covers the last 5,000 years (Athanasiadis et al., 2000). Francke et al. (2013) found that, during the Lateglacial period, high abundance of clay clasts, high mean grain size, an older shell fragment, and hydro-acoustic data indicated a low lake level and redeposition before ca. 12,100 cal yr BP. This was followed by a lake-level increase inferred from hydro-acoustic data, the absence of clay clasts, low mean grain size, and high potassium (K) concentration. During the earliest Holocene (ca. 11,500–10,700 cal yr BP), overall coarse sediment suggested high water inflow and high lake level, which led to high evaporation as indicated by high $\delta^{18}O_{carb}$ values; lower K concentration indicated lower erosion, and low organic matter content suggested low in-lake productivity. During the early Holocene (ca. 10,700–8,300 cal yr BP), decreasing $\delta^{18}O_{carb}$ values suggested increasing humidity; lower K concentration implied less clastic input while increasing organic matter content was from more allochthonous supply. During the mid Holocene, relatively high lake level was inferred from hydro-acoustic data and finer grain-size distribution. After ca. 2,800 cal yr BP, hydro-acoustic data and fine grain-size distribution suggested a lake-level lowstand apart from around 1,000 cal yr BP (Francke et al., 2013), while pollen data indicated changes in the catchment from a natural landscape to one modified by intensified human impact (Athanasiadis et al., 2000).

MATERIAL AND METHODS

In June 2011, based on a hydro-acoustic survey, a 717 cm-long core Co1260 was recovered from the deepest (6.7 m water depth), south-central part of Lake Dojran using UWITEC gravity and piston coring equipments (www.uwitec.at). The age model was established by Francke et al. (2013) by polynomial interpolation between the calibrated radiocarbon ages of six terrestrial plant macrofossils, one charcoal fragment, two bulk organic matter samples, and a regionally-correlated point of a $CaCO_3$ minimum. Three carbonate shell dates and one dislocated terrestrial plant date were not included into the calculations, and Francke et al. (2013) provided more detailed discussion of each age-control point (Table 1). This age model indicates that core Co1260 covers the last 12,500 years (Fig. 4), spanning the Younger Dryas and Holocene periods.

Table 1: Ages from core Co1260. The calibration of radiocarbon ages into calendar ages is based on Calib 7.0.2 (Stuiver and Reimer, 1993) and IntCal13 (Reimer et al., 2013) and on a 2 uncertainty

Core depth (cm)	Material	Radiocarbon age (^{14}C yr BP)	Calendar age (cal yr BP)
16.5	Bulk organic matter	140 ± 35	140 ± 140
53.3	Terrestrial plant remains	360 ± 70	410 ± 110
111.3	Terrestrial plant remains	840 ± 70	790 ± 120
253.0	Terrestrial plant remains	2,430 ± 30	2,520 ± 170
287.3	Terrestrial plant remains	3,080 ± 30	3,290 ± 80
309.1	Charcoal	3,560 ± 40	3,850 ± 130
404.9	Carbonate shell	6,410 ± 40	7,350 ± 80[a]
406.4	Terrestrial plant remains	8,020 ± 150	8,960 ± 440
460.9	Terrestrial plant remains	9,520 ± 160	10,820 ± 420[a]
502.9	Carbonate shell	9,840 ± 40	11,250 ± 60[a]
521.9	Terrestrial plant remains	9,330 ± 160	10,660 ± 440
635.0	Bulk organic matter	10,220 ± 70	11,920 ± 310
682.0	Carbonate shell	28,570 ± 170	32,830 ± 650[a]

[a]Indicating the dates excluded into the calculations of the age model.

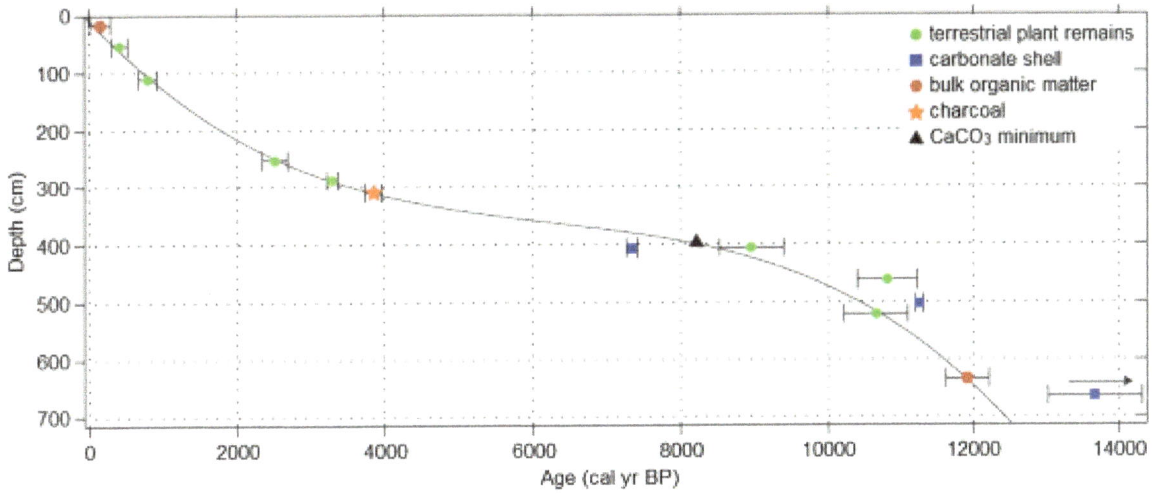

Figure 4: Age-depth model of core Co1260 (modified from Francke et al., 2013).

Diatom analysis was carried out on 107 subsamples, taken at 8 cm intervals but at a higher resolution of 4 cm for important phases comprising the bottom 30 cm, the 50 cm between ca. 11,800–11,400 cal yr BP, the 20 cm around ca. 8,200 cal yr BP, and the top 60 cm. The age resolution is ca. 70–150 years, except the top and bottom sections (ca. 30–40 years) and the middle section (ca. 200–500 years). The low age resolution between ca. 8,000–4,000 cal yr BP is correlated with low sediment accumulation rate indicated by the age model, and with a stable environment indicated by sedimentological and geochemical data (Francke et al., 2013). echniques in Battarbee et al. (2001) were adopted for diatom preparation of ca. 0.1 g dry weight subsamples, using 30% H_2O_2 to oxidise organics and a few drops of concentrated HCl to remove carbonates. Known quantities of microspheres were added to allow calculation of absolute valve concentration. Slides were mounted using Naphrax™. Diatom valves were counted at × 1000 magnification under oil immersion on an OLYMPUS BX51 light microscope. More than 500 valves per slide were counted, and around 100 valves for some slides where preservation was very poor. Diatom identification was based on a range of published literature (Krammer and Lange-Bertalot, 1986, Krammer and Lange-Bertalot, 1988, Krammer and Lange-Bertalot, 1991a, Krammer and Lange-Bertalot, 1991b, Lange-Bertalot, 2001, Krammer, 2002, Levkov et al., 2007 and Levkov, 2009) and diatoms of Lake Dojran (Levkov, unpublished data), adopting the nomenclature of the Catalogue of Diatom Names (on-line version) (Fourtanier and Kociolek, 2011). Following current taxonomic principles, *Stephanodiscus minutulus* (Kützing) Cleve & Möller and *S. parvus* Stoermer & Håkansson are merged into *S. minutulus/parvus* (Scheffler and Morabito, 2003, Cruces et al., 2010, Hobbs et al., 2011 and Bennion et al., 2012). Diatom percentages were displayed using Tilia version 1.7.16, and zone boundaries were defined according to Constrained Incremental Sum of Squares (CONISS) cluster analysis (Grimm, 2011). Unconstrained ordination techniques were used to explore the variance in the diatom data using Canoco for Windows 4.5 (Ter Braak and Šmilauer, 2002). Detrended correspondence analysis (DCA) gave the largest gradient length of 3.91 SD units, and thus this unimodal model is better than linear methods such as principal components analysis (PCA) (Ter Braak, 1995 and Lepš and Šmilauer, 2003). Quantitative diatom-inferred total phosphorus (DI-TP) and conductivity (DI-Cond) reconstructions were performed based on the Swiss TP and Combined Salinity training sets, respectively, available within the European Diatom Database (EDDI; Juggins, 2001), using classic weighted-averaging regression (WA-Cla) with bootstrapping cross-validation in C2 version 1.7.3 (Juggins, 2007). Both the Swiss TP and Combined Salinity training sets have higher determination coefficients (R^2) and lower maximum bias values in WA-Cla than weighted-averaging with inverse deshrinking (WA-Inv) and weighted-averaging partial least squares (WAPLS) component 2. WA-Cla also performs better than WAPLS component 2 in terms of root mean squared error of prediction (RMSEP), and has a slightly higher RMSEP than WA-Inv (bootstrapped RMSEP: 0.26 and 0.23, respectively, for log_{10}-TP, and 0.48 and 0.47, respectively, for log_{10}-conductivity). WA-Inv is susceptible to edge effects (bias at the gradient ends) (Juggins and Birks, 2012), and thus WA-Cla was chosen. Some potential problems have been identified with the application of DI-TP transfer functions (Bennion et al., 2001, Bennion et al., 2010, Sayer, 2001, Hall and Smol, 2010 and Juggins et al., 2013) and some have suggested there are some fundamental problems with underlying assumptions in the transfer function approach (Sayer et al., 2010 and Juggins, 2013), but the species TP optimum and tolerance estimates, particularly of planktonic taxa, may still be useful for trophic reconstruction. Among the EDDI TP transfer functions, the Swiss training set is predominantly from mesotrophic and eutrophic lakes (mean TP: 41.9 μg l^{-1}), the Central European training set is mainly from oligotrophic and mesotrophic lakes (mean TP: 23.5 μg l^{-1}), and the Combined TP training set spans the gradient from oligotrophic to hypereutrophic (mean TP: 98.6 μg l^{-1}) but incorporates many non-karstic, hypereutrophic British and Danish lakes (Juggins, 2001). Although improving the likelihood of finding modern analogues, the merging in the Combined TP training set tends to extend the apparent distribution of species unevenly along the TP gradient, which not only affects species optimum estimates but also widens species tolerance ranges and makes them poor trophic indicators. Griffiths et al. (2002) suggested that Lake Dojran was mesotrophic in the recent past, and as shown above, it is currently eutrophic and occasionally hypereutrophic. Thus the Swiss training set was selected as more appropriate, although oligotrophic and hypereutrophic states are likely to be somewhat underestimated as they lie at the gradient ends. Importantly, the shifts between the dominance of centric (*Cyclotella ocellata* Pantocsek and *S. minutulus/parvus*) and fragilariod species in core Co1260 are well represented by their distributions along the TP gradient in the Swiss training set (Lotter et al., 1998).

Results and Diatom Interpretation

Six major diatom assemblage zones can be defined, which correlate clearly with lithostratigraphic boundaries from Francke et al. (2013) in Fig. 5. Diatom preservation quality is high with >500 valves counted and $>10^7$ g^{-1} concentration throughout until the late Holocene (Zone D-6). DI-TP $\log_{10}$ values range from 1.02 to 2.18 (equivalent to 10.6–149.8 μg l^{-1}), estimated standard error of prediction (eSEP) ranges from 0.26 to 0.29, and the error range (2 × eSEP) varies from 0.51 to 0.58 (calculated as 13.7–210.8 μg l^{-1}). DI-Cond $\log_{10}$ values range from 2.39 to 3.62 (equivalent to 0.2–4.2 mS cm^{-1}), eSEP ranges from 0.47 to 0.67, and the error range varies from 0.93 to 1.34 (calculated as 0.8–13.4 mS cm^{-1}). We treat the DI-TP and DI-Cond reconstructions with caution.

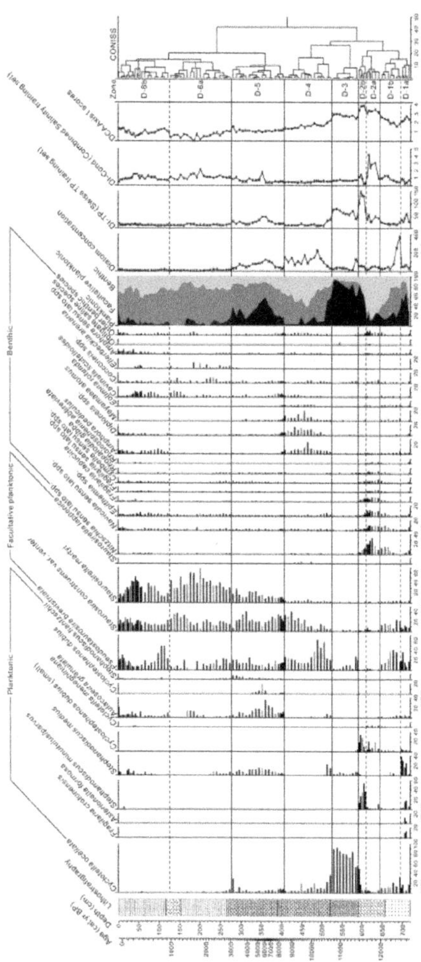

Figure 5: Lithostratigraphy (modified from Francke et al., 2013) and summary diatom diagram of core Co1260, showing diatom-inferred total phosphorus (DI-TP) and conductivity (DI-Cond) reconstructions based on the Swiss TP and Combined Salinity training sets (EDDI; Juggins, 2001), respectively, and Axis 1 scores from detrended correspondence analysis (DCA). The calibrated radiocarbon chronology is marked at 1000-year intervals, with an estimated age of 12,500 cal yr BP at the base of the sequence.

Zone D-1 (717–645 cm, ca. 12,500–12,000 cal yr BP)

In Subzone D-1a (717–696 cm, ca. 12,500–12,400 cal yr BP), the basal sample has relatively high abundance of benthic taxa (40%), possibly suggesting a lake-level lowstand at the base of the sequence, and this is followed at 713 cm depth by the dominance of planktonic taxa (50–70%), indicating relatively high lake level. Longer valves of *Fragilaria crotonensis* Kitton and *Asterionella formosa* Hassall have competitive advantages in deep waters for buoyancy and also prefer high nitrogen and silica concentrations (Bailey-Watts, 1986 and Saros et al., 2005), and *S. minutulus/parvus* prefers high phosphorus concentration (Bennion, 1995 and Kilham et al., 1996). Together, this also indicates a eutrophic state, presumably from high catchment runoff and nutrient erosion processes. The subsequent dominance of mesotrophic *Stephanodiscus medius* Håkansson and oligotrophic-mesotrophic *C. ocellata* suggests ongoing water inflow but declining nutrient enrichment, supported by declining DI-TP values. In Subzone D-1b (696–645 cm, ca. 12,400–12,000 cal yr BP), the increase in the relative abundance of facultative planktonic taxa, mainly comprising *Pseudostaurosira brevistriata* (Grunow) Williams & Round and *Staurosira construens*var. *venter* (Ehrenberg) Hamilton, and the clear increasing trend in benthic taxa from 667 cm depth suggest shallowing, starting with the peak concentration in the entire sequence at the 696 cm depth.

Zone D-2 (645–591 cm, ca. 12,000–11,500 cal yr BP)

Planktonic *C. ocellata* is at low abundance, and together with high relative abundance of benthos, this indicates a very low lake level. Facultative planktonic *Pseudostaurosira*, *Staurosira* and *Staurosirella* species are also rare. However, planktonic *Cyclotella meneghiniana* Kützing, *S. minutulus/parvus* and *Cyclostephanos dubius* (Fricke) Round (diameter <5 μm) are consistently present. *C. meneghiniana* lives in a wide range of habitats from shallow to deep waters (Gasse, 2002) and from fresh waters of high conductivity to saline conditions (Saros and Fritz, 2000). *S. minutulus/parvus* and *C. dubius* (small) are eutrophic freshwater taxa but tolerate oligosaline conditions (Fritz et al., 1993). Benthic taxa are remarkably diverse, most of which are eutrophic and halotolerant, such as *Nitzschia frustulum* (Kützing) Grunow, *Tryblionella constricta*(Kützing) Poulin, *Ctenophora pulchella* (Ralfs ex Kützing) Williams & Round, *Tabularia fasciculata* (Agardh) Williams & Round (Fritz et al., 1993, Reed, 1998a, Gasse, 2002 and Reed et al., 2012). Importantly, the presence of rare obligate saline taxa such as *Biremis circumtexta* (Meister ex Hustedt) Lange-Bertalot & Witkowski and *Campylodiscus clypeus* (Ehrenberg) Kützing indicates aridity in a closed basin due to enhanced evaporative concentration, since the groundwater influence on lake-water conductivity and salinity is insignificant. Maximum DCA Axis 1 scores also make this zone distinctive. In Subzone D-2a (645–611 cm, ca. 12,000–11,700 cal yr BP), eutrophic, halotolerant benthic taxa are dominant, and the DI-Cond reconstruction indicates a peak concentration of total dissolved solids (>3 mS cm^{-1} conductivity). In Subzone D-2b (611–591 cm, ca. 11,700–11,500 cal yr BP), *S. minutulus/parvus* and *C. dubius* (small) increase distinctly. Since obligate saline taxa and eutrophic, halotolerant taxa indicate the maintenance of low lake level, enhanced nutrient input is inferred, supported by the peak in DI-TP.

Zone D-3 (591–525 cm, ca. 11,500–10,700 cal yr BP)

Zone D-3 exhibits a marked transition to the low-diversity dominance of planktonic oligotrophic-mesotrophic *C. ocellata* (70–90%), initially as a co-dominant with *P. brevistriata* (ca. 25%). The abundance of facultative planktonic taxa is <10% thereafter. Planktonic mesotrophic *S. medius* is consistently present at ca. 5% abundance. The abundance of benthic taxa is consistently <10%. A deep, oligotrophic to mesotrophic state can be inferred. A moderate, nearly stable nutrient level is shown by the DI-TP reconstruction. This zone is also clearly distinguished by stable, intermediate DCA Axis 1 scores.

Zone D-4 (525–407 cm, ca. 10,700–8,500 cal yr BP)

A shallow condition prevails in Zone D-4, clearly indicated by a transition to the co-dominance of *C. ocellata*, *S. medius* and *F. brevistriata* (ca. 20% each), to the dominance of *F. brevistriata* (up to 60%) and then to the relatively high abundance of *S. construens*var. *venter* (30–40%). Heavily-silicified facultative planktonic *Staurosirella martyi*(Héribaud) Morales & Manoylov and *S. lapponica* (Grunow) Williams & Round are consistently present. Benthic taxa are at higher abundance, including smaller *Amphora pediculus* (Kützing) Grunow and *Mayamaea atomus* (Kützing) Lange-Bertalot, and heavier *Diploneis mauleri* (Brun) Cleve and *Eolimna rotunda* (Hustedt) Lange-Bertalot, Kulikovskiy & Witkowski. *A. pediculus* is tolerant of oligotrophic to eutrophic conditions, *D. mauleri* and *E. rotunda* live mostly in mesotrophic waters, and *M. atomus* is a eutrophic species. A relatively high trophic level can also be inferred; this is not clear in the DI-TP reconstruction but is supported by high diatom concentration in this phase.

Zone D-5 (407–277 cm, ca. 8,500–3,000 cal yr BP)

The relatively high abundance of planktonic eutrophic *Aulacoseira granulata* (Ehrenberg) Simonsen and mesotrophic *S. medius*, the consistent presence of eutrophic *Stephanodiscus hantzschii* Grunow, and the short-lived peak of eutrophic *C. dubius*around 353 cm depth (ca. 5,700 cal yr BP), indicate a relatively deep and turbid state. It is supported by relatively high DI-TP and diatom concentration, particularly between 377 and 329 cm depth (ca. 7,200–4,500 cal yr BP), where it correlates with a broad peak of plankton and high in-lake productivity. Low light availability in the deeper water would reduce the growth of benthic diatoms, while tychoplanktonic *Pseudostaurosira*, *Staurosira* and *Staurosirella* species (maintained at 40–60%) are commonly transported into the water column (Battarbee et al., 2001) and are tolerant of disturbed environments (Anderson, 2000). Larger and heavily-silicified planktonic diatoms sink rapidly in the absence of mixing, even in a deep lake with a long settling distance (Bennion et al., 2010 and Wolin and Stone, 2010), and in a shallow lake there is insufficient mixing to suspend these taxa for long growth periods (Bennion, 1995). Thus it is relatively high lake level and mixing that make robust planktonic *A. granulata*, *C. dubius*, *S. hantzschii* and *S. medius* and tychoplanktonic *S. martyi* remain in the photic zone. This supports an interpretation of the planktonic abundance both in terms of productivity and increased lake level.

Zone D-6 (277–0 cm, ca. 3,000 cal yr BP-Present)

Diatom concentration is low, with high dissolution, particularly between 209 and 177 cm depth (ca. 1,900–1,500 cal yr BP) and between 81 and 21 cm depth (ca. 600–100 cal yr BP), where the diatom count is < 300 valves and assemblages are dominated by poorly-preserved valves of robust taxa. More fragile taxa such as *P. brevistriata* and *A. pediculus*are possibly dissolved. Despite the dissolution, the dominance of facultative planktonic taxa and low abundance of planktonic taxa indicate a shallow environment, possibly with the expansion of emergent vegetation, because the base of emergent macrophytes is a major habitat for small fragilaroid species (Sayer, 2001). After ca. 1,000 cal yr BP (in Subzone D-6b) there is a slight increase in *Amphora* species, but no major ecological shift occurs in the recent past. The topmost sample is distinguished by the relatively high abundance of *A. granulata*, which is probably a reflection of the incorporation of living diatoms rather than an indication of recent accelerated eutrophication. In the light of the lake's modern eutrophic to hypereutrophic state, the ecologically-consistent diatom assemblages suggest that high trophic level possibly prevails in this zone, although nutrient reconstruction is not particularly sensitive here. There are several possible reasons for lack of diatom evidence for accelerated eutrophication in the recent past: 1) eurytopic diatoms with broad tolerance ranges tend to dominate at a high trophic state as an adaptation to wide fluctuations in water chemistry, compounding the aforementioned potential for poor reconstruction at the upper end of the nutrient gradient; 2) non-planktonic diatoms respond to water-column nutrient additions less directly than phytoplankton, and they are more sensitive to habitat availability as they can derive nutrients from sediments and macrophytes (Bennion et al., 2010 and Hall and Smol, 2010); 3) other algae (chlorophytes, cyanobacteria and dinoflagellates) are the most important primary producers rather than diatoms in this lake at a high trophic level; and 4) a more turbid state due to increased phytoplankton growth in turn reduces diatom growth, while chlorophytes and cyanobacteria are better competitors for light (Tilman et al., 1986).

Multi-Proxy, Regional- and Catchment-Scale Interpretation

The diatoms provide strong proxy data for Lateglacial and Holocene changes in lake levels and trophic status in Lake Dojran, and strengthen previous lake-level and palaeoclimate interpretation by comparison with extant sedimentological and geochemical data from the same core in Fig. 6. The diatom data are also important in disentangling regional climate effects from the influence of local catchment processes on Holocene hydrological variability, and we exploit extant regional pollen data with robust chronologies for vegetation change in the highlands and lowlands in Fig. 7. We compare with various proxy data from the northeastern Mediterranean, and assess the importance of seasonality in driving Holocene moisture availability by using the summer and winter latitudinal temperature gradient (LTG) to disentangle precipitation from temperature effects (Fig. 7).

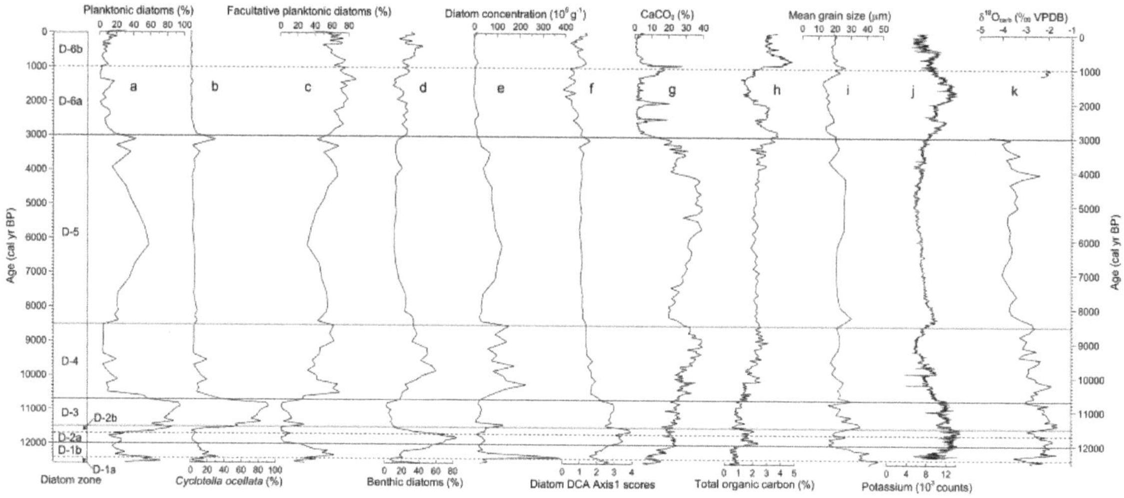

Figure 6: Comparison of diatom data with selected sedimentological and geochemical data (Francke et al., 2013) in Lake Dojran. (a)–(d) the relative abundance of planktonic diatoms, *C. ocellata*, facultative planktonic diatoms and benthic diatoms; (e) absolute diatom concentration; (f) diatom DCA Axis 1 scores; (g) and (h) carbonate and organic matter contents; (i) mean grain size; (j) K concentration; (k) carbonate oxygen stable isotope data.

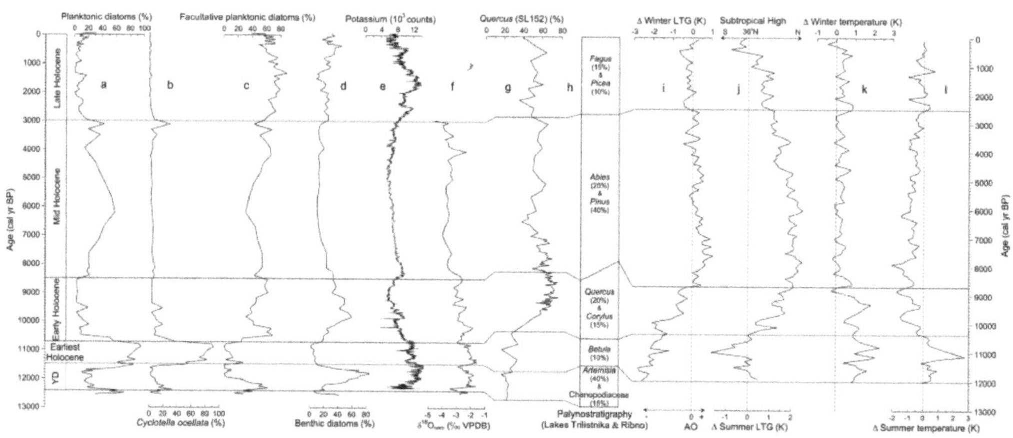

Figure 7: Comparison of Lake Dojran diatom data with key palaeoclimate data. (a)–(d) the relative abundance of planktonic diatoms, *C. ocellata*, facultative planktonic diatoms and benthic diatoms; (e) K concentration (Francke et al., 2013); (f) carbonate oxygen stable isotope data (Francke et al., 2013); (g) percentage of *Quercus* pollen in SL152 (northern Aegean Sea) (Kotthoff et al., 2008a and Kotthoff et al., 2008b); (h) palynostratigraphy with key pollen taxa and approximate average percentages from Lakes Trilistnika and Ribno (the Rila Mountain, southwestern Bulgaria) (Tonkov et al., 2008 and Tonkov et al., 2013); (i) and (j) the Holocene winter and summer latitudinal temperature gradient (LTG) between northern and southern Europe (Davis and Brewer, 2009); (k) and (l) the Holocene winter and summer temperature anomalies in southeastern Europe (Davis et al., 2003).

Younger Dryas (ca. 12,400–11,500 cal yr BP)

Francke et al. (2013) suggested that the presence of clay clasts, the occurrence of a 32,830 cal yr BP old shell fragment, and an undulated reflector in the hydro-acoustic data probably indicate a low lake level and redeposition between ca. 12,500–12,100 cal yr BP. This interpretation of a shallow state may be consistent with the relatively high abundance of benthic diatoms at the base of the sequence at ca. 12,500 cal yr BP. Subsequently in Subzone D-1a, even if the presence of eutrophic plankton may relate more to high productivity than lake level, the well-preserved diatom flora are important in indicating the presence of permanent water. A viable interpretation is that redeposition may occur, but that diatoms indicate lake refilling after desiccation, from a shallow state with the relatively high abundance of benthos to a relatively high lake level and eutrophic state with the dominance of plankton just above the core base. In Subzone D-1b (ca. 12,400–12,000 cal yr BP), diatom-inferred shallowing appears inconsistent with redeposition, since diatoms are well preserved, with a clear shift from the dominance of facultative planktonic taxa (50–65%) to increasing abundance of benthos, and peak concentration at the base of this subzone. Diatoms can be preserved during redeposition, as in the mass wasting deposit from Lake Ohrid (Zhang, unpublished data), but they would be present at extremely low concentration and with enhanced dissolution.

The diatom-inferred shallowing culminates in an extremely low lake level and eutrophic, oligosaline condition in the endorheic lake between ca. 12,000–11,500 cal yr BP (Zone D-2), indisputably interpreted as peak aridity of the Younger Dryas due to enhanced evaporative concentration.

Based on strong evidence for the presence of permanent water derived from hydro-acoustic data, grain-size composition and the absence of clay clasts, Francke et al. (2013) interpreted this zone as one of higher lake level than the shallow or even desiccated state at the base of the sequence. Since extremely shallow or even ephemeral lakes may be characterised by 'lacustrine' sediment (Reed, 1998a and Reed, 1998b), the data in conjunction indicate that this later phase (Zone D-2) can be interpreted as a stable and lacustrine state, but with the classic enhanced aridity of the Younger Dryas.

In the northeastern Mediterranean, there is growing palynological evidence for aridity during the Younger Dryas, commonly marked by a peak in steppic pollen taxa *Artemisia* and Chenopodiaceae throughout the altitudinal range (e.g. Lake Ribno, 2184 m a.s.l.,Tonkov et al., 2013; Lake Prespa, 849 m a.s.l., Panagiotopoulos et al., 2013; Valle di Castiglione, central Italy, 44 m a.s.l., Di Rita et al., 2013), and including a peak in *Ephedra* in marine records (e.g. Kotthoff et al., 2008a and Desprat et al., 2013). Pollen-based biome reconstructions (Allen et al., 2002 and Bordon et al., 2009) and quantitative temperature and/or precipitation reconstructions (e.g. Bordon et al., 2009 and Dormoy et al., 2009) also provide evidence of a cold, arid steppe environment. Our results are important in strengthening the sparse regional palaeohydrological dataset, which to date only comprises a distinct decrease in planktonic diatoms (particularly *C. ocellata*) in Lake Ioannina (Wilson et al., 2008), a peak of the eutrophic diatom species *A. granulata* and facultative planktonic *Staurosirella pinnata* (Ehrenberg) Williams & Round in Lake Prespa, Macedonia/Albania/Greece (Cvetkoska et al., 2014), a decrease in rubidium (Rb)/strontium (Sr) ratio in Lake Stymphalia, southern Greece (Heymann et al., 2013), and an increase in $\delta^{18}O_{carb}$ values and magnesium (Mg)/calcium (Ca) ratio in Lake Van, eastern Turkey (Wick et al., 2003 and Litt et al., 2009).

The Earliest Holocene (ca. 11,500–10,700 cal yr BP) (corresponding to the Preboreal Period)

Diatom-inferred high lake level and relatively low trophic state during the earliest Holocene (Zone D-3) strengthens the previously tentative interpretation of high $\delta^{18}O_{carb}$ values. Francke et al. (2013) suggested that evaporation was promoted during this period by large lake surface area that was accompanied by lake-level increase in such a flat-bottomed basin. The high lake level in Lake Dojran is in accord with abrupt isotopic depletion at the Younger Dryas–Holocene transition in Lake Ioannina, Lake Van and Eski Acıgöl (central Turkey) (Roberts et al., 2008 and references therein). It is also in accord with the inference of high humidity from increased Rb/Sr ratio in Lake Stymphalia (Heymann et al., 2013). However, it is not in complete agreement with vegetation development in the northeastern Mediterranean during the earliest Holocene. Non-steppic herb pollen increased rather than *Quercus* pollen in SL152 in the northern Aegean region at this time (Kotthoff et al., 2008a). Non-steppic herb pollen were abundant during this period although *Quercus* pollen increased rapidly at the onset of the Holocene in Nisi Fen (Lawson et al., 2005) and Lake Accesa (central Italy) (Drescher-Schneider et al., 2007). Non-steppic herb and steppe pollen were replaced by *Quercus* pollen gradually until its maximum around 10,500 cal yr BP in Lake Prespa (Panagiotopoulos et al., 2013) and Tenaghi Philippon (northeastern Greece) (Müller et al., 2011). Non-steppic herb pollen was dominant, with a gradual increase of *Quercus* pollen until ca. 10,500 cal yr BP in Eski Acıgöl (Roberts et al., 2001 and Turner et al., 2008) and MD04-2797 (Siculo-Tunisian Strait) (Desprat et al., 2013). Non-steppic herb pollen replaced steppe pollen gradually during this period, along with a slightly increasing trend in *Quercus* pollen in Lake Van (Litt et al., 2009). Non-steppic herb taxa were more important in these records than the percentage data implied, because they are mostly lower pollen producers than *Quercus* (Broström et al., 2008).

The high lake level in Lake Dojran and isotopic depletion in the northeastern Mediterranean suggest that the increase in humidity during the earliest Holocene would be attributed to increased precipitation, since Younger Dryas mountain glaciations did not develop widely in this region (Hughes et al., 2006 and Hughes, 2012) and meltwater input is an unlikely forcing function. However, the wide distribution of non-steppic herb taxa suggest that increased moisture availability was insufficient to support extensive forest development, since afforestation linked to soil development was asynchronous in this region. The limited increase in humidity is possibly the effect of high evaporation, corresponding to high pollen-inferred area-average summer and winter temperature in southeastern Europe (Davis et al., 2003) and high alkenone-inferred sea surface temperature (SST) at the beginning of the Holocene in the Mediterranean Sea (Abrantes et al., 2012). The comparison with the summer and winter LTG supports increased precipitation during the earliest Holocene. The negative (strong) winter LTG resulted in the negative phase of AO (Davis and Brewer, 2009), and promoted the penetration of more westerly moisture of North Atlantic origin in winter. The summer LTG suggests that the Subtropical High pressure was not strengthened and displaced northward (Davis and Brewer, 2009), and summer moisture was not scarce at this time. Thus the LTG and atmospheric moisture availability would contribute to the increase in precipitation during the earliest Holocene, modulated by high temperature-induced evaporation. Changes in catchment vegetation and erosion can have a major influence on lake hydrology and nutrient input and hence diatom composition (Fritz and Anderson, 2013). At the catchment scale, the steppe vegetation was replaced by birch forest at high altitudes (Lakes Ribno and Trilistnika, the Rila Mountain, southwestern Bulgaria) (Tonkov et al., 2008 and Tonkov et al., 2013) and by non-steppic herbs in the lowlands (SL152, northern Aegean Sea) (Kotthoff et al., 2008a). Despite increased precipitation and water inflow, the vegetation development restrained catchment erosion and nutrient input, which is supported by lower clastic input indicated by decreased K concentration (Francke et al., 2013). Together with the dilution effect of increased freshwater input, lower nutrient input resulted in the relatively low trophic level and in-lake productivity, which is supported by low organic matter content (Francke et al., 2013). In the northeastern Mediterranean, low trophic levels during the earliest Holocene were also indicated by pigment and diatom data in Lake Albano, central Italy (Guilizzoni et al., 2002). In all, increased precipitation was a major contributor to the high lake level in Lake Dojran and increased humidity in the northeastern Mediterranean during the earliest Holocene, although high evaporation reduced the effect of increased precipitation.

The Early Holocene (ca. 10,700–8,500 cal yr BP)

The diatom data show a low lake level and relatively high trophic state during the early Holocene (Zone D-4), which is not in accord with decreasing $\delta^{18}O_{carb}$ values in Lake Dojran. Francke et al. (2013) suggested that the littoral zone might extend during this phase. If valid, epiphytic and epipelic diatoms can be similarly facilitated, and they could reach the coring site through water mixing and/or sediment redistribution. However, water mixing would cause nutrients to become well distributed in the water column as well as in the pelagic zone, conflicting with the rather low relative abundance of planktonic taxa at the coring site; sediment disturbance would be unfavourable to the settlement and preservation of smaller, fragile valves, conflicting with the relatively high percentages of *A. pediculus* and *M. atomus*.

The diatom-inferred low lake level is also inconsistent with low bulk carbonate $\delta^{18}O$ values in Lake Pergusa (Sicily) (Sadori et al., 2008), authigenic carbonate $\delta^{18}O$ values in Lake Gölhisar (southwestern Turkey) (Eastwood et al., 2007) and ostracod $\delta^{18}O$ values in Lake Ioannina (Frogley et al., 2001). Roberts et al. (2008) suggested that the freshening of surface water in the eastern Mediterranean Sea in parallel with sapropel formation would affect the isotopic composition of precipitation during the early Holocene. However, the low lake level in Lake Dojran is in line with relatively high $\delta^{18}O$ values of authigenic carbonates in Lake Ohrid (Macedonia/Albania) (Leng et al., 2010), Lake Prespa (Leng et al., 2013) and Lake Van (Litt et al., 2009), and mollusc $\delta^{18}O$ values in Lake Frassino (northern Italy) (Baroni et al., 2006). The apparent discrepancy in isotope proxy data is probably due to the control of different hydroclimatic parameters (Jones and Roberts, 2008) and the influence of catchment factors on the specific hydrology of each lake (Leng and Marshall, 2004). The diatom-based inferences in Lake Dojran are also in accord with the inference of low lake levels from low Rb/Sr ratio in Lake Stymphalia (Heymann et al., 2013) and high Ca/titanium (Ti) ratio in Lake Iznik, northwest Turkey (Roeser et al., 2012) at this time.

Palynological data are also complex, with the occurrence of different ecological pollen groups in this region (e.g. Roberts et al., 2011b and Sadori, 2013) and even in the same record (e.g. Peyron et al., 2011 and Panagiotopoulos et al., 2013), which have been interpreted in different ways. The apparent discrepancy in the regional vegetation distribution was attributed to spatial and altitudinal variation in climatic factors (De Beaulieu et al., 2005 and Sadori, 2013), and Roberts et al. (2011b) suggested that increased seasonality of climatic factors is an important factor. Willis (1992a) invoked the distance from mountain refugia of different taxa, and Sadori et al. (2011) linked this to edaphic conditions and water retention capacity suitable for different plant growth. With respect to the combination of ecologically-incompatible pollen groups in the same record, Panagiotopoulos et al. (2013) attributed this to a more even distribution of annual precipitation, while Magny et al. (2013) invoked high seasonality of precipitation. Roberts et al. (2011b) cautioned against interpreting this pollen flora too closely in terms of modern climate analogues. Despite this complexity, the Lake Dojran diatom data are supported by pollen-based quantitative reconstructions of higher winter precipitation and lower or consistently low summer precipitation in SL152 (Dormoy et al., 2009), Lake Accesa (Peyron et al., 2011), Lake Pergusa (Magny et al., 2012 and Peyron et al., 2013) and MD04-2797 (Desprat et al., 2013). The low lake level and relatively high trophic state in Lake Dojran are probably driven by strong seasonal hydrological contrasts, and extreme summer aridity offset the effect of winter precipitation recharge. This is consistent with the wide distribution of *Quercus ilex* in the northern Aegean region rather than *Quercus* deciduous type (Kotthoff et al., 2008b), although moisture availability was sufficient to support tree growth.

The comparison with the summer and winter LTG supports this climatic interpretation as high seasonality. The negative (strong) winter LTG suggests that AO was in the negative phase (Davis and Brewer, 2009), the storm track was in a southerly path and moisture availability in winter was high during the early Holocene. The positive (weak) and increasing summer LTG suggests that the Subtropical High pressure was migrating northward (Davis and Brewer, 2009), blocking westerly moisture penetration in summer and leading to much drier summers than today. It is associated with the intensified African monsoon and the large number of lake records in the Sahara and Sahel (Lézine et al., 2011), and with sapropel formation in the eastern Mediterranean Sea since 10,800 cal yr BP (De Lange et al., 2008). Thus the LTG and high seasonality of moisture availability would contribute to the reduced lake level and increased nutrient level in Lake Dojran during the early Holocene.

At the catchment scale, the Lake Dojran diatom data coincide with extensive forest development, mainly comprising deciduous oak forest (*Quercus robur* and *Quercus cerris*) at high altitudes (Lakes Ribno and Trilistnika, southwestern Bulgaria) (Tonkov et al., 2008 and Tonkov et al., 2013) and evergreen oak forest (*Q. ilex*) in the lowlands (SL152, northern Aegean Sea) (Kotthoff et al., 2008a and Kotthoff et al., 2008b). According to the contrasting growth requirements of deciduous and evergreen oaks (Roberts et al., 2011b), it can be posited that the dense, thick vegetation would reduce runoff and erosion in the catchment throughout the year through soil absorption and retention, which may contribute to the low lake level, high shell abundance and lower K concentration during this phase (Francke et al., 2013). However, forest vegetation would enhance chemical weathering and nutrient supply through soil development, which may contribute to the relatively high trophic level in spite of the limited water inflow. This is supported by high diatom concentration, as well as higher organic matter and carbonate content (Francke et al., 2013), indicating high in-lake productivity. In all, the low lake level and relatively high trophic state in Lake Dojran may result climatically from high seasonality of precipitation and locally from dense forest development and limited, nutrient-rich catchment runoff.

The Mid Holocene (ca. 8,500–3,000 cal yr BP)

A relatively high lake level and maximum Holocene trophic level are inferred during the mid Holocene (Zone D-5). This is supported by decreased $\delta^{18}O_{carb}$ values. Francke et al. (2013) also discussed a relatively high lake level during this phase based on sedimentological data. The relatively high lake level in Lake Dojran is consistent with low $\delta^{18}O_{carb}$ values in Lake Prespa (Leng et al., 2013) and Lake Van (Litt et al., 2009), low mollusc $\delta^{18}O$ values in Lake Frassino (Baroni et al., 2006), and high lake levels in Lake Xinias, central Greece (Digerfeldt et al., 2007). However, in the northeastern Mediterranean, an aridification trend was shown by isotopic enrichment in Eski Acıgöl (Roberts and Jones, 2002), diatom succession in Lake Ioannina (Wilson et al., 2008 and Jones et al., 2013), and lithological changes in Lake Pergusa (Sadori and Narcisi, 2001). Sadori et al. (2011) and Roberts et al. (2011b) improved understanding of this aridification process after ca. 8,000 cal yr BP based on a regional synthesis of pollen and isotope data, respectively. The comparison with the summer and winter LTG does not give support to increased humidity during the mid Holocene. Both the summer and winter LTG were more positive (weaker) during this period (Davis and Brewer, 2009), and the positive phase of AO and the northern position of the Subtropical High pressure suggest that winter was not influenced by the northward-shifted storm track and summer was controlled mostly by the downdraught of dry air, respectively. This would lead to low winter precipitation and much drier summers than today, although several regional syntheses of temperature (Davis et al., 2003, Finné et al., 2011 and Abrantes et al., 2012) suggested cooling and decreased evaporation during the mid Holocene.

At the catchment scale, the relatively high lake level and high trophic state in Lake Dojran coincide with the dominance of coniferous forest (mainly firs) at high altitudes at ca. 8,500–7,800 cal yr BP (Lakes Ribno and Trilistnika, southwestern Bulgaria) (Tonkov et al., 2008 and Tonkov et al., 2013) and the opening of the oak forest with an increase in non-steppic herbs in the lowlands (SL152, northern Aegean Sea) (Kotthoff et al., 2008a and Kotthoff et al., 2008b). The reduction of forest density in the lowlands, most probably corresponding to the aridification process discussed above, was also clearly indicated by a distinct decrease in *Quercus* pollen concentration at ca. 8,400 cal yr BP (Kotthoff et al., 2008a), which would cause an increase in nutrient mobility. The expansion of firs at high altitudes is a regional signal across the southern Balkans, including Rezina Marsh (1800 m a.s.l.) at ca. 8,600 cal yr BP (Willis, 1992b). Fir trees developed in more humid and organic soils (Sadori et al., 2011), and the wetting of the highlands and forest development enhanced runoff and nutrient supply. In all, the wetting of the highlands and the drying of the lowlands and associated local vegetation succession enhanced catchment runoff and nutrient erosion, and resulted in the relatively high lake level and eutrophic state in Lake Dojran.

The Late Holocene (ca. 3,000–0 cal yr BP)

Despite low diatom concentration and high dissolution, a low lake level and high trophic state can be inferred during the late Holocene (Zone D-6). Late-Holocene aridity prevailed in the northeastern Mediterranean, indicated by isotope data (e.g. Roberts et al., 2008, Roberts et al., 2011b and Leng et al., 2013) and lake-level reconstructions (e.g. Harrison and Digerfeldt, 1993, Digerfeldt et al., 2007 and Magny et al., 2011). The comparison with the summer and winter LTG suggests that atmospheric moisture availability is not a determining factor, because the Subtropical High pressure regressed and AO index was slightly negative, which would not result in the obvious moisture deficit in this region. In contrast, the increase in regional temperature (Davis et al., 2003 and Finné et al., 2011) and high evaporation may contribute to the aridity during the late Holocene, although one model simulation suggests that anthropogenic deforestation may itself have caused a decrease in summer and winter temperature in southern Europe due to increased albedo effect (Strandberg et al., 2014).

A previous late-Holocene palynological study in the littoral zone of Lake Dojran revealed that the oak forest in the lowlands and the conifer and beech forests in the mountain region were replaced by herb vegetation and secondary trees at ca. 2,800 cal yr BP as a result of intensified human impact (Athanasiadis et al., 2000). At the catchment scale, beech and spruce forests expanded widely at high altitudes, and beech trees were favoured by human agricultural and stock-breeding activities and resultant soil deterioration (Marinova et al., 2012 and Tonkov et al., 2013). Non-steppic herbs expanded in the lowlands along with the further decline of the oak forest, and human disturbance was an important factor (Kotthoff et al., 2008a). Deforestation at mid-low altitudes during the late Holocene was also a regional event, for example, in Nisi Fen (475 m a.s.l.) the temperate forest declined at ca. 3,500 cal yr BP (Lawson et al., 2005 and Kotthoff et al., 2008a), and in Lake Ohrid (693 m a.s.l.) it was reflected by the reduction of the coniferous forest at ca. 2,500 cal yr BP (Wagner et al., 2009). Thus human impact is possibly an important factor not only leading to the high trophic state during the late Holocene by deforestation and enhanced nutrient erosion, but also contributing to the low lake level through agriculture and irrigation.

CONCLUSIONS

The Lake Dojran diatom data, supported by extant sedimentological and geochemical data from the same core and late-Holocene palynological data from a separate littoral core, give a new insight into Younger Dryas and Holocene changes in lake levels and trophic status in the northeastern Mediterranean. The Lake Dojran diatom data provide more robust evidence and strengthen previous lake-level interpretation based on sedimentological and geochemical data during the earliest, mid and late Holocene, and also clarify previous uncertainty in interpretation of Younger Dryas and early-Holocene lake-level change. Following a very shallow or even desiccated state at the core base at ca. 12,500 cal yr BP, indicated by sedimentological and hydro-acoustic data, diatoms indicate lake infilling, from a shallow state with abundant benthos to a plankton-dominated relatively high lake level and eutrophic state thereafter. Diatom-inferred shallowing between ca. 12,400–12,000 cal yr BP and a very low lake level and eutrophic, oligosaline state between ca. 12,000–11,500 cal yr BP provide clear evidence for aridity during the Younger Dryas. Although a slightly higher lake level was previously inferred during the second part of this period, the lacustrine state with permanent water does not conflict with the diatom-inferred salinity shift. Lake Dojran's water level increased markedly during the earliest Holocene (ca. 11,500–10,700 cal yr BP). A low lake level and relatively high trophic state are inferred during the early Holocene (ca. 10,700–8,500 cal yr BP), conflicting with the previous inference of increased humidity from decreasing $\delta^{18}O_{carb}$ values and sedimentological data. Lake Dojran was relatively deep and exhibited the maximum Holocene trophic level during the mid Holocene (ca. 8,500–3,000 cal yr BP), and it became shallow during the late Holocene (ca. 3,000–0 cal yr BP). Our results indicate that, being located at the juncture of the proposed boundaries between the west–east and north–south contrasting Holocene climatic domains, Lake Dojran cannot be classified simply into the western or eastern sector, or the northern or southern sector in the Mediterranean.

Our results are also important in disentangling regional climate effects from local catchment dynamics during the Holocene, and to this end we exploit extant regional palynological data for vegetation change in the highlands and lowlands. The importance of seasonality in driving Holocene climate change is assessed by reference to the summer and winter LTG model, linked to the Subtropical High pressure and AO, respectively. We suggest that diatom-inferred high lake level during the earliest Holocene was attributed to the increase in precipitation, in spite of high pollen-inferred, temperature-induced evaporation, and is coherent with high winter and summer atmospheric moisture availability inferred from the LTG. The relatively low trophic state at this time was possibly driven in part by vegetation development and reduced catchment erosion. The diatom-inferred early-Holocene lake-level reduction and increased trophic level may result climatically from high seasonality of precipitation, coherent with the contrasting summer and winter LTG, and locally from limited, nutrient-rich runoff in a densely-forested catchment. The relatively deep, eutrophic state in Lake Dojran during the mid Holocene shows strong affinity with palaeolimnological data from central Greece, northern Italy and eastern Turkey, but not with other records in the northeastern Mediterranean, where aridification is recognised. It is also not coherent with the LTG-inferred low atmospheric moisture availability. This may reflect local complexity of climate variability, but may also indicate that changes were driven more by local vegetation succession and associated changes in catchment processes than by climate change. During the late Holocene, diatom-inferred shallow and high trophic state is consistent with strong regional evidence for temperature-induced aridity, coupled with the influence of intensified human land use.

Overall, our study is important in strengthening existing multi-proxy interpretation of Lateglacial and Holocene palaeohydrology and associated shifts in nutrient status. This study improves understanding of Younger Dryas and Holocene climate change in the northeastern Mediterranean, providing a coherent interpretation which suggests the important role of the LTG on moisture availability during the Holocene and clarifies the influence of catchment processes on Holocene hydrological variability.

ACKNOWLEDGEMENTS

The study was funded by a University of Hull-China Scholarship Council (CSC) PhD Scholarship to X.S. Zhang. It is part of the project Scientific Collaboration on Past Speciation Conditions in Lake Ohrid (SCOPSCO), supported by the German Research Foundation (DFG) and the International Continental Scientific Drilling Program (ICDP). As part of SCOPSCO, this project contributes to the successful bid for UK's ICDP membership. The support of the UK's Natural Environment Research Council (NERC), via the British Geological Survey (BGS), is gratefully acknowledged in their direct contribution to funding of SCOPSCO drilling projects and in their investment in the UK's ICDP membership. We would like to thank Anne Böhm, Daniel Treu, Frank Schäbitz and Konstantinos Panagiotopoulos for assistance with fieldwork. Thanks are also due to Aleksandra Cvetkoska, Elena Jovanovska and Aleksandar Pavlov for fruitful discussions of diatom taxonomy. Neil Roberts and two anonymous reviewers are also thanked for providing detailed and constructive comments, suggestions and improvements on the manuscript.

REFERENCES

1. Abrantes, F., Voelker, A., Sierro, F.J., Naughton, F., Rodrigues, T., Cacho, I., Ariztegui, D., Brayshaw, D., Sicre, M.A., Batista, L., 2012. Paleoclimate variability in the Mediterranean region. In: Lionello, P. (Ed.), The Climate of the Mediterranean Region: from the Past to the Future. Elsevier, Amsterdam, pp. 1e86.
2. Allen, J.R.M., Watts, W.A., McGee, E., Huntley, B., 2002. Holocene environmental variabilityethe record from Lago Grande di Monticchio, Italy. Quat. Int. 88, 69e80.
3. Anderson, N.J., 2000. Diatoms, temperature and climatic change. Eur. J. Phycol. 35, 307e314.
4. Athanasiadis, N., Tonkov, S., Atanassova, J., Bozilova, E., 2000. Palynological study of Holocene sediments from Lake Doirani in northern Greece. J. Paleolimnol. 24, 331e342.
5. Aufgebauer, A., Panagiotopoulos, K., Wagner, B., Schaebitz, F., Viehberg, F.A., Vogel, H., Zanchetta, G., Sulpizio, R., Leng, M.J., Damaschke, M., 2012. Climate and environmental change in the Balkans over the last 17 ka recorded in sediments from Lake Prespa (Albania/ F.Y.R. of Macedonia/Greece). Quat. Int. 274, 122e135.
6. Bailey-Watts, A.E., 1986. The ecology of planktonic diatoms, especially Fragilaria crotonensis, associated with artificial mixing of a small Scottish loch in summer. Diatom Res. 1, 153e168.
7. Baroni, C., Zanchetta, G., Fallick, A.E., Longinelli, A., 2006. Mollusca stable isotope record of a core from Lake Frassino, northern Italy: hydrological and climatic changes during the last 14 ka. Holocene 16, 827e837.
8. Battarbee, R.W., Jones, V.J., Flower, R.J., Cameron, N.G., Bennion, H., 2001. Diatoms. In: Smol, J.P., Birks, H.J.B., Last, W.M. (Eds.), Tracking Environmental Change Using Lake Sediments, Terrestrial, Algal, and Siliceous Indicators, vol. 3. Kluwer Academic Publishers, Dordrecht, pp. 155e202.
9. Bennion, H., 1995. Surface-sediment diatom assemblages in shallow, artificial, enriched ponds, and implications for reconstructing trophic status. Diatom Res. 10, 1e19.
10. Bennion, H., Appleby, P.G., Phillips, G.L., 2001. Reconstructing nutrient histories in the Norfolk Broads, UK: implications for the role of diatom-total phosphorus transfer functions in shallow lake management. J. Paleolimnol. 26, 181e204.
11. Bennion, H., Sayer, C.D., Tibby, J., Carrick, H.J., 2010. Diatoms as indicators of environmental change in shallow lakes. In: Smol, J.P., Stoermer, E.F. (Eds.), The Diatoms: Applications for the Environmental and

Earth Sciences, second ed. Cambridge University Press, Cambridge, pp. 152e173.

12. Bennion, H., Carvalho, L., Sayer, C.D., Simpson, G.L., Wischnewski, J., 2012. Identifying from recent sediment records the effects of nutrients and climate on diatom dynamics in Loch Leven. Freshw. Biol. 57, 2015e2029.
13. Bordon, A., Peyron, O., Lezine, A.M., Brewer, S., Fouache, E., 2009. Impact of Late- glacial cold events on the northern Aegean region reconstructed from marine and terrestrial proxy data. Quat. Int. 200, 19e30.
14. Brayshaw, D.J., Rambeau, C.M.C., Smith, S.J., 2011. Changes in Mediterranean climate during the Holocene: insights from global and regional climate modelling. Holocene 21, 15e31.
15. Brewer, S., Guiot, J., Torre, F., 2007. Mid-Holocene climate change in Europe: a datamodel comparison. Clim. Past 3, 499e512.
16. Brostrom, A., Nielsen, A.B., Gaillard, M.J., Hjelle, K., Mazier, F., Binney, H., Bunting, J., € Fyfe, R., Meltsov, V., Poska, A., R€ as€ anen, S., Soepboer, W., von Stedingk, H., Suutari, H., Sugita, S., 2008. Pollen productivity estimates of key European plant taxa for quantitative reconstruction of past vegetation: a review. Veget. Hist. Archaeobot. 17, 461e478.
17. Combourieu-Nebout, N., Peyron, O., Bout-Roumazeilles, V., Goring, S., Dormoy, I., Joannin, S., Sadori, L., Siani, G., Magny, M., 2013. Holocene vegetation and climate changes in the central Mediterranean inferred from a high-resolution marine pollen record (Adriatic Sea). Clim. Past 9, 2023e2042.
18. Cruces, F., Rivera, P., Urrutia, R., 2010. Observations and comments on the diatom Stephanodiscus minutulus (Kützing) Cleve & Moller (Bacillariophyceae) found € for the first time in Chile from bottom sediments collected in Lake Laja. Gayana Bot. 67, 12e18.
19. Cvetkoska, A., Levkov, Z., Reed, J.M., Wagner, B., 2014. Late glacial to Holocene climate change and human impact in the Mediterranean: the last ca. 17 ka diatom record of Lake Prespa (Macedonia/Albania/Greece). Paleogeogr. Paleoclimatol. Paleoecol. 406, 22e32.
20. Davis, B.A.S., Brewer, S., Stevenson, A.C., Guiot, J., Data Contributors, 2003. The temperature of Europe during the Holocene reconstructed from pollen data. Quat. Sci. Rev. 22, 1701e1716.
21. Davis, B.A.S., Stevenson, A.C., 2007. The 8.2 ka event and EarlyeMid Holocene forests, fires and flooding in the Central Ebro Desert, NE Spain. Quat. Sci. Rev. 26, 1695e1712.
22. Davis, B.A.S., Brewer, S., 2009. Orbital forcing and role of the latitudinal insolation/ temperature gradient. Clim. Dyn. 32, 143e165.
23. De Lange, G.J., Thomson, J., Reitz, A., Slomp, C.P., Principato, M.S., Erba, E., Corselli, C., 2008. Synchronous basin-wide formation and redox-controlled preservation of a Mediterranean sapropel. Nat. Geosci. 1, 606e610.
24. De Beaulieu, J.L., Miras, Y., Andrieu-Ponel, V., Guiter, F., 2005. Vegetation dynamics in north-western Mediterranean regions: instability of the Mediterranean bioclimate. Plant Biosyst. 139, 114e126.
25. Desprat, S., Combourieu-Nebout, N., Essallami, L., Sicre, M.A., Dormoy, I., Peyron, O., Siani, G., Bout Roumazeilles, V., Turon, J.L., 2013. Deglacial and Holocene vegetation and climatic changes in the southern Central Mediterranean from a direct landesea correlation. Clim. Past 9, 767e787.
26. Di Rita, F., Anzidei, A.P., Magri, D., 2013. A Lateglacial and early Holocene pollen record from Valle di Castiglione (Rome): vegetation dynamics and climate implications. Quat. Int. 288, 73e80.
27. Digerfeldt, G., Sandgren, P., Olsson, S., 2007. Reconstruction of Holocene lake-level changes in Lake Xinias, central Greece. Holocene 17, 361e367.
28. Dormoy, I., Peyron, O., Combourieu Nebout, N., Goring, S., Kotthoff, U., Magny, M., Pross, J., 2009. Terrestrial climate variability and seasonality changes in the Mediterranean region between 15 000 and 4000 years BP deduced from marine pollen records. Clim. Past 5, 615e632.
29. Drescher-Schneider, R., de Beaulieu, J.L., Magny, M., Walter-Simonnet, A.V., Bossuet, G., Millet, L., Brugiapaglia, E., Drescher, A., 2007. Vegetation history, climate and human impact over the last 15,000 years at Lago dell'Accesa (Tuscany, Central Italy). Veg. Hist. Archaeobot. 16, 279e299.
30. Eastwood, W.J., Leng, M.J., Roberts, N., Davis, B., 2007. Holocene climate change in the eastern Mediterranean region: a comparison of stable isotope and pollen data from Lake Golhisar, southwest Turkey. J. Quat. Sci. 22, 327 € e341.
31. Felis, T., Rimbu, N., 2010.Mediterranean climate variability documentedin oxygenisotope records from northern Red Sea coralsea review. Glob. Planet. Change 71, 232e241.
32. Finne, M., Holmgren, K., Sundqvist, H.S., Weiberg, E., Lindblom, M., 2011. Climate in the eastern Mediterranean, and adjacent regions, during the past 6000 yearsea review. J. Archaeol. Sci. 38, 3153e3173.
33. Fletcher, W.J., Debret, M., Sanchez Goni, M.F., 2013. Mid-Holocene emergence of a ~ low-frequency millennial oscillation in western Mediterranean climate: implications for past dynamics of the North Atlantic atmospheric westerlies. Holocene 23, 153e166.
34. Fourtanier, E., Kociolek, J.P., 2011. Catalogue of Diatom Names. On-line Version (updated 19 Sep 2011). California Academy of Sciences, San Francisco.
35. Francke, A., Wagner, B., Leng, M.J., Rethemeyer, J., 2013. A Late Glacial to Holocene record of environmental change from Lake Dojran (Macedonia, Greece). Clim. Past 9, 481e498.
36. Fritz, S.C., 2008. Deciphering climatic history from lake sediments. J. Paleolimnol. 39, 5e16.
37. Fritz, S.C., Juggins, S., Battarbee, R.W., 1993. Diatom assemblages and ionic characterization of lakes of the northern Great Plains, North America: a tool for reconstructing past salinity and climate fluctuations.

Can. J. Fish. Aquat. Sci. 50, 1844e1856.

38. Fritz, S.C., Anderson, N.J., 2013. The relative influences of climate and catchment processes on Holocene lake development in glaciated regions. J. Paleolimnol. 49, 349e362.

39. Frogley, M.R., Griffiths, H.I., Heaton, T.H.E., 2001. Historical biogeography and Late Quaternary environmental change of Lake Pamvotis, Ioannina (north-western Greece): evidence from ostracods. J. Biogeogr. 28, 745e756.

40. Gasse, F., Barker, P., Gell, P.A., Fritz, S.C., Chalie, F., 1997. Diatom-inferred salinity in palaeolakes: an indirect tracer of climate change. Quat. Sci. Rev. 16, 547e563.

41. Gasse, F., 2002. Diatom-inferred salinity and carbonate oxygen isotopes in Holocene waterbodies of western Sahara and Sahel (Africa). Quat. Sci. Rev. 21, 737e767.

42. Giorgi, F., 2006. Climate change hot-spots. Geophys. Res. Lett. 33, L08707.

43. Gladstone, R.M., Ross, I., Valdes, P.J., Abe-Ouchi, A., Braconnot, P., Brewer, S., Kageyama, M., Kitoh, A., Legrande, A., Marti, O., Ohgaito, R., Otto-Bliesner, B., Peltier, W.R., Vettoretti, G., 2005. Mid-Holocene NAO: a PMIP2 model intercomparison. Geophys. Res. Lett. 32, L16707.

44. Guilizzoni, P., Lami, A., Marchetto, A., Jones, V., Manca, M., Bettinetti, R., 2002. Palaeoproductivity and environmental changes during the Holocene in central Italy as recorded in two crater lakes (Albano and Nemi). Quat. Int. 88, 57e68.

45. Griffiths, H.I., Reed, J.M., Leng, M.J., Ryan, S., Petkovski, S., 2002. The recent palaeoecology and conservation status of Balkan Lake Dojran. Biol. Conserv. 104, 35e49.

46. Grimm, E.C., 2011. Tilia Version 1.7.16. Illinois State Museum, Springfield.

47. Hall, R.I., Smol, J.P., 2010. Diatoms as indicators of lake eutrophication. In: Smol, J.P., Stoermer, E.F. (Eds.), The Diatoms: Applications for the Environmental and Earth Sciences, second ed. Cambridge University Press, Cambridge, pp. 122e151.

48. Harrison, S.P., Digerfeldt, G., 1993. European lakes as palaeohydrological and palaeoclimatic indicators. Quat. Sci. Rev. 12, 233e248.

49. Heymann, C., Nelle, O., Dor€ fler, W., Zagana, H., Nowaczyk, N., Xue, J., Unkel, I., 2013. Late Glacial to mid-Holocene palaeoclimate development of Southern Greece inferred from the sediment sequence of Lake Stymphalia (NE-Peloponnese). Quat. Int. 302, 42e60.

50. Hobbs, W.O., Fritz, S.C., Stone, J.R., Donovan, J.J., Grimm, E.C., Almendinger, E., 2011. Environmental history of a closed-basin lake in the US Great Plains: diatom response to variations in groundwater flow regimes over the last 8500 cal. yr BP. Holocene 21, 1203e1216.

51. Hughes, P.D., Woodward, J.C., Gibbard, P.L., 2006. Quaternary glacial history of the Mediterranean mountains. Prog. Phys. Geogr. 30, 334e364.

52. Hughes, P.D., 2012. Glacial history. In: Vogiatzakis, I.N. (Ed.), Mediterranean Mountain Environments. Wiley-Blackwell, Oxford, pp. 35e63.

53. Jones, M.D., Roberts, C.N., Leng, M.J., Türkes¸ , M., 2006. A high-resolution late Holocene lake isotope record from Turkey and links to North Atlantic and monsoon climate. Geology 34, 361e364.

54. Jones, M.D., Roberts, C.N., 2008. Interpreting lake isotope records of Holocene environmental change in the Eastern Mediterranean. Quat. Int. 181, 32e38.

55. Jones, T.D., Lawson, I.T., Reed, J.M., Wilson, G.P., Leng, M.J., Gierga, M., Bernasconi, S.M., Smittenberg, R.H., Hajdas, I., Bryant, C.L., Tzedakis, P.C., 2013. Diatom-inferred late Pleistocene and Holocene palaeolimnological changes in the Ioannina basin, northwest Greece. J. Paleolimnol. 49, 185e204.

56. Juggins, S., 2001. The European Diatom Database: User Guide. University of Newcastle, Newcastle.

57. Juggins, S., 2007. C2 Version 1.5 User Guide. Software for Ecological and Palaeoecological Data Analysis and Visualisation. University of Newcastle, Newcastle.

58. Juggins, S., 2013. Quantitative reconstructions in palaeolimnology: new paradigm or sick science? Quat. Sci. Rev. 64, 20e32.

59. Juggins, S., Birks, H.J.B., 2012. Quantitative environmental reconstructions from biological data. In: Birks, H.J.B., Lotter, A.F., Juggins, S., Smol, J.P. (Eds.), Tracking Environmental Change Using Lake Sediments, Data Handling and Numerical Techniques, vol. 5. Springer, Dordrecht, pp. 431e494.

60. Juggins, S., Anderson, N.J., Ramstack Hobbs, J.M., Heathcote, A.J., 2013. Reconstructing epilimnetic total phosphorus using diatoms: statistical and ecological constraints. J. Paleolimnol. 49, 373e390.

61. Kilham, S.S., Theriot, E.C., Fritz, S.C., 1996. Linking planktonic diatoms and climate change in the large lakes of the Yellowstone ecosystem using resource theory. Limnol. Oceanogr. 41, 1052e1062.

62. Kotthoff, U., Müller, U.C., Pross, J., Schmiedl, G., Lawson, I.T., van de Schootbrugge, B., Schulz, H., 2008a. Lateglacial and Holocene vegetation dynamics in the Aegean region: an integrated view based on pollen data from marine and terrestrial archives. Holocene 18, 1019e1032.

63. Kotthoff, U., Pross, J., Müller, U.C., Peyron, O., Schmiedl, G., Schulz, H., Bordon, A., 2008b. Climate dynamics in the borderlands of the Aegean Sea during formation of sapropel S1 deduced from a marine pollen record. Quat. Sci. Rev. 27, 832e845.

64. Kotthoff, U., Koutsodendris, A., Pross, J., Schmiedl, G., Bornemann, A., Kaul, C., Marino, G., Peyron, O., Schiebel, R., 2011. Impact of Lateglacial cold events on the northern Aegean region reconstructed from marine and terrestrial proxy data. J. Quat. Sci. 26, 86e96.

65. Krammer, K., Lange-Bertalot, H., 1986. Süsswasserflora von Mitteleuropa. Bacillariophyceae. 1. Teil: Naviculaceae, vol. 2/1. Gustav Fischer Verlag, Stuttgart.

66. Krammer, K., Lange-Bertalot, H., 1988. Süsswasserflora von Mitteleuropa. Bacillariophyceae. 2. Teil:

Epithemiaceae, Bacillariaceae, Surirellaceae, vol. 2/2. Gustav Fischer Verlag, Stuttgart.

67. Krammer, K., Lange-Bertalot, H., 1991a. Süsswasserflora von Mitteleuropa. Bacillariophyceae. 3. Teil: Centrales, Fragilariaceae, Eunotiaceae, vol. 2/3. Gustav Fischer Verlag, Stuttgart.

68. Krammer, K., Lange-Bertalot, H., 1991b. Süsswasserflora von Mitteleuropa. Bacillariophyceae. 4. Teil: Achnanthaceae, vol. 2/4. Gustav Fischer Verlag, Stuttgart.

69. Krammer, K., 2002. Cymbella. In: Lange-Bertalot, H. (Ed.), Diatoms of Europe, Diatoms of the European Inland Waters and Comparable Habitats, vol. 3. A.R.G. Gantner Verlag, Ruggell.

70. Lamy, F., Arz, H.W., Bond, G.C., Bahr, A., Patzold, J., 2006. Multicentennial-scale € hydrological changes in the Black Sea and northern Red Sea during the Holocene and the Arctic/North Atlantic Oscillation. Paleoceanography 21, PA1008.

71. Lange-Bertalot, H., 2001. Navicula sensu stricto, 10 genera separated from Navicula sensu lato, Frustulia. In: Lange-Bertalot, H. (Ed.), Diatoms of Europe, Diatoms of the European Inland Waters and Comparable Habitats, vol. 2. A.R.G. Gantner Verlag, Ruggell.

72. Lawson, I., Frogley, M., Bryant, C., Preece, R., Tzedakis, P., 2004. The Lateglacial and Holocene environmental history of the Ioannina basin, north-west Greece. Quat. Sci. Rev. 23, 1599e1625.

73. Lawson, I.T., Al-Omari, S., Tzedakis, P.C., Bryant, C.L., Christanis, K., 2005. Lateglacial and Holocene vegetation history at Nisi Fen and the Boras mountains, northern Greece. Holocene 15, 873e887.

74. Leng, M.J., Marshall, J.D., 2004. Palaeoclimate interpretation of stable isotope data from lake sediment archives. Quat. Sci. Rev. 23, 811e831.

75. Leng, M.J., Baneschi, I., Zanchetta, G., Jex, C.N., Wagner, B., Vogel, H., 2010. Late Quaternary palaeoenvironmental reconstruction from Lakes Ohrid and Prespa (Macedonia/Albania border) using stable isotopes. Biogeosciences 7, 3109e3122.

76. Leng, M.J., Wagner, B., Boehm, A., Panagiotopoulos, K., Vane, C.H., Snelling, A., Haidon, C., Woodley, E., Vogel, H., Zanchetta, G., Baneschi, I., 2013. Understanding past climatic and hydrological variability in the Mediterranean from Lake Prespa sediment isotope and geochemical record over the Last Glacial cycle. Quat. Sci. Rev. 66, 123e136. Leps, J., Smilauer, P., 2003. Multivariate Analysis of Ecological Data Using CANOCO. Cambridge University Press, Cambridge.

77. Lesoski, J., Zdraveski, N., Krstic, S., 2010. Preliminary results on cyanobacteria survey on Dojran Lakeethe beginning of revealing of the ultimate truth about the lake›s water quality. In: BALWOIS 2010, Ohrid.

78. Levkov, Z., Krstic, S., Metzeltin, D., Nakov, T., 2007. Diatoms of Lakes Prespa and Ohrid. In: Lange-Bertalot, H. (Ed.), Iconographia Diatomologica, vol. 16. A.R.G. Gantner Verlag, Ruggell.

79. Levkov, Z., 2009. Amphora sensu lato. In: Lange-Bertalot, H. (Ed.), Diatoms of Europe, Diatoms of the European Inland Waters and Comparable Habitats, vol. 5. A.R.G. Gantner Verlag, Ruggell.

80. Lezine, A.M., H ely, C., Grenier, C., Braconnot, P., Krinner, G., 2011. Sahara and sahel vulnerability to climate changes, lessons from Holocene hydrological data. Quat. Sci. Rev. 30, 3001e3012.

81. Lionello, P., Malanotte-Rizzoli, P., Boscolo, R., 2006. Mediterranean Climate Variability. Elsevier, Amsterdam.

82. Litt, T., Krastel, S., Sturm, M., Kipfer, R., Orcen, S., Heumann, G., Franz, S.O., € Ülgen, U.B., Niessen, F., 2009. 'PALEOVAN', International Continental Scientific Drilling Program (ICDP): site survey results and perspectives. Quat. Sci. Rev. 28, 1555e1567.

83. Lokoska, L., Jordanoski, M., Veljanoska-Sarafiloska, E., Tasevska, O., 2006. Water quality of Lake Dojran from biological and physical-chemical aspects. In: BALWOIS 2006, Ohrid.

84. Lotter, A.F., Birks, H.J.B., Hofmann, W., Marchetto, A., 1998. Modern diatom, cladocera, chironomid, and chrysophyte cyst assemblages as quantitative indicators for the reconstruction of past environmental conditions in the Alps. II. nutrients. J. Paleolimnol. 19, 443e463.

85. Magny, M., Vanniere, B., Calo, C., Millet, L., Leroux, A., Peyron, O., Zanchetta, G., La Mantia, T., Tinner, W., 2011. Holocene hydrological changes in south-western Mediterranean as recorded by lake-level fluctuations at Lago Preola, a coastal lake in southern Sicily, Italy. Quat. Sci. Rev. 30, 2459e2475.

86. Magny, M., Peyron, O., Sadori, L., Ortu, E., Zanchetta, G., Vanniere, B., Tinner, W., 2012. Contrasting patterns of precipitation seasonality during the Holocene in the south- and north-central Mediterranean. J. Quat. Sci. 27, 290e296.

87. Magny, M., Combourieu-Nebout, N., de Beaulieu, J.L., Bout-Roumazeilles, V., Colombaroli, D., Desprat, S., Francke, A., Joannin, S., Ortu, E., Peyron, O., Revel, M., Sadori, L., Siani, G., Sicre, M.A., Samartin, S., Simonneau, A., Tinner, W., Vanniere, B., Wagner, B., Zanchetta, G., Anselmetti, F., Brugiapaglia, E., Chapron, E., Debret, M., Desmet, M., Didier, J., Essallami, L., Galop, D., Gilli, A., Haas, J.N., Kallel, N., Millet, L., Stock, A., Turon, J.L., Wirth, S., 2013. Northesouth palaeohydrological contrasts in the central Mediterranean during the Holocene: tentative synthesis and working hypotheses. Clim. Past 9, 2043e2071.

88. Marinova, E., Tonkov, S., Bozilova, E., Vajsov, I., 2012. Holocene anthropogenic landscapes in the Balkans: the palaeobotanical evidence from southwestern Bulgaria. Veg. Hist. Archaeobot. 21, 413e427.

89. Martinson, D., Maslowski, W., Thompson, D., Wallace, J.M., 2000. Warm Phase/Cool Phase. National Geographic Magazine, Tampa.

90. Mauri, A., Davis, B.A.S., Collins, P.M., Kaplan, J.O., 2013. The influence of atmospheric circulation on the mid-Holocene climate of Europe: a data-model comparison. Clim. Past. Discuss. 9, 5569e5592.

91. Mittermeier, R.A., Gil, P.R., Hoffmann, M., Pilgrim, J., Brooks, T., Mittermeier, C.G., Lamoreux, J., da Fonseca,

G.A.B., 2004. Hotspots Revisited: Earth›s Biologically Richest and Most Endangered Terrestrial Ecoregions. Conservation International, Washington.

92. Müller, U.C., Pross, J., Tzedakis, P.C., Gamble, C., Kotthoff, U., Schmiedl, G., Wulf, S., Christanis, K., 2011. The role of climate in the spread of modern humans into Europe. Quat. Sci. Rev. 30, 273e279.

93. Myers, N., Mittermeier, R.A., Mittermeier, C.G., da Fonseca, G.A.B., Kent, J., 2000. Biodiversity hotspots for conservation priorities. Nature 403, 853e858.

94. Panagiotopoulos, K., Aufgebauer, A., Schabitz, F., Wagner, B., 2013. Vegetation and € climate history of the Lake Prespa region since the Lateglacial. Quat. Int. 293, 157e169.

95. Peyron, O., Goring, S., Dormoy, I., Kotthoff, U., Pross, J., de Beaulieu, J.L., DrescherSchneider, R., Vanniere, B., Magny, M., 2011. Holocene seasonality changes in the central Mediterranean region reconstructed from the pollen sequences of Lake Accesa (Italy) and Tenaghi Philippon (Greece). Holocene 21, 131e146.

96. Peyron, O., Magny, M., Goring, S., Joannin, S., de Beaulieu, J.L., Brugiapaglia, E., Sadori, L., Garfi, G., Kouli, K., Ioakim, C., Combourieu-Nebout, N., 2013. Contrasting patterns of climatic changes during the Holocene across the Italian Peninsula reconstructed from pollen data. Clim. Past 9, 1233e1252.

97. Popovska, C., Bonacci, O., 2008. Ecohydrology of Dojran Lake. In: Hlavinek, P., Bonacci, O., Marsalek, J., Mahrikova, I. (Eds.), Dangerous Pollutants (Xenobiotics) in Urban Water Cycle. Springer, Dordrecht, pp. 151e160.

98. Reed, J.M., 1998a. A diatom-conductivity transfer function for Spanish salt lakes. J. Paleolimnol. 19, 399e416.

99. Reed, J.M., 1998b. Diatom preservation in the recent sediment record of Spanish saline lakes: implications for palaeoclimate study. J. Paleolimnol. 19, 129e137.

100. Reed, J.M., Mesquita-Joanes, F., Griffiths, H.I., 2012. Multi-indicator conductivity transfer functions for Quaternary palaeoclimate reconstruction. J. Paleolimnol. 47, 251e275.

101. Reimer, P.J., Bard, E., Bayliss, A., Beck, J.W., Blackwell, P.G., Bronk Ramsey, C., Buck, C.E., Cheng, H., Edwards, R.L., Friedrich, M., Grootes, P.M., Guilderson, T.P., Haflidason, H., Hajdas, I., Hatte, C., Heaton, T.J., Hogg, A.G., Hughen, K.A., Kaiser, K.F., Kromer, B., Manning, S.W., Niu, M., Reimer, R.W., Richards, D.A., Scott, E.M., Southon, J.R., Turney, C.S.M., van der Plicht, J., 2013. IntCal13 and Marine13 radiocarbon age calibration curves 0e50,000 years cal BP. Radiocarbon 55, 1869e1887.

102. Roberts, N., Jones, M., 2002. Towards a regional synthesis of Mediterranean climatic change using lake stable isotope records. PAGES News 10, 13e15.

103. Roberts, N., Reed, J.M., Leng, M.J., Kuzucuoglu, C., Fontugne, M., Bertaux, J., Woldring, H., Bottema, S., Black, S., Hunt, E., Karabıyıkoglu, M., 2001. The tempo of Holocene climatic change in the eastern Mediterranean region: new highresolution crater-lake sediment data from central Turkey. Holocene 11, 721e736.

104. Roberts, N., Jones, M.D., Benkaddour, A., Eastwood, W.J., Filippi, M.L., Frogley, M.R., Lamb, H.F., Leng, M.J., Reed, J.M., Stein, M., Stevens, L., Valero-Garces, B., Zanchetta, G., 2008. Stable isotope records of Late Quaternary climate and hydrology from Mediterranean lakes: the ISOMED synthesis. Quat. Sci. Rev. 27, 2426e2441.

105. Roberts, N., Brayshaw, D., Kuzucuoglu, C., Perez, R., Sadori, L., 2011a. The mid- Holocene climatic transition in the Mediterranean: causes and consequences. Holocene 21, 3e13.

106. Roberts, N., Eastwood, W.J., Kuzucuoglu, C., Fiorentino, G., Caracuta, V., 2011b. Cli- matic, vegetation and cultural change in the eastern Mediterranean during the mid-Holocene environmental transition. Holocene 21, 147e162.

107. Roberts, N., Moreno, A., Valero-Garces, B.L., Corella, J.P., Jones, M., Allcock, S., Woodbridge, J., Morellon, M., Luterbacher, J., Xoplaki, E., Türkes ¸ , M., 2012. Palaeolimnological evidence for an eastewest climate see-saw in the Mediterranean since AD 900. Glob. Planet. Change 84e85, 23e34.

108. Roeser, P.A., Franz, S.O., Litt, T., Ülgen, U.B., Hilgers, A., Wulf, S., Wennrich, V., On, S.A., Viehberg, F.A., Ça € gatay, M.N., Melles, M., 2012. Lithostratigraphic and geochronological framework for the paleoenvironmental reconstruction of the last ~36 ka cal BP from a sediment record from Lake Iznik (NW Turkey). Quat. Int. 274, 73e87.

109. Sadori, L., Narcisi, B., 2001. The Postglacial record of environmental history from Lago di Pergusa, Sicily. Holocene 11, 655e670.

110. Sadori, L., Zanchetta, G., Giardini, M., 2008. Last Glacial to Holocene palaeoenvironmental evolution at Lago di Pergusa (Sicily, Southern Italy) as inferred by pollen, microcharcoal, and stable isotopes. Quat. Int. 181, 4e14.

111. Sadori, L., Jahns, S., Peyron, O., 2011. Mid-Holocene vegetation history of the central Mediterranean. Holocene 21, 117e129.

112. Sadori, L., 2013. Pollen records, postglacial, southern Europe. In: Elias, S.A. (Ed.), Encyclopedia of Quaternary Science, second ed. Elsevier, Amsterdam, pp. 179e188.

113. Saros, J.E., Fritz, S.C., 2000. Nutrients as a link between ionic concentration/ composition and diatom distributions in saline lakes. J. Paleolimnol. 23, 449e453.

114. Saros, J.E., Michel, T.J., Interlandi, S.J., Wolfe, A.P., 2005. Resource requirements of Asterionella formosa and Fragilaria crotonensis in oligotrophic alpine lakes: implications for recent phytoplankton community reorganizations. Can. J. Fish. Aquat. Sci. 62, 1681e1689.

115. Sayer, C.D., 2001. Problems with the application of diatom-total phosphorus transfer functions: examples from a shallow English lake. Freshw. Biol. 46, 743e757.

116. Sayer, C.D., Davidson, T.A., Jones, J.I., Langdon, P.G., 2010. Combining contemporary ecology and palaeolimnology to understand shallow lake ecosystem change. Freshw. Biol. 55, 487e499.

117. Scheffler, W., Morabito, G., 2003. Topical observations on centric diatoms (Bacillariophyceae, Centrales) of Lake Como (N. Italy). J. Limnol. 62, 47e60.
118. Sotiria, K., Petkovski, S., 2004. Lake Dojranean Overview of the Current Situation. Greek Biotope/ Wetland Centre (EKBY), Society for the Investigation and Conservation of Biodiversity and the Sustainable Development of Natural Ecosystems (BIOECO), Thermi.
119. Stojov, V., 2012. Hydrological state of Dojran Lake related to tectonic, climatic and human impacts. In: BALWOIS 2012, Ohrid.
120. Strandberg, G., Kjellstrom, E., Poska, A., Wagner, S., Gaillard, M.J., Trondman, A.K., € Mauri, A., Davis, B.A.S., Kaplan, J.O., Birks, H.J.B., Bjune, A.E., Fyfe, R., Giesecke, T., Kalnina, L., Kangur, M., van der Knaap, W.O., Kokfelt, U., Kunes, P., Latałowa, M., Marquer, L., Mazier, F., Nielsen, A.B., Smith, B., Seppa, H., Sugita, S., 2014. € Regional climate model simulations for Europe at 6 and 0.2 k BP: sensitivity to changes in anthropogenic deforestation. Clim. Past 10, 661e680.
121. Stuiver, M., Reimer, P.J., 1993. Extended 14C data base and revised Calib 3.0 14C age calibration program. Radiocarbon 35, 215e230.
122. Tasevska, O., Kostoski, G., Guseska, D., 2010. Rotifers based assessment of the Lake Dojran water quality. In: BALWOIS 2010, Ohrid.
123. Temponeras, M., Kristiansen, J., Moustaka-Gouni, M., 2000. Seasonal variation in phytoplankton composition and physical-chemical features of the shallow Lake Doïrani, Macedonia, Greece. Hydrobiologia 424, 109e122.
124. Ter Braak, C.J.F., 1995. Ordination. In: Jongman, R.H.G., Ter Braak, C.J.F., van Tongeren, O.F.R. (Eds.), Data Analysis in Community and Landscape Ecology. Cambridge University Press, Cambridge, pp. 91e173.
125. Ter Braak, C.J.F., Smilauer, P., 2002. CANOCO Reference Manual and CanoDraw for Windows User›s Guide: Software for Canonical Community Ordination (version 4.5). Microcomputer Power, Ithaca.
126. Thompson, D.W.J., Wallace, J.M., 1998. The Arctic Oscillation signature in the wintertime geopotential height and temperature fields. Geophys. Res. Lett. 25, 1297e1300.
127. Tilman, D., Kiesling, R., Sterner, R., Kilham, S.S., Johnson, F.A., 1986. Green, bluegreen and diatom algae: taxonomic difference in competitive ability for phosphorus, silicon and nitrogen. Arch. Hydrobiol. 106, 473e485.
128. Tonkov, S., Bozilova, E., Possnert, G., Velcev, A., 2008. A contribution to the postglacial vegetation history of the Rila Mountains, Bulgaria: the pollen record of Lake Trilistnika. Quat. Int. 190, 58e70.
129. Tonkov, S., Bozilova, E., Possnert, G., 2013. Postglacial vegetation history as recorded from the subalpine Lake Ribno (NW Rila Mts), Bulgaria. Cent. Eur. J. Biol. 8, 64e77.
130. Turner, R., Roberts, N., Jones, M.D., 2008. Climatic pacing of Mediterranean fire histories from lake sedimentary microcharcoal. Glob. Planet. Change 63, 317e324.
131. Tzedakis, P.C., 2007. Seven ambiguities in the Mediterranean palaeoenvironmental narrative. Quat. Sci. Rev. 26, 2042e2066.
132. Vanniere, B., Power, M.J., Roberts, N., Tinner, W., Carrion, J., Magny, M., Bartlein, P., Colombaroli, D., Daniau, A.L., Finsinger, W., Gil-Romera, G., Kaltenrieder, P., Pini, R., Sadori, L., Turner, R., Valsecchi, V., Vescovi, E., 2011. Circum-Mediterranean fire activity and climate changes during the mid-Holocene environmental transition (8500e2500 cal. BP). Holocene 21, 53e73.
133. Wagner, B., Lotter, A.F., Nowaczyk, N., Reed, J.M., Schwalb, A., Suipizio, R., Valsecchi, V., Wessels, M., Zanchetta, G., 2009. A 40,000-year record of environmental change from ancient Lake Ohrid (Albania and Macedonia). J. Paleolimnol. 41, 407e430.
134. Wanner, H., Bronnimann, S., Casty, C., Gyalistras, D., Luterbacher, J., Schmutz, C., € Stephenson, D.B., Xoplaki, E., 2001. North Atlantic oscillationeconcepts and studies. Surv. Geophys. 22, 321e382.
135. Wick, L., Lemcke, G., Sturm, M., 2003. Evidence of Lateglacial and Holocene climatic change and human impact in eastern Anatolia: high-resolution pollen, charcoal, isotopic and geochemical records from the laminated sediments of Lake Van, Turkey. Holocene 13, 665e675.
136. Willis, K.J., 1992a. The late Quaternary vegetational history of northwest Greece. III. a comparative study of two contrasting sites. New. Phytol. 121, 139e155.
137. Willis, K.J., 1992b. The late Quaternary vegetational history of northwest Greece. II. Rezina marsh. New. Phytol. 121, 119e138.
138. Wilson, G.P., Reed, J.M., Lawson, I.T., Frogley, M.R., Preece, R.C., Tzedakis, P.C., 2008. Diatom response to the Last GlacialeInterglacial Transition in the Ioannina basin, northwest Greece: implications for Mediterranean palaeoclimate reconstruction. Quat. Sci. Rev. 27, 428e440.
139. Wolin, J.A., Stone, J.R., 2010. Diatoms as indicators of water-level change in freshwater lakes. In: Smol, J.P., Stoermer, E.F. (Eds.), The Diatoms: Applications for the Environmental and Earth Sciences, second ed. Cambridge University Press, Cambridge, pp. 174e185.

CHAPTER 16

DEVELOPMENT OF AQUACULTURE IN COASTAL AREAS

Johann Hofherr, Fabrizio Natale, Pablo Trujillo

Keywords: GIS, Aquaculture, Europe, Coastal areas, Tourism

INTRODUCTION

While in the last three decades (1980–2010) world food fish production from aquaculture expanded at least tenfold, at an average annual growth rate of 9.5% (FAO, 2014), in the EU production is stagnating in the freshwater and molluscs segments and growing at a much lower rate of 4% in the case of marine aquaculture (own elaboration on the basis of FAO data). Following this low growth rate, the EU share on world aquaculture production fell from 2.3% in 2009 to 1.5% in 2011. Presently the EU only supplies 35% of its seafood demand and the remaining 65% is imported (STECF, 2014).

As regards the potential of growth in the EU, in its report of 2013 on the economic performance of the EU aquaculture sector, the Scientific, Technical and Economic Committee for Fisheries (STECF) stated: "Marine fish aquaculture is characterised by being generally capital intensive, with high input and high labour productivity. This segment has potential to compete on the increasingly globalised market but it faces constrains which hinder further expansion" (STECF, 2013).

Marine aquaculture developed in EU member states in the mid-1980s and 1990s, although with large differences between countries in the rate of growth and development (FAO, 2012). It was essentially an economic development within small and medium sized enterprises in remote areas where alternative employment was scarce. This has been particularly evident for Atlantic salmon in Scotland, Norway and Ireland, sea bass and sea bream in the Mediterranean and mussel farming by line or raft in Ireland, Spain and France (Fernandes et al., 2000).

Information from competent authorities and aquaculture associations in the main fish farming member states revealed that no or only very few (1 or 2) new farming licences were issued in the past 10–15 years for marine finfish in cages (Hofherr et al., 2012).

The very few new licences are indicative of problems of governance of the sector in addressing some common constrains. One reason for not expanding could be the economic performance of fish farming. Especially in the Mediterranean Sea, the segment of sea bass and sea bream production had a low profitability and a process of restructuration and consolidation can be observed (STECF, 2014). Despite favourable market conditions, also in the salmon production no licences for new sites have been issued (STECF, 2014). Analyses of the governance and regulatory systems for the EU aquaculture commissioned by the European Parliament (Hedley and Huntington, 2009) and evaluations of the sector's performance (OECD, 2010, Hofherr et al., 2012, STECF, 2013, European Commission, 2013, OECD, 2014 and STECF, 2014) indicate that EU aquaculture development is hindered by i) competition for space in coastal areas, ii) lack of clear priorities for the development of the sector, iii) fragmentation of competences for the authorization of aquaculture sites, and iv) diverging interpretations and applications of environmental legislations which is causing uncertainty for potential investors. Also for Turkey comparable constrains are described by Yucel-Gier et al. (2009). A recent in-depth analysis of conflicts in relation to the environmental justice theory confirms the complex set of claims of the various actors over finfish aquaculture in Europe, often aligning opposition from the tourist sector, small scale fisheries, local population and NGOs (Ertör and Ortega-Cerdà, 2015).

Citation: Johann Hofherr, Fabrizio Natale, Pablo Trujillo, 17. Development of aquaculture in coastal areas, Ocean & Coastal Management, Volume 116, November 2015, Pages 27-36, ISSN 0964-5691, http://dx.doi.org/10.1016/j.ocecoaman.2015.06.010.

Similar problems for the development of aquaculture are observed in coastal regions in the United States, eastern Australia and northern New Zealand (Gibbs, 2009). In these countries the recreational and amenity services provided by coastal regions, have become highly prioritised values, and aquaculture is often perceived to be a threat to these values. These values often are confused or mingled with other arguments regarding the overall sustainability of aquaculture activities. Where Gibbs sees a risk that prospective operators and administrative regulators are confronted with the need to demonstrate 'indefinite sustainability', other authors describe ways and criteria to assess ecological, economic and social aspects of aquaculture activities for a wide range of applications, e.g. Trujillo (2008) to have an objective tool to demonstrate long-term sustainability.

Differently to the isolated view on marine aquaculture, Coll et al. (2012) studied in a fine-scale analysis the spatial accumulation of human activities for the Mediterranean Sea. The findings show that the interaction between cumulative threats and areas of high marine biodiversity is mainly concentrated along certain coastal areas. Most of these areas are also used for aquaculture. Putting greater emphasis on the ecosystem approach, these findings could stimulate the tendency of moving aquaculture further offshore or in closed systems on land (recirculating aquaculture systems - RAS).

An extensive review of the literature on determinants for aquaculture siting listed approximately 20 bio-physical and 10 socio-cultural variables affecting the positioning of marine farms (Rennie, 2002). Among the bio-physical variables, water quality and sheltered conditions are considered key requirement for most farming systems. Over time there was lower relevance assigned to sheltered conditions which may be explained by the availability and adoption of technological solutions (i.e. submersible cages, mooring technologies) which allow farming in more open waters avoiding competition in areas close to shore. In many cases the difficulties encountered by aquaculture can be traced back to social conflicts arising from the incompatibility of the aquaculture activities with the social context, rather than with issues related to the biophysical environment. In an attempt to avoid conflicts, aquaculture enterprises, increasingly consider remoteness and distance from urban areas as key criteria for site selection (Rennie et al., 2009).

The kind of social conflicts and opposition faced by aquaculture development in coastal area is variegated and determined by local socio-political conditions. In some cases aquaculture enterprises are seen as outsiders to the local community and the allocation of licences for the establishment of aquaculture farms is seen as a form of expropriation of the common sea space used for traditional fishing activities by local groups (Pinkerton and Silver, 2011, Marshall, 2001 and Suryanata and Umemoto, 2003). In other cases it is the external touristic use of the coastline which is seen more hindering the aquaculture development. Communities are in general less motivated to embrace aquaculture if they see opportunities to generate local employment elsewhere and conflicts increase closer to urbanised areas and areas popular for recreation (Gibbs, 2009).

Science may play a manifold role in the debates around the siting of aquaculture activities. These debates are characterised by divergent sets of values favouring or contrasting aquaculture development using sustainability concerns as the main argument. The interplay between science and these values are seen at the opposite ends in the influences of "client-science" supporting the industry and "civic-science" supporting the preservation of local tradition from the establishment of new aquaculture activities (McGinnis and Collins, 2013).

The European Commission in the context of the new EU Common Fisheries Policy issued guidelines for the sustainable development of aquaculture to boost the growth of the sector (European Commission, 2013). The guidelines contain recommendations to improve governance systems and reducing bureaucracy. According to these guidelines, EU member states are expected to establish marine spatial plans in which the needs for the development of the sector are balanced against other uses of the marine space in coastal areas.

Despite the relevance of the issue of lack of space in coastal areas often indicated by the aquaculture industry and the high priority assigned to spatial planning for a better governance of the sector, information on the spatial characteristics and needs of marine aquaculture is limited, especially when zooming out from a very local geographical scale of specific coastal regions. On one side there is statistical and economic data collected through the EU Data Collection Framework which is highly aggregated at national level (European Commission, 2009) and on the other side there is spatial information on specific sites (European Commission, 2008a) which is used for spatial planning and environmental impact assessment at a local geographical scale. Both levels of spatial aggregation don›t allow appreciating the socio-economic factors which favoured the establishment of aquaculture in specific coastal regions in a country or form an EU wider perspective.

In a wide range of fields, Geographic Information Systems (GIS) and remote sensing techniques are employed for spatial planning, modelling and monitoring activities, as e.g. in the water resource management in agriculture (Rahimi et al., 2015), in the restoration of wetlands (Newcomer et al., 2014), in crop-monitoring (Wu et al., 2014), in pesticide exposure modelling for large areas (Wan, 2015) or in urban planning by automated sensing of ecological indicators (Behling et al., 2015).

Since 2008, EUROSTAT started to collect, along the production volume and value statistics, also data on surfaces occupied by aquaculture according to the Regulation (EC) 762/2008 (European Parliament and the Council, 2008). This information is still partial and collected as aggregated country statistics without geographical references. Geographical information on aquaculture sites is available in several national registers and maps but this information is dispersed and does not generally provide information on surfaces and production volumes.

At the supra-national level there have been recently three main initiatives to map marine aquaculture sites.

The Global Mariculture Database (GMD) in the context of the 'Sea Around Us' project represents the first global time-series compilation of production data of its kind. For each coastal country, historical mariculture production data were collected, and when necessary estimated. The datasets were then taxonomically and geographically disaggregated into distinct species and attributed to more than 600 different provinces/states in 112 coastal countries between 1950 and 2004. Nonetheless, GMD production figures are not assigned to specific geographical coordinates or the farm level. Instead they are generally spread across the Exclusive Economic Zone of the country in question.

FAO (Meaden et al., 2013 and Aguilar-Manjarrez and Crespi, 2013) set up exhaustive technical guidelines on GIS and remote sensing for fisheries and aquaculture to address the need for better marine spatial planning and ecosystem approaches through these techniques. The National Aquaculture Sector Overview – NASO maps which are a GIS tool published by FAO, should illustrate geographically where aquaculture is taking place. Key information features that could accompany the geographical locations are either by administrative units or individual farms and they include: cultured species, technology used, culture systems, environments, farm characteristics and respective production quantities, and main issues (credit, diseases, environmental impact, etc.). It is not clear until when all FAO member countries who are reporting aquaculture statistics to FAO and who also wish to inventory and monitor aquaculture in their respective countries will have fed the necessary information into the system.

Trujillo et al. (2012) mapped marine aquaculture sites in the Mediterranean Sea to estimate fish production. The Google Earth method intuitively makes use of freely available satellite images to locate finfish farms. Along with providing geographical coordinates to the fish farms, the images offer the possibility of estimating fish farm production volume via surface area of pens and cages at a certain point in time. These estimates can be used to verify the reliability of official statistics on aquaculture production by FAO and to disaggregate production data at lower geographical level than in national statistics. Yet the quality and resolution of the images will determine how accurate the proxy can be for calculating production by surface occupancy. Some difficulties that have to be considered in estimating production levels from cage surfaces detected from satellite images are related to the conversion of a production factor or density by volume into a production factor by surface, to the lack of information on the effective occupation of the cages and to the lack of information on the species and phase of production.

Although using a very similar approach of remote imaging, the present study puts its focus on the space occupancy of marine finfish aquaculture and its spatial relations with respect to coastline length and to touristic use. This spatial analysis is mostly concerned on social and policy aspects and considers the broader national and supra-national scale in respect of classical GIS studies on aquaculture considering bio-physical and environmental impacts at the very local level of specific sites or coastal areas (Rennie, 2002, Meaden et al., 2013, Perez et al., 2005, Pérez et al., 2003 and IUCN, 2009; ECASA project – Ecosystem Approach for Sustainable Aquaculture http://www.ecasa.org.uk/).

The mapping of aquaculture sites in this study has three main objectives. The first is to assess the issue of lack of space in coastal waters. Given the fact that lack of space is often invoked as the cause for the lack of growth of the sector in the EU it is important to understand if this is true in absolute terms or if this has to be rather considered in the context of conflicts with the other possible uses of space or other actors.

The second objective is to identify geographical clusters where aquaculture was able to establish and to develop. Theoretical explanations of the spatial clustering of economic activities in the economic literature can be grouped in five main approaches: the reduction for transportation and transaction costs, institutional thickness, agglomeration economies, innovative milieus and market conditions (Giuliani, 2010). In the case of marine aquaculture the fact that activities are taking place in a common resource space for which licencing and concessions have to be allocated poses specific challenges compared to other farming or industrial activities on land. The establishment of new aquaculture farms requires establishing property rights at sea and this aspect gives particular significance to local governance and social acceptance by coastal communities as determinants for the clustering of activities. Ideally, the clustering of aquaculture activities would also take account of socio-economic parameters. However, socio-economic data for the EU aquaculture are available on national levels only in aggregated form (STECF, 2013) and the remote imagery does not allow allocating cage aquaculture sites to specific enterprises. The identification of spatial clusters therefor can only represent a first step towards understanding which were these enabling conditions and identifying examples of best practices in governance.

The third objective is to provide evidence of the negative interactions between aquaculture and the touristic use of the coastline which has been often indicated in literature as a major problem for the expansion of the sector in developed countries, e.g. Ertör and Ortega-Cerdà, 2015 and Gibbs, 2009.

METHODS

The study covered marine finfish aquaculture in sea cages in ten EU member states (Cyprus, Spain, France, Greece, Croatia, Ireland, Italy, Malta, Slovenia and United Kingdom) and Turkey. It did not include inland structures for marine finfish production, or space occupied for the production of bivalves and algae. The position of the aquaculture sites was identified starting from coordinates available in national registers on aquaculture production established according to the EU veterinary legislation (European Commission, 2008a). This information was combined with a mapping exercise of aquaculture in the Mediterranean by Trujillo et al. (2012). In respect of the paper by Trujillo et al. the coverage was extended to other EU member states in terms of production and a temporal reference was added to each site considering all layers of historical images available in Google Earth.

The images in Google Earth of the years 2000–2012 were analysed by drawing at each site the boundary of the polygons surrounding interconnected groups of cages.

Several spatial analysis methods were applied on these polygons to define geographical cluster, calculate the sea surface occupation, the length of coastline affected and the interaction with other uses of the coastline. The analyses were performed using a combination of packages in the statistical software R (rgeos, sp, maptools, spatstat) and in ARCGIS.

The bounding polygons around the cages represent a minimum measure of surface occupation. A geographical cluster analysis was performed to identify larger areas which may more realistically represent allocated zones for aquaculture. The geographical clustering was carried out by applying a buffer around the polygons and grouping the sites which were subsequently intersecting. Several buffer sizes from 500 to 8500 m were tested to determine the most appropriate criteria for clustering. The selection of the best buffer size was done analysing silhouettes widths of the generated clusters (Rousseeuw, 1987). The optimal buffer size was chosen on the basis of the highest average silhouettes width of the generated cluster which is indicative of a clustering approach that minimises the average distances between observations in each cluster and maximises their average distances from observation in other clusters.

In addition to the geographical clustering each site was linked to the closest municipality on land (local administrative units at NUTS3 level from EUROSTAT) to compute a series of descriptive statistics by administrative units.

To calculate the length of coastline affected by aquaculture activities the centroids of the aquaculture sites were snapped on the closest point on the coastline, a buffer of 1500 m was created around each of these points, the resulting intersecting polygons were dissolved and the length of the section of coastline falling in the dissolved polygon was computed. Finally the length occupied by aquaculture was related to the total length of coastlines for each local administrative unit to assess the relevance of aquaculture activities on the coastline.

The interaction with other uses of the coastline was analysed considering the geographical distribution of hotels around the aquaculture sites. The position of the touristic structures was obtained using a web service provided by the Google Places API, which gives the possibility to extract programmatically the coordinates of points of interest a certain radius from a given location. A specific program was created in the R statistical software to interrogate the Google API and extracting the coordinates of business of the type "lodging" along the EU coastline (these category includes in addition to hotels other touristic infrastructures such as bed & breakfast and camping sites, but will be summarized further in the text under 'hotels'). From the individual coordinates it was possible to derive a density map of hotels and more specifically to calculate their frequencies at different distances from the aquaculture sites.

As a term of comparison a similar exercise was carried out in respect of the position of bathing water sampling sites. Information on bathing water quality is collected at EU level by the European Environmental Agency and is made available through the geo data portal 'Water Information System for Europe' – WISE. More than 22,000 bathing areas are monitored annually according to the provision of the bathing water directive (Directive 2006/7/EC). The monitoring takes place where most bathers or the greatest risk of pollution are expected. Based on these requirements the sites cover all main beaches along the coastline of the EU member states and their position can be considered as a proxy to represent centres of gravitation of touristic activities along the coastline. To account of different distributions of hotels is EU regions, frequencies of hotels were calculated in respect of bathing water sampling sites randomly selected in the same areas of the aquaculture sites and the results compared at an aggregate level.

RESULTS

In total 4257 polygons of aquaculture cages were identified from the analysis of the satellite images. These polygons represent sites at different temporal layers ranging from the year 2000–2012. The images of different years of the same coastlines indicated in some areas substantial changes in cage position, especially in Turkey, Spain and Italy. However, for the same site, images were not always available for all years or in some cases the quality was not sufficient to identify properly the aquaculture cages; therefore an analysis of the evolution of occupied surfaces and affected length of coastline by time was not possible. Instead of analysing data by year the results on area occupation and length of coastline were considered as maximum size over the entire period 2000–2012.

Table 1 shows the area occupation and the percentage of length of coastline affected by country. In all cases the surface and the portion of coastline affected by aquaculture are extremely limited. In the case of Greece for example, which has an aquaculture production accounting for more than 28% of the entire EU marine finfish aquaculture is occupying a surface of 230 ha and the portion of coastline affected by finfish aquaculture is 3% of the total coastline. The large differences between individual countries regarding the percentage of coastline affected by aquaculture are related to their differences in aquaculture activities and to the length of available coastline. Although France and Spain have a comparable length of coastline, France contributed 1.6% to the total EU marine finfish production in 2010, while Spain had a share of 11.1%. On the example of smaller countries, such as Malta and Slovenia it can be seen that a relatively small aquaculture sector can affect a high percentage of coastline.

Table 1: Area occupied and length of coastline affected by aquaculture (percentage of volume of production: own elaboration based on FAO data for 2010)

Country	Area occupied by aquaculture (ha)	Length of coastline (km)	Length of coastline affected by aquaculture (km)	Percentage of coastline affected by aquaculture (%)	Percentage of the volume of production in respect of the EU total in 2010 (%)
Croatia	69.98	5664	99	1.7	1.9
Cyprus	22.70	617	19	3.0	1.1
France	5.52	7330	38	0.5	1.6
Greece	230.32	15,147	466	3.0	28.1
Ireland	68.02	6437	195	3.0	4.5
Italy	37.16	9266	55	0.5	2.0

Malta	27.30	198	45	22.9	0.8
Slovenia	0.48	41	4	11.8	<0.1
Spain	135.19	7268	50	0.7	11.1
UK	34.18	19,717	280	1.4	44.1
Sum of above	630.85	71,685	1251	1.7	95.2
Turkey	170.38	8140	463	5.4	not applicable

Fig. 1 shows boxplots for the distances of the aquaculture sites from the closest point on the coastline by country. Cages are in general positioned very close to the coastline with median values of less than 1000 m in all countries. While there was no apparent trend by time, clear differences emerged among countries. Low average values are indicating that cages are placed very close to the coastline in the case of Greece (68 m), France (164 m), Turkey (184, m), while they are more distant in the case of Malta (592 m), Cyprus (628 m), Slovenia (766 m), Croatia (839 m), Italy (991 m) and Spain (1106 m). Intermediate values are present in UK (374 m) and Ireland (379 m).

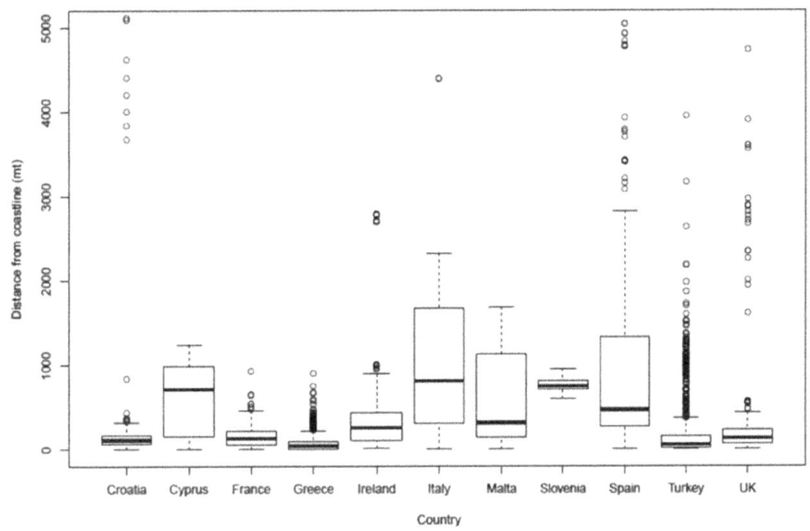

Figure 1: Distributions of distances of the aquaculture sites from the coastline.

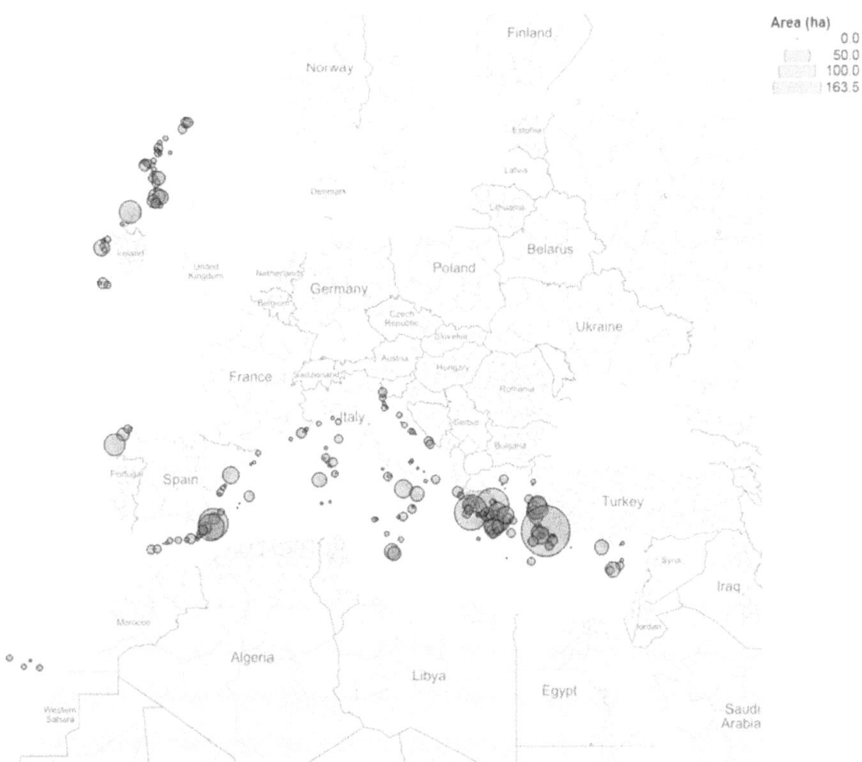

Figure 2: Clusters of aquaculture sites and area occupied (the size of the circles is proportional to the surface by cages in the cluster).

A high variability in the distribution of values and several examples of aquaculture at more than 2000 m from the coastline are present in Croatia, Ireland, Spain, Turkey and UK.

In the case of Croatia, Italy and Spain the high distance of the aquaculture sites from the coastline may be attributed to cages for the farming of tuna, while in the case of Ireland and UK there are cases of salmon farming in more exposed waters developing in particular in recent years.

In the clustering exercise the best approach was obtained by using a buffer of 4500 m. This buffer size resulted in the highest value of a silhouette width of 0.80 and allowed to identify a total of 217 clusters. The map in Fig. 2 shows the location of these clusters with symbols proportional to their surface. The 10 largest clusters in terms of surfaces were in Ireland (Donegal), Spain (Murcia), Greece (Dytiky Ellada, Stereá Ellåda, Peloponnisos) and Turkey (Izmir and Muğla).

Another approach for grouping the aquaculture sites was to relate them to the closest local administrative unit on land. In total 257 local administrative units in the ten EU member states have aquaculture sites along their coastline. Fig. 3 shows for each of them the percentage of affected coastline by length of the coastline, while the size of the symbols is proportional to the maximum cage surface across all years. The three largest clusters have a surface of 47 ha (in Spain), 44 ha and 30 ha (in Greece) and occupy respectively 33%, 7% and 45% of the coastline of the local administrative units. The local administrative units which have a medium to high percentage of their coastline occupied by aquaculture, considerable surfaces of aquaculture and at the same time a total length of the coastline of around 20–80 km stand out for the relevance the aquaculture activities may have for the respective local coastal communities. In these local administrative units aquaculture must have found local enabling conditions and social acceptance which allowed its establishments and development up to the point of representing a relevant portion of the available coastline.

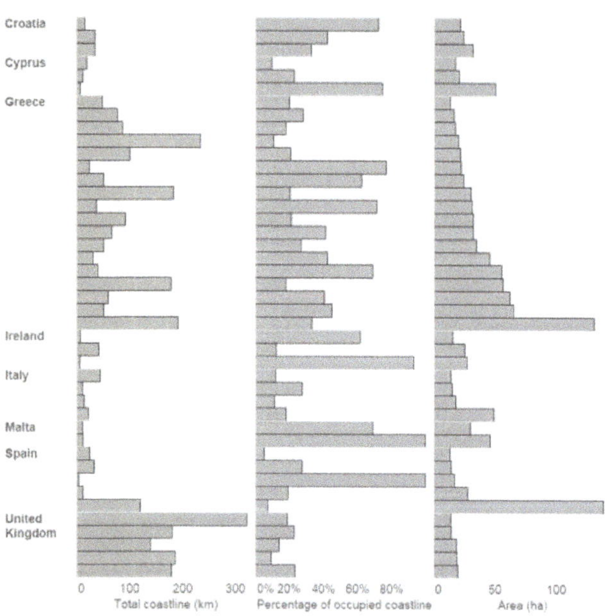

Figure 3: Total length of the coastline of local administrative units with aquaculture and percentage of coastline and area occupied by aquaculture (showing the top 20 local administrative units on the basis of the area of aquaculture sites).

Further in the analysis the interactions between aquaculture sites and tourism was evaluated. Fig. 4 shows how the density of hotels within 5 km to the nearest point of the coast along the EU coastlines varies between member states, with the lowest density in the Northeast of the Baltic Sea (around 0.04 hotels/km²) and highest along the French, Spanish and Italian coasts (around 0.1 hotels/km²). Based on these densities the problem of conflicting use between aquaculture and tourism may be considered less relevant in countries like Sweden, Denmark and Finland which a have a very low density of hotels. Aquaculture expansion could be expected to incur in fewer conflicts with tourism in Scotland, western Ireland and Greece in respect of France, Spain and Italy.

Figure 4: Density of hotels along the EU coastlines (number of hotels/km²).

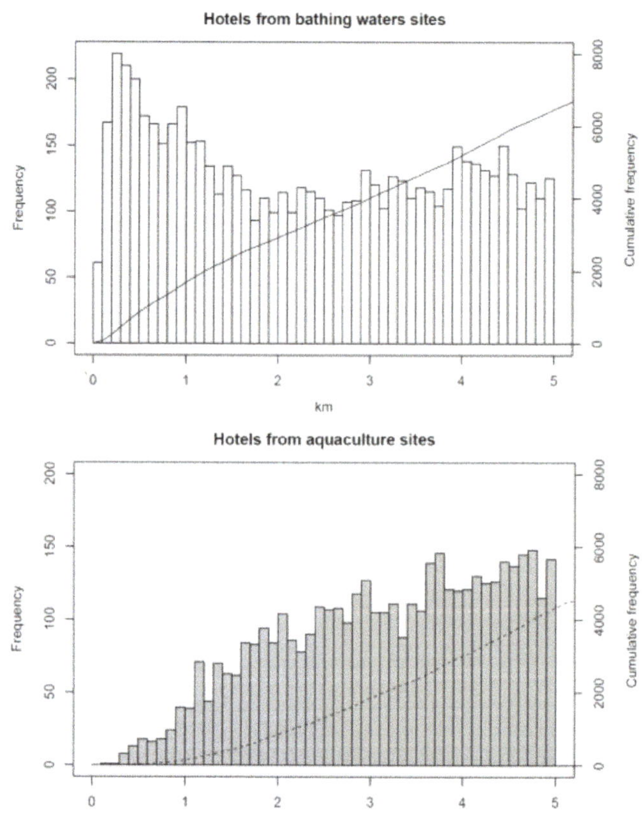

Figure 5: Frequency (columns) and cumulative frequency (lines) of hotels at increasing distances from bathing water sites (above) and aquaculture sites (below).

The analysis also considered how the density of hotels varies in respect of the presence of aquaculture sites and as term of comparison in respect of bathing water sampling sites. Fig. 5 plots the frequencies and cumulative frequencies of hotels by increasing distance from the aquaculture and bathing water sampling sites. A total of 4550 hotels were identified within 5 km of the aquaculture sites while 5905 were identified at the same distance from the bathing water sampling sites. For lower distances, the frequency of hotels in respect of the aquaculture sites has an opposite distribution pattern in respect of the bathing water sampling sites. Very few hotels can be found in the close proximity of the aquaculture sites while they tend to be concentrated at few hundred meters in respect of bathing water sampling sites. The two frequency distributions start to become parallel at around 3 km.

DISCUSSION

This study used GIS methods to examine the characteristics of marine aquaculture sites in the EU. In particular it provided the first map of marine aquaculture at EU level for some ten member states, it quantified the used surface and affected coastline and it identified a series of geographical and administrative clusters where aquaculture development was particularly significant. Finally the study considered the interaction between aquaculture sites and the touristic uses of the coastline looking at densities of hotels around the aquaculture sites and, as a term of comparison, around water quality sampling sites. This study differs from previous GIS analyses applied to the aquaculture sector since it considers a larger, supra-national scale and other factors than bio-physical and environmental for site selection.

The lack of space is often cited by the industry as one of the main hindering factors for further expansion of the marine finfish aquaculture. This study showed that only around 3% of EU coastline is affected by aquaculture and that the marine finfish sector occupies a negligible surface of marine waters. In the ten member states analysed; some 630 ha of cages correspond to the production of around 95% of the entire EU marine finfish by volume in 2010. These occupied surface values seem extremely low especially if compared with on land farming activities (for example 1335 agriculture farms in the UK have an average size of 2416 ha per farm (EUROSTAT 18/2011)). The high relevance assigned to the issue of lack of space as a hindering factor for aquaculture development seems to relate more to the common nature of the sea space and the amenity values of the sea landscape in western countries as Gibbs (2009) points out. While property rights on land are historically established and land farming activities are often even seen as contributing to the amenity value of terrestrial landscapes, marine aquaculture is a relatively new activity which has to negotiate its space demands in the not fully anthropic coastal environment against much stronger economic interests such as those represented by tourism.

Not each stretch of coastline is equally suitable and accessible for aquaculture. The observed clustering of sites in specific regions may be explained by the favourable specific physical, biological and environmental conditions of local stretch of the coastline. But obviously similar suitable locations e.g. in Scotland or in Greece show no or much less aquaculture activities. The siting of aquaculture especially when zooming out from the local perspective may be traced back to similar institutional and/or social acceptance of the sector by some coastal communities (Ertör and Ortega-Cerdà, 2015). This acceptance seems particularly significant in respect of other determinants for the clustering of economic activities normally considered in the economic geography literature such as agglomeration economies and market conditions, for which not sufficient disaggregated data are available for the aquaculture sector. One peculiarity which may explain this difference is that aquaculture has to go through complex and in most cases cumbersome authorisation procedures for the allocation of site licences (Hedley and Huntington, 2009, OECD, 2014 and OECD, 2010).

One factor which emerges from the analysis of the location of existing sites is that in most of the cases aquaculture finds its space in areas hidden from the visibility range of touristic infrastructures. The analyses on the interaction with touristic use of the coastline showed that aquaculture sites are clearly located to avoid interference with hotels. A negative spatial interaction was proved by comparing densities of hotels around aquaculture sites with water quality sampling sites. Water quality sampling sites were chosen as proxy for centre of touristic attraction along the coastline being a quite comprehensive set of spatial data on beaches and bathing places. The densities of hotels are increasing with distance in the case of aquaculture sites and decreasing in the case of water quality sampling sites. The differences in distributions which are expressions of positive and negative interactions disappear at a distance of around 3 km. This distance can be considered as the threshold at which the negative and positive effects disappear and the density of hotels is starting to assume the characteristics of a random distribution. This findings support with quantitative spatial information the observations of Gibbs (2009) regarding the conflicts affecting the establishment of marine aquaculture in coastal regions with very high recreational and amenity values in USA, Canada, Australia and New Zealand. In return, it could be assumed that in coastal regions with less touristic use the chances for establishing new aquaculture sites should be higher. The plot of hotel density on EU coasts suggests a higher potential for finding new aquaculture sites especially on the northern coasts of the Baltic Sea, the eastern North Sea, but also around Scotland, Ireland, northern Denmark and in Greek waters.

Although, as the above analysis shows, marine aquaculture affects only very limited stretches of EU coastline, sustainable coastal aquaculture requires adequate consideration of the interactions among the social, economic and ecological components (Primavera, 2006). What stands out in the spatial approaches is the heterogeneity of the coastal conditions where marine finfish aquaculture activities currently exist, raising the question on whether certain coastal systems may be better or ill-suited for finfish aquaculture developments. In the EU, to bring these loose ends together, integrated coastal zone management – ICZM and maritime spatial planning – MSP are considered suitable instruments addressing the conflict between different uses of coastal zones, increasing demand for aquaculture products and environmental conservation, taking into account the heterogeneity of the coastal conditions (Kaiser and Stead, 2002 and Primavera, 2006). A spatial analysis of conflicting claims of users in the Bay of Izmir (Yucel-Gier et al., 2010) backs the view that fish farming is often a minor claimant for marine space and that aquaculture developments should therefore be planned in the broader integrated framework of ICZM and MSP to become a recognised partner. ICZM is described as a dynamic planning and coordinating process whereby decisions are made for the sustainable use of coastal areas and resources. ICZM addresses both coastal waters and land and takes into account the interests of the various coastal stakeholders (European Parliament and the Council, 2002). MSP is seen as a tool for improved decision-making. It provides a framework for arbitrating between competing human activities and managing their impact on the marine environment. Its objective is to balance sectorial interests and achieve sustainable use of marine resources (European Commission, 2008b).

ICZM and MSP activities are in various stages of implementation, e.g. Kelly et al., 2014 and Scarff et al., 2015. A number of projects have been carried out to give guidance or to identify best practices for ICZM and MSP (e.g. http://iczm.ucc.ie/, http://www.coexistproject.eu/, http://ec.europa.eu/ourcoast/). In some regions, coastal zone planning has strong emphasis on aquaculture development, e.g. (IUCN, 2009 and Grant, 2010). In Greece, on the basis of the spatial planning provisions, aquaculture development areas were defined in various regions, in which areas of organised development of aquaculture activities – POAY can be established (Joint Ministerial Decision, 2011). Although it will take some years to evaluate the effects the Greek spatial planning provisions will have on their national aquaculture industry, the integration of aquaculture activities in the spatial planning process and defining aquaculture development areas is seen to have multiple beneficial effects compared to the present situation: it guarantees the integration of aquaculture in the coastal zones, identifies suitable areas in socio-economic and environmental terms, minimises the conflicts for the use of space with other stakeholders, encourages investments and local production by common and simpler licencing rules, reduces and mitigates environmental impacts by an integrated monitoring system (Argyrou and Papaioannou, 2008).

Mapping of geographical clusters in relation to local administrative units show where aquaculture production is particularly significant. The example of licences for new marine aquaculture sites in some regions of Spain (Hofherr et al., 2012), albeit the generally high recreational use of coasts, indicate that there are further enabling factors. The differences in aquaculture occupation between administrative units could point to different regional/local administrative conditions or planning approaches and would merit further research for understanding the local enabling conditions which favoured the expansion of aquaculture and at the same time satisfied the participation of stakeholders in these areas. It may thus allow identifying examples of best practices for the governance of the sector as also the results of Ertör and Ortega-Cerdà (2015) on conflicts with finfish aquaculture indicate frequently problems regarding even distribution of burdens and benefits, recognition of relevant stakeholders, effective participation process in which all actors have access to adequate and transparent information and the capacity to influence the decision-making.

Further efforts to improve our knowledge regarding coastal attributes are encouraged if EU marine aquaculture developmental goals to increase production in a sustainable way are feasible future outcomes.

It is acknowledged that there are a number of other, equally important aspects affecting the growth of marine aquaculture and sites selection which have not been addressed by this study. E.g., the findings of Coll et al. (2012) regarding anthropogenic threats to biodiversity may help to direct the industry's future development to coastal zones with lower biodiversity accounts (such as the south-eastern Mediterranean waters) where finfish production may be perceived less controversial. Another aspect which will determine acceptance of marine aquaculture is its long-term sustainability, where common criteria for ecological, economic and social aspects of aquaculture can be applied to show distinctive differences between countries, independent of areas and species farmed (Trujillo, 2008). Recent work highlighting the overall condition and sustainable use of the global ocean evidenced that mariculture (all species combined) in coastal zones was perceived as performing poorly in this context, and garnered widespread concern from the public, policy makers and resource managers (Halpern et al., 2012). Such concerns are not only of spatial character, as e.g. the use of pelagic fish for aqua feed, although a clear trend for a substantial reduction of fish-based ingredients in commercial fish diets can be observed (Tacon and Metian, 2008; Sarker et al., 2013). The growth of marine aquaculture with mainly piscivorous species will therefore also be influenced by the ability to further substitute fishmeal and fish oil in the diets and improving feed conversion ratios (Pelletier et al., 2009).

ACKNOWLEDGEMENTS

We wish to thank Marco Ferretti for the development of the software used for the interrogation of the Google APIs.

REFERENCES

1. Aguilar-Manjarrez, J., Crespi, V., 2013. National Aquaculture Sector Overview map collection. User manual./Vues gen erales du secteur aquacole national (NASO) (Manuel de l'utilisateur).
2. Argyrou, I.N., Papaioannou, E.A., 2008. Aquaculture in Greece: sustainable development in the coastal zone. In: Presented at the International Conference & Trade Exhibition "Aquaculture Europe 2008-Resource Management," Cracow, Poland.
3. Behling, R., Bochow, M., Foerster, S., Roessner, S., Kaufmann, H., 2015. Automated GIS-based derivation of urban ecological indicators using hyperspectral remote sensing and height information. Ecol. Indic. 48, 218e234. http://dx.doi.org/ 10.1016/j.ecolind.2014.08.003.
4. Coll, M., Piroddi, C., Albouy, C., Ben Rais Lasram, F., Cheung, W.W.L., Christensen, V., Karpouzi, V.S., Guilhaumon, F., Mouillot, D., Paleczny, M., Palomares, M.L., Steenbeek, J., Trujillo, P., Watson, R., Pauly, D., 2012. The Mediterranean Sea under siege: spatial overlap between marine biodiversity, cumulative threats and marine reserves: the Mediterranean Sea under siege. Glob. Ecol. Biogeogr. 21, 465e480. http://dx.doi.org/10.1111/j.1466-8238.2011.00697.x.
5. Ertor, I., Ortega-Cerd € a, M., 2015. Political lessons from early warnings: marine fin- fish aquaculture conflicts in Europe. Mar. Policy 51, 202e210. http://dx.doi.org/ 10.1016/j.marpol.2014.07.018.
6. European Commission, 2013. Strategic Guidelines for the Sustainable Development of EU Aquaculture. COM,

229 final.

7. European Commission, 2009. COMMISSION DECISION 2010/93/EU Adopting a Multiannual Community Programme for the Collection, Management and Use of Data in the Fisheries Sector for the Period 2011-2013. OJ L 41, p.8 (16.02.10.).
8. European Commission, 2008a. DECISION 2008/392/EC Implementing Council Directive 2006/88/EC as Regards an Internet-based Information Page to Make Information on Aquaculture Production Businesses and Authorised Processing Establishments Available by Electronic Means. OJ L 138, p.12 (28.05.08).
9. European Commission, 2008b. Roadmap for Maritime Spatial Planning: Achieving Common Principles in the EU e Commission Communication. COM, 791 final.
10. European Parliament and the Council, 2002. Recommendation 2002/413/EC of 30 May 2002 Concerning the Implementation of Integrated Coastal Zone Management in Europe. OJ L 148, p. 24 (06.06.02.).
11. European Parliament and the Council, 2008. Regulation (EC) 762/2008 of 9 July 2008, on the Submission by Member States of Statistics on Aquaculture and Repealing Council Regulation (EC) 788/96. OJ L 218, p.1 (13.08.08.).
12. FAO, 2012. The State of World Fisheries and Aquaculture 2012. Food and Agriculture Organization of the United Nations. Eurospan [distributor], Rome; London.
13. Fernandes, T.F., Miller, K.L., Read, P.A., 2000. Monitoring and regulation of marine aquaculture in Europe. J. Appl. Ichthyol. 16, 138e143. http://dx.doi.org/10.1046/ j.1439-0426.2000.00267.x.
14. FAO, 2014. The State of World Fisheries and Aquaculture 2014: Opportunities and Challenges. Rome.
15. Gibbs, M.T., 2009. Implementation barriers to establishing a sustainable coastal aquaculture sector. Mar. Policy 33, 83e89. http://dx.doi.org/10.1016/j.marpol.2008.04.004.
16. Giuliani, D., 2010. Methods for Detecting Spatial Clustering of Economic Activities Using Micro-geographic Data. Thesis. University La Sapienza, Rome.
17. Grant, J., 2010. Coastal communities, participatory research, and far-field effects of aquaculture. Aquac. Environ. Interact. 1, 85e93. http://dx.doi.org/10.3354/aei00009.
18. Halpern, B.S., Longo, C., Hardy, D., McLeod, K.L., Samhouri, J.F., Katona, S.K., Kleisner, K., Lester, S.E., O'Leary, J., Ranelletti, M., Rosenberg, A.A., Scarborough, C., Selig, E.R., Best, B.D., Brumbaugh, D.R., Chapin, F.S., Crowder, L.B., Daly, K.L., Doney, S.C., Elfes, C., Fogarty, M.J., Gaines, S.D., Jacobsen, K.I., Karrer, L.B., Leslie, H.M., Neeley, E., Pauly, D., Polasky, S., Ris, B., St Martin, K., Stone, G.S., Sumaila, U.R., Zeller, D., 2012. An index to assess the health and benefits of the global ocean. Nature 488, 615e620. http://dx.doi.org/ 10.1038/nature11397.
19. Hedley, C., Huntington, T., 2009. Regulatory and Legal Constraints for European Aquaculture (Study No. PE 431.568). European Parliament, Internal Policy DG, Policy Departement B: Structural and Cohesion Policies, Fisheries.
20. Hofherr, J., Natale, F., Fiore, G., European Commission, Joint Research Centre, Institute for the Protection and the Security of the Citizen, 2012. An Approach towards European Aquaculture Performance Indicators Indicators for Sustainable Aquaculture in the European Union. Publications Office, Luxembourg.
21. IUCNeThe World Conservation Union, Spain, Federation of European Aquaculture Producers, 2009. Aquaculture Site Selection and Site Management, Guide for the Sustainable Development of Mediterranean Aquaculture. IUCN; in Collaboration with the Spanish Ministry of the Environment and Rural and Marine Affairs, and the European Federation of Aquaculture Producers (FEAP). Gland, Switzerland: Malaga, Spain.
22. Joint Ministerial Decision, 2011. Decision Nr. 31722/4-11-2011 of 4 November 2011: Adoption of a Special Framework for Spatial Planning and Sustainable Development for Aquaculture and its Strategic Environmental Impact.
23. Kaiser, M., Stead, S.M., 2002. Uncertainties and Vvalues in European Aaquaculture: Ccommunication, Mmanagement and Ppolicy Iissues in Ttimes of "changing Ppublic perceptions." Aquac. Int. 10, 469e490.
24. Kelly, C., Gray, L., Shucksmith, R., Tweddle, J.F., 2014. Review and evaluation of marine spatial planning in the Shetland Islands. Mar. Policy 46, 152e160. http:// dx.doi.org/10.1016/j.marpol.2014.01.017.
25. Marshall, J., 2001. Landlords, leaseholders & sweat equity: changing property regimes in aquaculture. Mar. Policy 25, 335e352.
26. McGinnis, M.V., Collins, M., 2013. A race for marine space: science, values, and aquaculture planning in New Zealand. Coast. Manag. 41, 401e419. http:// dx.doi.org/10.1080/08920753.2013.822284.
27. Meaden, G.J., Aguilar-Manjarrez, J., Food and Agriculture Organization of the United Nations, 2013. Advances in Geographic Information Systems and Remote Sensing for Fisheries and Acquaculture.
28. Newcomer, M.E., Kuss, A.J.M., Ketron, T., Remar, A., Choksi, V., Skiles, J.W., 2014. Estuarine sediment deposition during wetland restoration: a GIS and remote sensing modeling approach. Geocarto Int. 29, 451e467. http://dx.doi.org/ 10.1080/10106049.2013.798356.
29. OECD, 2014. Measurement and Reduction of Administrative Burdens in 13 Sectors in Greece (Final Report Fisheries). OECD.
30. OECD, 2010. Advancing the aquaculture agenda. In: Workshop Proceedings. Ed. OECD Publishing, Paris/ France.
31. Pelletier, N., Tyedmers, P., Sonesson, U., Scholz, A., Ziegler, F., Flysjo, A., Kruse, S., Cancino, B., Silverman, H., 2009. Not all Salmon are created equal: life cycle assessment (LCA) of global Salmon farming systems. Environ. Sci. Technol. 43, 8730e8736. http://dx.doi.org/10.1021/es9010114.
32. Perez, O.M., Telfer, T.C., Ross, L.G., 2005. Geographical

information systems-based models for offshore floating marine fish cage aquaculture site selection in Tenerife, Canary Islands. Aquac. Res. 36, 946e961. http://dx.doi.org/10.1111/ j.1365-2109.2005.01282.x.

33. Perez, O., Ross, L., Telfer, T., del Campo Barquin, L., 2003. Water quality requirements for marine fish cage site selection in Tenerife (Canary Islands): predictive modelling and analysis using GIS. Aquaculture 224, 51e68. http://dx.doi.org/ 10.1016/S0044-8486(02)00274-0.
34. Pinkerton, E., Silver, J., 2011. Cadastralizing or coordinating the clam commons: can competing community and government visions of wild and farmed fisheries be J. reconciled? Mar. Policy 35, 63e72. http://dx.doi.org/10.1016/j.marpol.2010.08.002.
35. Primavera, J.H., 2006. Overcoming theimpacts of aquaculture on the coastal zone. Ocean Coast. Manag. 49, 531e545. http://dx.doi.org/10.1016/j.ocecoaman.2006.06.018.
36. Rahimi, S., Gholami Sefidkouhi, M.A., Raeini-Sarjaz, M., Valipour, M., 2015. Estimation of actual evapotranspiration by using MODIS images (a case study: Tajan catchment). Arch. Agron. Soil Sci. 61, 695e709. http://dx.doi.org/10.1080/ 03650340.2014.944904.
37. Rennie, H.G., 2002. A Geography of Marine Farming Rights in New Zealand: Some Rubbings of Patterns on the Face of the Sea. Thesis. The University of Waikato.
38. Rennie, H.G., White, R., Brabyn, L., 2009. Developing a conceptual model of marine farming in New Zealand. Mar. Policy 33, 106e117. http://dx.doi.org/10.1016/ j.marpol.2008.05.001.
39. Rousseeuw, P.J., 1987. Silhouettes: a graphical aid to the interpretation and validation of cluster analysis. J. Comput. Appl. Math. 20, 53e65.
40. Sarker, P.K., Bureau, D.P., Hua, K., Drew, M.D., Forster, I., Were, K., Hicks, B., Vandenberg, G.W., 2013. Sustainability issues related to feeding salmonids: a Canadian perspective. Rev. Aquac. 5, 199e219. http://dx.doi.org/10.1111/raq.12013.
41. Scarff, G., Fitzsimmons, C., Gray, T., 2015. The new mode of marine planning in the UK: aspirations and challenges. Mar. Policy 51, 96e102. http://dx.doi.org/ 10.1016/j.marpol.2014.07.026.
42. STECF, 2014. The Economic Performance of the EU Aquaculture Sector (STECF 14e18) (JRC Scientific and Technical Reports No. JRC 93169,). Scientific, Technical and Economic Committee for Fisheries (STECF). Publications Office of the European Union, Luxembourg.
43. STECF, 2013. Economic Performance of the EU Aquaculture Sector - 2012 Exercise (STECF-owp-13e03) (JRC Scientific and Technical Reports No. JRC 81620). Scientific, Technical and Economic Committee for Fisheries (STECF).
44. Suryanata, K., Umemoto, K.N., 2003. Tension at the nexus of the global and local: culture, property, and marine aquaculture in Hawai'i. Environ. Plan. A 35, 199e213.
45. Tacon, A.G.J., Metian, M., 2008. Global overview on the use of fish meal and fish oil in industrially compounded aquafeeds: trends and future prospects. Aquaculture 285, 146e158. http://dx.doi.org/10.1016/j.aquaculture.2008.08.015.
46. Trujillo, P., 2008. Using a mariculture sustainability index to rank countries' performance (Fisheries Center research reports No. 16(7)). In: Alder, J., Pauly, D. (Eds.), A Comparative Asessment of Biodiversity, Fsheries and Aquaculture in 53 Countries' Exclusive Economic Zones. University of British Columbia.
47. Trujillo, P., Piroddi, C., Jacquet, J., 2012. Fish farms at sea: the ground truth from google Earth. PLoS One 7, e30546. http://dx.doi.org/10.1371/journal.pone.0030546.
48. Wan, N., 2015. Pesticides exposure modeling based on GIS and remote sensing land use data. Appl. Geogr. 56, 99e106. http://dx.doi.org/10.1016/j.apgeog.2014.11.012.
49. Wu, B., Meng, J., Li, Q., Yan, N., Du, X., Zhang, M., 2014. Remote sensing-based global crop monitoring: experiences with China's crop watch system. Int. J. Digit. Earth 7, 113e137. http://dx.doi.org/10.1080/1753 8947.2013.821185.
50. Yucel-Gier, G., Arisoy, Y., Pazi, I., 2010. A spatial analysis of fish farming in the context of ICZM in the Bay of Izmir-Turkey. Coast. Manag. 38, 399e411. http://dx.doi.org/10.1080/08920753.2010.498111.
51. Yucel-Gier, G., Uslu, O., Kucuksezgin, F., 2009. Regulating and monitoring marine finfish aquaculture in Turkey. J. Appl. Ichthyol. 25, 686e694. http://dx.doi.org/ 10.1111/j.1439-0426.2009.01367.x.

CHAPTER 17 FORMIN IS ASSOCIATED WITH LEFT-RIGHT ASYMMETRY IN THE POND SNAIL AND THE FROG

Angus Davison, Gary S. McDowell, Jennifer M. Holden, Harriet F. Johnson, Georgios D. Koutsovoulos, M. Maureen Liu, Paco Hulpiau, Frans Van Roy, Christopher M. Wade, Ruby Banerjee, Fengtang Yang, Satoshi Chiba, John W. Davey, Daniel J. Jackson, Michael Levin, Mark L. Blaxter

RESULTS AND DISCUSSION

Bilaterian animals are more or less symmetrical about the midline that divides left and right, but internally most organs are asymmetric in location or shape. How is symmetry broken during development if the "right" and "left" sides are essentially arbitrary? A longstanding model posits that a chiral "F molecule" is orientated relative to the anteroposterior and dorsoventral axes [9]. This asymmetric molecular reference then determines left-right (LR) differentiation at the cellular and organismal level.

Given the importance of chiral patterning in the three bilaterian superphyla, Deuterostomia, Ecdysozoa, and Lophotrochozoa, a continuing problem is a lack of knowledge of the first symmetry-breaking steps in the Lophotrochozoa, even though the first described locus that reverses the whole body structure of an animal was from the pond snail *Lymnaea* [10 and 11]. Recently, commonalities between different species have been discovered [12, 13, 14 and 15]. For example, in both vertebrates (Deuterostomia) and snails (Lophotrochozoa), *nodal* and *pitx* encode key signaling molecules required for the establishment of LR asymmetry, suggesting that these genes may have been used in the last common ancestor of Bilateria, but lost in Ecdysozoa [1, 2 and 16]. However, neither *nodal* nor *pitx* is the earliest symmetry-breaking determinant in snails, ultimately limiting a knowledge of whether this represents deep conservation or convergent use of the same genes.

Within Lophotrochozoa, snails are unique in that they exhibit genetically tractable, natural variation in chirality [17] and so may aid in understanding of the establishment and conservation of LR asymmetry. Here, we use genetics, genomics, and pharmacological inhibition to show that the *Lymnaea stagnalis* chirality gene is a scaffolding component of the cytoskeleton. We present strong evidence that this same molecule is one component of an early chiral cytoskeletal structure that is involved in the earliest symmetry-breaking steps across the Bilateria.

A Gene that Is Associated with Chirality in Pond Snails

The gastropod mollusc *L. stagnalis* is naturally variable in left-right asymmetry, outwardly visible in the chirality of the spiral shell, and under the control of a single maternally expressed locus. In *L. stagnalis* (Figure 1A), maternal *D* alleles dominantly determine a clockwise ("dextral") twist in offspring [18 and 19]. Specifically, during the third cleavage in dextral embryos, four micromeres simultaneously emerge from four macromeres and twist clockwise ("spiral deformation"; [19]). In homozygous recessive (*dd*) sinistral embryos the four micromeres initially emerge neutrally, without rotation, with a later counterclockwise twist taking place during furrow ingression [19]. These differences are presaged by the orientation of metaphase-anaphase spindles. In dextral embryos, micromere spindle orientation is chiral ("spindle inclination"; [19]), while in sinistral embryos, the spindles are positioned radially, such that they do not exhibit chirality.

Citation: Angus Davison, Gary S. McDowell, Jennifer M. Holden, Harriet F. Johnson, Georgios D. Koutsovoulos, M. Maureen Liu, Paco Hulpiau, Frans Van Roy, Christopher M. Wade, Ruby Banerjee, Fengtang Yang, Satoshi Chiba, John W. Davey, Daniel J. Jackson, Michael Levin, Mark L. Blaxter, Formin Is Associated with Left-Right Asymmetry in the Pond Snail and the Frog, Current Biology, Volume 26, Issue 5, 7 March 2016, Pages 654-660, ISSN 0960-9822, http://dx.doi.org/10.1016/j.cub.2015.12.071.

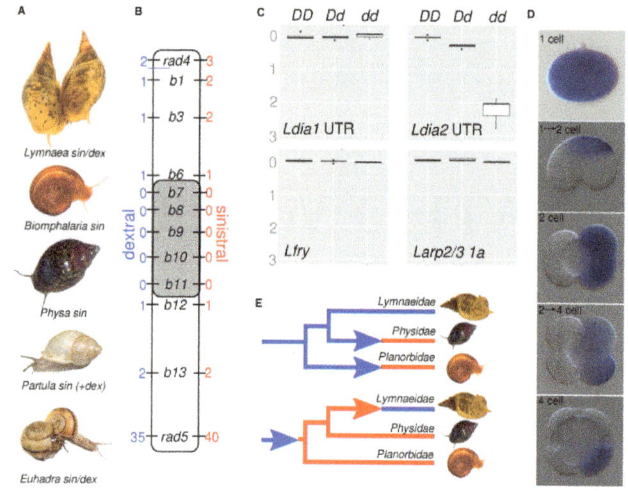

Figure 1: Mapping the Formin Gene, Maternal Expression, and Evolution of Chirality

(A) The snail genera used in this study (image credits: *Lymnaea* [E. de Roij], *Biomphalaria* and *Physa*[creative commons], *Partula* and *Euhadra* [A.D.]).

(B) 3,403 offspring were used to infer the recombination breakpoints that bound the *D* locus. Numbers of mapped recombinants for 1,507 sinistral (*dd*) snails are shown on the right and for 1,896 dextral (*DD* or *Dd*) on the left. The sinistral mutation must be between loci b6 and b12 (shaded), a region that spans 267 kb (not to scale).

(C) Boxplots show normalized relative quantities (NRQs), on log scale, of quantitative real-time PCR assays of transcripts of three candidate genes and one control (*Larp2/3 1a*) in single-cell egg samples from dextral homozygote (*DD*), dextral heterozygote (*Dd*), and sinistral recessive homozygote (*dd*) individuals. Significant differences in expression were detected for *Ldia2* only (*DD*:*dd*, $p = 0.002$; *DD*:*Dd*and *Dd*:*dd*, $p = 0.004$).

(D) WMISH of maternal *Ldia* transcripts in early, dextral *L. stagnalis* embryos.

(E) Schematic showing two hypotheses for the evolution of chirality in three snail families (dextral = blue; sinistral = red). Either sinistrality evolved once from a dextral ancestor, with the ancestral Lymnaeid reverting to dextral (bottom), or sinistrality evolved twice (top).

See also Table S1 for the mapping data; Figure S3 for further WMISH and quantitative real-time PCR data; and Figure S2C for the full snail phylogeny.

Previously we defined a ~0.4 Mb region of the 1 Gb *L. stagnalis* genome that must contain the chirality locus [18]. To identify the chirality-determining gene, we generated two new resources: a sequenced bacterial artificial chromosome (BAC) clone walk across the chirality locus interval and draft genome sequences of *DD* homozygote and *Dd* heterozygote individuals. The BAC clone walk was oriented using three-color fiber fluorescent in situ hybridization (FISH). We used the offspring of a large cross to recombination breakpoint map the position, orientation, and haplotype origin (*D*or *d*) of each BAC clone relative to the original restriction associated DNA sequencing (RAD-seq) markers (rad4 and rad5) and the chirality locus *D* (Table S1). The chirality locus was located between markers b6 and b12, a region of 267 kb (Figure 1B). We predicted genes across the scaffolded BAC walk assembly and mapped *DD* and *Dd*whole-genome sequencing data to the assembly to identify haplotype-specific variation.

Only six genes were in perfect linkage with *D* in our mapping cross, all contained on a single BAC scaffold: lysosomal pro-x carboxypeptidase (*Lprcp*), furry (*Lfry*), a tudor domain-containing protein (*Ltud*), a major facilitator superfamily domain-containing protein (*Lmfsd*), and a pair of tandemly duplicated diaphanous-related formin genes (*Ldia1* and *Ldia2*; Figure S2A). In sinistral (*dd*) snails, *Ldia2* was found to contain a homozygous, single base deletion in the 5′ of the coding region that causes a frameshift. The mutation was confirmed by cloning and sequencing *Ldia* transcripts from both sinistral and dextral snails (KU341302–KU341305). No disabling mutations were discovered in the coding sequences of the other genes, including the adjacent paralog *Ldia1*.

Maternal expression of genes was assessed by quantitative real-time PCR in single-cell embryos from *DD*, *Dd*, and *dd* mothers. Of the six candidate genes, only *Ldia2* showed significant differences in expression associated with genotype (Figures 1C and S3A). *Ldia2* transcripts were readily detected in dextral embryos from *DD* mothers. However, in sinistral embryos from *dd* mothers, *Ldia2* transcript levels were ~0.6% that of *DD*embryos, whereas in embryos from heterozygous *Dd*mothers, they were ~50%. These differences in expression are consistent with frameshifted $Ldia2^{sin}$ transcripts being degraded by nonsense mediated decay.

The tandem duplication perhaps explains why a $Ldia2^{sin}$ mutation is not simply lethal [20, 21 and 22], in that *Ldia1* and *Ldia2* may have overlapping roles in embryonic development, albeit with some specialized function: while we found that mRNAs for both *Ldia1* and *Ldia2* were present in equal quantities in the dextral 1-cell embryo, *Ldia2* was ~3-fold enriched in 1-cell dextral embryos relative to somatic tissues, whereas *Ldia1* was ~10-fold depleted (~30-fold difference overall).

Asymmetric Expression of Maternal Genes Precedes Asymmetric Morphology

Expression of candidate loci was examined in early *L. stagnalis* embryos using whole-mount in situ hybridization (WMISH; [23]). The first two embryonic cleavages in *L. stagnalis* are equal, so that the four macromeres that are formed from the first two cleavages are indistinguishable [3, 4 and 5], until contact between the third quartet of micromeres induces one of the macromeres to become the future D blastomere (but see 24]). We expected that mRNA transcripts would initially be equally distributed between the four macromeres. Surprisingly, however, we found that *Ldia* mRNA was asymmetric at the 2-cell stage and largely confined to one macromere by the 4-cell stage (Figure 1D), with *Lfry* sometimes showing an asymmetric pattern, albeit less striking, and with more variation between individuals.

While we do not know whether the *Ldia2*-positive macromere is the D blastomere, individual macromeres must have an identity as early as the two-cell stage, contrary to the expectation based on accepted models [3, 4 and 5] of development in equal cleaving snails. This also further emphasizes the general point that asymmetry is determined very early and intracellularly—asymmetry of molecules precedes visible morphological asymmetry, although not necessarily in a causative manner (see [4 and 25] for comparison with unequal cleaving embryos). The results also suggest an explanation for the reduced viability of sinistral embryos [20, 21 and 22]—as chiral rotation of micromeres in sinistrals is sometimes incomplete or error prone, inviability may be caused by a conflict in identity between individual cells. Further investigation is necessary, but the sinistral embryos seem to show more generalized, less obviously asymmetric staining, consistent with this explanation (perhaps because of reduced transport on actin microfilaments, [4]).

Pharmacological Inhibition of Formin in Early Snail Embryos Mimics the Sinistral Phenotype

As transgenic approaches are not yet established in *L. stagnalis*, and microinjection is usually lethal [22], we tested formin involvement in chirality using SMIFH2, an FH2 domain inhibitor [26]. SMIFH2 prevents formin nucleation and processive elongation of filamentous actin by decreasing the affinity of formin for the barbed end. Micromolar concentrations of SMIFH2 disrupt the formation of formin-dependent, but not Arp2/3 complex-dependent, actin cytoskeletal structures [26].

When 100 μM SMIFH2 was added to genetically dextral 4-cell embryos shortly after completion of the second cleavage, relatively few embryos (~30%–60%) reached third cleavage at ~80 min (Figure 2). However, in ~25%–35% of those that did, all four micromeres emerged neutrally, with no chiral twist (SMIFH2 treated at 0 min, n = 6 experiments, 274 embryos, compared to n = 11 control experiments, 219 embryos; $p < 0.001$, $U = 5.5$, Mann-Whitney U), with up to ~45% of individual micromeres emerging neutrally (Figure 2). SMIFH2 treatment of dextral embryos thus phenocopies normal sinistral embryos. We also visualized spindles during the third cleavage of *L. stagnalis* by indirect immunofluorescence with anti-β-tubulin antibody. In line with previous findings [19], control dextral embryos showed the characteristic spindle inclination, especially in the latter stages of mitosis. In SMIFH2-treated dextral embryos, the spindles were more frequently radially symmetric, resembling untreated sinistral embryos (Figure 3).

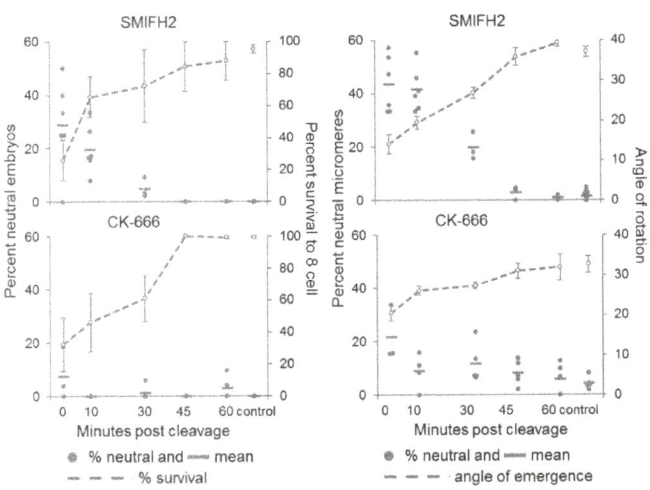

Figure 2: Impact of Drug Treatment upon 4-Cell Snail Embryos.

When applied shortly after the second cleavage had completed, both SMIFH2 and CK-666 reduced the proportion of embryos that survived to the 8-cell stage (left-hand graphs). Following SMIFH2 treatment (top left), a high proportion of the viable embryos emerged neutrally, without a chiral twist. In contrast, the proportion of neutral embryos following CK-666 treatment (bottom left) was low. Both drugs reduce the angle of rotation as the micromeres emerge (right-hand graphs). Mean values for each experiment and SE are shown.

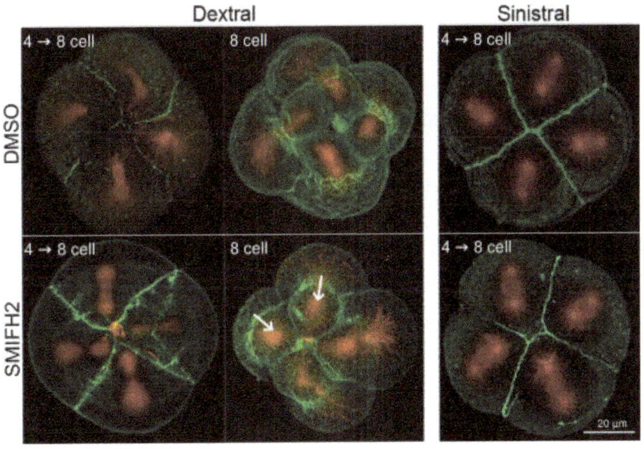

Figure 3: Tubulin and Actin Staining of Control and Drug-Treated Embryos.

Embryos were fixed and stained with Cy3-b-tubulin (red) and 488-phalloidin (green) to highlight the spindle microtubules and filamentous actin, respectively. DMSO-treated embryos predominantly showed spindle inclination (left image, 4-cell stage), with the micromeres usually emerging with a dextral twist (right image, 8-cell stage). A minority of SMIFH2 treated embryos had mitotic spindles that showed a radial orientation (left image), an arrangement that was not observed in DMSO control dextral embryos. In the SMIFH2 right-hand image (8-cell stage), the top middle and middle left micromeres are emerging neutrally (arrows), with the other two showing a partial rotation. Addition of SMIFH2 did not influence spindle orientation in 4-cell DMSO control or sinistral embryos, with spindles typically showing a radial orientation.

To rule out a non-specific effect of SMIFH2, we compared the effect of another inhibitor of actin assembly, CK-666, which acts specifically on Arp2/3 complex-dependent actin patches [27] and not on formin-dependent actin cables. CK-666 also had a lethal effect when applied to genetically dextral early four-cell embryos and also tended to reduce the average angle of emergence of micromeres. However, an achiral phenotype was observed in rather few CK-666-treated embryos (Figure 2; CK-666 treated 0 min, n = 3 experiments, 196 embryos, compared to n = 5 control experiments, 190 embryos; $p = 0.107$, $U = 2.5$, Mann-Whitney U). Therefore, the differential effects of SMIFH2 and CK-666 suggest that formin-mediated actin assembly may be a critical factor in determining chiral cleavage orientation between the second and third cleavages.

In SMIFH2-treated, genetically dextral embryos that continued to develop following neutral emergence of micromeres (i.e., like sinistrals), the direction of the subsequent twist was dextral, rather than sinistral (6/6; Movie S1). The individual micromeres of genetic sinistrals also sometimes twisted dextrally (~0 to 4% in our experiments; the fourth time-lapse in Movie S1 shows an embryo in which all four micromeres twist dextrally after neutral emergence; see also [22]). This later twist is also actin dependent [19], which suggests dextrality may be the default pathway, independent of FH2 domain function.

A Sinistral Ancestral Pond Snail?

We used genomic and transcriptomic resources, and new sequences, to explore links between diaphanous formin and chiral evolution. First, we found that the chromosomal region that contains the *L. stagnalis* chirality locus is deeply conserved, exhibiting synteny of *dia*, *fry*, and *tud* between *L. stagnalis*, *Biomphalaria glabrata* (planorb snail, sinistral cleaving), and *Capitella telata* (polychaete annelid, dextral cleaving). Second, while the *dia* duplication is also evident in *Lymnaea truncula*ta, and so must pre-date its divergence from *L. stagnalis*, single-copy *dia* genes in *B. glabrata* and *Physa acuta* (both sinistral snails: Figure 1A) are more similar to *Ldia1* than *Ldia2* . The *B. glabrata dia gene* and *Ldia1* also share a non-repetitive UTR element, absent from *Ldia2*. Thus, *Ldia2* is likely the derived paralog, which may have evolved specific function in the embryo, and for which loss leads to a sinistral phenotype.

We mapped chirality onto a new phylogeny of the Hygrophila. The predominantly dextral Lymnaeidae and the sinistral Physidae cluster together, so either sinistrality evolved once, with a sinistral ancestral Lymnaeid subsequently reverting to dextral, or else, sinistrality evolved on two separate occasions (Figure 1E). Both explanations are equally parsimonious. However, only the first is consistent if the duplication was involved in enabling dextrality in the Lymnaeidae.

As no natural variation in chirality has been described in *Biomphalaria* or *Physa*, it is difficult to further test the function of *dia* in relation to copy number. Instead, we sampled *dia* orthologs in two other chirally variable snail genera, *Euhadra*, a Japanese land snail, and *Partula*, an endangered species from Polynesia (Figure 1A; [28 and 29]). In both of these genera, chirality was not associated with variation in a single-copy *dia* that was recovered. Molecular understanding of chiral variation in these species is therefore likely to reveal additional components of the LR asymmetry pathway, including variants in genes that enable chiral evolution without negative pleiotropic effects upon fitness.

Formin Also Regulates LR Patterning in the Frog

A popular model of LR symmetry breaking in vertebrates relies on chiral flow of extracellular fluid during neurulation. However, this mechanism cannot be universal as many phyla lack the ciliated structures required or achieve correct LR patterning prior to the ciliated structure differentiation (reviewed in [8, 14 and 15]). Inherent asymmetry in the cytoskeleton could provide an ancient, well-conserved mechanism used by vertebrate embryos at the earliest stages of development to initiate the LR pattern and instruct the entire body plan [8, 30 and 31]. We investigated whether formin inhibition in early embryos, before the neurula ciliary flow, could affect chirality in the vertebrate model *Xenopus laevis*.

SMIFH2 and CK-666 drug treatments were carried out with *X. laevis* embryos at different stages of development and scored by measuring heterotaxia (independent organ LR inversion) in tadpoles (Figure 4; Table S3). In early embryos (stages 1–6), during which the cytoskeleton instructs LR patterning [30], treatment with 50 µM SMIFH2 had a strong effect on LR patterning (13% heterotaxia, X^2, $p < 0.001$), with a smaller but significant effect with CK-666 (7% heterotaxia, X^2, $p < 0.001$), whereas embryos treated at neurula stages (stages 19–21), when ciliary flow is present, showed a strong effect for both 50 µM SMIFH2 (12%, X^2, $p < 0.001$) and 50 µM CK-666 (16%, X^2, $p < 0.001$). Treatments spanning late blastula to gastrula (stages 8–14) had a reduced (3%–4%) but still significant effect (Table S3) on organ heterotaxia, showing that the efficacy of early treatment cannot be due to remnant drug persisting to cilia stages.

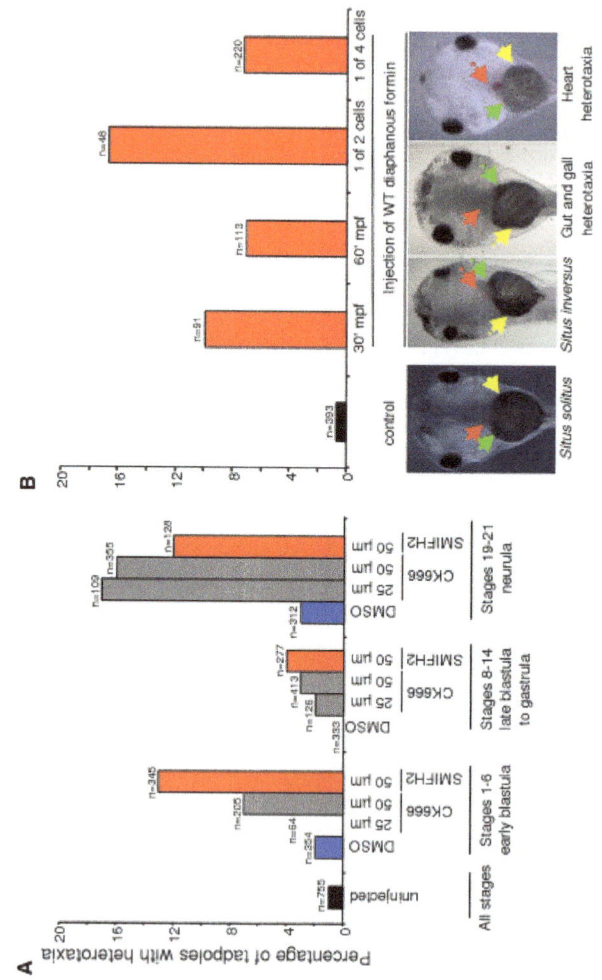

Figure 4: Effect of Drug Treatment and Microinjection of Overexpressed Formin on Chirality in the Frog

(A) Embryos were treated with DMSO, CK-666, or SMIFH2 at the concentrations indicated, allowed to develop, and scored for visceral organ chirality at stage 45.

(B) Embryos were injected into the animal pole with mRNA encoding mouse *dia1* formin and scored for visceral organ *situs* at stage 45. Images: Examples of organ *situs* for experimental microinjection with wild-type mouse *dia1* mRNA. The control shows a wild-type (*situs solitus*) tadpole, ventral view, demonstrating the normal arrangement of the stomach (yellow arrowhead), heart apex (red arrowhead), and gall bladder (green arrowhead). Heterotaxic tadpoles (ventral view) resulting from formin overexpression show reversal of all three organs, i.e., *situs inversus*; the gut position and looping and gall bladder; or the heart.

For a more specific gain-of-function test of formin function in *Xenopus*, mouse *dia1*mRNA was injected into the animal pole of frog embryos and organ *situs* assessed at stage 45. Similar to the in vitro finding that the spontaneous counterclockwise alignment of actin bundles can be reversed by overexpression of alpha-actinin [7], we found that overexpression of *dia1* resulted in a high and significant proportion of heterotaxia, whether injected 30 or 60 min post-fertilization, or in one of two or four cells (Figures 4and S4). In addition, targeted injection into the dorsal left (DL) or ventral right (VR) blastomeres at the 4-cell stage showed that while significantly different from uninjected controls (X^2, $p < 0.001$), there is no significant difference between DL and VR in the effect on heterotaxia (n = 85 DL, 5% heterotaxia; n = 135 VR, 9% heterotaxia; p = 0.75 Student's t test), which would be expected if the effect were at the point of ciliary flow, as the left side of the embryo is required for ciliary flow affecting LR patterning [32].

Formin, an F Molecule?

The implication of a key cytoskeletal protein in LR patterning of both molluscan and vertebrate embryos is consistent with a view of asymmetry as a highly conserved, ancient property in which diverse body plans leverage asymmetry from the same intracellular chiral elements. Bilaterian LR asymmetry may be dependent upon the physical orientation of the actin cytoskeleton, which, by exerting mechanical stresses on the cell, results in helical rotation [6 and 7]. While multiple elements potentially contribute to the establishment of this asymmetry, formins may have pivotal roles in coordinating functions that depend upon both the actin and microtubule cytoskeleton [7, 30, 33 and 34]. Pond snails are now an experimentally tractable, comparative model in which to integrate understanding of the action of downstream patterning genes, such as *nodal*, and the dynamics of cellular interaction and movement in the embryo to generate handedness.

Author Contributions

Conceptualization, A.D., M.L., and M.L.B; Methodology and Investigation, A.D. and C.M.W. (phylogenetics); Methodology and Investigation, H.F.J., A.D. (quantitative real-time PCR), J.M.H., and A.D. (snail microscopy and pharmacology); Methodology and Investigation, D.J.J. (WMISH), R.B., and F.Y. (FISH, with help from A.D. and M.M.L.); Methodology and Investigation, A.D. and M.M.L. (BAC walk, with BioS&T); Methodology and Investigation, G.S.M. and M.L. (laboratory - frog); Methodology and Investigation, A.D. and S.C. (Japan fieldwork); Methodology and Investigation, A.D., C.M.W., D.J.J., G.D.K., J.W.D., M.L.B., P.H., and F.V.R. (bioinformatics); Writing – Original draft, A.D. and M.L.B.; Writing – Review & Editing, A.D., C.M.W., D.J.J., G.S.M., G.D.K., H.F.J., J.M.H., M.M.L., M.L., and M.L.B. Supervision and Data Analysis, A.D., M.L., and M.L.B; Discovered the mutation, A.D. The order in which G.S.M. and J.M.H. appear in the author list was decided by a "best of three" coin toss.

ACKNOWLEDGMENTS

The work was principally funded by BBSRC grant BB/F018940/1 to A.D., M.L.B., and Aziz Aboobaker, with additional funding provided by the Universities of Edinburgh and Nottingham, the Wellcome Trust Sanger Institute (WT098051), BBSRC grants G00661X and F021135, MRC grant (G0900740), NERC grant (R8/H10/56) to M.L.B., Leverhulme Trust grant F/00114U to C.M.W., grants from the G. Harold and Leila Y. Mathers Charitable Foundation and the Physical Science Oncology Center supported by Award Number U54CA143876 from the National Cancer Institute to M.L., and DFG funding to D.J.J. (JA2108/1-1), with additional funding from the JSPS and the Daiwa Foundation to A.D. G.D.K. and H.F.J. were supported by BBSRC PhD studentships. The authors would like to thank BioS&T/Areti Karadimos for BAC library construction; Edinburgh Genomics staff for sequencing support; Nigel P. Carter for support in initiating the FISH assay; Takahiro Asami, Edmund Gittenberger, and Gerhard Falkner for the initial finding and breeding of sinistral snails; and Aziz Aboobaker, William Brown, John Armour, and David Lambert for providing plentiful, useful advice. Nebis Navarro, Paul Richards, Cendrine Hudelot, and Sheila Keeble assisted with some laboratory work; Yuta Morii and Kazuki Kimura assisted with fieldwork. *Partula mooreana* were kindly supplied by Ross Poulter at the Zoological Society of Scotland. Helpful reviews were provided by Edmund Gittenberger and two anonymous referees. For *Xenopus laevis*, this study was carried out in strict accordance with the recommendations in the Guide for the Care and Use of Laboratory Animals of the NIH.

REFERENCES

1. Grande, C., and Patel, N.H. (2009). Nodal signalling is involved in left-right asymmetry in snails. Nature 457, 1007–1011.
2. Hierck, B.P., Witte, B., Poelmann, R.E., Gittenberger-de Groot, A.C., and Gittenberger, E. (2005). Chirality in snails is determined by highly conserved asymmetry genes. J. Molluscan Stud. 71, 192–195.
3. Freeman, G., and Lundelius, J.W. (1992). Evolutionary implications of the mode of D quadrant specification in coelomates with spiral cleavage. J. Evol. Biol. 5, 205–247.
4. Lambert, J.D., and Nagy, L.M. (2002). Asymmetric inheritance of centrosomally localized mRNAs during embryonic cleavages. Nature 420, 682–686.
5. Martindale, M.Q. (1986). The 'organizing' role of the D quadrant in an equalcleaving spiralian, Lymnaea stagnalis as studied by UV laser deletion of macromeres at intervals between third and fourth quartet formation. International Journal of Invertebrate Reproduction and Development 9, 229–242.
6. Mogilner, A., and Fogelson, B. (2015). Cytoskeletal chirality: swirling cells tell left from right. Curr. Biol. 25, R501–R503.
7. Tee, Y.H., Shemesh, T., Thiagarajan, V., Hariadi, R.F., Anderson, K.L., Page, C., Volkmann, N., Hanein, D., Sivaramakrishnan, S., Kozlov, M.M., and Bershadsky, A.D. (2015). Cellular chirality arising from the self-organization of the actin cytoskeleton. Nat. Cell Biol. 17, 445–457.
8. Vandenberg, L.N., Lemire, J.M., and Levin, M. (2013).

It's never too early to get it right: a conserved role for the cytoskeleton in left-right asymmetry. Commun. Integr. Biol. 6, e27155.

9. Brown, N.A., and Wolpert, L. (1990). The development of handedness in left/right asymmetry. Development 109, 1–9.
10. Sturtevant, A.H. (1923). Inheritance of direction of coiling in Limnaea. Science 58, 269–270.
11. Boycott, A.E., and Diver, C. (1923). On the inheritance of sinistrality in Limnaea peregra. Proc. R. Soc. Lond. B Biol. Sci. 95, 207–213.
12. Coutelis, J.-B., Gonza´ lez-Morales, N., Ge´ minard, C., and Noselli, S. (2014). Diversity and convergence in the mechanisms establishing L/R asymmetry in metazoa. EMBO Rep. 15, 926–937.
13. Blum, M., Feistel, K., Thumberger, T., and Schweickert, A. (2014). The evolution and conservation of left-right patterning mechanisms. Development 141, 1603–1613.
14. Okumura, T., Utsuno, H., Kuroda, J., Gittenberger, E., Asami, T., and Matsuno, K. (2008). The development and evolution of left-right asymmetry in invertebrates: lessons from Drosophila and snails. Dev. Dyn. 237, 3497– 3515.
15. Spe´ der, P., Petzoldt, A., Suzanne, M., and Noselli, S. (2007). Strategies to establish left/right asymmetry in vertebrates and invertebrates. Curr. Opin. Genet. Dev. 17, 351–358.
16. Grande, C., Martı´n-Dura´ n, J.M., Kenny, N.J., Truchado-Garcı´a, M., and Hejnol, A. (2014). Evolution, divergence and loss of the Nodal signalling pathway: new data and a synthesis across the Bilateria. Int. J. Dev. Biol. 58, 521–532.
17. Schilthuizen, M., and Davison, A. (2005). The convoluted evolution of snail chirality. Naturwissenschaften 92, 504–515.
18. Liu, M.M., Davey, J.W., Banerjee, R., Han, J., Yang, F., Aboobaker, A., Blaxter, M.L., and Davison, A. (2013). Fine mapping of the pond snail left-right asymmetry (chirality) locus using RAD-Seq and fibre-FISH. PLoS ONE 8, e71067.
19. Shibazaki, Y., Shimizu, M., and Kuroda, R. (2004). Body handedness is directed by genetically determined cytoskeletal dynamics in the early embryo. Curr. Biol. 14, 1462–1467.
20. Davison, A., Barton, N.H., and Clarke, B. (2009). The effect of coil phenotypes and genotypes on the fecundity and viability of Partula suturalis and Lymnaea stagnalis: implications for the evolution of sinistral snails. J. Evol. Biol. 22, 1624–1635.
21. Utsuno, H., Asami, T., Van Dooren, T.J.M., and Gittenberger, E. (2011). Internal selection against the evolution of left-right reversal. Evolution 65, 2399–2411.
22. Kuroda, R. (2014). How a single gene twists a snail. Integr. Comp. Biol. 54, 677–687.
23. Hohagen, J., Herlitze, I., and Jackson, D.J. (2015). An optimised whole mount in situ hybridisation protocol for the mollusc Lymnaea stagnalis. BMC Dev. Biol. 15, 19.
24. Meshcheryakov, V.N., and Beloussov, L.V. (1975). Asymmetrical rotations of blastomeres in early cleavage of Gastropoda. Wilhelm Rouxs Archives of Developmental Biology 177, 193–203.
25. Lyons, D.C., and Weisblat, D.A. (2009). D quadrant specification in the leech Helobdella: actomyosin contractility controls the unequal cleavage of the CD blastomere. Dev. Biol. 334, 46–58.
26. Rizvi, S.A., Neidt, E.M., Cui, J., Feiger, Z., Skau, C.T., Gardel, M.L., Kozmin, S.A., and Kovar, D.R. (2009). Identification and characterization of a small molecule inhibitor of formin-mediated actin assembly. Chem. Biol. 16, 1158–1168.
27. Nolen, B.J., Tomasevic, N., Russell, A., Pierce, D.W., Jia, Z., McCormick, C.D., Hartman, J., Sakowicz, R., and Pollard, T.D. (2009). Characterization of two classes of small molecule inhibitors of Arp2/3 complex. Nature 460, 1031–1034.
28. Davison, A., Chiba, S., Barton, N.H., and Clarke, B. (2005). Speciation and gene flow between snails of opposite chirality. PLoS Biol. 3, e282.
29. Lee, T., Burch, J.B., Jung, Y., Coote, T., Pearce-Kelly, P., and O Foighil, D. (2007). Tahitian tree snail mitochondrial clades survived recent mass extirpation. Curr. Biol. 17, R502–R503.
30. Lobikin, M., Wang, G., Xu, J., Hsieh, Y.-W., Chuang, C.-F., Lemire, J.M., and Levin, M. (2012). Early, nonciliary role for microtubule proteins in left-right patterning is conserved across kingdoms. Proc. Natl. Acad. Sci. USA 109, 12586–12591.
31. McDowell, G.S., Lemire, J.M., Pare, J., Cammarata, G., Lowery, L.A., and Levin, M. (2016). Conserved roles for cytoskeletal components in determining laterality. Integrative Biology, 10.1039/C5IB00281H.
32. Vick, P., Schweickert, A., Weber, T., Eberhardt, M., Mencl, S., Shcherbakov, D., Beyer, T., and Blum, M. (2009). Flow on the right side of the gastrocoel roof plate is dispensable for symmetry breakage in the frog Xenopus laevis. Dev. Biol. 331, 281–291.
33. Pohl, C., and Bao, Z.R. (2010). Chiral forces organize left-right patterning in C. elegans by uncoupling midline and anteroposterior axis. Dev. Cell 19, 402–412.
34. Naganathan, S.R., Middelkoop, T.C., Fu¨rthauer, S., and Grill, S.W. (2016). Actomyosin-driven left-right asymmetry: from molecular torques to chiral self organization. Current Opinion in Cell Biology 38, 24–30.

CHAPTER 18

HOUSING AND HEALTH INEQUALITIES

Marcia Gibson, Mark Petticrew, Clare Bambra, Amanda J. Sowden, Kath E. Wright, Margaret Whitehead

Keywords: Housing, Health inequalities, Systematic review, Intervention, Evidence

INTRODUCTION

Health inequalities persist in developed countries, despite general improvements in health outcomes across the population (Department of Health, 2008). The unequal distribution of health has led to a growing awareness that health is socially determined by factors originating in different levels of society, ranging from the individual to the structural, as represented by models such as Dahlgren and Whitehead's well-known 'rainbow' model (Dahlgren and Whitehead, 2007). At the individual level, biological and behavioural factors influence health, and are amenable to healthcare or behavioural interventions—these are known as 'downstream' interventions because they are located closest to the apparent source of health problems. At the other end of the spectrum, structural factors such as employment conditions and education also influence health, and because they are unevenly distributed, play a role in the creation and maintenance of health inequalities. Interventions aimed at these 'upstream determinants' are therefore required to tackle health inequalities (Acheson, 1998 and Graham, 2004). Housing and neighbourhood conditions are such an upstream determinant. Thus, attempts to improve housing and neighbourhood conditions are seen as upstream interventions with the potential to tackle health inequalities, the importance of which was recently highlighted by the WHO report of the Commission for Social Determinants of Health (2008).

Factors linked to housing and neighbourhood conditions that influence health can be grouped into three broad categories, or pathways (Acevedo-Garcia et al., 2004): area characteristics, internal housing conditions and housing tenure, all of which have been shown to have independent effects on health (Shaw, 2004). Within each of these pathways, there are numerous specific mechanisms, which may influence health. Area characteristics may impact on health in many ways. Deprived areas may experience higher levels of crime and social disorder, making it more stressful and dangerous to live in them. Access to amenities may be worse than in more affluent areas, and there may be fewer jobs available. Cultures influencing health behaviours and employment may differ, and both residents and outsiders may perceive neighbourhoods negatively (Kling et al., 2007). Many aspects of internal housing conditions have the potential to influence health. In particular, cold and damp conditions may cause or exacerbate respiratory health conditions. Poisoning may be caused by lead piping, lead paint or carbon monoxide, and injuries may be caused by accidents related to high-rise housing or lack of safety equipment, particularly among children and the elderly. Lack of smoke alarms, fire extinguishers and sprinklers may exacerbate the risk of injury from fire (Shaw, 2004). There is some evidence that housing design and street layout can impact on psychosocial outcomes (Gibson et al., in press). Housing tenure may have psychosocial impacts on health; owning one's one home may confer greater feelings of security or prestige than social or private renting, and is often used as an indicator of greater long-term command over resources. Conversely, the burden of debt involved may lead to anxiety and worry. Renting a home may be seen as socially inferior in some national or historical contexts but not in others (Macintyre et al., 2003). It is difficult to disentangle the impacts of these pathways; for instance, it would appear that much of the relationship between housing tenure and health is explained by the worse internal housing conditions and greater level of area problems associated with social housing (Macintyre et al., 2003). There is therefore a need to identify and collate the evidence on

Citation: Marcia Gibson, Mark Petticrew, Clare Bambra, Amanda J. Sowden, Kath E. Wright, Margaret Whitehead, Housing and health inequalities, Health & Place, Volume 17, Issue 1, January 2011, Pages 175-184, ISSN 1353-8292, http://dx.doi.org/10.1016/j.healthplace.2010.09.011.

interventions, which aim to improve health by addressing each of these multi-factorial pathways linking housing to health: (1) area characteristics, (2) internal housing conditions and (3) housing tenure.

The use of systematic reviews to locate, appraise and synthesise evidence in fields such as housing and public health has become increasingly common, as the growing number of studies on given topics necessitate a thorough overview to make these studies accessible and to contextualise conflicting findings. However, the recent proliferation of systematic reviews has been such that it has now become difficult to keep pace with their output (Petticrew et al., 2008). Thus, so-called meta- or 'umbrella' reviews have emerged as a means of identifying, appraising and synthesising the evidence contained in systematic reviews and of making findings accessible to policy-makers and practitioners (Bialy et al., 2006 and Egan et al., 2008). Presenting the overarching findings of reviews of 'upstream' interventions is arguably increasingly important given the growing recognition of the importance of structural factors in the social patterning of health. For these reasons, we conducted a systematic overview of systematic reviews of housing and community interventions with three main aims:

- to identify what types of housing and neighbourhood interventions have been reviewed systematically and how these relate to the different pathways between housing and health;
- to establish what gaps exist in the systematic review evidence base on housing interventions; and
- to consider what existing reviews can tell us about the impact of housing and neighbourhood interventions on health and health inequalities.

METHODS

Systematic review methods were used to locate and evaluate published and unpublished systematic reviews of studies that evaluated the impact of housing and community interventions on health and health inequalities. We sought reviews of housing interventions, which focused on health outcomes, and particularly on health inequalities, whether via the targeting of interventions at disadvantaged groups or by reporting differential impacts according to social sub-groups.

Search Strategy

The searches were designed and conducted by an experienced Information Scientist from the Centre for Reviews and Dissemination (CRD). We searched the CRD Wider Public Health database manually from 2000–2002. In addition, we conducted electronic searches of the Cochrane Database of Systematic Reviews, the Criminal Justice Abstracts database (2000–2007) and the Database of Abstracts of Reviews of Effects (DARE; 2002–2007), and hand searched the Campbell Collaboration Database and the Evidence for Policy and Practice Information and Coordinating Centre database (2002–2007). (An example of the search strategy used to use search the DARE database can be viewed at Appendix A.) Bibliographies and relevant websites were searched, and experts were contacted. Four journals were hand searched: American Journal of Public Health, American Journal of Preventive Medicine, Journal of Epidemiology and Community Health, Social Science and Medicine (2002–2007). Through our own contacts, we were also aware of one review (Thomson et al., 2009) conducted outwith the search time frame, which we included because it represents a major contribution to the evidence base on internal housing conditions and health.

Inclusion and Exclusion Criteria

Only reviews from 2000–2007 of adult participants (16+) or the general population in OECD countries (North America, Europe, Australasia, Japan) were eligible for inclusion. Reviews of studies evaluating interventions aimed at altering housing or neighbourhood conditions, which collected data on health or well-being outcomes were included. Systematic reviews had to meet what were then the two mandatory criteria of the Database of Abstracts of Reviews of Effects; that there is a defined review question and that an effort has been made to identify all the relevant literature. A minimum of one or more named databases needed to be searched, in conjunction with either reference checking, hand-searching, citation searching or contact with authors in the field to meet the second criterion.

Data Extraction and Critical Appraisal

All titles and abstracts (n=1694) were independently screened by two reviewers (MG/CB), and relevant reviews (n=84) were retrieved and assessed for inclusion. The flowchart is provided in Fig. 1. Discrepancies were resolved by consensus, or referred to a third reviewer if necessary (MP). Data relating to the review methods (search strategy, inclusion criteria, synthesis) were extracted along with information about the intervention, participants, outcomes, results (including number of studies and study design), authors' conclusions and research recommendations. Each systematic review was critically appraised using a checklist list adapted from DARE. The checklist was used to highlight variations in the reviews and to assess their reliability and validity (see Box 1).

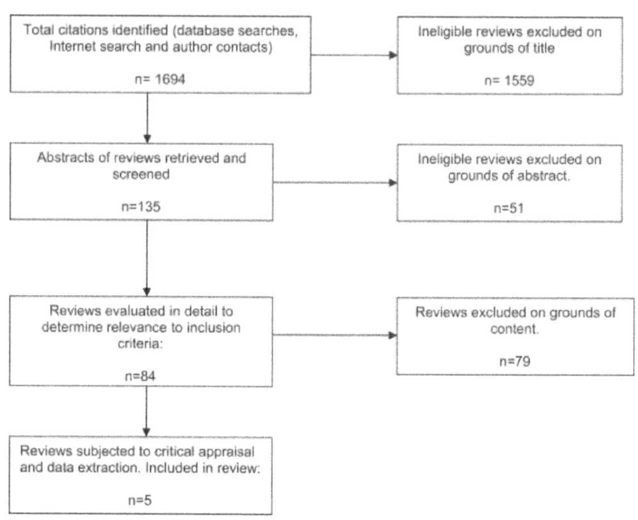

Figure 1: Flow diagram of included reviews.

Box 1–Methodological quality checklist.
Source: adapted from Database of Abstracts of Reviews of Effects (DARE), (Main et al., 2008; Egan et al., 2008)

1. Is there a well defined question?
2. Is there a defined search strategy?
3. Are inclusion/exclusion criteria stated?
4. Are the primary study designs and number of studies clearly stated?
5. Have the primary studies been quality assessed?
6. Have the studies been appropriately synthesised?
7. Has more than one author been involved at each stage of the review process?

RESULTS

Five reviews met the criteria for inclusion, containing a total of 130 studies of relevance to this overview of reviews, although the number of unique studies is smaller, as there is some overlap between the reviews. Three of the systematic reviews we included were aimed at addressing pathway one—the impact of area characteristics. Two of these were reviews of US interventions aimed at reducing the concentration of poverty by using various means to relocate families living in high poverty areas to more affluent areas (Acevedo-Garcia et al., 2004 and Anderson et al., 2003). The third assessed area-based urban regeneration interventions in the UK (Thomson et al., 2006). One review addressed pathway two, internal housing conditions and included studies of warmth and energy efficiency interventions, housing improvement, refurbishment and relocation (Thomson et al., 2009). The fifth review included a range of housing interventions aimed at several of the pathways linking housing to health, including rehousing, injury prevention and behaviour change interventions (Saegert et al., 2003). No reviews relating to the third pathway, altering housing tenure, were located. The results are summarised in Table 1, Table 2 and Table 3, and a narrative synthesis is presented below. We also relate these findings to a recent literature review from the WHO Regional Office for Europe, which summarised the evidence of the extent of social inequities in risk associated with housing and residential location (Braubach and Fairburn, 2010). This review considered risks derived from seven sources: (i) housing and indoor environments, (ii) fuel poverty and thermal comfort, (iii) indoor environmental exposures and overcrowding, (iv) water and sanitation, (v) outdoor environments and residential location, (v) neighbourhood deprivation, safety and physical activity (vi) noise and (vii) pollution and environmental deprivation. This allows us to consider whether the systematic reviews we uncovered (which assess the effectiveness of interventions) tell us how to intervene within the main risk categories identified in the Braubach and Fairburn (2010) review.

Table 1: Summary of area effects reviews (*n*=3)

	Anderson et al. (2003)	Acevedo-Garcia et al. (2004)	Thomson et al. (2006)
Review objective	"to address mixed income housing programs"	"to summarize the research evidence that U.S. housing mobility policies may help improve health outcomes"	"to synthesise data on the impact on health and key socio-economic determinants of health and health inequalities reported in evaluations of national UK regeneration programmes"
Review inclusion criteria	Evaluation of mixed income or rental assistance intervention; before and after or no-intervention comparison; include housing hazards, neighbourhood safety, youth risk behaviour or physical/ mental health outcomes.	Empirical evaluation of housing intervention; include--> ≥1 health outcome; have a comparison group	'Evaluations that reported achievements or impacts drawing on data from at least two target areas of a national ABI programme in the UK.'
Intervention(s)	A. Mixed income housing developments in low SES neighbourhoods. B. Tenant based rental assistance programmes (housing vouchers allowing low SES access to more expensive areas). Participants required to remain in higher SES are for 1 year. Counselling also provided in some interventions.	Tenant based rental assistance programmes (rent subsidies in private sector requiring participants to move from high to low poverty areas). Participants required to remain in higher SES are for 1 year. Counselling also provided in some interventions.	Area-based initiatives (ABIs)—urban regeneration programmes.
Population	Low SES families with children	Low SES families with children	No restrictions, but most interventions aimed at deprived areas.
Total/Nof included studies	Not provided	Not provided	Not provided

Health outcomes	Community health; residential stability (family moves, crowded living conditions, homelessness); physical and mental health, youth behavioural problems, violence and injuries; community cohesion and civil engagement	Mental or physical health; experience of violence; substance abuse, medical care	Quality of life, well-being, health, morbidity, mortality, use of or satisfaction with local health services. Also, housing, income, education, training or employment
Relevant primary study*N*	12	13	3
Study designs	experimental, controlled and uncontrolled prospective before and after	randomised and non-randomised experimental studies	prospective cohorts, prospective repeat cross-sectional
Database*N*	10	8	8
Location	USA	USA	UK
Synthesis method	Narrative with median change across studies reported for some outcomes	Narrative	Narrative
Main findings	No studies included for intervention A.	Overall health and 'calmness' (1 study), distress and anxiety (1 study), depression (1 study), problem drinking and substance abuse (2 studies) improved significantly in the experimental groups. Exposure to violence decreased at follow-up (6 studies).	Impact of interventions on reported outcomes was highly variable.
	Intervention B:	Residential mobility programmes have the potential to improve health	Self-reported health: 1 before and after evaluation found deteriorations in 3 out of 4 measures of self-reported health (+ 3.8%).
	Statistically sig. improvement in 28/30 measures of crime and social disorder across 10 studies.		Mortality: 2 case study area evaluations reported improvements in standardised mortality rate: paper A=131v 114; paper B=122v 118.
	2 good quality MTO studies reported sig. improvements in 14 mental health measures and sig. improvements in self-reported health status; 11% increase in SR good/excellent health in New York, 12% increase in Boston		Authors conclude that there is 'little evidence to demonstrate the impacts on health or socio-economic outcomes' of ABIs, although 'a small overall positive impact is suggested.' However, adverse impacts of ABIs are also possible.
	1 MTO study of less good quality reported a sig. decrease of 28%-53%. in health and safety risks (e.g. rodent infestation, inadequate plumbing)		
	Tenant based rental assistance programs (2) improve household safety (crime, social disorder).		
Quality appraisal	1, 2, 3, 4, 5, 6	1, 2, 3, 4, 5, 6, 7	1, 2, 3, 4, 6, 7
Quality of included studies	Unclear	1 classed as 'good', 5 as 'fair', and 7 as 'limited'	Not provided: authors comment overall study quality is poor

Table 2: Summary of housing conditions review

	Thomson et al. (2009)
Review objective	"To conduct a systematic review of the health impacts of housing improvement"
Intervention(s)	Warmth and energy efficiency (insulation, improved central heating, improved flued heat source. Some included additional measures, e.g. light bulbs, domestic repairs 15 studies); area-based rehousing/refurbishment w/out neighbourhood renewal (11 studies); rehousing from slum conditions (4 studies).
Review inclusion criteria	Studies of housing improvement, which involved improving the physical attributes of the housing infrastructure'; any health or illness-related outcome measure
Population	All but 4 studies aimed at low-income households
Total *N* of included studies	Not provided
Health outcomes	General health, respiratory health, mental health, illness/symptoms
Relevant primary study N	30
Study designs	RCTs
	Prospective controlled
	Qualitative
Database N	42
Location	USA, UK, New Zealand, Europe
Synthesis method	Narrative with some results pooled
Main findings	Warmth and energy efficiency: 2 RCTs reported sig. improvements in general and respiratory health;
	2 prospective controlled studies reported sig. improvements in general health. Impacts on general and respiratory health in the remaining studies were unclear.
	6 studies reported sig. improvements in mental health.
	10 studies reported diverse and inconsistent impacts on other illness/symptoms.
	Rehousing/refurbishment: 3 better quality studies reported small insignificant improvements in general health. 1 poor study reported sig. increase in adult's poor health (+12.3%). 3 studies found little evidence of improvement in respiratory health outcomes; in some cases, outcomes were better for the control group. 3 good quality studies found no clear impact on SF-36 mental health outcomes. A further 6 studies of lesser quality reported sig. positive impacts on a range of mental health measures. 3 studies reported mixed impacts on other illness/symptoms.
	Rehousing from slums: 4 studies conducted over 40 years ago reported no sig. impacts.
Quality appraisal	1, 2, 3, 4, 5, 6, 7
Quality of included studies	8 studies graded 'A', 13 graded 'B', and 21 graded 'C'

Table 3: Summary of multiple pathways review

	Saegert et al. (2003)
Review objective	"to characterize and to evaluate the success of current public health interventions related to housing"
Review inclusion criteria	US housing interventions to improve health; published in peer reviewed journal between January 1990 and December 2001.
Intervention(s)	Housing improvements: rehousing, changes in physical infrastructure, changes in indoor equipment or furniture, changes in housing policy
Population	USA housing residents (31% of studies related to low SES
Total N of included studies	Not provided
Health outcomes	Any health outcomes
Relevant primary study *N*	72
Study designs	44 controlled, 35 randomised studies

DatabaseN	10
Location	
	USA
Synthesis method	
	Narrative, quantitative content analysis
Main findings	49/72 studies reported a sig. improvement in health outcomes (unspecified).
Quality appraisal	1, 2, 3, 7
Quality of included studies	Quality not assessed

Pathway 1: Reviews of Interventions Aimed at Improving Area Characteristics

Tenant based Rental Assistance Programmes

Tenant based rental assistance programmes aim to influence area characteristics by moving disadvantaged people from areas of high poverty to areas of low poverty. We found two reviews, which considered the health impacts of these kinds of intervention. It should be noted though that in many cases the primary aim of the interventions was economic, that is to increase labour market activity and economic self-sufficiency.

A US review of interventions designed to address area characteristics (Anderson et al., 2003) searched for evaluations of mixed income housing developments in areas of high poverty, but found none of sufficient quality for inclusion. They also searched for interventions providing low-income families with housing vouchers to allow them to move to more affluent areas. Twelve studies were located evaluating four such programmes: the Housing Allowance Experiment, Housing and Urban Development (HUD), Section 8 and Gautreaux and Moving to Opportunity (MTO). They included randomised controlled trials (RCTs; the MTO studies) and controlled and uncontrolled prospective studies. The programmes subsidised housing costs for families with income below 50% of the area median, allowing them to seek accommodation in the private rental market. Intervention group participants were required to move to a lower poverty area and to remain in that area for a minimum of a year post-intervention. The families then contributed 30% of their monthly income to the rental costs, with the remainder being subsidised as part of the intervention. The authors analysed outcome data by summing similar outcome measures within and between studies, then calculating a median effect across all of the outcomes. From these measures, it was not possible to deduce anything about which study reported the measure, the study quality, sample size or whether any studies reported adverse effects. Data were presented for each outcome measure in the Appendix. Thirty measures of crime and social disorder across 10 studies were reported. The majority of these improved significantly in favour of the intervention group. A significant decrease in four health and safety risks was found by one study (MTO) of lower quality (range 28–53% in rodent infestation, peeling paint, inadequate plumbing and broken or missing locks). In two good quality studies (Boston and New York MTO), significant improvements in 14 measures of adult mental health were found, and there were significant improvements in two measures of self-reported health (11% increase in self-reported good or excellent health in New York, effect size 32%, $p=0.1$ and 12% increase in Boston, effect size 20% $p=0.05$). The authors conclude that there is insufficient evidence on physical and mental health outcomes on which to base any policy recommendations. However, they recommend such interventions as a means to improve household safety. This review met six out of seven critical appraisal criteria; it was unclear whether more than one author had been involved at each stage of the review process. The publications in the review were quality appraised, but the appraisal tool was not provided, and the manner of describing individual study quality did not permit identification of each study's quality, or of the numbers attaining a given level of quality. The role of study quality in interpreting the findings of the review was not discussed.

Another US review (Acevedo-Garcia et al., 2004) included evaluations of US housing policies aimed at addressing area characteristics, which included at least one health outcome (mental or physical health, experience of violence, health behaviours and medical care). Included studies had to have a comparison group. This review examined a wider range of housing mobility policies with a closer focus on health outcomes. They located 13 such studies, of five housing mobility programmes. Three of these (Gautreaux, Section 8 [this reported only child outcomes and is not reported here], and MTO) used various combinations of Section 8 vouchers alone (to allow families to move to private rented accommodation in more affluent areas), or Section 8 vouchers plus counselling. Two further 'scattered-site' interventions (Yonkers and Cincinnati) involved building new public housing units in low poverty areas.

Quasi-experimental studies of the Gautreaux programme found that tenants who moved within the city reported greater satisfaction with medical care than at baseline, while those who moved to the suburbs reported lower satisfaction than did city movers. The MTO programme (all studies were RCTs) was delivered in five US metropolitan sites, and included treatment (Section 8 voucher limited to low poverty area plus counselling), Section 8 (unrestricted voucher without counselling), and control (no voucher, but remain eligible for public housing) groups. After 2–3 years, the Boston MTO reported a small but significant decrease in personal crime victimisation, and significant increases in both overall health and 'feeling calm and peaceful' for both treatment and Section 8 groups. The New York MTO treatment group reported significant decreases in distress symptoms and exposure to violence. There were also indications of positive health and social outcomes for children and adolescents across all the sites, although these were mediated by gender. An interim report published in 2003 (on all of the MTO study sites 4–7 years post-intervention) did not report site-specific results, but found that there were large and significant reductions in obesity for the treatment and Section 8 groups, and in mental health problems for the treatment group. No differences in outcomes by ethnicity, socio-economic status (SES) or gender were reported. Of the 'scattered-site' interventions, one Cincinnati study reported a significant reduction in personal attacks and robberies, and an increase in employer provided healthcare benefits, while two Yonkers studies found that those who relocated reported significantly lower depression symptoms, problem drinking, marijuana use and experience of violent or traumatic events. No data on effect sizes or sample sizes were reported.

The review authors conclude that both tenant- and unit-based residential mobility programmes have the potential to improve health and health behaviours, but since there are few experimental or quasi-experimental studies, more research is needed to confirm this. They also note the methodological limitations of many of the studies, notably the risk of selection bias even in the randomised studies; there is some evidence to suggest that individuals with existing health problems were more likely to remain in their area of origin. The authors also point out that changes in internal housing conditions, rather than area characteristics, may account for improvements in health outcomes for those who moved. They argue that hypothesised causal pathways need to be elucidated more clearly and tested more rigorously via collection of qualitative data, data on internal housing conditions, and more detailed neighbourhood data including validated measures of neighbourhood quality. This review met all of the critical appraisal criteria. The included studies were quality assessed, and a table was provided, which allowed clear identification of each study's quality. The full quality appraisal tool was not provided, but a reference to the tool used was provided, with some explanation of the parameters for assessing quality. In addition, the authors discussed potential threats to validity in each study and considered how these influenced interpretation of the evidence as a whole.

Urban regeneration

The health and social impacts of UK urban regeneration programmes, or area-based initiatives (ABIs), which aim to tackle area characteristics by improving deprived areas, were the focus of a review by Thomson et al. (2006). Such programmes are aimed at the entire population of the target area, which are often mixed in their socio-economic composition; thus, those affected by the intervention are not necessarily personally disadvantaged. Three uncontrolled evaluations of ABIs reported data on health outcomes: two evaluated the Single Regeneration Budget (SRB), a multi-agency ABI aimed at improving employment, training, economic growth, housing, crime, environment and quality of life; and one evaluated New Life for Urban Scotland, which was also a multi-agency ABI targeted at housing, the environment, service provision, training and employment. One panel study of the SRB found that three out of four measures of self-reported health got worse, by up to 3.8% over three years. Two evaluations of case study areas using routine data found that standardised mortality rates improved (131 v 114 in the New Life programme, 122 v 118 in the SRB programme; significance unclear). Thus, impacts on health were variable, and in some cases negative. The authors conclude that there is evidence of small positive health impacts, but that adverse impacts are also possible. However, they comment that the majority of UK ABI evaluations collected data solely on outputs rather than outcomes, and that therefore the scope for ABIs to tackle health inequalities is unknown. This review met six of the critical appraisal criteria; although study quality was discussed, a summary was not provided for each study, nor was the tool used to assess quality provided.

The evidence we found in this pathway maps onto the category (v), in the WHO review (outdoor environments and residential location). The risks in this category relate to poor neighbourhood quality, and perceived safety in deprived neighbourhoods as a source of inequality. The evidence from intervention studies suggests that residential mobility programmes may impact on these risks (through reducing exposure to crime). However the potential of urban regeneration programmes to impact on these risks is largely unknown, because of the lack of outcome evaluations.

Pathway 2: Review of Interventions Aimed at Internal Housing Conditions

One review included evaluations of a range of interventions designed to improve internal housing conditions (Thomson et al., 2009). This is an update of an earlier review of such interventions (Thomson et al., 2001). The authors identified 40 evaluation studies, which collected data on direct health impacts of internal housing improvements, 30 of which were relevant to this overview (4 included only child outcomes, and 6 were set in developing countries). The majority of these studies reported effects on low-income households. Interventions to improve housing conditions included warmth and energy efficiency (15 studies), rehousing and refurbishment with or without neighbourhood renewal (11 studies, 10 of UK neighbourhood renewal interventions), and rehousing from slums (4 studies). Data on general health outcomes, respiratory health, mental health and other illnesses or symptoms were extracted.

Nine studies of warmth and energy efficiency interventions reported impacts on general health outcomes; two non-randomised prospective controlled trials (one German and one UK) and two New Zealand RCTs reported significant improvements in several general health measures. Five less rigorous studies reported unclear impacts. Respiratory health impacts were reported in eleven studies of warmth interventions. The two NZ RCTs reported significant improvements in a range of measures; the remainder of the studies reported positive, unclear or conflicting impacts. Of seven studies which assessed mental health impacts, all but one reported positive effects. No consistent results were reported for the impact of warmth and energy efficiency interventions on other illnesses or symptoms.

Ten studies of rehousing or refurbishment reported findings from UK neighbourhood renewal interventions. Of six studies reporting impacts of rehousing or refurbishment on general health, three good quality studies reported small improvements, which were not statistically significant, and one study of lower quality reported a large significant increase in poor health among adults (12.3%). Three studies included respiratory health impacts, showing little evidence of improvement and reporting better outcomes for the control group on some measures. Three rigorous studies reported unclear impacts on SF-36 mental health outcomes; a further six lower quality studies reported significant positive impacts on mental health measures. One US study of refurbishment and affordable home ownership reported a significant impact on mental health, but no impacts on other outcomes. Studies that included illness or symptom impacts (3) reported no clear impacts in either direction. Four older studies reported the impacts of rehousing slum dwellers. No significant impacts were reported in any of the studies.

Overall, warmth and energy efficiency interventions seemed to have the clearest positive impacts on health. In the review the authors note that the interventions that reported the largest effects were targeted at vulnerable groups, including those with existing health conditions and the elderly. The evidence for impact of rehousing or refurbishment on physical health outcomes is as yet unclear; however, the impacts of these area-focused interventions may be diluted because area-level deprivation indices may mask considerable socio-economic heterogeneity at the individual level; thus many of those affected by area-focused interventions may not individually be in need. In most studies the intervention actually delivered to individual households varied considerably within the sample, which hampered reliable estimation of their impacts. The authors argue that there is still a need for more large-scale controlled and quasi-experimental studies, which provide more information on intervention integrity, and which permit assessment of impacts on different social groups. Study follow-up periods may be too short to detect impacts, which may take some years to manifest. The authors also comment that poor housing conditions are interlinked with other forms of deprivation; tackling internal housing conditions alone may not have major impacts on health while other determinants such as poverty and unemployment remain unchanged. All of the critical appraisal criteria were met by this review. The full quality assessment was supplied for each study, and consideration was given to the potential impact of sources of bias on the interpretation of the results.

Three of the WHO risk categories are particularly relevant to Thomson et al. (2009) (i) housing and indoor environments, (ii) fuel poverty and thermal comfort and (iii) indoor environmental exposures and overcrowding. In all of these there appears to be evidence of effective interventions, with the clearest evidence relating to category (ii). The first WHO category also includes environmental tobacco smoke in the home, which was not considered in the reviews we examined. However other evidence suggests that the evidence here is inconsistent (Edwards et al., 2009 and Hyland et al., 2009).

Review of Interventions Aimed at Multiple Pathways

One further review included US interventions aimed at a number of pathways linking housing conditions to health (Saegert et al., 2003; the WHO literature review did not report on combinations of risks and is not discussed further here). These included rehousing and changes to: indoor equipment or furniture; respondents' knowledge or behaviour; community norms or collective behaviour; housing policy or regulatory practices, and health practitioners' behaviour. Seventy-two studies were included, of which 36% were interventions aimed at lead paint hazards, 35% at other safety hazards and 29% at asthma triggers or air quality hazards. Thirty-one per cent of these were aimed at low SES groups. Randomised studies (49%), and studies with a comparison group (60%), which provided data on any health outcomes, were included in the review. Forty-nine out of 72 studies reported a significant improvement in health outcomes; however, since neither the nature of the interventions, which yielded such improvements, nor the specific health outcomes, which improved are reported in the review, it is difficult to attribute any effects to specific interventions. The authors argue that policy interventions (of an unspecified nature) are relatively cost-effective, and also that ecological interventions, which target multiple levels (i.e. individuals, households, housing and neighbourhoods) are most likely to be successful. They also observe that few studies provided information on respondent SES, or on the content of interventions. They conclude that housing interventions have the potential to address health inequalities. This review met four of the seven quality appraisal criteria; the study designs were not clearly stated, the primary studies were not quality appraised, and due to the difficulty in attributing impacts to given interventions, it was deemed that the studies were not appropriately synthesised. No reference is made to assessment of study quality in this review; therefore, it is not possible to tell whether the studies were quality appraised, or whether they were considered to be of high or low quality.

DISCUSSION

We conducted a systematic overview of housing interventions, finding five systematic reviews, which met our inclusion criteria. In total, these reviews contained 130 relevant publications evaluating the health impacts of housing interventions. The majority of the interventions included in these reviews could be considered to address health inequalities by targeting disadvantaged groups. Overall, there is evidence to suggest that interventions aimed at altering disadvantaged participants' neighbourhood conditions by moving them to areas of lower poverty can lead to reductions in the percentage of participants reporting depression and increases in the proportion reporting good or excellent health (Acevedo-Garcia et al., 2004 and Anderson et al., 2003). There is some evidence of positive impact for area effects interventions designed to improve high poverty areas, although adverse impacts can also result from these (Thomson et al., 2006). From Thomson et al. (2009), there is strong evidence that warmth and energy efficiency interventions have positive impacts on health, although the evidence on general improvements to housing conditions remains unclear. The studies synthesised by Saegert et al. (2003) led them to conclude that multiple level housing and neighbourhood interventions were most likely to be successful. However, the level of detail included in their review hampers identification of the interventions that had positive impacts.

Area Characteristics

It appears that interventions aimed at addressing area characteristics by moving people from high to low poverty neighbourhoods can improve mental health, reduce obesity, and impact positively on some wider determinants of health, such as respondents' experience of social disorder. It seems they have the potential to alleviate health inequalities by improving the health of disadvantaged groups. Importantly, these results lend some support to hypotheses regarding the impact of area effects on health, in that for individuals who moved to lower poverty areas, there were improvements in mental health and in obesity. However, Acevedo-Garcia et al. (2004) highlight the need for greater understanding of the specific mechanisms linking these interventions to such outcomes. Later work on the Moving to Opportunity programme has attempted to address the issue of mechanisms linking area change to health improvement; data from the nested qualitative studies suggest that moving away from an area in which there is a constant threat of random violence is the most likely explanation for the observed improvements in mental health, which may in turn lead to reductions in obesity (Kling et al., 2007), perhaps because respondents were more able to be active in a less threatening outdoors environment. It should also be noted that the MTO interventions and studies were highly complex and debate continues around certain aspects of them, notably the degree to which the findings are influenced by selection bias (Clampet-Lundquist and Massey, 2008 and Sampson, 2008) and the impact of the interventions on the concentration of poverty in the sending and receiving areas (Galster, 2003 and Sampson, 2008).

In relation to the policy implications of these findings, in an overview of all of the MTO studies, Kling et al. (2007) observe that it is cheaper to provide such subsidised housing vouchers than to build new public housing units, whilst also noting that the evidence on the role of area effects provided by MTO suggests that interventions to improve distressed areas may have similar impacts. Sampson (2008) suggests that area-level interventions may be more cost-effective than moving individuals to better areas, a point which has been argued by others who also point out that area-level interventions may benefit the community as a whole (Jackson et al., 2009 and Leventhal and Brooks-Gunn, 2003). Thomson et al.'s (2006) review of UK ABIs investigates the impact of such interventions. Unfortunately, it seems that existing evaluations of such interventions can tell us little about their effects on health or health inequalities, although there is evidence of small positive health impacts on the residents of these deprived areas. A recent commentary by Thomson (2008) reflects on some of the obstacles to effective evaluation of such interventions, including difficulties with tracking individuals in regeneration areas, identifying suitable comparison groups or areas, and monitoring all components of these complex interventions in order to identify mechanisms leading to health changes. A further issue with area-level interventions, which focus on improving deprived areas is the difficulty of reliably assessing impacts when many of those receiving the intervention are not in the greatest need.

Internal Housing Conditions

With regards to interventions aimed at improving internal housing conditions, the review of warmth and energy efficiency interventions, housing refurbishment and relocation (Thomson et al., 2009) found strong evidence that improvements in warmth and energy efficiency have positive impacts on the health of low-income groups, particularly where these are targeted at the elderly or people with existing health conditions. Evidence on the impact of housing refurbishment and relocation remains inconclusive, although this may be due to short follow-up times, low intervention integrity or the impact of other factors such as socio-economic deprivation on intervention participants. Thus, interventions carefully targeted at those in greatest need may hold the most promise for improving health.

RESEARCH RECOMMENDATIONS

The authors of the systematic reviews made a range of methodological recommendations. Several called for more, and more robust, prospective controlled studies (Thomson et al., 2009 and Saegert et al., 2003), which collect data on intervention integrity and health impacts on different socio-economic groups (Thomson et al., 2009). The need to develop a better understanding of the mechanisms at work in housing interventions was highlighted by Acevedo-Garcia et al. (2004), who recommended qualitative methods as a means of doing so. In their earlier review of housing improvement interventions Thomson et al. (2001) also recommended qualitative methods as part of a more holistic approach to understanding pathways linking housing and health. Improved specification of intervention components and hypothesised causal pathways linking components to outcomes would lead to improved evaluations (Acevedo-Garcia et al., 2004 and Thomson et al., 2009).

In terms of substantive research topics, based on the findings of this synthesis of systematic reviews we would recommend further research on the health impacts of housing interventions aimed at housing tenure, including mixed tenure interventions and shared equity programmes, which promote affordable home ownership for low-income groups. A systematic review of mixed tenure interventions in the UK has recently been conducted and will add significantly to understanding of the evidence base on this topic (Bond et al., in press).

Methodologically, we would strongly echo calls for the greater use of mixed method studies to investigate specific mechanisms linking interventions to health outcomes; much of the evidence on mechanisms in the MTO interventions comes from the qualitative research that was conducted alongside the surveys. Indeed, health emerged as an outcome of interest during qualitative interviews conducted early in the MTO study (Acevedo-Garcia et al., 2004). This points to the increasing importance of qualitative research as a critical component of evaluation studies. We would also advocate the analysis of secondary longitudinal data to evaluate changes in structural determinants of health. An explicit equity focus should be adopted when evaluating housing interventions, so that differential impacts across different sub-groups can be assessed.

The interventions included in these reviews illustrate the difficulties inherent in disentangling the evidence on the various pathways linking health with housing and neighbourhood conditions; the studies of interventions to move people to lower poverty areas did not collect data on housing conditions, which are likely to have altered as a result of moving, and thus represent a confounding factor which may have altered the interventions' impacts on health outcomes (Acevedo-Garcia et al., 2004). Similarly, interventions targeted at deprived areas often include measures to improve housing conditions (Thomson et al., 2006) and interventions to improve housing conditions often occur within a context of neighbourhood renewal (Thomson et al., 2009). This underscores the need for clarity in specifying intervention components and hypothesised causal pathways.

STRENGTHS AND LIMITATIONS

This synthesis of systematic reviews provides an overview of current evidence on the health effects of interventions aimed at different pathways linking housing and health. By bringing together evidence from many published reports of intervention evaluations, it provides insight into what is currently known about the most promising means of improving health through altering housing conditions. There are however some limitations of the review, which should be acknowledged.

It is recognised that searching for primary studies or systematic reviews of complex social interventions, including those in the field of housing, is particularly challenging (Jackson and Waters, 2004 and Ogilvie et al., 2005). Where the reviews of interest may include a wide range of interventions, study designs and outcomes, designing a search that identifies all relevant papers without generating an unmanageable number of references can be very difficult. To overcome these challenges, our search strategy was developed iteratively, piloted and revised wherever necessary to ensure maximum relevance. An experienced information scientist from the Centre for Reviews and Dissemination conducted the searches. We also hand searched appropriate journals and contacted authors. Notwithstanding these efforts, we cannot rule out the possibility that we were unable to locate some relevant reviews. Another issue is a loss of detail in the progress of information from primary studies to systematic reviews and then to systematic overviews; it is clearly not possible to provide the same level of detail in an overarching summary of reviews.

Any such synthesis of systematic reviews is also inevitably limited by the level of detail reported in the original reviews. Here we have found that reporting of intervention information is often limited, and few review authors discuss implementation issues. Reporting of quality appraisal processes and study assessment was also lacking in some cases. Where statistically significant findings were noted, it was not clear whether the review authors judged these to be clinically or socially significant. Few reviews indicated what sample sizes were available for all analyses, so it was not possible to identify whether analyses were sufficiently powered. This emphasises the importance of adhering to existing guidelines on the reporting of systematic reviews, such as those provide by PRISMA (Preferred Reporting Items for Systematic Reviews and Meta-Analyses, http://www.prisma-statement.org/). However, we feel that these issues do not detract from the utility of a systematic overview of the existing evidence base on the health impacts of housing interventions.

CONCLUSION

Five systematic reviews of the health effects of interventions aimed at different pathways linking housing and health were identified. Three of these reviews included studies aimed at improving area characteristics, one included studies aimed at internal housing conditions, and one included interventions aimed at a range of pathways. The lack of systematic reviews of the health impact of housing interventions aimed at altering housing tenure represents a significant gap in the systematic review evidence base on pathways linking housing and health. The findings of this systematic overview indicate that attempting to address area characteristics by moving disadvantaged people to lower poverty area appears to have some success in improving health outcomes for those who move. However, although it is cheaper than focusing investment on deprived areas, it does not help to improve conditions in these areas, thus leaving the remaining residents to contend with the existing problems. Focusing investment on deprived areas may assist all of the residents and thus be more cost-effective. However, it is difficult to gather robust evidence of impact for area-level interventions aimed at improving either area characteristics or internal housing conditions, in part because impacts may be diluted by benefiting many who are not personally disadvantaged. There are strong indications that warmth and energy efficiency interventions which are targeted at those in most need deliver at least short-term improvements in health, suggesting that interventions to improve internal housing conditions which are targeted at the most vulnerable individuals within a disadvantaged area may yield the best results. However, improved evaluation of area-level interventions may demonstrate that these also have the potential to improve health.

ACKNOWLEDGEMENTS

The authors undertook the work as part of the Public Health Research Consortium. The Public Health Research Consortium is funded by the Department of Health Policy Research Programme. Marcia Gibson is funded by the Chief Scientist Office of the Scottish Government Health Directorates. The views expressed in the publication are those of the authors and not necessarily those of the DH. Information about the wider programme of the PHRC is available from www.york.ac.uk/phrc. Funded by MRCUK.

REFERENCES

1. Acevedo-Garcia, D., Osypuk, T.L., Werbel, R.E., Meara, E.R., Cutler, D.M., Berkman, L.F., 2004. Does housing mobility policy improve health? Housing Policy Debate 15 49–98.
2. Acheson, D., 1998. Independent Inquiry into Inequalities in Health (the Acheson Report). HMSO, London.
3. Anderson St, L.M., Charles, J., Fullilove, M.T., Scrimshaw, S.C., Fielding, J.E., Normand, J., 2003. Providing affordable family housing and reducing residential segregation by income—a systematic review. American Journal of Preventive Medicine 24, 47–67.
4. Bialy, L., Smith, M., Bourke, T., Becker, L., 2006. The Cochrane Library and bronchiolitis: an umbrella review. Evidence Based Child Health 1, 939–947.
5. Bond, A., Sautkina, E., Kearns, A., in press. Mixed messages about mixed tenure: do reviews tell the real story? Housing Studies, doi:10.1080/02673037.2010.512752.
6. Braubach, M., Fairburn, J., 2010. Social inequities in environmental risks associated with housing and residential location—a review of evidence. European Journal of Public Health 20, 36–42.
7. Clampet-Lundquist, S., Massey, Douglas S., 2008.

Neighborhood effects on economic self-sufficiency: a reconsideration of the Moving to Opportunity experiment. American Journal of Sociology 114, 107–143.

8. Commission on Social Determinants of Health, 2008. Closing the Gap in a Generation: Health Equity through Action on the Social Determinants of Health. World Health Organisation, Geneva.
9. Dahlgren, G., Whitehead, M., 2007. European Strategies for Tackling Social Inequities in Health: Levelling Up. Part 2. WHO Regional Office for Europe Copenhagen.
10. Department of Health, 2008. Health Inequalities: Progress and Next Steps.
11. Department of Health, London.
12. Edwards, R., Gifford, H., Waa, A., Glover, M., Thomson, G., Wilson, N., 2009. Beneficial impacts of a national smokefree environments law on an indigenous population: a multifaceted evaluation. International Journal of Equity in Health 8.
13. Egan, M., Tannahill, C., Petticrew, M., Thomas, S., 2008. Psychosocial risk factors in home and community settings and their associations with population health and health inequalities: a systematic meta-review. BMC Public Health 8, 239.
14. Galster, G., 2003. MTO's impact on sending and receiving neighborhoods. In: Goering, J., Feins, J.D. (Eds.), Choosing a Better Life? Evaluating the Moving to Opportunity Social Experiment. The Urban Institute Press, Washington, D.C.
15. Gibson, M., Thomson, H., Kearns, A., Petticrew, M., in press. 'It's just a better way of living'; qualitative insights into the psychosocial impact of housing type on well-being. Housing Studies.
16. Graham, H., 2004. Social determinants and their unequal distribution: clarifying policy understandings. The Milbank Quarterly 82, 101–124.
17. Hyland, A., Hassan, L., Higbee, C., Boudreau, C., Fong, G., Borland, R., Cummings, K., Yan, M., Thompson, M., Hastings, G., 2009. The impact of smokefree legislation in Scotland: results from the Scottish ITC: Scotland/UK longitudinal surveys. European Journal of Public Health 19.
18. Jackson, L., Langille, L., Lyons, R., Hughes, J., Martin, D., Winstanley, V., 2009. Does moving from a high-poverty to lower-poverty neighborhood improve mental health? A realist review of Moving to Opportunity'. Health and Place 15, 961–970.
19. Jackson, N., Waters, E., 2004. The challenges of systematically reviewing public health interventions. Journal of Public Health-UK 26, 303–307.
20. Kling, J.R., Liebman, J.B., Katz, L.F., 2007. Experimental analysis of neighborhood effects. Econometrica 75, 83–119.
21. Leventhal, T., Brooks-Gunn, J., 2003. Moving to Opportunity: an experimental study of neighborhood effects on mental health. American Journal of Public Health 93, 1576–1582.
22. Macintyre, S., Ellaway, A., Hiscock, R., Kearns, A., Der, G., McKay, L., 2003. What features of the home and the area might help to explain observed relationships between housing tenure and health? Evidence from the west of Scotland. Health & Place 9, 207–218.
23. Main, C., Thomas, S., Ogilvie, D., Stirk, L., Petticrew, M., Whitehead, M., Sowden, A., 2008. Population tobacco control interventions and their effects on social inequalities in smoking: placing an equity lens on existing systematic reviews. BMC Public Health 8.
24. Ogilvie, D., Hamilton, V., Egan, M., Petticrew, M., 2005. Systematic reviews of health effects of social interventions: 1. Finding the evidence: how far should you go? Journal of Epidemiology and Community Health 59 804–808.
25. Petticrew, M., Bambra, C., Gibson, M., Sowden, A.J., Whitehead, M., Wright, K., 2008. Tackling the wider social determinants of health and health inequalities: evidence from systematic reviews. Public Health Research Consortium, University of York, York.
26. Saegert, S.C., Klitzman, S., Freudenberg, N., Cooperman-Mroczek, J., Nassar, S., 2003. Healthy housing: a structured review of published evaluations of US interventions to improve health by modifying housing in the United States, 1990–2001. American Journal of Public Health 93.
27. Sampson, Robert J., 2008. Moving to inequality: neighborhood effects and experiments meet social structure. American Journal of Sociology 114, 189–231.
28. Shaw, M., 2004. Housing and public health. Annual Review of Public Health 25, 397–418.
29. Thomson, H., Petticrew, M., Morrison, D., 2001. Health effects of housing interventions: a systematic review of intervention studies. British Medical Journal 323, 187–190.
30. Thomson, H., 2008. A dose of realism for healthy urban policy: lessons from area-based initiatives in the UK. Journal of Epidemiology and Community Health 62, 932–936.
31. Thomson, H., Atkinson, R., Petticrew, M., Kearns, A., 2006. Do urban regeneration programmes improve public health and reduce health inequalities: a synthesis of the evidence from UK policy and practice (1980-2004). Journal of Epidemiology and Community Health 60, 108–115.
32. Thomson, H., Thomas, S., Sellstrom, E., Petticrew, M., 2009. The health impacts of housing improvement: a systematic review of intervention studies from 1887 to 2007. American Journal of Public Health 99, S681–692.

INDEX